Human Resources and Change Management for Safety Professionals

Occupational Safety and Health Guide Series

Series Editor: Thomas D. Schneid, Eastern Kentucky University, Richmond, USA

The aim of this series is to develop books to address the myriad of unique topics which are encompassed within the areas of responsibility of a safety professional. The scope of the series is broad, addressing traditional occupational safety and health topics as well as emerging and correlating topic areas such as ergonomics, security, labor, school safety, employment, workplace violence, and other topics. Safety professionals often assume a broad spectrum of responsibilities, depending on the organization, which go beyond the traditional safety professional's role. This series should become the "one stop" source for safety professionals to locate texts on traditional as well as emerging and correlating subjects related to the safety and health function.

Human Resources and Change Management for Safety Professionals
Thomas D. Schneid and Shelby L. Schneid

Safety Law
Legal Aspects in Occupational Safety and Health
Thomas D. Schneid

Physical Hazards of the Workplace
Barry Spurlock

Loss Prevention and Safety Control
Terms and Definitions
Dennis P. Nolan

Discrimination Law Issues for the Safety Professional
Thomas D. Schneid

Labor and Employment Issues for the Safety Professional
Thomas D. Schneid

The Comprehensive Handbook of School Safety
Edited by E. Scott Dunlap

Creative Safety Solutions
Thomas D. Schneid

Physical Security and Safety
A Field Guide for the Practitioner
Edited by Truett A. Ricks, Bobby E. Ricks and Jeffrey Dingle

Workplace Safety and Health
Assessing Current Practices and Promoting Change in the Profession
Thomas D. Schneid

For more information about this series, please visit: https://www.crcpress.com/Occupational-Safety--Health-Guide-Series/book-series/CRCOCCSHGS

Human Resources and Change Management for Safety Professionals

Thomas D. Schneid and Shelby L. Schneid

CRC Press
Taylor & Francis Group
Boca Raton London New York

CRC Press is an imprint of the
Taylor & Francis Group, an **informa** business

CRC Press
Taylor & Francis Group
6000 Broken Sound Parkway NW, Suite 300
Boca Raton, FL 33487-2742

Printed on acid-free paper

International Standard Book Number-13: 978-1-4987-7025-5 (Hardback)

Library of Congress Cataloging-in-Publication Data

Names: Schneid, Thomas D., author. | Schneid, Shelby L., author.
Title: Human resources and change management for safety professionals / authored by Thomas D. Schneid and Shelby L. Schneid.
Description: Boca Raton : Taylor & Francis, CRC Press, 2019. | Series: Occupational safety and health guide series
Identifiers: LCCN 2018043916| ISBN 9781498770255 (hardback : alk. paper) | ISBN 9780429029660 (e-book)
Subjects: LCSH: Industrial safety--Psychological aspects. | Safety regulations. | Safety education. | Organizational change. | Corporate culture.
Classification: LCC T55.3.B43 S36 2019 | DDC 363.10068/3--dc23
LC record available at https://lccn.loc.gov/2018043916

Visit the Taylor & Francis Web site at
http://www.taylorandfrancis.com

and the CRC Press Web site at
http://www.crcpress.com

Contents

Preface

Safety and human resource professionals are often consumed with their day-to-day activities of compliance, drug and alcohol programs, hiring, worker' s compensation, security, legal, and other activities. Change is one aspect of all organizations that must not be lost in day-to-day activities but rather carefully managed to succeed. Organizational change touches every part of the company including structure, technology, and people. From managing relationships, spans of control, work processes, equipment, people, expectations, and perceptions, successful change requires time, resources, and energy, but will ultimately lead your organization to success.

Safety and human resource professionals should not be afraid of change, but should rather be excited to move the company and its employees forward. Being a change agent is a key piece of the change management process. Understanding the change and changing safely is the responsibility of safety professionals. Changing safely not only provides a smoother transition but prevents near-misses and recordable injuries.

In this text, the authors hope to help you master the change management process and how it embodies short-term and long-term planning from various different angles. The authors hope you are able to understand the importance of change and how to execute change no matter your organization, to lead to not only personal success but allow you to make a widespread impact of successful change.

Disclaimer

The authors have attempted in every way to ensure that the information contained in this text is accurate and up-to-date; however, laws, regulations, and case decisions are being modified and changing daily. It is imperative that safety professionals research and acquire the assistance of local counsel prior to depending solely on the information in this text.

This text is designed to serve as a "starting point" for students and safety professionals to identify the potential sources of the law or regulation that may be applicable to their specific circumstances. Safety professionals and students should be aware that not all areas of potential legal liability are covered in this text. It is imperative that safety professionals appropriately research their issue at hand and review all matters with local counsel.

The authors have attempted to identify the issues that occur with the greatest frequency within the pages of this text. The authors provide no warranty, either expressed or implied, as to the accuracy of the law, standards, regulations, or other information in this text. Although suggestions and recommendations are offered, the authors do not intend for this text to provide specific legal counsel regarding individual circumstances. Competent legal counsel should always be sought to assist with specific circumstances and situations.

Authors

Thomas D. Schneid, PhD, is a tenured professor in the Department of Safety, Security, and Emergency Management at Eastern Kentucky University and serves as the Chair of the Department. Dr. Schneid earned a BS in Education, MS and CAS in Safety, an MS in International Business, and a PhD in Environmental Engineering, as well as his Juris Doctor (JD in law) from West Virginia University and LLM (Graduate Law) from the University of San Diego. He is a member of the bar for the US Supreme Court, Sixth Circuit Court of Appeals, and a member of the bars of a number of federal districts as well as the Kentucky and West Virginia state bar associations. Dr. Schneid has authored and coauthored 20 texts and more than 100 articles on various safety, fire, and medical service (EMS), disaster management, and legal topics.

Shelby L. Schneid is a human resource manager with experience in a number of Fortune 500 companies. Schneid earned an MS in Management (concentrating in human resources and change management) from the University of Central Florida, an MS in Safety, Security, and Emergency Management, and a BA in Public Relations from the Eastern Kentucky University. Schneid has authored a number of articles on various human resource and safety topics including interview preparation, successful thinking, and organizational behavior.

1 Human Resource Function and Interaction with the Safety and Health Professional

***

"A smooth sea never made a skilled mariner."

English Proverb

"There are few, if any, jobs in which ability alone is sufficient. Needed, also, are loyalty, sincerity, enthusiasm and team play."

William B. Givens, Jr.

***

The safety and health function does not work in a vacuum. In most organizations, the safety and health professional is part of the management team and serves in a leadership role to guide the safety and health efforts of the organization. However, in performing the functions required to ensure the safety and health of all employees, the safety and health function must interact with the laws, regulations, and policies of other functions within the organization, as well as outside of the organization. Given the fact that the safety and health function works closely with employees at all levels within the organizational structure, the safety and health function often works closely with the human resource or personnel function on a daily basis. This frequent interaction between the safety and health function and the human resource function is essential in order to provide the appropriate leadership and guidance to employees with regards to not only workplace safety and health issues but also to operational, quality, and related expectations for each employee by the organization.

The location of the safety and health function within the organizational structure of the company is often unique to the individual company and has developed throughout the history of the company. For more established companies, injecting formalized safety and health after 1970, the safety and health function often started as a compliance ensuring entity and was aligned with production or quality. As the safety and health function expanded beyond compliance to become a vital component within the organizational structure, companies often designed an internal hierarchy for the safety and health function within the overall organizational structure ranging from operational-level through corporate-level safety and health functions.

Given the expanded role of the safety and health professional and the daily interaction with the human resource function, many companies aligned their reporting structure, especially at operational levels, to permit the safety and health function to report directly to the human resource function with a secondary reporting line to a corporate safety and health function.

To ensure the efficiency and effectiveness of the safety and health function, many companies found that it made operational "sense" to align the safety and health function with the human resource function due to the overlapping responsibilities and expertise requirements. In order to achieve a proactive environment which prevents not only accidents but also violations of specific laws relevant to the human resource function, alignment of the reporting structure merged not only the functions but also the experience and expertise of the functions, permitting overlapping compliance requirements to be addressed in an efficient and effective manner. However, education and training are essential for the safety and health function to become knowledgeable with the laws, regulations, and policies within the human resource function and the human resource function to become knowledgeable about the laws, regulations, and policies within the safety and health function.

Often as the result of difficult economic or organizational circumstances, which can result in downsizing within the operations, safety and health professionals can be tasked with peripheral functions in addition to their safety and health responsibilities, such as security or environmental functions. These peripheral functions were often individual functions under the human resource umbrella prior to the economic downturn, however, they could be consolidated for cost-saving reasons. These "slash" positions (i.e., Safety/Security Coordinator) often require new functions to integrate within the human resource function for operational purposes. Although these commingling functions often require a minimalist approach simply due to the responsibilities and time limitations, the survival of the company is often at risk and the safety and health professional can integrate and apply his/her managerial skills and abilities to the correlating function.

The safety and health professional is often the unofficial "eyes and ears" of the human resource function within the operational setting. Safety and health professionals are frequently working within the operations conducting assessments, observations, and related activities within the safety and health function. However, while working within the operations, the safety and health professional often observes employee activities which may impact laws, regulations, or policies within the human resource function. It is imperative that the safety and health professional possess the knowledge and ability to be able to recognize the potential violation and the ability to communicate this information to the human resource function in order that preventative action can be initiated. For example, while the safety and health professional is conducting his/her compliance-related inspection, an employee is upset within the operation. In a discussion, the employee explains that she has a very ill family member and asked her supervisor for time off work. The supervisor denied the request. The safety and health professional, with knowledge of the Family and Medical Leave Act (FMLA) and identifying that this could be an FMLA issue, informs the human resource function, who can further investigate and take the appropriate action to avoid potential liability for the company.

It is imperative that safety and health professionals possess a working knowledge of the laws, regulations, and policies within the human resource function. Additionally, if it is a unionized operation, it is essential that the safety and health professional has a working knowledge of the collective bargaining agreement and, in nonunion operations, knowledge of the employee handbook. In general, safety and health professionals should possess a working knowledge and be able to recognize situations involving potential discrimination, including pregnancy, age, religion, gender, disability, race, and national origin. Additionally, safety and health professionals should be able to recognize issues involving leave from the workplace including FMLA, short-term and long-term disability (by policy), Americans with Disabilities Act (ADA) and workers compensation overlap, and related return-to-work issues. Safety and health professionals should have a firm grasp of the issues related to employee discipline, employee involuntary termination due to direct impact on the safety and health function, and controlled substance testing.

Safety and health professionals are often tasked with the management of an individual state's workers compensation function. Although the workers compensation function is reactive (i.e., the injury or illness has already occurred), wherein the safety and health function is proactive, many organizations tend to commingle these functions within their corporate structure. Safety and health professionals should be aware that each state possesses individual state laws, regulations, and systems, and these laws can overlap with other state and federal laws, as well as internal human resource or company protocols. Given the nature of most state workers compensation systems, the state legislatures create the laws and an administrative structure is established within state government which establishes rules and regulations for virtually every aspect of the workers compensation function from rates to duration to care. Safety and health professionals tasked with this responsibility must adhere to these laws and regulations while assisting the injured or ill employee and balancing this with compliance with other laws and regulations. As noted previously, it is essential that safety and health professionals have a working knowledge of these laws and regulations so not to create situations involving noncompliance. For example, an employee is injured on-the-job on a weekend. He/she goes to the emergency room after work and is instructed by the emergency room physician to remain off work for a week. The employee remains at home without notifying his/her supervisor or the company for three days and is subsequently terminated in accordance with the company absenteeism policy (No Call-No Show). Should the doctor's directive or the company policy take precedence? If the employee is terminated, how will this impact the company's workers compensation costs? Following this case further, the employee has reached maximum medical recovery and has rated at a permanency rating of 40 percent of a whole person. The employee is released to return to work and requests an accommodation to be able to perform the job. Is the employee qualified under the Americans with Disabilities Act? Is the company required to provide accommodation to the employee?

As noted, the safety and health function is definitely not "siloed" within the function and every situation and decision possesses the potential to impact other laws and regulations depending on the circumstances and facts of the situation. Given the fact that safety and health professionals work with employees and the human resource function is the primary managing function for employees in many companies, the potential for conflict among and between functions exists in virtually every situation involving employees. It is essential that safety and health professionals become knowledgeable regarding the policies, protocol, and personnel within the human resource function and investigate each situation thoroughly before finalizing a decision which may have human resource implications. With virtually every decision made by the safety and health professional impacting employees directly or indirectly, careful analysis of potential conflicts with human resource laws, regulations, and policies is essential.

In general, safety and health professionals should possess a working knowledge of the major laws and regulations impacting the safety and health function as well as an expert knowledge of the specific laws and regulations that are within his/her sphere of direct responsibilities. In general, safety and health professionals should possess an expert level of knowledge with regards to the Occupational Safety and Health Act and the numerous standards. This expert level must be maintained on a daily basis with all changes, modifications, or decisions being noted with application within the safety and health structure. Safety and health professionals with concurrent environmental or workers compensation responsibilities should possess expert knowledge in various laws, regulations, and court decisions. However, in addition to the expert level within these specific spheres of direct responsibilities, safety and health professionals should have a working knowledge of the potential impacting laws and regulations which can be embedded within the specific situation and decision-making process. In general, these laws and regulations include, but are not limited to, the areas of discrimination in the workplace, labor laws, ADA and laws governing handicap or disability, leave laws (such as FMLA), hiring and recruitment laws, confidentiality laws, and employee record laws.

In the area of hiring, most safety and health professionals do not possess direct responsibility for the hiring and placement of employees; however, this function, or the impact from this function, can have a direct correlation to the safety and health function. Hiring, recruitment, and placement errors can result in an increased probability of workplace accidents and injuries. For example, if an individual is hired for a very physical job function when the individual does not possess the physical capabilities of performing the job function. Although the new employee may be able to perform the job function successfully over a short period of time, the risk probabilities are substantially increased that the new employee may incur a work-related injury if appropriate job modifications or adjustments are not made to the job function.

The laws and regulations surrounding the hiring and recruitment function have substantially changed as a result of new laws and regulations as well as technology. Areas such as pre-employment background checks, especially in the areas of criminal background checks and financial background checks, have changed as a result of not only federal laws but also individual state laws. Questions which were historically asked within the interview process may now be prohibited and can serve as the foundation for discrimination action. The laws and regulations regarding medical screening, controlled substance testing, and confidentiality of records have substantially changed. Technology, especially in the area of social media, as well as recent legislation and court decisions, have impacted the hiring and recruitment process. At a minimum, safety and health professionals should be aware of the best practices in this important area and ensure that compliance is assured.

As noted previously, the wrong placement of new employees within the operational structure can detrimentally impact the safety and health performance. Safety and health professionals should take an active role in the development of written job descriptions, which are key within the hiring and placement process. The written job description should address the specific activities required within the job, appropriate and specific physical demands, specific job duties and responsibilities, and, above all, be completely accurate and up-to-date. As safety and health professionals are well aware, as soon as the candidate becomes an employee, he/she is your responsibility and should be appropriately placed and prepared for long-term success within the job position.

Given the fact that progressive disciplinary action is one of the primary methods through which behavior modification is achieved in the workplace, safety and health professionals should be knowledgeable of the process and impacts of each level within this important human resource policy. Under most progressive disciplinary policies or protocols, disciplinary action for failure to comply with a safety and health requirement is usually part and parcel of the policy. Whether prescribed in a policy and employee handbook within nonunionized operations or within the collective bargaining agreement in unionized operations, the safety and health professional is often actively involved in disciplinary investigations as well as throughout the progressive disciplinary process.

An area where a confluence of many different laws and regulations impacting a situation can amass is the area of restricted duty return to work programs. Safety and health professionals should be aware of the many possible protections afforded to the injured employee beyond simply workers compensation coverage. Although many companies have initiated restricted duty programs to reduce costs within the category of workers compensation time loss payments to the employee and maintaining the employee within the routine of the job function, mismanagement of this function can create impacts not only within the workers compensation area but also in the areas of ADA and state protections, internal short- and long-term disability policies, and related laws and regulations.

Safety and health professionals seldom are disciplined or terminated as a result of their direct efforts within the safety and health realm. However, when a safety and health professional "steps on his/her tail" in areas encompassed by other laws, companies are often placed in the awkward position of being required to act against a management team member. Safety and health professionals should be aware that although they are a valued member of the management team, he/she is still an employee of the company and required to adhere to all laws, regulations, and company policies. Areas of specific focus for safety and health professionals often include, but are not limited to, violations of discrimination laws and policies, relationships in violation of company policies, improper initiation of actions in violation of a law, or inappropriate or unprofessional conduct within the scope of the position.

In summation, the safety and health professional often works hand-in-hand with the human resource function on a daily basis within many operations. Safety and health professionals should recognize this relationship and be aware that the safety and health function does not "work in a vacuum" and is impacted, directly or indirectly, by the laws, regulations, and policies with which the human resource function is tasked with ensuring compliance. Safety and health professionals should not only be experts in their field but also be knowledgeable and be able to recognize the various laws, regulations, and policies which could impact their function. It is the laws and regulations which the safety and health professional is not aware of or has not recognized that will create the greatest difficulties. Preparation, knowledge, and recognition are key for safety and health professionals to avoid missteps or conflicts in the effective and efficient management of a safety and health program.

2 Safety and Health Professional's Guide to Labor and Employment Laws

***

"Don't condescend to unskilled labor. Try it for half a day first."

Brooks Atkinson

"Excellence, in any department, can now be attained, only by the labors of a lifetime. It is not purchased at a lesser price."

Johnson

***

There are three basic labor and employment structures under which employees work in the United States today. First is the "At Will Employment Doctrine." In general, most safety and health professionals working for companies are salaried management team members and thus fall within the "At Will Employment Doctrine." This means that the safety and health professional can be terminated with or without cause for any reason at any time. There are exceptions to this general rule; however, these exceptions are narrow and state specific. Second, safety and health professionals working as consultants or other structured capacities within companies usually work under the terms and conditions of a specified contract. A valid contract would consist of an offer, acceptance, and "consideration" or payment for work activities. When a safety and health professional is working under a contract, the contract terms and conditions are binding and the safety and health professional is considered a "contractor" and not an employee. The third structure is often focused on a group of employees performing similar work activities and usually being paid on an hourly basis. Under the laws and regulations to be discussed, employees are permitted to join together to collectively bargain with the employer regarding wages, hours, and conditions of employment. The result of this negotiations is often called a "collective bargaining agreement" or "union contract." Under this structure, the wages, hours, and conditions of employment take precedent and bind the employees and employer during the term of the contract.

In general, unionization in the United States has depended upon the actions or inactions by employers with regards to wages, working conditions, and related workplace situations. In essence, if the employer is providing a good wage, appropriate hours, and safe and healthful working conditions, employees often saw no need for a third party to represent and negotiate on their collective behalf.

However, where employers provide substandard wages and hours, poor working conditions, and related conditions, employees have been empowered under the law for join and form unions in order to collectively band together to create a greater negotiating power than individual employees may possess. Over the decades, this pendulum has swung from a large number of companies being unionized in the 1930s and 1940s to a substantially lower percentage in the 2000s.

In the early years of this century, the working conditions changed as a result of the emerging industrialization of the United States and events such as the Great Depression, World War I, and the New Deal. Unsafe working conditions, low wages, and other workplace conditions created an environment which was favorable for the growth of unions and the creation of worker's rights in the industrial workplace. The current labor relations structure was created by the laws enacted during this turbulent period of time, namely the Norris-LaGuardia Act of 1932; the National Labor Relations Act (also known as the Wagner Act of 1935); the Labor Management Relation Act of 1947 (also known as the Taft-Hartley Act); and the Labor-Management Reporting and Disclosure Act of 1959 (also known as the Landrum-Griffin Act). Although there were a number of earlier laws which laid the groundwork, including the Sherman Antitrust Act of 1890, the Clayton Act of 1914, the Railroad Labor Act of 1926, and the National Industrial Recovery Act of 1933, the framework for today's labor-management relations and collective bargaining process was established by the combination of the National Labor Relations Act, the Labor-Management Relations Act, and the Labor-Management Reporting and Disclosure Act. Although there is a myriad of other laws which impact the labor-management relations area in various ways today, including such laws as the Occupational Safety and Health Act and the Civil Rights Act, the basic foundation and framework of labor law in the United States remains the Wagner Act, the Taft-Hartley Act, and the Landrum-Griffin Act.

Safety and health professionals are not usually directly involved in the collective bargaining process; however, a base level knowledge of the important aspects that do impact the safety function is essential. A labor organization, also commonly called a union, can be the representative of the employees who choose or are required to become members for the purposes of collective bargaining or negotiating with the company or organization over mandatory subjects including wages, hours, and conditions of employment. In addition to the federal labor laws, many states have passed "right-to-work" laws, which outlaw compulsory union membership and the required dues paid by employees.

Following the specific labor laws identified above, as well as the procedures established by the National Labor Relation Board, employees can form or join a labor organization. If elected, the labor organization becomes the employee's representative. The labor organization and the company or organization are required to negotiate to impasse over the wages, hours, and conditions of employment (which includes safety and health). This negotiation often leads to a collective bargaining agreement, or union contract, which contractually establishes the rules of the workplace. Members of the labor organization often pay a membership fee to join and a periodic payment, usually referred to as dues, to maintain membership.

Safety and Health professionals should be aware that there are different types of union security arrangements which affect the payment of dues by employees. First, safety and health professionals should be aware of the Taft-Hartley Act made “closed shops” or workplaces requiring membership in the union prior to employment, illegal. However, several other types of union security arrangements are permitted to be included within a collective bargaining agreement. In a “union shop,” the company or organization is free to hire whomever they choose; however, employees hired for positions covered by the collective bargaining agreement are required to join the labor organization after a specified probationary period. The “union shop” is the strongest relationship between labor and management currently permitted under law. An “agency shop” is a union security arrangement where nonunion employees are required to pay the union monies equal to the union dues or fees as a condition of continuing employment with the company. The purpose of this payment of union dues despite not being a union member is to compensate the union for their collective bargaining work and the company’s desire to make union membership voluntary. The most common and the least desirable form of union security arrangements is an “open shop.” In an open shop, membership in the union is voluntary and employees choosing not to belong to the union are not required to pay dues.

In very general terms, safety and health professionals should be aware of the process governed by the National Labor Relations Board (NLRB) through which employees can vote to elect or not to elect a labor organization to represent them in the workplace. This general process includes the following steps:

1. Employees contact the labor organization.
2. The labor organization files a petition with the NLRB.
3. Bargaining unit determined by the NLRB.
4. Union authorization cards signed by employees. (Note: If over 50 percent, the union can ask for voluntary recognition of the union without an election.)
5. Company notified. Excelsior list of employees usually requested.
6. NLRB established “laboratory conditions” in the workplace.
7. Very specific rules during the pre-election period.
8. Picketing, solicitation, boycotts, and campaigning.
9. Unfair labor practice charges filed by company and union.
10. NLRB hold the election.
11. If the labor organization wins the election, collective bargain negotiations begin.
12. If the company wins, the union is barred for a period of time.

Of particular importance to safety and health professionals is the impact an election can have on their safety and health programs and efforts. Historically, the time involved from employee contact with a union through the election was one or more months. However, under the recent “fast track” methodology employed by the NLRB, this time period has been condensed to as little as three weeks. During this time period, special rules and requirements will be installed by the NLRB to ensure “laboratory conditions and safety and health professionals be aware that their actions or inactions can negatively impact the process.”

The foundational element that safety and health professionals should provide special attention within the Labor Management Relations Act and the Wagner Act is employees' Section 7 (of the Labor Management Relations Act) rights, which provide protection for employees to form, join, or assist a labor organization; to bargain collectively through representatives of their choosing; and to engage in concerted activities for mutual aid and protection. Section 7 also provides that employees have the right to refrain from any or all such activities except to the extent that such a right may be affected by an agreement requiring membership in a labor organization as a condition of employment. Most of the provisions of Section 7 of the Labor Management Relations Act are designed to protect the rights of the employee identified above.

Safety and health professionals should pay particular attention to the actions and/or inactions which constitute an unfair labor practice under Section 8 of the LMRA. Safety professionals, as agents for the company or organization, can be charged with an unfair labor practice against the company or organization if they:

> (1) interfere with, restrain or coerce employees in the exercise of the rights guaranteed under Section 7; (2) dominate or interfere with the formation or administration of any labor organization or contribute financial or other support to it; discriminate in regards to hire or tenure of employment or any term or condition of employment to encourage or discourage membership in any labor organization; (4) discharge or otherwise discriminate against an employee because he has filed charges or given testimony under the Act; and/or (5) refuse to bargain collectively with representatives of his employees.[1]

Safety and health professionals should be aware of the activities that constitute unfair labor practices and even if he/she is acting in good faith, the performance of the safety function can often place the safety professional in a position where an unfair labor practice charge may be filed by the employee or labor organization.

Safety and health professionals should be aware that some companies or organizations go to great lengths to avoid unionization. Safety and health professionals, as an agency of the company or organization, can easily be "caught in the middle" during a union organizing campaign with unfair labor practice charges being filed against the company or organization for his/her actions or inactions. Some of the common unfair labor practice charges filed by employees or labor organizations include, but are not limited to: interrogation of employees; polling employees; investigative interviews (in violation of the Weingarten rule); threats, promises, and reprisals; granting of benefits; and spying. Some of the activities and statements which the NLRB could find to be in violation of Section 8(a)(1) include the following:

THREATS

- A safety and health professional threatens an employee with violence because of the employee's union activities.
- A safety and health professional tells an employee, "the company will never sign a contract with that union."
- A safety and health professional tells an employee that the company knows who signed the authorization cards and will "get them."

INTERROGATION

- A safety and health professional asks an employee about who is attending a union meeting.
- A safety and health professional asks an employee how she feels about the union.
- A safety health professional asks an employee how other employees feel about the union.

PROMISES

- A safety and health professional states an employee will get a raise if she votes against the union.
- A safety and health professional tells an employee he will get a promotion if he votes against the union.

SPYING

- A safety and health professional implies surveillance when she tells an employee that he had a lot to say at the union meeting the previous evening.
- A safety and health professional visits an employee's home for the purposes of ascertaining her union support.

The above are only examples of possible violations of Section 8(a)(1) and should not be construed in any way as all inclusive. Safety and health professionals with responsibilities within the areas of workers compensation and security as well as safety and health should exercise caution and acquire legal guidance before embarking into any activities or conversations which could constitute an unfair labor practice.

Safety and health professionals should be aware that unfair labor practices can also be committed by the labor organization under Section 8(b). Labor organizations and their agents are prohibited from restraining or coercing employees in the exercise of their Section 7 rights or employers in the selection of their representatives for the purposes of collective bargaining or grievances. Additionally, it is an unfair labor practice to attempt or cause an employer to discriminate against an employee with respect to membership in an organization being denied or terminated on the grounds of failure to provide periodic dues or initiation fees uniformly required as a condition of acquiring or maintaining membership.

Safety and health professionals should be aware that most private sector safety functions fall within Section 2(11) as "supervisors" and are considered management and thus outside of the collective bargaining unit. A "supervisor" means the following:

> any individual having authority, in the interest of the employer, to hire, transfer, suspend, layoff, recall, promote, discharge, assign, reward, or discipline other employees, or responsible to direct them, or to adjust their grievances, or effectively to recommend such action, if in connection with the foregoing exercise of such authority is not of a merely routine or clerical nature, but requires the use of independent judgment.[2]

As a representative of management, safety and health professionals should exercise caution when contacted by a representative of a labor organization. Labor organizations often employ tactics in an attempt to acquire recognition from the company or organization before an election. For example, a union representative may contact the safety professional to ask for an opportunity to discuss alleged complaints regarding safety violations and other safety issues. Prudent safety professionals should contact their human resource department or legal counsel immediately if this happens. Generally, safety and health professionals are not required to comply with the union's request until and unless the union becomes the designated representative of the employees. If the union is not recognized by the company or organization or designated by the NLRB as the employees' representative (usually after the election), the safety and health professional, as the agent for the company or organization, is not required to bargain with the union on any subject, including safety and health.

Safety and health professionals should exercise caution whenever a union organizing campaign is underway. Guidance should always be sought from the human resource department or legal counsel. Generally, safety and health professionals should not look at any lists of employees provided by the labor organization or look at any authorization cards or letters from the labor organization. This could constitute recognition of the labor organization. Additionally, safety professionals should not accept any registered mail or any written documents to the safety professional from the labor representative.

A labor organizing campaign can be very disruptive to safety and health program efforts. Safety and health professionals should be aware that the labor organization must acquire signed authorization cards from 30 percent of the employees in the identified collective bargaining unit in order for the NLRD to order an election. The NLRB conducts a secret ballot, usually onsite, and employees who signed the authorization cards are not bound by the card to vote for the union. If the majority of the employees vote for a union, the NLRB would recognize the union and order collective bargaining negotiations with the company or organization to begin. If the union does not receive a majority of the votes from the collective bargaining unit employees, the NLRB has the ability to recognize the union and order the company or organization to bargain with the union. However, before the NLRB can order bargaining, the union must show that the employer improperly denied the union's pre-election bargaining requests or the union lost its majority as a result of unfair labor practices by the company or organization. If the company or organization committed serious unfair labor practices, the NLRB can order the company or organization to bargain with the union if the union obtained a majority of the authorization cards without even holding a secret ballot.

Given the visibility and functions of the safety and health professional within the company or organization, the safety and health professional can often be a focal point for the filing of unfair labor practices by the labor organization. One of the first tactics often employed is to contact the Occupational Safety and Health Administration (OSHA) to initiate an inspection in order that any alleged violations could potentially be utilized during the labor organization's organizing campaign. The unfair labor practice charges, which usually are filed by both the labor organization and the company, play an important role in the organization process. If the labor organization can prove unfair labor practices by the company or organization, the NLRB can order the company to bargain with the union even if the union lost the election. Additionally, in the event of a strike or lockout situation, unfair labor practices play a key role in determining whether the strike was an economic strike or an unfair labor practice strike, and thus, whether the striking employees or replacement employees return to the jobs.

A labor organizing campaign can be a minefield for the safety and health professional. Safety and health programs, or lack thereof, may be one of the reasons why the employees initially contacted the labor organization. The organizing campaign can be very disruptive to the overall safety and health efforts within the company or organizations while the company or organization and the union battle for the hearts and minds of the voting employees. Safety professionals, as a member of the management team, should exercise caution in their daily job activities and be extremely careful in their interaction and communication with employees and the labor organization representatives. The safety and health professional's actions or inactions directly reflect on and impact the company or organization. Prudent safety and health professionals should always receive direction and advice from their human resource or legal counsel when encountering a labor organizing campaign.

NOTES

1. NLRA Section 8(a).
2. NLRA Section 2(11).

3 Human Resources and Safety in a Unionized Environment

***

"A man's greatest enemies are his apathy and stubbornness."

Frank Tyger

"Nothing can stop the man with the right mental attitude from achieving his goal; nothing on earth can help the man with the wrong mental attitude."

W.W. Ziege

***

The conceptual aspects of safety and health as applicable to a unionized and non-union operation are the same; however, the procedural implementation and management of the safety and health function in a unionized operation can be significantly different from an organization where employees are not represented by a labor union. For safety and health professionals, the differences in managing the safety and health function often involve a more formalized structure and require "steps" for implementation provided by the collective bargaining agreement as well as the third-party representation of an individual group and/or all employees within the operations. In short, the safety and health concept or OSHA requirement is the same, nevertheless, the methodology for implementation within the operation may be significantly elongated within a unionized operation due to structural requirements.

Unionization in the United States is a pendulum which swings with the state of the workplace for the American worker. When employees believe they are receiving adequate wages, working appropriate hours, receiving appropriate benefits, and are working in a safe and healthful environment, there is usually no need to band collectively to force change. However, where employees perceive that they are underpaid, overworked, lack benefits, and the working conditions are unsafe or unhealthy, employees often look to a union organization to fight for change. Over the decades, this pendulum has swung from nearly half of all US-based companies being organized to where approximately 10 percent of all private sector companies are covered under a collective bargaining agreement today. This pendulum can swing back if the variables of wages, hours, benefits, AND workplace safety and health are not continuously monitored and maintained by the company. For the safety and health professional, creating and maintaining a work environment which is safe, healthful, and which employees are actively involved and take "ownership" of the safety function is essential. Where the safety and health function has faltered and employees feel their safety and health could be impacted on the job, there is a greater probability that the employees will require a third party to intervene to force corrective actions by the company. Safety and health issues are often one of, if not "the" reason, for employees to choose to pursue union representation.

The primary authority which impacts the human resource and safety functions in a unionized environment is the collective bargaining agreement. The collective bargaining agreement, often referred to as the "union contract," is a negotiated document which addresses wages, hours, and conditions of employment, which include workplace safety and many of the functions prescribed to human resources. Although safety and human resource functions are included within the four corners of the collective bargaining agreement, areas and issues not addressed are usually left to management's discretion.

Within the human resource function, the collective bargaining agreement identifies and proscribes the primary areas of wages and hours, i.e., how much employees are paid per hour and how long they have to work. The negotiated rates are usually in excess of the federal and state minimum wage levels and the hour requirements usually parallel these laws. However, within the wage and hour perimeters prescribed within the collective bargaining agreement, numerous issues can arise which require the expertise of the human resource manager. The "golden rule" within the human resource function is to ensure employees are paid correctly for the time worked and paid on time. Failure to pay an employee the correct wages or failure to pay on time can create substantial issues within the human resource function. Human resource and safety professionals should remember that failure to provide appropriate wages in a timely manner has a detrimental effect not only on the employee but also his/her family. Many employees live from paycheck to paycheck and any delay or inaccuracy will have an impact on the employee and his/her family.

Within a unionized workplace, the tracking of attendance by employees is a daily activity within the human resource function. Within the collective bargaining agreement is usually a number of prescribed days which can be excused or unexcused absences as well as the disciplinary impact of such absences. For example, under the terms negotiated within the collective bargaining agreement a "no call-no show" (i.e., the employee did not call to report off and did not show up for work), the prescribed disciplinary action would be a one-day suspension on the first offense. Within an organization in which employees are not represented by a labor organization, the employee handbook and/or attendance policies would proscribe the disciplinary action levels. Although this type of progressive disciplinary action is common, the types and levels of disciplinary action will vary among and between collective bargaining agreements and organizations.

Safety and health professionals should also be cognizant of the probationary period often prescribed for new employees within the organization. Although the probationary period often varies with the organization, the probationary period is a time in which the organization can assess the skills, abilities, and attitude of a new employee to ensure that he/she is a "good fit" for the organization. Probationary periods can be as short as 30 days and as long as a year; however, during this period of time, the new employee is usually provided training and assessments conducted to carefully assess the new employee's abilities to become a long-term employee within the organization. One of the major assessments, as well as training, is in the area of safety and health. During this probationary period, new employees can usually be terminated from employment without cause and are provided no protection under the collective bargaining agreement.

Under virtually all collective bargaining agreements, seniority is the governing rule. The longer time an employee has with the company or in the union, the greater his/her ability to bid for positions with higher pay, to bump positions of employees with less seniority during difficult times, and to be provided recognition within the organization and union for the time he/she has served within the organization. Seniority is the "coin of the realm" within the collective bargaining system. For safety and health professionals, senior employees know their jobs better than anyone else within the organization and know safety and health is essential within the workplace. Safety and health professionals often rely on senior employees to serve on their safety committees, conduct specific equipment training, and other important safety and health activities.

Attendance, or lack thereof, is essential for production, quality, and safety. Although virtually all collective bargaining agreements provide a prescribed progressive disciplinary system in the area of attendance, employees often have holidays, personal holidays, FMLA leave of absence, excused absences, and other time away from the workplace which does not impact the progressive disciplinary system. For safety and health professionals, one of the major areas in which employee absences impact the function is in the area of safety and health training. Employees who are absent when an important safety and health training is scheduled often require a second or multiple additional training sessions to ensure that all employees receive the required training. This duplication of effort often impacts the safety and health professional's ability to conduct other essential work within the workplace.

Within the required areas of negotiations, namely wages, hours, and conditions of employment, safety and health is usually considered a condition of employment and thus a negotiable item within the collective bargaining process. Although there is a wide variation among and between collective bargaining agreements, safety and health professionals working within a unionized operation should be cognizant of the safety and health requirements under the collective bargaining agreement and recognize that new programs, new policies, and other aspects of the safety and health program may be a negotiable issue. Safety and health professionals should work closely with the human resource function and legal counsel to identify and acknowledge the process through which safety and health programs, policies, safety equipment, and other safety and health issues must be negotiated with employee representatives within the labor organization.

Under most collective bargaining agreements, a prescribed method is negotiated through which employees can grieve and/or file a grievance against the company for violation of the collective bargaining agreement or workplace issue. Although the grievance procedure often varies from a one-step grievance hearing before a panel of company and union representatives to a multiple-step process including professional arbitrators. The grievance process for employees often includes discussions with the union representative or a written document identifying the basis for the grievance against the company. Safety and health professionals should be aware that there is a high probability of employees filing grievances when change is enacted to improve the safety and health function within the operations. Even with the vetting of the safety and health program or activity with the labor organization, employees usually have the right to file a grievance addressing their individual situations. For example, the installation of a personal protective equipment program is enacted to protect employees from injury or illness within the operations. Overall, employees recognize and support a greater level of safety and health in their job functions. However, the PPE may create unique difficulties for individual employees, such as a loss of dexterity with gloves or fogging of safety glasses, which the employee may file a grievance if not addressed in a timely manner by the safety and health professional.

Safety and health professionals should also be aware that when disciplinary action is utilized to change unsafe behavior by an employee, there is a high probability that the employee will grieve the disciplinary action. It is essential that all employees follow the prescribed safety policies, programs, and procedures in order to safeguard all employees within the workplace. The level of disciplinary action for failure to follow the prescribed safety policy, program, or procedure is usually provided within the collective bargaining agreement. The burden of proof is usually on the organization to prove the violation. Safety and health professionals, as representatives of the organization, often have to prove the violation through documentation or other evidence within the grievance process. Strict adherence with safety and health programs, policies, and procedures are essential and the enforcement process is usually a negotiated process with various levels of appeal.

Depending on the organization, the safety and health professional may report directly to the human resource director or indirectly while reporting directly to a corporate function. No matter what the reporting structure is for the organization, the safety and health professional will usually work closely with the human resource function due to the overlapping issues involving employees. One area in which the human resource function, safety and health function, and labor organization often work together is after an employee incurs a work-related injury or illness.

Safety and health professionals often have responsibility for the management or supervision of the workers' compensation function through an administrator or insurance carrier. The laws and regulations governing workers compensation are unique to each state. Additionally, within the collective bargaining agreement, additional benefits may have been negotiated which may be applicable as well as absence from the workplace. Although most state laws proscribe the compensation and benefit amounts within the workers' compensation structure, issues, such as timely payment and return to work, can create conflicts resulting in grievances under the collective bargaining agreement as well as individual state agency involvement.

The human resource function is often the organization's representative and the shop steward is the onsite representative of the labor organization. Although the organization and labor organization often have larger teams of representatives in the organizing and negotiation process, the human resource manager and onsite labor representative are the "day-to-day" administrators of the collective bargaining agreement. Daily enforcement of company policies, such as attendance, are usually conducted by the human resource function. The shop steward or onsite union representative is usually the first step for an employee to file a grievance against enforcement of the policy against him/her. The levels of appeal for the grievance process are usually prescribed within the collective bargaining agreement.

The general trend in the past decade has been an increase in public sector union membership and a reduction in private sector union membership. With safety and health being a primary issue for most employees and a lack of or negligent safety and health efforts often being a major reason why employees seek union representation, safety and health professionals must have a firm grasp of not only compliance requirements and programs but also must go "over and above" to create and maintain a safe and healthful work environment for all workers. The pendulum in the private sector has swung in favor of the companies at this point in time because workers do not need third-party representation and the power of collective bargaining to achieve a safe workplace, fair wages, appropriate benefits, and other necessities. However, when the companies do not maintain safety and health as a high priority and reduce wages, benefits, and other conditions within the workplace, the pendulum can quickly shift and unionization can increase. The safety and health of ALL employees are paramount and it is the safety and health professional's job to create and maintain all aspects of this environment within the workplace.

Modified for the Purposes of this Text.
NEWS RELEASE—US Bureau of Labor Statistics
For release 10:00 a.m. (EST) Friday, January 19, 2018
USDL-18-0080 Technical information: (202) 691-6378
• PressOffice@bls.gov

UNION MEMBERS—2017

The union membership rate—the percent of wage and salary workers who were members of unions—was unchanged at 10.7 percent in 2017, the US Bureau of Labor Statistics reported today. The number of wage and salary workers belonging to unions, at 14.8 million in 2017, edged up by 262,000 from 2016. In 1983, the first year for which comparable union data are available, the union membership rate was 20.1 percent and there were 17.7 million union workers. The data on union membership are collected as part of the Current Population Survey (CPS), a monthly sample survey of about 60,000 eligible households that obtains information on employment and unemployment among the nation's civilian noninstitutional[1] population age 16 and over. For more information, see the Technical Note in this news release.

Highlights from the 2017 data:

- The union membership rate of public-sector workers (34.4 percent) continued to be more than five times higher than that of private-sector workers (6.5 percent).
- Workers in protective service occupations and in education, training, and library occupations had the highest unionization rates (34.7 percent and 33.5 percent, respectively).
- Men continued to have a higher union membership rate (11.4 percent) than women (10.0 percent).

- Black workers remained more likely to be union members than White, Asian, or Hispanic workers.
- Nonunion workers had median weekly earnings that were 80 percent of earnings for workers who were union members ($829 versus $1,041). (The comparisons of earnings in this release are on a broad level and do not control for many factors that can be important in explaining earnings differences.)
- Among states, New York continued to have the highest union membership rate (23.8 percent), while South Carolina continued to have the lowest (2.6 percent).

Industry and Occupation of Union Members In 2017, 7.2 million employees in the public sector belonged to a union, compared with 7.6 million workers in the private sector. Although the union membership rate for private-sector workers edged up by 0.1 percentage point in 2017, their unionization rate continued to be substantially lower than that for public-sector workers (6.5 percent versus 34.4 percent).

Within the public sector, the union membership rate was highest in local government (40.1 percent), which employs many workers in heavily unionized occupations, such as teachers, police officers, and firefighters. Private-sector industries with high unionization rates included utilities (23.0 percent), transportation and warehousing (17.3 percent), telecommunications (16.1 percent), and construction (14.0 percent). Low unionization rates occurred in finance (1.1 percent), food services and drinking places (1.4 percent), and professional and technical services (1.7 percent).

Among occupational groups, the highest unionization rates in 2017 were in protective service occupations (34.7 percent) and in education, training, and library occupations (33.5 percent). The rate for workers in education, training, and library occupations continued to decline in 2017. Unionization rates were lowest in sales and related occupations (3.2 percent); farming, fishing, and forestry occupations (3.4 percent); food preparation and serving related occupations (3.8 percent); and in computer and mathematical occupations (3.9 percent).

Selected characteristics of union members in 2017: The union membership rate continued to be higher for men (11.4 percent) than for women (10.0 percent); over the year, the rate for men edged up, and the rate for women edged down. However, the gap between their rates has narrowed considerably since 1983 (the earliest year for which comparable data are available), when rates for men and women were 24.7 percent and 14.6 percent, respectively. Among major race and ethnicity groups, Black workers continued to have a higher union membership rate in 2017 (12.6 percent) than workers who were White (10.6 percent), Asian (8.9 percent), or Hispanic (9.3 percent). By age, union membership rates continued to be highest among workers ages 45 to 64. In 2017, 13.2 percent of workers ages 45 to 54 and 13.5 percent of those ages

55 to 64 were union members. In 2017, the union membership rate for full-time workers was about twice the rate for part-time workers (11.8 percent versus 5.7 percent). Union representation in 2017: 16.4 million wage and salary workers were represented by a union. This group includes both union members (14.8 million) and workers who report no union affiliation but whose jobs are covered by a union contract (1.6 million).

Earnings among full-time wage and salary workers: Union members had median usual weekly earnings of $1,041 in 2017, while those who were not union members had median weekly earnings of $829. In addition to coverage by a collective bargaining agreement, this earnings difference reflects a variety of influences, including variations in the distributions of union members and nonunion employees by occupation, industry, age, firm size, or geographic region.

Union membership by state in 2017: 27 states and the District of Columbia had union membership rates below that of the US average, 10.7 percent, while 22 states had rates above it and 1 state had the same rate. All states in both the East South Central and West South Central divisions had union membership rates below the national average, while all states in the New England, Middle Atlantic, and Pacific divisions had rates above it. Union membership rates increased over the year in 25 states and the District, decreased in 21 states, and were unchanged in 4 states. Nine states had union membership rates below 5.0 percent in 2017, with South Carolina having the lowest rate (2.6 percent). The next lowest rates were in North Carolina (3.4 percent) and Utah (3.9 percent). Two states had union membership rates over 20.0 percent in 2017: New York (23.8 percent) and Hawaii (21.3 percent). The largest numbers of union members lived in California (2.5 million) and New York (2.0 million). Over half of the 14.8 million union members in the United States lived in just 7 states (California, 2.5 million; New York, 2.0 million; Illinois, 0.8 million; Michigan and Pennsylvania, 0.7 million each; and New Jersey and Ohio, 0.6 million each), though these states accounted for only about one-third of wage and salary employment nationally.[2]

NOTES

1. Unemployment Rate LNS14000000, Bureau of Labor Statistics Data, December 12, 2018.
2. News Release, USDL 18-0080, January 19, 2018.

4 Collective Bargaining and the Impact on the Safety Function

***

"Liberty, when it begins to take root, is a plant of rapid growth."

George Washington

"Conformity is the jailer of freedom and the enemy of growth."

John F. Kennedy

***

Collective bargaining, "as contemplated by the National Labor Relations Act, is a procedure looking toward making a collective bargaining agreement between employer and accredited representative of employees concerning wages, hours, and other conditions of employment, and requires the parties deal with each other with open and fair minds and sincerely endeavor to overcome obstacles existing between them to the end that employment relations may be stabilized and obstruction to free flow of commerce prevented."[1]

A collective bargaining unit is defined as "all of the employees of a single employer unless the employees of a particular department or division have voted otherwise."[2] A collective bargaining agreement, or often referred to as the "union contract," is "an agreement between the employer and a labor union which regulates terms and conditions of employment. The joint and several contract of members of a union made by officers of the union as their agents. Such is enforceable by and against the union in matters which affect all members alike or large classes of members, particularly those who are employees of other party to contract."[3]

The safety and health of employees is a condition of employment and this a mandatory bargaining issue. Safety and health professionals may or may not be directly involved in the collective bargaining negotiations process depending on your company or organization. However, safety and health professionals should be aware that safety and health issues arising from the workplace are often a major issue within the negotiation process and they are considered a mandatory subject for bargaining. Whether the safety and health professional is "at the negotiation table" or working behind the scenes providing safety and health-related information to the company spokesperson, safety and health professionals are often a key member of the negotiations team.

The primary times in the life of the collective bargaining process which may impact the safety and health function often includes the initial campaign by the labor organization, the negotiation of a collective bargaining agreement, and the expiration of a collective bargaining agreement. Safety and health professionals should be aware of the status of the labor situation within their operations. In companies or organizations where the labor organization was already present, the situation may involve a collective bargaining agreement that is expiring and is being re-negotiated. In other situations, the labor organizations may have been recognized by the company or organization upon request or before the certification of the results of an NLRB election. The most common is when the labor organization is recognized and negotiations are required after a secret ballot election and certification of the results of the election identifying the labor organization as the representative of the employees within the bargaining unit. In any of these situations, the safety and health function often incurs decreased focus, decreased vigilance, and increased risk in the workplace.

Section 8(b) of the NLRA provides the duty to both the company or organization and the labor organization to bargain collectively and includes the mutual obligation to meet at reasonable times and to negotiate in good faith over the subjects of wages, hours, and other terms and conditions of employment, including safety. This section also identified that the company and labor organization are required to execute a written contract including any/all agreements reached by the parties. However, Section 8(d) does not require or compel the company or labor organization to agree to the other side's proposals or to make concessions in order to reach an agreement. Safety and health professionals should ensure that all aspects of the safety and health function requiring negotiations and agreement as a condition of employment are identified and brought to the table.

Safety and health professionals should also be aware that Section 8(a)(3) identified that it is an unfair labor practice for the company or organization to refuse to bargain with the recognized labor organization. And under Section 9(a), the labor organization who won the majority of the votes in the election is the exclusive representative of the employees within the identified collective bargaining unit and the company or organization is required to negotiate in good faith over wages, hours, and other conditions of employment. It should be noted that Section 9(a) does not prevent individual employees from taking grievances directly to the safety and health professional, as a representative of the company, as long as this communication is not inconsistent with the terms of the collective bargaining agreement. However, safety and health professionals should exercise extreme caution and acquire guidance from legal counsel in order to avoid any circumstances which could possibly be construed as an unfair labor practice.

Under Section 8(a)(5), a company or organization can commit a violation if the company or organization refuses to negotiate with the recognized union over the mandatory bargaining subjects of wages, hours, and conditions of employment.[4] This violation can be applicable even if the company or organization bargains in good faith and desires to reach an agreement. A company or organization is required to bargain only over the mandatory bargaining issues of wages, hours, and conditions of employment and the company or organization is not required to bargain over any subject or issues that are not considered a mandatory subject.

Although the lines between a mandatory subject and a remote or nonmandatory subject can often get blurred, safety professionals should be aware that safety rules have been found to be a mandatory subject requiring negotiations.[5] Other safety-related issues which have also been found to be mandatory subjects include bonuses or gifts,[6] job transfers,[7] work schedules,[8] and anti-discrimination clauses.[9] Safety professionals should be aware that although negotiations are not limited to mandatory subjects, insistence by the company or organization or labor organization to include nonmandatory subjects in the collective bargaining agreement as a condition of agreement or signing can constitute an unfair labor practice.[10]

Safety and health professionals should be aware that labor organizations are under the same duty to bargain in good faith under Section 9(a). Labor organizations possess the correlating duty to meet and confer with the company or organization in good faith with the desire to reach an agreement. A labor union can commit an unfair labor practice by refusing to bargain to impasse, refusing to sign a negotiated agreement, or demanding to negotiate for a unit that has not been recognized. Safety professionals should be aware that it is also an unfair labor practice for a labor organization to bargain to impasse over nonmandatory subjects of bargaining.[11]

As can be seen, there is a multitude of different rules and regulations governing the collective bargaining process and the requirements may change depending on the circumstances. The negotiations can take a variety of forms from first-time negotiations, to the continuation of an existing collective bargaining agreement, to a successor employer who purchased the company or organization, or to a multi-employer bargaining unit. Most safety and health professionals will work "behind the scenes" to assist the spokesperson or legal counsel who is representing the company or organization at the negotiation table with specific safety or related issues. However, it is important that the safety professional be properly prepared to provide guidance and assistance before, during, and after the negotiation process.

In preparing for pending negotiations, safety and health professionals should be cognizant that any adversarial climate which developed between the company or organization and the labor organization before and during the election process usually continues at the negotiating table. This is often the case with initial contract negotiations where the company or organization and the labor organization were combative during the organizing campaign with the company or organization losing the election. The "tactics" often employed by the company as well as the labor organizations during the organizing campaign create a level of animosity on both sides which carries over to the negotiating table.

Safety and health professionals should be aware that preparing for collective bargaining negotiations takes substantial time, research, training, and resources. Although safety professionals should be aware that until a collective bargaining agreement is reached and ratified, supervisors and management team members retain all management authority they possessed prior to the union certification. Although the working environment should be the "status quo," in many circumstances, the organizing campaign and election results have created changes in the workplace. Safety and health professionals should be aware that this changing work environment may require supervisors and members of the management team to undergo significant behavioral, ego, practical, and attitudinal changes that require substantial re-training.

In preparing for the collective bargaining negotiations, safety and health professionals may be asked to prepare and share information with the labor organization. Although all requests should be approved by the lead negotiator or legal counsel prior to providing any information to the labor organization, the Taft-Hartley amendments to the NLRA and orders from the NLRB can permit the labor organization to request specific information from the company or organization prior to negotiations. Safety and health professionals should be aware that information often requested by labor organizations, in addition to wage date, hours worked, EEO data, and related information, includes safety inspections, safety audits, safety compliance programs, injury and illness data, training records, workers compensation date, and insurance information.

Safety and health professionals should be prepared to conduct a substantial amount of research, case studies, and case and data analysis for each safety or related issue in preparing for negotiations. The information acquired is often reviewed and assembled into proposals and counterproposals for each identifiable issue and submitted to the labor organization at the bargaining table. Conversely, the labor organization will conduct their own research and develop proposals and counterproposals for each identifiable issue. Each proposal or counterproposal should be carefully analyzed with an appropriate justification or rationale to justify the proposal or counterproposal. Proposals and counterproposals should be carefully worded and provided in clear language as well as being consistent with the position and strategy of the company or organization.

One of the most difficult challenges for safety and health professionals is the preparation of various analysis of data to determine the estimated cost of each of the various proposals or counterproposals. For safety and health professionals, the cost of various safety efforts can possess a wide variable but often include, but not limited to, equipment costs, replacement costs, training costs, overtime costs, insurance costs, and related costs. For larger companies or organizations, simply the training costs of x number of employees multiplied by the employee's rate of pay multiplied by the number of training hours can reach a substantially large amount of money.

The collective bargaining process usually involved four (4) basic steps; namely, the initiation of the actual bargaining, the establishment of the rapport between the negotiators, the negotiation and consolidation of the proposals, and the completion of bargaining and consolidation of the agreements into the collective bargaining contract. The safety and health professional, no matter whether he/she is seated at the negotiating table or working behind the scenes, is an instrumental member of most negotiating teams. Safety and health professionals should be well prepared at all times and be ready at a moment's notice to produce information to support any safety-related proposals or counterproposals. Safety and health professionals should also be aware that the negotiating process can be a grueling and time-consuming process often stretching for weeks or even months. The location of the negotiation sessions is often at locations away from the operations, thus, safety and health professionals actively involved as part of the negotiating team should remember to make appropriate arrangement within the company or organization to fulfill the required duties and responsibilities in the safety and health professional's absence.

Safety and health professionals should be aware that the results of the collective bargaining negotiations can move in various directions. Although both the labor organization and company or organization possess a duty to bargain in good faith, sometimes an impasse is reached resulting in a lockout of employees in the bargaining unit by the company or organization. This can be followed by the replacement of the employees or the labor organization can call for a strike by employees in the bargaining unit. During this time period, safety and health professionals should exercise caution because the intensity of the situation will increase dramatically and the number of unfair labor practice charges filed by the company or organization and the labor organization can often increase dramatically. In general, if the strike is found to be a purely economic strike, the replacement workers usually maintain their jobs after the conclusion of the strike. If the strike is found to be the result of an unfair labor practice by the company or organization, the striking employees usually return to their jobs and the replacement workers lose their jobs.

In most negotiations, an agreement between the company or organization and the labor organization is reached before impasse. The results of the various negotiated proposals and the agreed-upon wages, hours, and conditions of employment which were negotiated are codified in a document usually called the collective bargaining agreement or union contract. Section 8(b) of the NLRA provides a duty to the company or organization as well as the labor organization to include "the execution of a written contract incorporating any agreement reached if requested by either party."[12] If the labor organization or company or organization refuses to sign the written contract containing the terms, conditions, and provisions that were agreed to between the company or organization and the labor organization, this refusal would constitute a violation of Section 8(b)(3) of the NLRA.

Safety and health professionals should be aware that once signed and ratified, the collective bargaining agreement becomes the "law of the workplace." Most collective bargaining agreements require some form of grievance process wherein any alleged violation of the collective bargaining agreement can result in a grievance being filed by an employee or the labor organization resulting in hearings up to and including arbitration. Safety and health professionals should carefully analyze the collective bargaining agreement and appropriately adjust or modify existing safety programs, policies, and procedures to adhere to the terms and conditions negotiated in the collective bargaining agreement. Safety and health professionals should be aware that any disagreements with regards to virtually any aspect of the safety and health programs, policies, or procedures usually result in formal grievances being filed by the labor organization or individual employees.

Although this chapter is a "broad brush" overview of the various aspects of the collective bargaining process and the areas which impact many safety and health professionals, this chapter in no way encompasses all of the various laws and regulations that are involved in the NLRA and related labor laws. Safety and health professionals are not usually expected to be labor law experts and are seldom the primary negotiator in the collective bargaining process; however, they are often key members of the negotiation team. Safety and health, as a condition of employment, is a mandatory subject of bargaining in any collective bargaining negotiations and it is a vital issue in any company or organization. Safety and health is not only an important issue for safety and health professionals; it is also of the highest priority for employees and their labor representatives. Safety and health professionals should maintain their focus on the priorities of the function and ensure that the working environment is maintained in a manner in which the safety and health of all employees is the highest priority.

NOTES

1. *Black's Law Dictionary.* Also see NLRA Section 8(5), 29 USCA Section 158(5).
2. *Black's Law Dictionary.*
3. Ibid.
4. *NLRB v. Ins. Agents' Int. Union*, 361 US 477 (1960).
5. *NLRB v. Miller Brewing Co.*, 408 F.2d 12 (CA-9, 1969).
6. *NLRB v. Electric Steam Radiator Corp.*, 321 F.2d 733 (CA-6, 1063).
7. *Rapid Roller Co. v. NLRB*, 126 F.2d 452 (9CA-7, 1942).
8. *Houghton v. NLRB*, 18 F.2d 486 (CA-3, 1941).
9. *United Packinghouse, Food & Allied Workers Int. Union v. NLRB*, 135 App. DC 111, 416 F.2d 1126 (1969).
10. *NLRB v. Wooster Div. of Borg-Warner Corp.*, 356 US 342 (1958).
11. NLRA Section 8(b)(3).
12. NLRA Section 8(b).

5 Engaging and Empowering Employees within the Safety Function

***

"Our character is what we do when we think no one is looking."

H. Jackson Brown, Jr.

***

Everyone wants a safe workplace. Employees want to go home after their workday in the same physical and mental condition as when they started at the beginning of the day. Employers want to create and maintain a safe workplace because it is smart business, it is the law, and workplace injuries and illnesses have costs. However, despite a common goal, achieving and maintaining a safe, healthful, and friendly workplace on a consistent basis is often a substantial challenge and costs money. Saying you want a safe and healthful workplace and actually achieving and maintaining a safe and healthful workplace on a daily basis are significantly different.

To create and maintain a truly safe and healthful work environment which empowers employees, the organization must have the commitment of the top tier of management. Top-level managerial commitment should include all facets of commitment, ranging from necessary funding, necessary manpower and resources, making safety and health a top priority, and a complete "buy-in" to create and maintain a safe and healthful workplace. Often, the top level of organizations have "talked the talk" for safety, however, they have failed to provide adequate funding and support to truly create and maintain a safe and healthful workplace. Top-level management support and funding are essential to create an environment where employees can truly be engaged and empowered to accept responsibility and become an active participant within the safety and health function.

The human resource function correlating with the safety and health function is key in creating the environment where employees can engage with management and fellow employees as well as learn to accept responsibility for their safety and the safety and health of their fellow employees. The safety and health manager or the management team member with direct responsibility for compliance and the safety operations should create a transparent system which actively involves employees in all facets of the safety function. Although there may be times, such as individual disciplinary actions, which must be compartmentalized for confidentiality purposes, the safety function should be open and encourage active involvement by all employees.

However, as a result of the recent economic recession and the resulting impacts, there may be a loss of trust between organizations and employees. Historically, it was not uncommon that an employee would remain employed by one particular organization for decades and was rewarded with a defined pension paid by the company and a "gold watch" at the end of their career. Today, it is widespread for organizations to downsize as part of a corporate strategy and severing employees from their jobs without much forethought or compassion. Additionally, organizations have shifted much of the retirement risk to individual employees through the elimination of defined pension plans and substitution of 401K plans. Many of the employees who maintained their jobs within the organization and weathered the recent storms of the economic recessions have lost their home, vehicles were repossessed, savings disappeared, and otherwise suffered substantial financial setbacks. Their children, many of who are or have entered the workplace today as employees, have experienced the detrimental impacts on their family and may harbor an inherent distrust of organizations. Generally, the employee of today and in the future will be more entrepreneurial in nature, may possess an inherent distrust for organizations, possesses a higher probability of being employed for 36 months rather than 36 years, and are far more demanding for not only a safe and healthful work environment but an accommodative environment than previous generations.

As many organizations have realized, turnover in personnel costs money in terms of productivity, quality, and training. As organizations have devalued longevity in employment, many employees have also devalued loyalty to the organization and no longer "ride for the brand." This lack of trust between the organization and employee has created turnover as well as increased risk within the workplace. In today' s workplace, the trust between the organization operating on a quarter-to-quarter basis and who may downsize a workforce to meet a quarterly goal has created an environment in which trust and loyalty within their employee ranks are absent. Conversely, the entrepreneurial workforce of today, with transportable 401Ks and knowledge of their parents' displaced trust, have little or no loyalty to the organization and will voluntarily move to acquire a better workplace, more money, or greater satisfaction or happiness.

Before an organization can truly engage and empower their employees to accept responsibility for the safety functions, organizations must restore a level of trust within their workforce. This trust must be earned by the management of the organization through their actions and decisions. Within the safety realm, the top management must truly be committed to workplace safety and provide the necessary resources and support over an extended period of time to achieve success. Middle management must take an active role and prioritize safety and health within their daily activities. Employees must see, through daily actions, that their direct supervisor values their safety and health. The safety and health professional must "take the lead" to design and develop a safety system which integrates and empowers employees within the safety function while providing the "tools" to successfully engage and assume ownership of the safety and health function. HR managers play a pivotal role in creating and maintaining the environment which empowers individual employees while maintaining compliance with organizational policies within the workplace.

Every workplace is its own ecosystem. There are formal leaders and informal leaders. There are managers whose employees possess a comfort level and "get things done" attitude while there are other managers whose employees avoid work because of a lack of results, a confrontational nature, or other negative aspects. In many organizations, there is a mentality of separation, superiority, and even hostility between management and employees. Safety and health professionals, working with the human resource function and management team, must change this dynamic and create an inclusive environment with minimal delineation between arbitrary designations and levels among and between employees and create a trust level which can permit a team environment to evolve. If this cultural shift in the workplace ecosystem is absent, the status quo of old-line safety and health will be maintained and engagement and employee empowerment will not be a viable option.

To change the workplace ecosystem, safety and health professionals should consider activities to regain and/or build trust as well as activities to build employees' confidence in accepting responsibility for their safety and the safety of their fellow employees. Foundational activities can include the following:

1. Get to know jobs and individual employees: New safety and health professionals should allot sufficient time to meet with each employee at every role, requesting the employee explain/exhibit their job functions and activities. Not only will this be an exceptional learning activity for the safety and health professional but it will also provide an opportunity to meet with each individual employee "on his/her turf" and permit the employee to showcase his/her expertise within the job function. Who knows the specific job better than the employee who is performing the job on a daily basis!
2. Build competency through education and training: Safety and health professionals should consider methods through which to educate and train all employees regarding the requirements, methods, policies, and other components of the overall safety and health function. Specialized education and training activities should be considered for employees on the safety committee, serving as safety and health liaisons, and related functions.
3. Stand by your word: Safety and health professionals should be the onsite expert within the safety and health function and should "do their homework" prior to responding to questions and inquiries. Safety and health professionals should not be afraid of making an error and should stand behind their "word." If the response to the specific decision is "no," safety and health professionals should provide an explanation as to the decision and identify the mitigating safeguards being taken regarding the situation. For example, the appropriate corrective action would be to modify the particular machine which will take six months of lost production and cost $1 million in repairs. The organization's decision was to defer the repairs until the operations shut down for a period the following year and continue to operate with the current safeguards. Most employees can accept the decision if the safety and health professional can provide an appropriate explanation.

4. Inspire the employee and praise positive behavior: Positive reinforcement is one of the most cost-effective methods to emphasize.
5. Involvement in decision making.
6. Show you care: Recognize positive reinforcement.
7. Near miss programs.
8. Empower to stop work: Work with others.

Specifically for safety and health professionals, see the following from *Workplace Safety and Health*:

1. **Be enthusiastic**. Focus on a positive and winning attitude.
2. **Plan your work and work your plan**. Each and every level of your education, training and accountability process should be mapped and analyzed. Be creative and challenge the status quo.
3. **Show respect to all employees at all levels**. Who knows the job better than the individual who performs the job daily?
4. **Be accountable**. Your employees, managers and executives accountable for their safety and health performance. And the safety and health professional should hold himself/herself accountable. If you say you are going to do something, do it!
5. **Listen intently to your employees**. Safety and health professionals should stop what they are doing and listen intently to any comments or ideas which any employee should bring forward. Safety and health professionals should acknowledge and thank the employee for his/her active involvement in safety and health.
6. **Be a good example**. As the coach, safety and health professionals should always set the standard for safety and health by always wearing the appropriate PPE and always following the safety and health rules. "Do what I say and not what I do" doesn't work in safety and health.
7. **Speak and communicate safety and health daily**. Safety and health professionals should be current on all new regulations, standards and trends within the function and communicate this information throughout the organization. Safety and health doesn't take a day off thus the safety and health professional should be actively communicating and promoting safety and health on a daily basis.
8. **Open your door to employees**. Safety and health professionals should be readily available for employees to communicate with them. Safety and health professionals should strive to create a level of comfort for all employees to become actively involved and take ownership of the safety and health function.
9. **Be visible on the shop floor**. Although safety and health professionals are often overwhelmed with their many duties and responsibilities, it is important to schedule time each day to be on the shop floor. The visibility of the safety and health professional reinforces the importance of safety and health to employees.

10. **Communicate clearly and concisely**. Safety and health professionals should strive to ensure that all communications are clear, concise and with no "grey areas" which may be subject to interpretation. Safety and health professionals should ensure that communications are provided at the appropriate educational levels for each function within their organization and with substantial clarity to ensure no ambiguity.
11. **Pat employees on the back**. Safety and health professionals should consistently recognize their safety and health efforts of their employees and provide incentives to spur further involvement and personal ownership in the safety and health function. Incentives do not need to be a formalized incentive program but activities such as a "pat on the back," a hand-written note, or other personalized recognition can be effective.
12. **Avoid taking credit for employee's accomplishments**. Through empowerment of employees, the safety and health function becomes a team effort. Safety and health professionals should always acknowledge their employee's accomplishments in safety and health and should never claim credit for the work of the team.
13. **Leaders without titles**. Safety and health professionals should recognize that many organizations have leaders who function without the requisite title or position. These leaders can be instrumental in the success of your empowerment efforts as well as your overall safety and health program. Safety and health professionals should recognize these leaders and cultivate their leadership skills and abilities within the overall safety and health function.[1]
14. **Ensure ongoing education and training**. The safety and health function is a marathon rather than a sprint. Ongoing education and training empowers employees with the skills and abilities to take true ownership over the safety and health function.
15. **Acknowledge employee accomplishments**. Continuous and pertinent feedback as to the achievements of the safety and health efforts is essential. Safety and health professionals should recognize and reward the team's efforts within the safety and health function.
16. **Eliminate "I" from your language**. Safety and health professionals should recognize that the ownership of the safety and health function now belongs to the team. The safety and health professional, as the coach, should recognize this transition and always recognize the team's efforts. The simple change from "I" to "we" can go a long way in creating ownership of the safety and health function.
17. **Invite feedback and respond**. Safety and health professionals should create comfortable situations where employees are invited to provide their thoughts and feedback with regards to safety and health issues. Safety and health professionals should recognize that the comfort level and time for acceptance of these new duties and responsibilities may vary among and between individual employees.
18. **Be willing to do the job you are asking them to do**. Safety and health professionals should make the time to spend at least a few minutes at each

and every job station within the operation and talk with the employee performing the job function. Safety and health professionals should not ask any employee to perform any task which they themselves would be unwilling to perform.

19. **Be willing to be flexible**. Although safety and health standards can sometimes be restrictive, it is important for safety and health professionals to listen to their team members' ideas and be flexible in their assessments.
20. **Be fair and consistent**. In holding team members accountable for their safety and health responsibilities, safety and health professionals should ensure that they are always fair and consistent in each and every decision.
21. **Be a great coach**. Safety and health professionals, as the coach of the team, should strive to bring out the best safety and health performance in each employee in each job. The safety and health professional should challenge and inspire the team members and encourage top-level safety and health performance.[2]

Safety and health professionals working in ISO rated organizations should be aware of the new ISO 45001 that addresses specific requirements for an occupational safety and health program or system. The purpose of ISO 45001 is the establishment and maintenance of occupational safety and health management systems to improve safety and health performance, eliminate hazards and risks, take advantage of occupational safety and health opportunities, and address occupational safety and health programs and systems nonconformities. Of particular importance is the specific element of employee engagement and empowerment of employees within the safety function.

ISO 45001: OCCUPATIONAL HEALTH AND SAFETY

Over 7600 people die each day from work-related accidents or diseases – that's over 2.78 million every year*.

The burden of occupational injuries and diseases is significant, both for employers and the wider economy, resulting in losses from early retirements, staff absence and rising insurance premiums.

To combat the problem, ISO has developed a new standard, ISO 45001, *Occupational health and safety management systems – Requirements,* that will help organizations reduce this burden by providing a framework to improve employee safety, reduce workplace risks and create better, safer working conditions, all over the world.

The standard was developed by a committee of occupational health and safety experts, and follows other generic management system approaches such as ISO 14001 and ISO 9001. It will take into account other International Standards in this area such as OHSAS 18001, the International Labour Organization's ILO-OSH Guidelines, various national standards and the ILO's international labour standards and conventions.

Safety and health professionals should be aware that ISO 45001 applies to organizations regardless of size, type, and activities, and ISO 45001 does not state or require specific criteria for safety and health performance. Additionally, safety and health professionals should be aware that ISO 45001 does not address product safety or environmental related issues.

In conclusion, active employee engagement and empowerment within the safety and health function is an essential element to the achievement of safety and health objectives beyond compliance. Depending on the stage of the development of your safety and health program or system, safety and health professionals should consider appropriate and customized methodologies to earn/maintain employee trust as well as stimulate employee engagement. In an ideal circumstance, employees will learn and accept "ownership" for the safety and health function and be able to achieve levels far beyond the norms within the safety and health function.

NOTES

1. See Sampson, S.J., *Leaders Without Titles*, HRD Press, Inc. (2011).
2. Schneid, T.D., *Workplace Safety and Health*, CRC Press (2014).

6 Applicable Federal and State Laws

***

"The more laws, the less justice."

German Proverb

"Nobody has a more sacred obligation to obey the law than those who make the law."

Jean Anouilh

***

The safety and health function does not work in a vacuum. Although the Occupational Safety and Health Act is the primary federal law utilized by safety and health professionals, there are a number of other laws which may impact the situation or the decision-making process. These laws may include, but are not limited to, the following:

LABOR

National Labor Relation Act
Labor Management Relation Act
Labor Management Reporting and Disclosure Act
Labor Arbitration
Railway Labor Act
Federal Service Labor-Management Relations Act
Norris-LaGuardia Act
Federal Mediation and Conciliation Act
Postal Service Labor Relations
Hobbs Act (anti-racketeering)
Byrnes Act (anti-strikebreaker)

ANTI-DISCRIMINATION LAWS

Pregnancy Discrimination Act (PDA)
Title VII of the Civil Rights Act of 1964 (title VII)
Rehabilitation Act of 1973
Americans with Disabilities Act (ADA)
Equal Pay Act (EPA)
Age Discrimination in Employment Act (ADEA)

Genetic Information Nondiscrimination Act (GINA)
Immigration Reform and Control Act of 1986 (INCA)
Federal Anti-Discrimination Executive Orders 11246

FEDERAL WAGE AND HOUR LAWS

Fair Labor Standards Act (FLSA)
Walsh-Healy Act
David-Bacon Act
Government Service Contract Act
The Copeland Act
The Miller Act
Consumer Credit Protection Act
Contract Work Hours and Safety Standards Act

FEDERAL RETIREMENT, WELFARE, AND PRIVACY LAWS

Employee Retirement income Security Act (ERISA)
Social Security Act
Railroad Retirement Act
Health Maintenance Organizations Act (HMO)
Federal Unemployment Tax Act
Railroad Unemployment Insurance Act
Federal Employee Compensation Act
Longshoremen's and Harbor Workers' Compensation Act
The Jones Act

LAWS ENFORCED BY THE EQUAL EMPLOYMENT OPPORTUNITY COMMISSION

Title VII of the Civil Rights Act of 1964 (Title VII): This law makes it illegal to discriminate against someone on the basis of race, color, religion, national origin, or sex. The law also makes it illegal to retaliate against a person because the person complained about discrimination, filed a charge of discrimination, or participated in an employment discrimination investigation or lawsuit. The law also requires that employers reasonably accommodate applicants and employees sincerely held religious practices, unless doing so would impose an undue hardship on the operation of the employer's business.

The Pregnancy Discrimination Act: This law amended Title VII to make it illegal to discriminate against a woman because of pregnancy, childbirth, or a medical condition related to pregnancy or childbirth. The law also makes it illegal to retaliate against a person because the person complained about discrimination, filed a charge of discrimination, or participated in an employment discrimination investigation or lawsuit.

The Equal Pay Act of 1963 (EPA): This law makes it illegal to pay different wages to men and women if they perform equal work in the same workplace.

The law also makes it illegal to retaliate against a person because the person complained about discrimination, filed a charge of discrimination, or participated in an employment discrimination investigation or lawsuit.

The Age Discrimination in Employment Act of 1967 (ADEA): This law protects people who are 40 or older from discrimination because of age. The law also makes it illegal to retaliate against a person because the person complained about discrimination, filed a charge of discrimination, or participated in an employment discrimination investigation or lawsuit.

Title I of the Americans with Disabilities Act of 1990 (ADA): This law makes it illegal to discriminate against a qualified person with a disability in the private sector and in state and local governments. The law also makes it illegal to retaliate against a person because the person complained about discrimination, filed a charge of discrimination, or participated in an employment discrimination investigation or lawsuit. The law also requires that employers reasonably accommodate the known physical or mental limitations of an otherwise qualified individual with a disability who is an applicant or employee, unless doing so would impose an undue hardship on the operation of the employer's business.

Sections 102 and 103 of the Civil Rights Act of 1991: Among other things, this law amends Title VII and the ADA to permit jury trials and compensatory and punitive damage awards in intentional discrimination cases.

Sections 501 and 505 of the Rehabilitation Act of 1973: This law makes it illegal to discriminate against a qualified person with a disability in the federal government. The law also makes it illegal to retaliate against a person because the person complained about discrimination, filed a charge of discrimination, or participated in an employment discrimination investigation or lawsuit. The law also requires that employers reasonably accommodate the known physical or mental limitations of an otherwise qualified individual with a disability who is an applicant or employee, unless doing so would impose an undue hardship on the operation of the employer's business.

The Genetic Information Nondiscrimination Act of 2008 (GINA) Effective November 21, 2009: This law makes it illegal to discriminate against employees or applicants because of genetic information. Genetic information includes information about an individual's genetic tests and the genetic tests of an individual's family members, as well as information about any disease, disorder or condition of an individual's family members (i.e., an individual's family medical history). The law also makes it illegal to retaliate against a person because the person complained about discrimination, filed a charge of discrimination, or participated in an employment discrimination investigation or lawsuit.[1]

Summary of the Major Laws of the Department of Labor

- Wages and Hours
- Workplace Safety and Health
- Workers' Compensation

- Employee Benefits
- Unions and Their Members
- Employee Protection
- Uniformed Services Employment and Reemployment Rights Act
- Employee Polygraph Protection Act
- Garnishment of Wages
- The Family and Medical Leave Act
- Veterans' Preference
- Government Contracts, Grants, or Financial Aid
- Migrant and Seasonal Agricultural Workers
- Mine Safety and Health
- Construction
- Transportation
- Plant Closings and Layoffs
- Posters
- Related Agencies

The Department of Labor (DOL) administers and enforces more than 180 federal laws. These mandates and the regulations that implement them cover many workplace activities for about 10 million employers and 125 million workers.

The following is a brief description of many of DOL's principal statutes most commonly applicable to businesses, job seekers, workers, retirees, contractors, and grantees. This brief summary is intended to acquaint you with the major labor laws and not to offer a detailed exposition. For authoritative information and references to fuller descriptions on these laws, you should consult the statutes and regulations themselves.

Rulemaking and Regulations provides brief descriptions and various sources of information on DOL's rulemaking activities and regulations.

Wages and Hours

The **Fair Labor Standards Act (FLSA)** prescribes standards for wages and overtime pay, which affect most private and public employment. The act is administered by the Wage and Hour Division. It requires employers to pay covered employees who are not otherwise exempt at least the federal minimum wage and overtime pay of one-and-one-half-times the regular rate of pay. For nonagricultural operations, it restricts the hours that children under age 16 can work and forbids the employment of children under age 18 in certain jobs deemed too dangerous. For agricultural operations, it prohibits the employment of children under age 16 during school hours and in certain jobs deemed too dangerous.

The Wage and Hour Division also enforces the labor standards provisions of the **Immigration and Nationality Act (INA)** that apply to aliens authorized to work in the United States under certain nonimmigrant visa programs (H-1B, H-1B1, H-1C, H2A).

Workplace Safety and Health

The **Occupational Safety and Health (OSH) Act** is administered by the Occupational Safety and Health Administration (OSHA). Safety and health conditions in most private industries are regulated by OSHA or OSHA-approved state programs, which also cover public sector employers. Employers covered by the OSH Act must comply with the regulations and the safety and health standards promulgated by OSHA. Employers also have a general duty under the OSH Act to provide their employees with work and a workplace free from recognized, serious hazards. OSHA enforces the Act through workplace inspections and investigations. Compliance assistance and other cooperative programs are also available.

Workers' Compensation

If you worked for a private company or a state government, you should contact the workers' compensation program for the state in which you lived or worked. The US Department of Labor, Office of Workers' Compensation Programs, does not have a role in the administration or oversight of state workers' compensation programs.

The **Longshore and Harbor Workers' Compensation Act (LHWCA)**, administered by The Office of Workers Compensation Programs (OWCP), provides for compensation and medical care to certain maritime employees (including a longshore worker or other person in longshore operations, and any harbor worker, including a ship repairer, shipbuilder, and shipbreaker) and to qualified dependent survivors of such employees who are disabled or die due to injuries that occur on the navigable waters of the United States, or in adjoining areas customarily used in loading, unloading, repairing, or building a vessel.

The **Energy Employees Occupational Illness Compensation Program Act (EEOICPA)** is a compensation program that provides a lump-sum payment of $150,000 and prospective medical benefits to employees (or certain of their survivors) of the Department of Energy and its contractors and subcontractors as a result of cancer caused by exposure to radiation, or certain illnesses caused by exposure to beryllium or silica incurred in the performance of duty, as well as for payment of a lump-sum of $50,000 and prospective medical benefits to individuals (or certain of their survivors) determined by the Department of Justice to be eligible for compensation as uranium workers under Section 5 of the **Radiation Exposure Compensation Act (RECA)**.

The **Federal Employees' Compensation Act (FECA)**, 5 U.S.C. 8101 et seq., establishes a comprehensive and exclusive workers' compensation program which pays compensation for the disability or death of a federal employee resulting from personal injury sustained while in the performance of duty. The FECA, administered by OWCP, provides benefits for wage loss compensation for total or partial disability, schedule awards for permanent loss or loss of use of specified members of the body, related medical costs, and vocational rehabilitation.

The **Black Lung Benefits Act (BLBA)** provides monthly cash payments and medical benefits to coal miners totally disabled from pneumoconiosis ("black lung disease") arising from their employment in the nation's coal mines. The statute also provides monthly benefits to a deceased miner's survivors if the miner's death was due to black lung disease.

Employee Benefit Security

The **Employee Retirement Income Security Act (ERISA)** regulates employers who offer pension or welfare benefit plans for their employees. Title I of ERISA is administered by the Employee Benefits Security Administration (EBSA) (formerly the Pension and Welfare Benefits Administration) and imposes a wide range of fiduciary, disclosure, and reporting requirements on fiduciaries of pension and welfare benefit plans and on others having dealings with these plans. These provisions preempt many similar state laws. Under Title IV, certain employers and plan administrators must fund an insurance system to protect certain kinds of retirement benefits, with premiums paid to the federal government's Pension Benefit Guaranty Corporation (PBGC). EBSA also administers reporting requirements for continuation of healthcare provisions, required under the **Comprehensive Omnibus Budget Reconciliation Act of 1985 (COBRA)** and the healthcare portability requirements on group plans under the **Health Insurance Portability and Accountability Act (HIPAA)**.

Unions and Their Members

The **Labor-Management Reporting and Disclosure Act (LMRDA)** of 1959 (also known as the Landrum-Griffin Act) deals with the relationship between a union and its members. It protects union funds and promotes union democracy by requiring labor organizations to file annual financial reports, by requiring union officials, employers, and labor consultants to file reports regarding certain labor relations practices, and by establishing standards for the election of union officers. The act is administered by the Office of Labor-Management Standards (OLMS).

Employee Protection

Most labor and public safety laws and many environmental laws mandate whistleblower protections for employees who complain about violations of the law by their employers. Remedies can include job reinstatement and payment of back wages. OSHA enforces the whistleblower protections in most laws.

Uniformed Services Employment and Reemployment Rights Act

Certain persons who serve in the armed forces have a right to reemployment with the employer they were with when they entered service. This includes those called up from the reserves or National Guard. These rights are administered by the Veterans' Employment and Training Service (VETS).

Employee Polygraph Protection Act

This law bars most employers from using lie detectors on employees but permits polygraph tests only in limited circumstances. It is administered by the Wage and Hour Division.

Garnishment of Wages

Garnishment of employee wages by employers is regulated under the **Consumer Credit Protection Act** (CCPA), which is administered by the Wage and Hour Division.

The Family and Medical Leave Act

Administered by the Wage and Hour Division, the **Family and Medical Leave Act (FMLA)** requires employers of 50 or more employees to give up to 12 weeks of unpaid, job-protected leave to eligible employees for the birth or adoption of a child or for the serious illness of the employee or a spouse, child, or parent.

Veterans' Preference

Veterans and other eligible persons have special employment rights with the federal government. They are provided preference in initial hiring and protection in reductions in force. Claims of violation of these rights are investigated by the Veterans' Employment and Training Service (VETS).

Government Contracts, Grants, or Financial Aid

Recipients of government contracts, grants or financial aid are subject to wage, hour, benefits, and safety and health standards under:

- The **Davis-Bacon Act**, which requires payment of prevailing wages and benefits to employees of contractors engaged in federal government construction projects;
- The **McNamara-O'Hara Service Contract Act**, which sets wage rates and other labor standards for employees of contractors furnishing services to the federal government; and
- The **Walsh-Healey Public Contracts Act**, which requires payment of minimum wages and other labor standards by contractors providing materials and supplies to the federal government.

Administration and enforcement of these laws are by The Wage and Hour Division. The Office of Federal Contract Compliance Programs (OFCCP) administers and enforces three federal contract-based civil rights laws that require most federal contractors and subcontractors, as well as federally assisted construction contractors, to provide equal employment opportunity. The Office of the Assistant Secretary for Administration and Management's (OASAM) Civil Rights Center administers and enforces several federal assistance based civil rights laws requiring recipients of federal financial assistance from Department of Labor to provide equal opportunity.

Migrant and Seasonal Agricultural Workers

The **Migrant and Seasonal Agricultural Worker Protection Act (MSPA)** regulates the hiring and employment activities of agricultural employers, farm labor contractors, and associations using migrant and seasonal agricultural workers. The Act prescribes wage protections, housing, and transportation safety standards, farm labor contractor registration requirements, and disclosure requirements. The Wage and Hour Division administers this law.

The **Fair Labor Standards Act (FLSA)** exempts agricultural workers from overtime premium pay, but requires the payment of the minimum wage to workers employed on larger farms (farms employing more than approximately seven full-time workers). The Act has special child-labor regulations that apply to agricultural employment; children under 16 are forbidden to work during school hours and in certain jobs deemed too dangerous. Children employed on their families' farms are exempt from these regulations. The Wage and Hour Division administers this law. OSHA also has special safety and health standards that may apply to agricultural operations.

The **Immigration and Nationality Act (INA)** requires employers who want to use foreign temporary workers on H-2A visas to get a labor certificate from the Employment and Training Administration certifying that there are not sufficient, able, willing, and qualified US workers available to do the work. The labor standards protections of the H-2A program are enforced by The Wage and Hour Division.

Mine Safety and Health

The **Federal Mine Safety and Health Act of 1977** (Mine Act) covers all people who work on mine property. The Mine Safety and Health Administration (MSHA) administers this Act.

The Mine Act holds mine operators responsible for the safety and health of miners; provides for the setting of mandatory safety and health standards, mandates miners' training requirements; prescribes penalties for violations; and enables inspectors to close dangerous mines. The safety and health standards address numerous hazards including roof falls, flammable and explosive gases, fire, electricity, equipment roll-overs and maintenance, airborne contaminants, noise, and respirable dust. MSHA enforces safety and health requirements at more than 13,000 mines, investigates mine accidents, and offers mine operators training, technical and compliance assistance.

Construction

Several agencies administer programs related solely to the construction industry. OSHA has special occupational safety and health standards for construction; The Wage and Hour Division, under Davis-Bacon and related acts, requires payment of prevailing wages and benefits; The Office of Federal Contract Compliance Programs enforces **Executive Order 11246**, which requires federal construction contractors and subcontractors, as well as federally assisted construction contractors, to provide equal employment opportunity; the anti-kickback section of the **Copeland Act** precludes a federal contractor from inducing any employee to sacrifice any part of the compensation required.

Transportation

Most laws with labor provisions regulating the transportation industry are administered by agencies outside the Department of Labor. However, longshoring and maritime industry safety and health standards are issued and enforced by OSHA. The **Longshoring and Harbor Workers' Compensation Act (LHWCA)** requires employers to assure that workers' compensation is funded and available to eligible employees. In addition, the rights of employees in the mass transit industry are protected when federal funds are used to acquire, improve, or operate a transit system. Under the **Federal Transit law**, the Department of Labor is responsible for approving employee protection arrangements before the Department of Transportation can release funds to grantees.

Plant Closings and Layoffs

Such occurrences may be subject to the **Worker Adjustment and Retraining Notification Act (WARN)**. WARN offers employees early warning of impending layoffs or plant closings. The Employment and Training Administration (ETA) provides information to the public on WARN, though neither ETA nor the Department of Labor has administrative responsibility for the statute, which is enforced through private action in the federal courts.

Posters

Some of the statutes and regulations enforced by DOL require that notices be provided to employees and/or posted in the workplace. DOL provides free electronic and printed copies of these required posters.

The elaws Poster Advisor can be used to determine which poster(s) employers are required to display at their place(s) of business. Posters, available in English and other languages, may be downloaded and printed directly from the Advisor. If you already know which poster(s) you are required to display, see below to download and print the appropriate poster(s) free of charge.

Please note that the elaws Poster Advisor provides information on federal DOL poster requirements.[2]

Federal Agency Websites

Occupational Safety and Health Administration	www. osha.gov
US Department of Labor	www. dol.gov
National Labor Relation Board	www.nlrb.gov
Equal Employment Opportunity Commission	www.eeoc.gov
Environmental Protection Agency	www.epa.gov

Applicable State Laws

Disclaimer: Selected state laws that could have an impact on the safety and health and human resource functions were identified for the purposes of this text. These selected laws are not intended to be all inclusive in any manner and are subject to change. Prudent safety and HR professionals are advised to seek legal counsel where appropriate.

Alabama

{66.01} Labor Relations Laws.

Collective Bargaining Rights of Firefighters.

Code of Alabama, Sections 11-43-143 and 11-43-144.

Right to Work Statue.

Code of Alabama, Section 25-7-30, et seq.

Code of Alabama, Section 25-7-32.

Code of Alabama, Section 25-7-34.

{66.02} Strikes, Picketing, and Boycott Laws.

Conspiracy to Interfere with Business.

Code of Alabama, Section 13A-11-122.

Garner v. Teamsters Union, 346 US 485 (1953).
Interference with Employment Relationship.
Code of Alabama, Section 27-7-9.

{66.03} Mediation and Arbitration Laws.
Code of Alabama, Section 25-7-50.

{66.04} Regulation of Union Activities.
Code of Alabama, Sections 25-7-1, et seq.

{66.05} Regulation of Employment Practices.
Jury Duty.
Code of Alabama, Section 12-16-8.
Blacklisting.
Code of Alabama, Section 13A-11-123.
Resident Workmen Requirements.
Code of Alabama, Section 39-3-2.

{66.06} Wage and Hour Laws.
Child Labor.
Code of Alabama, Section 25-8-5.
Code of Alabama, Sections 25-8-4 and 25-8-2, Sections 25-8-2 and 25-8-11.
Code of Alabama, Sections 25-8-13 and 25-8-14.
Voting.
Code of Alabama, Sections 17-23-10 and 17-23-11.
Wage Claims.
Code of Alabama, Section 25-3-2.

{66.07} Safety and Health Laws.
Code of Alabama, Section 25-1-1.

{66.08} Unemployment Compensation Laws.
Code of Alabama, Section 25-4-78.
Code of Alabama, Section 25-4-78.

{66.09} Workers' Compensation Laws.
Code of Alabama, Sections 25-5-1, et seq.
Code of Alabama, Section 25-5-51.

{66.10} Employment-at-Will Developments.
"Independent Consideration" Exception.
Scott v. Lane, 409 S.2d 791 (Ala. 1982).
Peters v. Alabama Power Co., No. 82-248 (Ala. S. Ct. 1931).
Intentional Infliction of Emotional Distress Exception.
Rice v. United Ins. Co. of America, Case No. 83-84 (Ala. S.Ct., 1984).

Alaska

{67.01} Labor Relations Laws.
Public Employment Relations Act.
Section 23.40.070, et seq.
Alaska Statutes, Section 23.40.080.
Alaska Statutes, Section 23.40.70 through 23.40.220.
Teachers Collective Bargaining Rights.
Section 14.20.550, et seq.
Alaska Statutes, Section 14.20.550 through 14.20.610.
Nonright-to-Work Police.
Section 14(b) of the Labor Management Relations Act

{67.02} Strikes, Picketing, and Boycott Laws.
Riots
Alaska Statutes, Section 11.45.020.
Unlawful Assembly.
Alaska Statutes, Section 11.45.020.

{67.03} Mediation and Arbitration Laws.
Alaska Statutes, Section 09.43.010.
Alaska Statutes, Section 09.43.010.

{67.04} Regulation of Union Activities.
Alaska Statutes, Section 23.40.20.
Alaska Public Acts, S.B. 123, Section 1 (L.1959).
Alaska Public Acts, S.B. 123, Section 1 (L.1959).

{67.05} Regulation of Employment Practices.
Anti-Discrimination Laws.
Alaska Statutes, Section 18.80.220.
Polygraph Restrictions.
Alaska Statutes, Section 23.10.37.
False Representations to Procure Employees.
Alaska Statutes, Section 23.10.015.
Equal Pay.
Alaska Statutes, Section 18.80.220.

{67.06} Wage and Hour Laws.
Child Labor.
Alaska Statutes, Section 23.10.340.
Alaska Statutes, Section 23.10.350.
Alaska Statutes, Section 23.10.332.
Hourly Rate of Pay.
Alaska Statutes, Section 23.10.065.
Payment of Wages Upon Termination.
Alaska Statutes, Section 23.05.140.

Voting Time.
Alaska Statutes, Section 15.56.100.
Garnishment.
Alaska Statutes, Section 47.23.070 (b).

{67.07} Safety and Health Laws.
General Provisions.
Alaska Statutes, Section 18.60.075.
Toxic Substances-Right to Know.
Alaska Statutes, Section 18.60.066.
Alaska Statutes, Section 18.60.068.
Alaska Statutes, Section 18.60.067.

{67.08} Unemployment Compensation Laws.
Alaska statutes, Section 23.20.350.

{67.09} Workers Compensation Laws.
Alaska Statutes, Section 23.30.005, et seq.
Alaska Statutes, Section 23.30.005, et seq.
Alaska Statutes, Section 23.30.040.

{67.10} Employment-at-Will Developments.
Covenant of Good Faith and Dealing Exception.
Mitford v. lasala, 666 p.2d 1000 (Alaska, 1983).
"Permanent Employment" Exception.
Eales v. Tanana Valley Medical Surgical Group, Inc., 663 p .2d 958 (Alaska, 1983).

Arizona

New Law: Minimum Wage.
AZ ST Section 23-363 Arizona Revised Statutes Annotated Title 23. Labor.

{68.01} Labor Relations Laws.
Collective Bargaining Rights of Agricultural Employees.
Arizona Revised Statutes, Section 23-1381, et seq.
Arizona Revised Statutes, Section 23-1385.
Arizona Revised Statutes, Section 23-1385 and 23-1389.

Right-to-Work Law.
Arizona Revised Statutes, Section 23-1301, et seq.
"Yellow-Dog" Contracts.
Arizona Statutes, Section 23-1342.

{68.02} Strikes, Picketing, and Boycott Laws.
Anti-Injunction Statute.
Arizona Revised Statutes, Section 23-1393.

Unlawful Assembly.
Arizona Statutes, Section 13-2902.
Secondary Boycotts.
Arizona Statutes, Section 23-1322.

{68.03} Mediation and Arbitration Laws.
Arizona Revised Statutes, Section 12-1517 (although such agreements are recognized by the federal labor policy).

{68.04} Regulation of Union Activities.
Arizona Revised Statutes, Section 44-1453.
Arizona Revised Statutes, Section 16-471.

{68.05} Regulation of Employment Practices.
Anti-Discrimination Laws.
Arizona Statutes, Section 41-1463.
Jury Duty.
Arizona Revised Statutes, Section 21-236.
Backlisting.
Arizona Constitution, Article 18, Section 8.
Employment Under False Pretenses.
Arizona Revised Statutes, section 23-201 and 23-202.
Protection of Employees Political and Voting Freedom.
Arizona Revised Statutes, Section 16-1012.
Employee Right of Access to Consumer Reports.
Arizona Revised Statutes, Section 44-1693.
National Guard Member Protection.
Arizona Revised Statutes, Section 26-167 and 26-168.
Equal Pay.
Arizona Revised Statutes, Section 23-340 and 23-341.

{68.06} Wage and Hour Laws.
Child Labor.
Arizona Revised Statutes, Section 23-231 and 23-233.
Payment of Wages.
Arizona Revised Statutes, Section 23-351.
Payment Upon Termination.
Arizona Revised Statutes, Section 23-353.
Voting Time.
Arizona Revised Statutes, Section 16-897.

{68.07} Safety and Health Laws.
Arizona Revised Statutes, Section 23-403 and 23-404.

{68.08} Unemployment Compensation Laws.
Arizona Revised Statutes, Section 23-601, et seq.

See generally Arizona Revised Statutes, Section 23-601, et seq.
See generally Arizona Revised Statutes, Section 23-771 and 23-776.

{68.09} Workers Compensation Laws.
Arizona Revised Statutes, Section 23-901, et seq.
Arizona Revised Statutes, Section 23-902.
See, for example, Arizona Revised Statutes, Section 23-1044 through 24-1046.

{68.10} Employment- at- will Developments.
Public Policy Exception.
Vermillion v. AAA Pro Moving and Storage, Case No. CA-CIV 5297 (Ariz. Ct. of App., 1985).
Implied Contract Exception.
Wagenseller v. Scottsdale Memorial Hosp., Case No. 17646-PR (Ariz. S. Ct., 1985).

Arkansas

{69.01} Labor Relations Laws.
Right-to-Work Law.
Arkansas Statutes, Section 81-202.
Collective Bargaining Policy.
Arkansas Statutes, Section 81-201.
See Arkansas Attorney General Opinion No. 77-99.

{69.02} Strike, Picketing, and Boycott Laws.
Interference with Lawful Employment.
Arkansas Statutes, Section 81-206.
Interference with Railroad Trains During Picketing.
Arkansas Statutes, Section 81-214.

{69.03} Mediation and Arbitration Laws.
Arkansas Statutes, Section 81-107.

{69.04} Regulation of Union Activities.
Arkansas Statutes, Section 71-201.

{69.05} Regulation of Employment Practices.
Anti-Discrimination Laws.
Arkansas Statutes, Section 6-1506 and 81-405
Interference with Employment.
Arkansas Statutes, Section 81-210.
Blacklisting.
Arkansas Statutes, Section 81-211.
Wrongful Discharge.
Arkansas Statutes, Section 81-310.
Jury Service.
Arkansas Statutes, Section 39-103.

Discrimination Against Political and Voting Activates.
Arkansas Statutes, Section 3-1105(0).
Protection of Public Employees.
Arkansas Public Acts, no .46, Section 1 (L.1961).
Arkansas Public Acts, no .406, Section 2 (L.1973).
Medical Examination Payments.
Arkansas Statutes, Section 81-212.
Equal Pay.
Arkansas Statutes, Section 81-624.

{69.06} Wage and Hour Laws.
Child Labor.
Arkansas Statutes, Section 81-701.
Arkansas Statutes, Section 81-702 and 81-703.
Arkansas Statutes, Section 81-708 through 81-711.
Payment of Wages Upon Termination.
Arkansas Statutes, Section 81-308.
Access to Medical Reports.
Arkansas Statutes, Section 81-212.
Voting Time.
Arkansas Statutes, Section 3-1602.
Medical Insurance Continuance.
Arkansas Public Laws Acts 8115:854 (L.1985).

{69.07} Safety and Health Laws.
Arkansas Statutes, Section 81-108.

{69.08} Unemployment Compensation Laws.
Arkansas Statutes, Section 81.1101, et saq.
See for example: Arkansas Statutes, Section 81-1105, et seq.

{69.09} Workers Compensation Laws.
Arkansas Statutes, Section 81-1301, et seq.

{69.10} Employment-at-Will Developments.
Public Policy Exception.
Lucas v. Brown & Root, Inc., 736 f.2d 1202 (Ca-8,1984).
Intentional Infliction of Emotional Distress Exception.
M.B.M.Co., Inc. v. Counce, 268 Ark. 269, 596 s. w.2d 681(1980).

California

New Law: Section 1182.12. **Minimum Wage**; scheduled **Increases**; adjusted **Minimum Wage**; temporary suspension of **Increases.**

{70.01} Labor Relations Laws.
Collective Bargaining Rights of Public Employees.

California Government Code, Section 3502.
California Government Code, Section 3502.5.
California Government Code, Section 3505.
Collective Bargaining Rights of Agricultural Employees.
See California Labor Code, Section 1152, et seq.
California Labor Code, Section 1152-1154.
California Labor Code, Section 1156, et seq.
California Labor Code, Section 1160.

Higher Education Employees Bargaining Rights.
See California Government Code, Section 3560, et seq.
Public School Employees Bargaining Rights.
California Government Code, Section 3540.
See a discussion of the California law regulating the collective bargaining rights of public employees at {70.01}.
Firefighters Bargaining Rights.
California Code, Section 1960, et seq.
Nonright-to-Work Police.
Section 14(b) of the Labor Management Relations Act.
"Yellow-Dog" Contracts.
California Code, Section 920, et seq.

{70.02} Strikes, Picketing, and Boycott Laws.
Striker Replacements and Strikebreakers.
California Labor Code, Section 973-974.
California Labor Code, Section 1134.
California Labor Code, Section 1112-1113.
California Penal Code, Section 12590(a)(1).
Jurisdictional Strikes.
California Labor Code, Section 1115-1116.
Unlawful Assembly.
California Penal Code, Section 407.
Anti-Injunction Statutes.
See California civil procedure Code, Section 527, et seq.

{70.03} Mediation and Arbitration Laws.
Code Section 1280, et seq.

{70.04} Regulation of Union Activities.
California business and professions Code, Section 9972.
California business and professions Code, Section 9988.
California Labor Code, Section 1010, et seq.
California Labor Code, Section 1017.

{70.05} Regulation of Employment Practices.
California Government Code, Section 12940, et seq.
California Government Code, Section 12941, et seq.

California Government Code, Section 12940(I).
California Government Code, Section 12943 and 12945.
Protection of Political Freedom.
California Labor Code, Section 1101, et seq.
Registration of Employment Applications.
California Labor Code, Section 430, et seq.
Jury Duty.
California Labor Code, Section 230.
Sterilization.
California Government Code, Section 12945.5.
Investigative Consumer Reports.
California civil Code, Section 1786.
Arrest Record.
California Labor Code, Section 432.7.
Access to Personnel Records.
California Labor Code, Section 1198.5.
Confidentiality of Medical Information.
California civil Code, Section 56.20.
Employment Under False Pretense.
California Labor Code, Section 970.
Polygraph Restrictions.
California Labor Code, Section 432.2.
California Penal Code, Section 637.
Protection of Military Personnel.
California Code, Section 394.
California military and veterans Code, Section 394.
Blacklisting, Photographing, and Fingerprinting.
California Labor Code, Section 1051.
Whistle-Blowing Statute.
California Government Code, Section 10545.

Cancer Recovery.
California Government Code, Section 12926.
Alcoholic Rehabilitation.
California Labor Code, Section 1025, et seq.
Medical Examination Payments.
California Labor Code, Section 222.5.
Equal Pay.
California Labor Code, Section 1197.

{70.06} Wage and Hour Laws.
Child Labor.
California Labor Code, Section 1290.
California Labor Code, Section 1308.
Hours of Work.
California Labor Code, Section 552, et seq.

Payment of Wages.
California Labor Code, Section 204-210.
Payment Upon Termination.
California Labor Code, Section 201-203.
Garnishment.
California Labor Code, Section 2929.
Medical Insurance Conversion.
California Health and Safety Code, Section 1373.6.
Voting Time.
California Elections Code, Section 14350.
Anti-reprisal Statute.
California Labor Code, Section 98.6.

{70.07} Safety and Health Laws.
General Provisions.
California Labor Code, Section 6400, et seq.
California Labor Code, Section 6399.7.
Toxic Substances-Right to Know.
California Labor Code, Section 6399, et seq.

{70.08} Unemployment Compensation Laws.
California Unemployment Ins. Code, Section 1, et seq.
California Unemployment Ins. Code, Section 2601, et seq.

{70.09} Workers Compensation Laws.
California Labor Code, Section 3201, et seq.
California Labor Code, Section 132a.

{70.10} Employment-at-Will Developments.
Implied Contract Exception.
Walker v. Northern San Diego County Hosp. Dist.,135 cal. App. 3d 896, 185 Cal. Rptr. 617 (1982).
Covenant of Good Faith and Fair Dealing Exception.
Pugh v. See's Candies, Inc., 116 Cal. App. 3d 311, 171 Cal. Rptr. 917 (1981), see also: *Cancellier v. Federated Department Stores*, 672 f. 2d1312 (CA-9, 1982).
"Independent Consideration" Exception.
Alvarez v. Dart Industries, Inc., 55 Cal. App. 3d 91, 127 Cal. Rptr. 222 (1976).
Public Policy Exception.
Tameny v. Atlantic Richfield Co., 27 Cal. 3d 167, 610 p. 2d 1330, 164 Cal. Rptr. 839(1980); *Petermann v. International Brotherhood of Teamsters*, 174 Cal. App. 2d 184, 344, p.2d 25 (1959); *Crossen v. Foremost McKesson, Inc.*, 537 F. Supp. 1067 (ND Cal., 1982).

Colorado

{71.01} Labor Relations Laws.
Collective Bargaining Laws.
Colorado Revised Statutes, Section 8-13-101, et seq.

Colorado Revised Statutes, Section 80-3-106, et seq.
Nonright-to-Work Policy.
Section 14(b) of the Labor Management Relations Act.

"Yellow-Dog" Contracts.
Colorado Revised Statutes, Section 8-3-108.

{71.02} Strikes, Picketing, and Boycott Laws.
Prohibition of Injunctions to Restrain Strikes.
Colorado Revised Statutes, Section 8-3-118.
Interference with Employment.
Colorado Revised Statutes, Section 8-2-101, et seq.
Unlawful Assembly.
Colorado Revised Statutes, Section 18-9-108.
Secondary Boycotts.
Colorado Revised Statutes, Section 8-3-108(g).
Colorado Revised Statutes, Section 8-2-112.
Striker Replacements.
Colorado Revised Statutes, Section 12-24-109.
Colorado Revised Statutes, Section 8-2-104, et seq.
Strikebreakers.
Colorado Revised Statutes, Section 8-2-104.

{71.03} Mediation and Arbitration Laws.
Colorado Revised Statutes, Section 8-3-12, et seq.
Colorado Revised Statutes, Section 8-3-113.

{71.04} Regulation of Union Activities.
Colorado Revised Statutes, Section 7-70-101, et seq.

{71.05} Regulation of Employment Practices.
Anti-Discrimination Laws.
Colorado Revised Statutes, Section 24-34-402.
Colorado Revised Statutes, Section 8-2-116.
Employment Under False Pretenses.
Colorado Revised Statutes, Section 8-2-104, et seq.
Protection of Political Freedom.
Colorado Revised Statutes, Section 1-13-719.
Military Leaves of Absence.
Colorado Revised Statutes, Section 28-3-609, et seq.

Jury Duty.
Colorado Revised Statutes, Section 13-71-118.
Blacklisting.
Colorado Revised Statutes, Section 8-2-111.

Medical Examination Payments.
Colorado Revised Statutes, Section 8-2-118.
Equal Pay.
Colorado Revised Statutes, Section 8-5-101, et seq.

{71.06} Wage and Hour Laws.
Child Labor.
Colorado Revised Statutes, Section 8-12-101, et seq.
Colorado Revised Statutes, Section 8-12-110.
Colorado Revised Statutes, Section 8-12-111 through 8-12-113.
Payment of Wages.
Colorado Revised Statutes, Section 8-4-105 and Section 8-4-106.
Payment Upon Termination.
Colorado Revised Statutes, Section 8-4-104.
Voting Time.
Colorado Revised Statutes, Section 31-10-603.
Garnishments.
Colorado Revised Statutes, Section 5-5-106, et seq.

{71.07} Safety and Health Laws.
Colorado Revised Statutes, Section 8-11-101, et seq.
Colorado Revised Statutes, Section 8-11-108.

{71.08} Unemployment Compensation Laws.
Colorado Revised Statutes, Section 8-70-103, et seq.
Colorado Revised Statutes, Section 8-70-101, et seq.

{71.09} Workers Compensation Laws.
Colorado Revised Statutes, Section 8-40-101, et seq.

{71.10} Employment-at-Will Developments.
Lampe v. Presbyterian Medical Center, 41 Colo. App. 465, 590 P.2d 513 (1978); *Rawson v. Sears Roebuck and Co.*, 530 F. Supp. 776 (D. Colo. 1982).
Brooks v. Trans World Airlines, Inc., 574 F. Supp.805 (D. Colo. 1983).

Connecticut

{72.01} Labor Relations Laws.
Connecticut Labor Relations Act.
Connecticut General Statutes, Section 31-101, et seq.
Connecticut General Statutes, Section 31-101.
State Employees Bargaining Rights.
Connecticut General Statutes, Section 5-270, et seq.
Connecticut General Statutes, Section 5-272.
Connecticut General Statutes, Section 5-271.
Municipal Employees Bargaining Rights.
Connecticut General Statutes, Section 7-467, et seq.

Teachers Bargaining Rights.
Connecticut General Statutes, Section 10-153.
Nonright-to-Work Policy
Section 14(b) of the Labor Management Relations Act.
"Yellow-Dog" Contracts.
Connecticut General Statutes, Section 31-90.

{72.02} Strikes, Picketing, and Boycott Laws.
Limitation of Anti-Strike Injunctions.
Connecticut General Statutes, Section 31-113.
Unlawful Assembly.
Connecticut General Statutes, Section 53a-177.
Striker Replacements.
Connecticut General Statutes, Section 31-121.

Strikebreakers.
Connecticut General Statutes, Section 31-121, et seq.
Use of Municipal Police Officers During Strike.
Connecticut Public Acts, No. 77 (L.1981).

{72.03} Mediation and Arbitration Laws.
Connecticut General Statutes, Section 52-408, et seq.

{72.04} Regulation of Union Activities.
Connecticut General Statutes, Section 31-77.

{72.05} Regulation of Employment Practices.
Anti-Discrimination Laws.
Connecticut General Statutes, Section 46a-51, et seq.
Inquiries Regarding Arrest Record.
Connecticut General Statutes, Section 31-51(I).
Employment of Illegal Aliens.
Connecticut General Statutes, Section 31-51(k).
Polygraphs Prohibited.
Connecticut General Statutes, Section 31-51(g).
Electronic Surveillance Forbidden.
Connecticut General Statutes, Section 31-48 (b).
Protection of Political Freedom.
Connecticut General Statutes, Section 9-365.
Jury Duty.
Connecticut General Statutes, Section 51-247(a).
Witness Duty.
Connecticut General Statutes, Section 54-85(b).

Military Leaves of Absence.
Connecticut General Statutes, Section 27-33, et seq.

Access to Personnel Records.
Connecticut General Statutes, Section 31-128a, et seq.
See Connecticut General Statutes, Section 31-128a, et seq.

Whistle-Blowing Statute.
Connecticut Public Acts, No.289 (L.1982).
Blacklisting.
Connecticut General Statutes, Section 31-51.
Plant Closures.
Connecticut Public Acts, No.451 (L.1983).
Equal Pay.
Connecticut General Statutes, Section 31-75.

{72.06} Wage and Hour Laws.
Child Labor.
Connecticut General Statutes, Section 10-189 and 10-193.
Hours of Work.
Connecticut General Statutes, Section 31-21.
Payment of Wages.
Connecticut General Statutes, Section 31-71a, et seq.
Payment Upon Termination.
Connecticut General Statutes, Section 31-71c.
Garnishments.
Connecticut Public Laws, No.83-581 (L.1983).
Medical Insurance Conversion.
Connecticut General Statutes, Section 38-262(d).

{72.07} Safety and Health Laws.
General Provisions.
Connecticut General Statutes, Section 31-370.
Toxic Substances-Right to Know.
Connecticut General Statutes, Section 31-40j, et seq.

{72.08} Unemployment Compensation Laws.
Connecticut General Statutes, Section 31-222, et seq.

{72.09} Workers Compensation Laws.
Connecticut General Statutes, Section 31-275, et seq.
Connecticut General Statutes, Section 31-284.

{72.10} Employment at-will Developments.
Violation of Public Policy.
Sheet v. Teddy's Frosted Foods, Inc., 179 Conn. 472 A.2d 385 (1980).
Covenant of Good Faith and Fair Dealing.
Magnan v. Anaconda Industries, Inc., 37 Conn. Super. 38 (1980).
"Independent Consideration" Exception.
Fisher v. Jackson, 142 Conn. 734,118 A.2d 316 (1955).

Delaware

{73.01} Labor Relations Laws.
Public Employees Bargaining Rights.
Delaware Code, Sections 19-1301.
Delaware Code, Sections 19-1305 through 19-1308.
School Employees Bargaining Rights.
Delaware Code, Sections 14-4001, et seq.
Delaware Code, Sections 14-4003.
Nonright-to-Work Policy.
Section 14(b) of the Labor Management Relations Act.

{73.02} Strikes, Picketing, and Boycott Laws.
Unlawful Assembly.
Delaware Code, Sections 11-1302, et seq.

Strikebreakers.
Delaware Code, Sections 19-704.
Delaware Code, Sections 19-703.

{73.03} Mediation and Arbitration Laws.
Delaware Public Acts, S.B.No.425 (L.1972).

{73.04} Regulation of Union Activities.
Delaware Code, Sections 6-3331.
Delaware Code, Sections 6-3305.

{73.05} Regulation of Employment Practices.
Anti-Discrimination Laws.
Delaware Code, Sections 19-711, et seq.
Polygraph Restrictions.
Delaware Code, Sections 19-705.
Protection of Political Freedom.
Delaware Code, Sections 15-5302.
National Guard Members.
Delaware Code, Sections 20-905.
Whistle-Blowing Statute.
Delaware Code, Sections 29-5115.
Access to Personnel Records.
Delaware Public Acts, Chapter 473 (L.1984).

{73.06} Wage and Hour Laws.
Delaware Code, Sections 19-512.
Delaware Code, Sections 19-542.
Delaware Code, Sections 19-541.
Payment of Wages.
Delaware Code, Sections 19-1102.

Payment on Termination.
Delaware Code, Sections 19-1103.

Garnishments.
Delaware Code, Sections 1-3509.

{73.07} Safety and Health Laws.
General Provisions.
Delaware Code, Sections 19-106.
Toxic Substances-Right to Know.
Delaware Public Laws, S.B. 436 (L.1984).

{73.08} Unemployment Compensation Laws.
Delaware Code, Sections 19-3301, et seq.

{73.09} Workers Compensation Laws.
Delaware Code, Sections 19-2101, et seq.
Delaware Code, Sections 19-2301, et seq.

{73.10} Employment-at-Will Developments.
Heideck v. Kent General Hospital, Inc., 446 A.2d 1095 (Del. 1982).

District of Columbia

{74.01} Labor Relations Laws.
Public Employees Bargaining Rights
D.C. Code, Section 1-618.1, et seq.
D.C. Code, Section 1-618.6.
D.C. Code, Section 1-618.4.
D.C. Code, Section 1-618.5.

{74.02} Strikes, Picketing, and Boycott Laws.
Unlawful Assembly
D.C. Code, Section 22-1107.
Threats of Violence
D.C. Code, Section 22-507.

{74.03} Mediation and Arbitration.
D.C. Code, section 16-4301, et seq.

{74.05} Regulation of Employment Practices.
Anti-Discrimination Laws.
D.C. Code, Section 1-2512, et seq.
D.C. Code, Section 1-2502(17), 1-2512.

Access to Personnel Records.
D.C. Code, Section 3105.

Polygraph Restrictions.
D.C. Code, Section 36-801, et seq.

{74.06} Wage and Hour Laws.
Child Labor.
D.C. Code, Section 36-501.
D.C. Code, Section 36-502.
D.C. Code, Section 36-507, et seq.
Payment of Wages.
D.C. Code, Section 36-102.
Payment Upon Termination.
D.C. Code, Section 36-103.

{74.07} Safety and Health Laws.
D.C. Code, Section 36-228.

{74.08} Unemployment Compensation.
D.C. Code, Section 46-101, et seq.
D.C. Code, section 46-110.

{74.09} Workers Compensation Laws.
The Longshoremen and Harbor Workers Compensation Act is discussed in chapter 62.
D.C. Code, Section 36-301, et seq.

{74.10} Employment-at-Will Developments.
Prouty v. National R.R. Passenger Corp., 572 F.Supp.200 (D. D.C. 1982).
Little v. Evening Star Newspaper Co., 120 F.2d 36 (D.C. Cir. 1941).
Kitzmiller v. Washington Post, 115 LRRM 3015 (D.D.C. 1984).

Florida

{75.01} Labor Relations Acts.
Employees Rights of Self-Organization.
Florida Statutes, Section 447.03.
Public Employees Bargaining Rights.
Florida Statutes, Section 447.209, et seq.
Florida Statutes, Section 447.501.
See generally Florida Statutes, Section 447.201, et seq.
Right-to-Work Statute.
Florida Revised Constitution Section 6, Art. 1.

{75.02} Strikes, Picketing, and Boycott Laws.
Regulation of Strike Activity.
Florida Statutes, Section 447.09(11) through (13).
Unlawful Assembly
Florida Statutes, Section 870.01, et seq.

Striker Replacements.
Florida Statutes, Section 449.07.

{75.03} Mediation and Arbitration Laws.
Florida Statutes, Section 453.01, et seq.
Florida Statutes, Section 682.02.

{75.04} Regulation of Union Activities.
Florida Statutes, Section 447.01.
Florida Statutes, Section 447.09.
Florida Statutes, Section 506.06.

{75.05} Regulation of Employment Practices.
Anti-Discrimination Laws.
Florida Statutes, Section 760.01, et seq.
Florida Statutes, Section 448.075.
Jury Duty.
Florida Statutes, Section 40.271.
Employment of Illegal Aliens.
Florida Statutes, Section 448.09.
Protection of Political Freedom.
Florida Statutes, Section 104.081.
Protection of National Guard Members.
Florida Statutes, Section 250.48.
Blacklisting.
Florida Statutes, Section 448.045.
Equal pay.
Florida statutes, Section 448.07.

{75.06} Wage and Hour Laws.
Child Labor.
Florida Statutes, Section 450.021.
Florida Statutes, Section 450.061.
Payment of Wages.
Florida Statutes, Section 532.01.
Medical Insurance Conversion.
Florida Statutes, Section 627.6675.

{75.07} Safety and Health Laws.
Florida Statutes, Section 440.56.
Toxic Substances-Right to Know.
Florida Public Laws, Chapters 84-223 (L. 1984).

{75.08} Unemployment Compensation Laws.
Florida Statutes, Section 443.011, et seq.

{75.09} Workers Compensation Laws.
Florida Statutes, Section 440.01, et seq.
Florida Statutes, Section 440.205.

{75.10} Employment-at-Will Developments.
See, for example: *Muller v. Stromberg Carlson Corp.*, 427 So.2d 266 (2d Dist. Ct. Fla. App. 1983).
Public Police Exception.
Smith v. Piezo Technology, 427 So.2d 182 (Fla. 1983).
"Independent Consideration" Exception.
Chatelier v. Robertson, 118 So.2d 241(2d Dist. Ct. Fla. App. 1960).

Georgia

{76.01} Firefighters Bargaining Rights.
Georgia Public Acts, H.B. 569 (L.1971).
Right-to-Work Statute.
Section 34-6-23 of the Georgia code.

{76.02} Strikes, Picketing, and Boycott Laws.
Anti-picketing Statute.
Section 34-6-2 of the Georgia code.
Unlawful Assembly.
Georgia Code Annotated, Section 16-11-30, et seq.
Interference with Employment Right.
Georgia Code Annotated, Section 34-6-4.
Mass Picketing Prohibited.
Georgia Code Annotated, Section 34-6-5.
Prohibition Against Public Employee Strikes.
Georgia Code Annotated, Section 45-19-1, et seq.

{76.03} Mediation and Arbitration Laws.
Georgia Code Annotated, Section 9-9-30, et seq.
Georgia Code Annotated, Section 34-2-6.

{76.04} Regulation of Union Activities.
Georgia Code Annotated, Section 10-1-451.
Georgia Code Annotated, Section 34-6-8.

{76.05} Regulation of Employment Practices.
Anti-Discrimination Laws (Public Sector).
Georgia Code Annotated, Section 45-19-29.
Georgia Code Annotated, Section 34-1-2.
Georgia Code Annotated, Section 34-6A-4.
Unlawful Enticement of Employees.
Georgia Code Prohibited, Section 60-9904, et seq.

Forged Letters of Reference Restrictions.
Georgia Code Annotated, Section 34-10-14.
Lie Detector Restrictions.
Georgia Code Annotated, Section 43-36-14.
Military Leave (Public Employees).
Georgia Code Annotated, Section 38-2-279.
Equal Pay.
Georgia Code Annotated, Section 45-19-29.

{76.06} Wage and Hour Laws.
Child Labor.
See, Georgia Child Labor Regulations, Section 300-7-1, et seq.
Garnishments.
Georgia Code forbids, Section 18-4-7.
Terms of Employment.
Georgia Code Annotated, Section 34-7-1.
Voting Time.
Georgia Code Annotated, Section 21-1-404.
Payment of Wages.
Georgia Code Annotated, Section 34-7-2.

{76.07} Safety and Health Laws.
Georgia Code Annotated, Section 34-2-10.

{76.08} Unemployment Compensation Laws.
Georgia Code Annotated, Section 34-8-1, et seq.

{76.09} Worker Compensation Laws.
Georgia Code Annotated, Section 34-9-1, et seq.

{76.10} Employment-at-Will Developments.
See, for example: *Troy v. Interincisal, Inc.*, 320 S.E.2d 872 (Ga. Ct. of App., 1984).
Nelson v. M and M Products Co., 168 Ga. App. 280, 308 S.E.2d 607 (Ga. Ct. of App., 1983).
Cox v. Brazo, 303 S.E.2d 71 (Ga. Ct. of App.,1983).
Beavers v. Johnson, 145 S.E.2d 776 (Ga. Ct. of App.,1965).
Smith v. Rich's, Inc., 104 Ga .App. 883, 123 S.E.2.d 316 (Ga. Ct. of App.,1961).

Hawaii

{77.01} Labor Relations Laws.
Hawaii Employment Relations Act.
Hawaii Revised Statutes, Section 377-1, et seq.
Public Employee's Bargaining Rights
Hawaii Revised Statutes, Section 89-3.
Hawaii Revised Statutes, Section 89-7 and 89-8.
Hawaii Revised Statutes, Section 89-6.
Hawaii Revised Statutes, Section 89-12.

Nonright-to-Work Policy
Section 14(b) of the Labor Management Relations Act.

{77.02} Striker, Picketing, and Boycott Laws.
Unlawful Assembly.
Hawaii Revised Statutes, Section 711-1104.
Picketing of residence or Dwelling.
Hawaii Revised Statutes, Section 379-A-1.
Striker Replacements and Strikebreakers.
Hawaii Revised Statutes, Section 379-3.
Hawaii Revised Statutes, Section 379-2.

Anti-Injunction Laws.
Hawaii Revised Statutes, Section 380-1, et seq.
Interference with Ingress or Egress.
Hawaii Revised Statutes, Section 711-1105.
Restrictions in Stevedore Strikes.
Hawaii Revised Statutes, Section 382-1, et seq.

{77.03} Mediation and Arbitration Laws.
Hawaii Public Acts, No. 146 (L.1949).
Hawaii Public Acts, No. 146, section 4163 (L.1949).

{77.04} Regulation of Union Activities.
Hawaii Revised Statutes, Section 482-4.
Hawaii Revised Statutes, Section 337-14.
Hawaii Revised Statutes, Section 377-10.

{77.05} Regulation of Employment Practices.
Anti-Discrimination.
Hawaii Revised Statutes, Section 378-1, et seq.
Unlawful Enticement.
Hawaii Public Acts, Chapter 70 (L.1911).
Jury Pay.
Hawaii Revised Statutes, Section 612-25. See also: Sections 95-26 and 79-14.
Polygraph Restrictions.
Hawaii Revised Statutes, Section 378-21.
Witness Appearances.
Hawaii Public Acts, No. 621 (L.1978).
Criminal and Arrest Records (State Employees).
Hawaii Revised Statutes, Section 378-2.
Military Leave.
Hawaii Revised Statutes, Section 79-23.
Hawaii Revised Statutes, Section 121-43.
Equal Pay.
Hawaii Revised Statutes, Section 387-4.

{77.06} Wage and Hour Laws.
Child Labor.
Hawaii Revised Statutes, Section 390-1, et seq.
Voting Time.
Hawaii Revised Statutes, Section 11-95.
Payment of Wages.
Hawaii Revised Statutes, Section 388-2.
Payment Upon Termination.
Hawaii Revised Statutes, Section 388-3(a).
Hawaii Revised Statutes, Section 388-3(b).
Garnishments.
Hawaii Revised Statutes, Section 378-32(1).

{77.07} Safety and Health Laws.
General Provisions.
Hawaii Revised Statutes, Section 396-6(a) and (c).
Hawaii Revised Statutes, Section 396-8 (e) (1).
Toxic Substances-Right to Know.
Hawaii Revised Statutes, Section 396-7.

{77.08} Unemployment Compensation Laws.
Hawaii Revised Statutes, Section 383-1, et seq.

{77.09} Worker Compensation Laws.
Hawaii Revised Statutes, Section 386-10, et seq.
Hawaii Revised Statutes, Section 386-1, et seq.
Hawaii Revised Statutes, Section 378-32(2).

{77.10} Employment-at-Will Developments.
Lim v. Motor Supply Co., 45 Hawaii 111, 364 P.2d 38 (1961).
Public Policy Exception.
Parnar v. Americana Hotels, Inc., 652P.2d 625 (Hawaii, 1982).
Promissory Estoppel Exception.
Ravelo v. County of Hawaii, 66 Hawaii 197, 658 P.2d 883 (1983).

Idaho

{78.01} Labor Relations Laws.
Agricultural Employees Bargaining Rights.
Idaho Code, Section 22-4101, et seq.
Teachers Bargaining Rights.
Idaho Code, Section 33-1271.
Idaho Code, Section 33-1274.
Municipal Employees' and Firefighters' Bargaining Rights.
Idaho Code, Section 44-1801, et seq.
Idaho Code, Section 50-901.
Right-to-Work Statute.
Idaho Code, Section 44-2001, et seq.

"Yellow-Dog."
Idaho Code, Section 44-901.

{78.02} Strikes, Picketing, and Boycott Laws.
Anti-Injunction Laws.
Idaho Code, Section 44-701, et seq.
Unlawful Assembly.
Idaho Code, Section 18-6402.
Strikebreakers.
Idaho Code, Section 18-712.

{78.03} Mediation and Arbitration Laws.
Idaho Code, Section 44-106.
Idaho Code, Section 7-901, et seq.

{78.04} Regulation of Union Activities.
Idaho Code, Section 44-601, et seq.

{78.05} Regulation of Employment Practices.
Anti-Discrimination Laws.
Idaho Code, Section 67-5909, et seq.
Freedom of Political Activities.
Idaho Code, Section 18-2319.
Jury or Witness Duty.
Idaho Code, Section 2-218.
Employment Under False Pretense.
Idaho Code, Section 18-3101.
National Guard Duty.
Idaho Code, Section 46-224.
Polygraph Restrictions.
Idaho Code, Section 44-903.
Equal Pay.
Idaho Code, Section 44-1702.

{78.06} Wage and Hour Laws.
Child Labor.
Idaho Code, Section 44-1301, et seq.
Idaho Code, Section 44-1301.
Idaho Code, Section 44-1302.
Idaho Code, Section 44-1301, et seq.
Payment of Wages.
Idaho Code, Section 45-610.
Payment Upon Termination.
Idaho Code, Section 45-606.
Garnishments.
Idaho Code, Section 28-35-106.

{78.07} Safety and Health Laws.
Idaho Code, Section 44-104.
Idaho Code, Section 72-1101.

{78.08} Unemployment Compensation.
Idaho Code, Section 72-1361, et seq.

{78.09} Workers Compensation Laws.
Idaho Code, Section 72-101, et seq.

{78.10} Employment-at-Will Developments.
Implied Contract Exception.
MacNeil v. Minidoka Memorial Hospital, 701 P. 2d 208 (Idaho S. Ct.,1985).
Fraud.
Verway v. Blinkcoe Packing Co., Case No. 15189 (Idaho Ct. of App., 1985).

Illinois

New Law (HB 3554/ PA 99-0762). Amends Illinois Wage and Collection Act.

{79.01} Labor Relations Laws.
Public Employees Bargaining Rights.
Illinois Executive Order No. 6 of 1973.
Bargaining Rights of Teachers.
Illinois Public Acts, No. 1014 (L.1984).
Nonright-to-Work Policy
Section 14(b) of the Labor Management Relations Act.
"Yellow-Dog" Contracts.
Illinois Revised Statutes, Ch. 48, Section 2b.

{79.02} Strikes, Picketing, and Boycott Laws.
Interference with Employment
Illinois Revised Statutes, Chapter 38, section 12.6
Unlawful Assemble.
Illinois Revised Statutes, Chapter 38, section 25.1.
Striker Replacements.
Illinois Revised Statutes, Chapter 48, section 2d.

Strikebreakers.
Illinois Revised Statutes, Chapter 48, Section 2f.
Anti-Injunction Statute.
Illinois Revised Statutes, Chapter 48, Section 2a.
Picketing of Residence or Dwelling.
Illinois Revised Statutes, Chapter 38, Section 21.1-1.
Inciting Strikes.
Illinois Revised Statutes, Chapter 38, Section 201.11.
Restriction of Railroad Picketing
Illinois Revised Statutes, Chapter 114, Section 101, et seq.

{79.03} Mediation and Arbitration Laws.
Illinois Revised Statutes, Chapter 10, Section 101, et seq.
Illinois Revised Statutes, Chapter 10, Section 20, et seq.

{79.04} Regulation of Union Activities.
Illinois Revised Statutes, Chapter 38, Section 242.
Illinois Public Acts, S.B. 192 (L.1961).

{79.05} Regulation of Employment Practices.
Anti-Discrimination Laws.
Illinois Revised Statutes, Chapter 68, Section 1-102, et seq.
Military Personnel Leave.
Illinois Revised Statutes, Chapter 126 ½ Section 31, et seq.
Arrest Records.
Illinois Revised Statutes, Chapter 48, Section 2-103.
Jury and Witness Duty.
Illinois Revised Statutes, Chapter 78, Section 4.1.
Whistle-Blowing Statute (Public Employees).
Illinois Public Acts, No.82-734 (L.1981).
Polygraph Restrictions.
Illinois Revised Statutes, Chapter 111, Section 2415.1.
Access to Personnel Records.
Illinois Revised Statutes, Chapter 48. Section 2001.
Illinois Revised Statutes, Chapter 48, Section 2001, et seq.
Medical Examination Payments.
Illinois Revised Statutes, Chapter 48, Section 172d.
Equal Pay.
Illinois Revised Statutes, Chapter 48, Section 4(a).

{79.06} Wage and Hour Laws.
Child Labor.
Illinois Revised statutes, Chapter 48, Section 31, et seq.
Payment of wages.
Illinois Revised Statutes, Chapter 48, Section 39m.
Payment on Termination.
Illinois Revised Statutes, Chapter 48, Section 39m.
Garnishments.
Illinois Revised Statutes, Chapter 110, Section 12-818.

{79.07} Safety and Health Laws.
General Provisions.
Illinois Revised Statutes, Chapter 48, Section 137.3.

Toxic Substances-Right to Know.
Illinois Revised Statutes, Chapter 48, Section 1401, et seq.

{79.08} Unemployment Compensation Laws.
Illinois Revised Statutes, Chapter 48, Section 300, et seq.
Illinois Revised Statutes, Chapter 48, Section 401, et seq.

{79.09} Workers Compensation Laws.
Illinois Revised Statutes, Chapter 48, Section 138, et seq.
Illinois Revised Statutes, Chapter 48, Section 138h.

{79.10} Employment-at-Will Developments.
Public Policy Exception.
Kelsay v. Motorola, Inc., 74 Ill.2d 172,384 N.E.2d 353 (1978).
Palmateer v. International Harvester, 85 Ill. App. 2d 124, 421 N.E.2d 876 (1981).
"Independent Consideration" Exception.
Martin v. Federated Life Insurance Co., 109 Ill. App.3d 596, 440 N.E.2d 998 (1982).
Intentional Infliction of Emotional Distress Exception.
Harris v. First Federal Savings and Loan Association of Chicago, 473 N.E.2d 457(Ill. Ct. of App., 1984).

Indiana

New Law—HB 1183 Employee Paid Sick Leave

{80.01} Labor Relations Laws.
Teachers' Bargaining Rights.
Indiana Statutes, Section 20-7 .5-1-1.
Indiana Statutes, Section 20-7 .5-1-7.
Indiana Statutes, Section 20-7 .5-1-10.
Indiana Statutes, Section 20-7 .5-1-14.
Right to Organize.
Indiana Public Acts, Senate Bill No. 251 (1957).
Nonright-to-Work Policy.
Section 14(b) of the Labor Management Relations Act.
"Yellow-Dog."
Indiana Statutes, Section 35-15-3-1.

{80.02} Strikes, Picketing, and Boycott Laws.
Anti-Injunction Statute.
Indiana Statutes, Section 22-6-1, et seq.
Striker Replacements.
Indiana Statutes, Section 25-16-1-12.
Unlawful Interference with Employment.
Indiana Statutes, Section 35-15-4-2.

{80.03} Mediation and Arbitration Laws.
Indiana Statutes, Section 22-1-1-8.
Indiana Statutes, Section 34-4-1-1.

{80.04} Regulation of Union Activities.
Indiana Statutes, Section 23-116

{80.05} Regulation of Employment Practices.
Anti-Discrimination Laws.
Indiana Statutes, Section 22-9-1-2, et seq.
Freedom of Political Activities.
Indiana Statutes, Section 3-4-7-3.
Service Letter.
Indiana Statutes, Section 22-6-3-1.
Whistle-Blowing Statute.
Indiana Statutes, Section 4-15-10-4.
Military Personnel Leaves.
Indiana Statutes, Section 10-5-9-2.
Blacklisting.
Indiana Statutes, Section 22-5-3-2.
Jury Service.
Indiana Statutes, Section 35-44-3-10.
Equal Pay.
Indiana Statutes, Section 22-2-2-4.

{80.06} Wage and Hour Laws.
Child Labor.
Indiana Statutes, Section 20-8-1-4-1, et seq.
Indiana Statutes, Section 20-8.1-4-21.
Indiana Statutes, Section 20-8.1-4-27.
Voting Time.
Indiana Statutes, Section 3-1-21-7.
Payment of Wages.
Indiana Statutes, Section 22-2-4-1.
Payment Upon Termination.
Indiana Statutes, Section 22-2-5-1.
Garnishments.
Indiana Statutes, Section 24-4.5-5-106.

{80.07} Safety and Health Laws.
Indiana Code Annotated, Section 22-8-1.1-1.
Indiana Code Annotated, Section 22-8-1.1-4.

{80.08} Unemployment Compensation Laws.
Indiana Code Annotated, Section 22-2-14-1, et seq.
Indiana Code Annotated, Section 22-5-14-3.

{80.09} Workers Compensation Laws.
Indiana Code Annotated, Section 22-3-2-1 et seq.
Indiana Code Annotated, Section 22-3-2-1, et seq.
Indiana Code Annotated, Section 22-3-7-1, et seq.

{80.10} Employment-at-Will Developments.
Public Policy Exception.
Frampton v. Central Indiana Gas Co., 297 N, E .2d 425 (Ind. 1973).

Perry v. Hartz Mountain Corp., F. Supp. 1387 (S.D. Ind. 1982).
Promissory Estoppel Exception.
Eby v. York Division, Borg Warner, 455 N.E.2d 623 (Ind. App. 1983); *Pepsi-Cola General Bottlers, Inc. v. Woods*, 440 N.E.2d 969 (Ind. App. 1982).

Iowa

{81.01} Labor Relations Laws.
Public Employees Bargaining Rights.
Code of Iowa, Section 20.8.
Code of Iowa, Section 20.7.
Code of Iowa, Section 20.10.
Code of Iowa, Section 20.15.
Code of Iowa, Section 20.12.
Right-to-Work-Statute.
Code of Iowa, Section 731.1.

{81.02} Strikes, Picketing, and Boycott Laws.
Unlawful Assembly.
Code of Iowa, Section 723.2.
Secondary Boycotts.
Code of Iowa, Section 732.1.
Jurisdictional Disputes.
Chapter 732.3 of the Iowa Public Acts.
Strikebreakers.
Code of Iowa, Section 732.6.

{81.03} Mediation and Arbitration Laws.
Code of Iowa, Section 679.1. et seq.
Code of Iowa, Section 90.1.

{81.04} Regulation of Union Activity.
Code of Iowa, Section 548.10.

{81.05} Regulation of Employment Practices.
Anti-Discrimination Statutes.
Code of Iowa, Section 601A.6.
Military Leaves of Absence.
Code of Iowa, Section 29A.43.
Drunk Driving Rehabilitation.
Code of Iowa, Section 321.283(8).
Blacklisting.
Code of Iowa, Section 730.2.
Whistle-Blowing Statute (Public Employees).
Code of Iowa, Section 19A.19.
Polygraph Restrictions.
Code of Iowa, Section 730.4(2).

Comparable Work Law (Public Employees).
Code of Iowa, Section 79.18.

{81.06} Wage and Hour Laws.
Child Labor.
Iowa Public Acts, Section 2.1, H. B. 1251{1970}.
Iowa Public Acts, Section 4, H. B. 1251{1970}.
Iowa Public Acts, Section 5, H. B. 1251{1970}.
Voting Time.
Code of Iowa, Chapter 49 Section 109.
Payment of Wages.
Code of Iowa, Chapter 91A Section 3(1).
Payment upon Termination.
Code of Iowa, Section 91A.4.
Garnishments.
Code of Iowa, Chapter 642 Section 21.

{81.07} Safety and Health Laws.
General Provisions
Code of Iowa, section 88.4.
Toxic Substances-Right to Know
Code of Iowa, section 455D.1.

{81.08} Unemployment Compensation Laws.
Code of Iowa, Chapter 96 section 1, et seq.

{81.09} Workers Compensation Laws.
Code of Iowa, section 85.1, et seq.

{81.10} Employment-at-Will Developments.
Allen v. Highway Equipment Co., 239 N.W.2d 135 (Iowa S. Ct., 1976).
Abrisz v. Pulley Freight Lines, Inc., 270 N.W.2d 454 (Iowa S. Ct., 1978).
Covenant of Good Faith and Fair Dealing Excepting.
High v. Sperry Crop., 581 F. Supp. 1246 (S.D. Iowa, 1984).
"Independent Consideration" Exception
Collins v. Parsons College, 203 N. W2d 594 (Iowa, 1973).

Kansas

{82.01} Labor Relations Laws.
Kansas LMRA.
Section 44-801, et seq. of the Kansas Statutes.
Agricultural Employees Bargaining Rights.
Kansas Statutes Annotated, Section 44-818, et seq.
Teachers' Bargaining Rights.
Kansas Statutes Annotated, Section 72-5411, et seq.
Kansas Statutes Annotated, Section 72-5414.

Kansas Statutes Annotated, Section 72-5430.
Public Employees' Bargaining Rights.
Kansas Statutes Annotated, Section 75-4321, et seq.

Right-to-Work Statute.
Kansas Statutes Annotated, Section 44-831.

{82.02} Strikes, Picketing, and Boycott Laws.
Anti-Injunction Laws.
Kansas Statutes Annotated, Section 60-904, et seq.
Unlawful Assembly
Kansas Statutes Annotated, Section 21-1001.
Interference with Railroads
Kansas Statutes Annotated, Section 21-1901.
Kansas Statutes Annotated, Section 21-1901, et seq.
Strikebreakers
Kansas Statutes Annotated, Section 21-1616, et seq.
Secondary Boycotts and Jurisdictional Disputes
Kansas Statutes Annotated, Section 44.809(a).

{82.03} Mediation and Arbitration Laws.
Kansas Statutes Annotated, Section 5-201, et seq.
Kansas Statutes Annotated, Section 44-603.

{82.04} Regulation of Union Activities.
Kansas Statutes Annotated, Section 81-105, et seq.
Kansas Statutes Annotated, Section 44-809(6).

{82.05} Regulation of Employment Practices.
Anti-Discrimination Laws.
Kansas Statutes Annotated, Section 44-1001, et seq.
Protection of Political Freedom
Kansas Statutes Annotated, Section 25-418.
Service Letter
Kansas Statutes Annotated, Section 44-808(3).
Hiring of Illegal Aliens.
Kansas Statutes Annotated, Section 21-4409.
Whistle-Blowing Statute (State Employees)
Kansas Public Laws, H.B. 2621 (L.1984).

Blacklisting
Kansas Statutes Annotated, Section 44-118.
Equal Pay
Kansas Statutes Annotated, Section 44-1210(b).

{82.06} Wage and Hour Laws.
Child Labor
Kansas Statutes Annotated, Section 38-614.

Voting Time
Kansas Statutes Annotated, Section 25-418.
Payment of Wages
Kansas Statutes Annotated, Section 44-314.
Payment Upon Termination
Kansas Statutes Annotated, Section 44-315 (a).
Garnishments
Kansas Statutes Annotated, Section 60-2311(a).
Medical Insurance Conversion
Kansas Statutes Annotated, Section 40-2209.

{82.07} Safety and Health Laws.
Kansas Statutes Annotated, Section 44-636.

{82.08} Unemployment Compensation Laws.
Kansas Statutes Annotated, Section 44-701, et seq.

{82.10} Employment-at-Will Developments.
Public Policy Exception
Murphy v. City of Topeka-Shawnee County Dept., 630 P.2d 186 (Kan., 1981).
Promissory Estoppel Exception
Lorson v. Falcon Coach, Inc., 522 P.2d 449 (Kan 1974).
Implied Contract Exception
Owens v. City of Derby, 586 F. Supp. 37 (D. Kan, 1984).
Rouse v. People's National Gas Co., 116 LRRM 2875 (D. Kan., 1984).

Kentucky

{83.01} Labor Relations Laws.
Employees' Right of Self-Organization
Kentucky Revised Statutes, Section 345.010, et seq.
Firefighters and Police Officers
Kentucky Public Acts, H.B.217 (L.1972).
Kentucky Revised Statutes, Section 340.050.
Nonright-to-Work Policy
Section 14(b) of the Labor Management Relations Act.

{83.02} Strikes, Picketing, and Boycott Laws.
Unlawful Assembly
Kentucky Revised Statutes, Section 437.010.
Striker Replacements
Kentucky Revised Statutes, Section 340.050.
Interference with Transportation
Kentucky Revised Statutes, Section 433.370.

{83.03} Mediation and Arbitration Laws.
Kentucky Revised Statutes, Section 417.010.
Kentucky Revised Statutes, Section 336.140.

{83.04} Regulation of Union Activities.
Kentucky Revised Statutes, Section 336.170.
Kentucky Revised Statutes, Section 365.120
Kentucky Public Acts, H. B. 362(1962).

{83.05} Regulation of Employment Practices.
Anti-Discrimination Laws
Kentucky Revised Statutes, Section 344.040.
Kentucky Revised Statutes, Section 207.130.
Kentucky Revised Statutes, Section 344.030.
Jury Duty
Kentucky Public Acts, H. B. 23-x (L.1976).
Unlawful Enticement
Kentucky Revised Statutes, Section 433.310.
Protection of Political Activities
Kentucky Revised Statutes, Section 121.310.
Employment of Convicted Persons
Kentucky Public Acts, H. B. 100(1978): see also Kentucky Revised Statutes, Section 335B.2(1).
National Guard Members
Kentucky Revised Statutes, Section 38.460.
Equal Pay
Kentucky Revised Statutes, Section 337.423.

{83.06} Wage and Hour Laws.
Child Labor
Kentucky Revised Statutes, Section 339.210, et seq.
Kentucky Revised Statutes, Section 339.230.
Voting Time
Kentucky Revised Statutes, Section 118.035.
Payment of Wages
Kentucky Revised Statutes, Section 337.020.
Payment Upon Terminating
Kentucky Revised Statutes, Section 337.055.
Garnishments
Kentucky Revised Statutes, Section 427.140.
Medical Insurance Conversion
Kentucky Revised Statutes, Section 304.18-110.

{83.07} Safety and Health Laws.
Kentucky Revised Statutes, Section 338.031.

{83.08} Unemployment Compensation Laws.
Kentucky Revised Statutes, Section 341.350, et seq.

{83.09} Workers Compensation Laws.
Kentucky Revised Statutes, Section 342.630:342.650.

{83.10} Employment-at-Will Developments.
Public Policy Exception.
Firestone Textile Co. Division v. Meadows, 666 S.W.2d &30 (Ky., 1983).
Implied Contract Exception.
Shah v. American Synthetic Rubber Crop., 655 S.W.2d 489 (Ky., 1983).
Presumption of Term Exception
Moore v. Young Women's Christian Ass'n, Case No. 84-CA-1508-MR (Ky, Ct of App., 1985).

Louisiana

{84.01} Labor Relations Laws.
General Right to Organize.
Louisiana Revised of the Section 23:822.
Public Transit Employees Bargaining Rights.
Louisiana public acts No. 127, Section 4 (1964).
Louisiana public acts No. 127, Section 5 (1964).
Right-to-Work Statute.
Louisiana Revised Statutes, Section 23:983.
"Yellow-Dog" Contracts.
Louisiana Revised Statutes, Section 23:823.

{84.02} Strikes, Picketing, and Boycott Laws.
Anti-Injunction Laws.
Louisiana Revised Statutes, Section 23:841.
Strikebreakers.
Louisiana Revised Statutes, Section 23:898.
Louisiana Revised Statutes, Section 23:901.
Obstruction of Commerce or Passageways.
Louisiana Revised Statutes, Section 14:100.
Picketing of Courts.
Louisiana Revised Statutes, Section 14:401.

{84.03} Mediation and Arbitration Laws.
Louisiana Revised Statutes, Section 23:6.

{84.04} Regulation of Union Activities.
Louisiana Revised Statutes, Section 51:211, et seq.

{84.05} Regulation of Employment Practices.
Anti-Discrimination Laws.
Louisiana Revised Statutes, Section 23:1006.
Louisiana Revised Statutes, Section 23:973.
Louisiana Revised Statutes, Section 46:2254(1).
Hiring of Illegal Aliens.
Louisiana Revised Statutes, Section 23:992.
Whistle-Blowing Statutes (Environmental Conditions).
Louisiana Revised Statutes, Section 30:1074.1.

Reemployment Right of Veterans.
Louisiana Revised Statutes, Section 29:28.
Protection of Political Activities.
Louisiana Revised Statutes, Section 23:962.
Jury Duty.
Louisiana Revised Statutes, Section 23:965.
Prior Criminal Convictions.
Louisiana Revised Statutes, Section 37:2950.
Access to Medical Records.
Louisiana Revised Statutes, Section 23:1125.

Medical Examination Payments.
Louisiana Revised Statutes, Section 23:897.

{84.06} Wage and Hour Laws.
Child Labor.
Louisiana Revised Statutes, Section 23:151, et seq.
Louisiana Revised Statutes, Section 23:161.
Payment of Wages.
Louisiana Revised Statutes, Section 23:633.
Payment Upon Termination.
Louisiana Revised Statutes, Section 23:631, et seq.
Garnishments.
Louisiana Revised Statutes, Section 23:731(c)

{84.07} Safety and Health Laws.
General Provisions.
Louisiana Revised Statutes, Section 23:8.
Toxic Substances-Right to Know.
Louisiana Revised Statutes, Section 23:1126.

{84.08} Unemployment Compensation Laws.
Louisiana Revised Statutes, Section 23:1600, et seq.

{84.09} Workers Compensation Laws.
Louisiana Revised Statutes, Section 23:1021, et seq.

{84.10} Employment-at-Will Developments.
Gil v. Metal Service Crop., 412 So.2d 706 (La. Ct. of App., 1982).
Williams v. Delta Haven, Inc., 416 So.2d 637 (La. Ct. of App., 1928).

Maine

New Law Section 664. Minimum wage; overtime rate.

{85.01} Labor Relations Laws.
State Employees Bargaining Rights.
Maine Revised Statutes Annotated, Title 26, Section 979.

Maine Revised Statutes Annotated, Title 26, Section 979-E and F.
Maine Revised Statutes Annotated, Title 26, Section 979-K.
Public Employees Bargaining Rights.
Maine Revised Statutes Annotated, Title 26, Section 963, et seq.
Maine Revised Statutes Annotated, Title 26, Section 964.
University Employees Bargaining Rights.
Maine Revised Statutes Annotated, Title 26, Section 10.21, et seq.
Bargaining Rights to Judicial Employees.
Maine public acts, Chapter 702 (L.1984).
General Right to Organize Statute.
Maine Revised Statutes Annotated, Title 26, Section 911.
Nonright-to-Work Policy.
Section 14(b) of the Labor Management Relations Act.

{85.02} Strikes, Picketing, and Boycott Laws.
Anti-Injunction Laws.
Maine public acts, Chapter 620 (L.1972).
Unlawful Assembly.
Maine Revised Statutes Annotated, Title 17, Section 3352.
Interference with Railroads.
Maine Revised Statutes Annotated, Title 17, Section 3605.
Strikebreakers.
Maine Revised Statutes Annotated, Title 26, Section 851.
Striker Replacement.
Maine Revised Statutes Annotated, Title 26, Section 921.
Mass Picketing.
Maine Revised Statutes Annotated, Title 17, Section 3606.
Cancellation of Health Insurance During Strike.
Maine Revised Statutes Annotated, Title 24-A, Section 2894.
Maine Revised Statutes Annotated, Title 26, Section 634.
Interference with Public Utilities Employment.
Maine Revised Statutes Annotated, Title 17, Section 3601.

{85.03} Mediation and Arbitration Laws.
Maine Revised Statutes Annotated, Title 14, Section 1151, et seq.
Maine Revised Statutes Annotated, Title 26, Section 891.

{85.04} Regulation of Union Activities.
Maine Revised Statutes Annotated, Title 16, Section 1341.

{85.05} Regulation of Employment Practices.
Anti-Discrimination Laws.
Maine Revised Statutes Annotated, Title 5, Section 4572-A.
Blacklisting.
Maine Revised Statutes Annotated, Title 17, Section 401.

Polygraph Restrictions.
Maine Revised Statutes Annotated, Title 32, Section 7166.
National Guard Members.
Maine Revised Statutes Annotated, Title 26, Section 811.
Military Leaves of Absence.
Maine Revised Statutes Annotated, Title 26, Section 811.
Hiring of Illegal Aliens.
Maine Revised Statutes Annotated, Title 26, Section 871.
Plant Closures.
Maine Revised Statutes Annotated, Title 26, Section 625 B (6).
Maine Revised Statutes Annotated, Title 26, Section 625 B (2).
Disclosure of Consumer Reports.
Maine Revised Statutes Annotated, Title 10, Section 1320.
Access to Employment Records.
Maine Revised Statutes Annotated, Title 26, Section 631.

Whistle-Blowing Statute.
Maine Revised Statutes Annotated, Title 26, Section 831, et seq.
Service Letter.
Maine Revised Statutes Annotated, Title 26, Section 630.
Comparable Pay.
Maine Revised Statutes Annotated, Title 26, Section 628.
Medical Examination Payments.
Maine Revised Statutes Annotated, Title 26, Section 592.

{85.06} Wage and Hour Laws.
Child Labor.
Maine Revised Statutes Annotated, Title 26, Section 711, et seq.
Payment of Wages.
Maine Revised Statutes Annotated, Title 26, Section 621.

{85.07} Safety and Health Laws.
General Provisions.
Maine Revised Statutes Annotated, Title 26, Section 561, et seq.
Toxic Substances-Right to Know.
Maine Revised Statutes Annotated, Title 26, Section 1701, et seq.

{85.08} Unemployment Compensation Laws.
Maine Revised Statutes Annotated, Title 26, Section 1192.

{85.09} Workers Compensation Laws.
Maine Revised Statutes Annotated, Title 39, Section 1, et seq.
Merrill v. Western Union Tele. Co., 2 A.847 (Maine, 1886).
MacDonald v. Eastern Fine Paper Inc., 485 A.2d 228 (Maine, 1984).
Larrabee v. Penobscot Frozen Foods, Inc., 486 A.2d 97 (Maine, 1984).

Maryland

{86.01} Labor Relations Laws.
General Right to Organize.
Annotated code of Maryland, art. 100, Section 63.
Public School Employees' Bargaining Rights.
Annotated code of Maryland, art. 77, Section 160A, et seq.
Nonright-to-Work Policy
Section 14(b) of the Labor Management Relations Act.
"Yellow-Dog" Contracts
Annotated code of Maryland, art. 100, Section 64.

{86.02} Strikes, Picketing, and Boycott Laws.
Anti-Injunction Laws
Annotated code of Maryland, art. 100, Section 70 and 71.
Strikebreakers
Annotated code of Maryland, art. 100, Section 51A.

{86.03} Mediation and Arbitration Laws.
Annotated code of Maryland, art. 64A, Section 52, et seq.
Annotated code of Maryland, art. 723 (L.1978).

{86.04} Regulation of Union Activities.
Annotated code of Maryland, art. 27, Section 187.

{86.05} Regulation of Employment Practices.
Anti-Discrimination Laws.
Annotated code of Maryland, art. 49A, Section 16.
Polygraph Restrictions.
Annotated code of Maryland, art. 100, Section 95.
Discrimination Against Volunteer Public Service Forbidden.
Annotated code of Maryland, art. 100, Section 109.
Reemployment Rights of Military Personnel.
Annotated code of Maryland, art. 65, Section 32A.

Whistle-Blowing Statute (Public Employees).
Annotated code of Maryland, art. 64, Section 12(G).
Arrest Records.
Annotated code of Maryland, art. 27, Section 740(A).
Medical Records.
Annotated code of Maryland, art. 100, Section 95A
Jury Duty.
Maryland courts and judicial proceedings Code Annotated, Section 8-105.
Comparable Pay.
Annotated code of Maryland, art. 100, Section 55A.

{89.06} Wage and Hour Laws.
Child Labor.
Annotated code of Maryland, art. 100, Section 23(a)-(6).

Voting Time.
Annotated code of Maryland, art. 33, Section 24-26.
Payment of Wages.
Annotated code of Maryland, art. 100, Section 94.
Payment Upon Termination.
Annotated code of Maryland, art. 100, Section 94E.
Garnishments.
Annotated code of Maryland, art. 15, Section 606.
Medical Insurance Conversion.
Annotated code of Maryland, art. 95A, Section 5A.

{86.07} Safety and Health Laws.
Annotated code of Maryland, art. 89, Section 32A.

{86.08} Unemployment Compensation Laws.
Annotated code of Maryland, art. 95A, Section 1, et seq.

{86.09} Workers Compensation Laws.
Annotated code of Maryland, art. 101, Section 36, et seq.

{86.10} Employment-at-Will Developments.
Public Policy Exception.
Adler v. American Standard Crop., 583 F. Supp. 572(D. Md., 1982); *Adler v. American Standard Corp.*, 432 A.2d 464 (Md. Ct. of App., 1981).
Intentional Infliction of Emotional Distress Exception.
Harris v. Jones, 380 A.2d 611(Md. Ct. of App., 1977).
Implied Contract Exception.
Staggs v. Blue Cross of Maryland, Inc., Case No. 538 (Md. Ct. of Sp. App., 1985).

Massachusetts

{87.01} Labor Relations Laws.
Massachusetts Labor Relations Law.
Massachusetts General laws, Chapter 150A, Section 3.
Massachusetts General laws, Chapter 150A, Section 1, et seq.
Housing Employees Bargaining Rights.
Massachusetts General laws, Chapter 121B, Section 29.
Public Employees Bargaining Rights.
Massachusetts General laws, Chapter 150E, Section 2.
Massachusetts General laws, Chapter 150E, Section 10.
Massachusetts General laws, Chapter 150E, Section 1, et seq.
State Employee's Grievance Rights.
Massachusetts General laws, Chapter 30, Section 53.
Prohibition Against Pay-Offs.
Massachusetts General laws, Chapter 149 Section 20D.
Collective Bargaining Successor Clauses.

Massachusetts General laws, Chapter 149, Section 179C.
Nonright-to-Work Policy.
Section 14(b) of the Labor Management Relations Act.
"Yellow-Dog" Contracts.
Massachusetts General laws, Chapter 149, Section 20A.

{87.02} Strikes, Picketing, and Boycott Laws.
Anti-Injunction Act.
Massachusetts General laws, Chapter 214, Section 6.
Unlawful Assembly.
Massachusetts General laws, Chapter 269 Section 1.
Unlawful Interference with Employment.
Massachusetts General laws, Chapter 149, Section 19 and 18.
Picketing of Courts.
Massachusetts General laws, Chapter 268 Section 13a.
Declaration of Emergency During Transit Strikes.
Massachusetts General laws, Chapter 544, Section 19A.
Strikebreakers.
Massachusetts General laws, Chapter 150D, Section 1, et seq.
Massachusetts General laws, Chapter 802, Section 22A.
Striker Replacements.
Massachusetts General laws, Chapter 149 Section 22.
Use of Police Officers and Firefighters During Strike.
Massachusetts General laws, Chapter 149, Section 23B.
Massachusetts General laws, Chapter 48, Section 88.

{87.03} Mediation and Conciliation Laws.
Massachusetts General laws, Chapter 150, Section 1-10.
Massachusetts General laws, Chapter 150B, Section 1-7.
Massachusetts public Acts, Chapter 154 (L-1979).
Massachusetts General laws, Chapter 161A, Section 19.

{87.04} Regulation of Union Activities.
Massachusetts General laws, Chapter 612, Section 1-5.
Massachusetts General laws, Chapter 110, Section 7-11 and chapter 266, section 71A.
Massachusetts General laws, Chapter 151D, Section 1, et seq.
Massachusetts General laws, Chapter 149, Section 150B.

{87.05} Regulation of Employment Practices.
Anti-Discrimination Laws.
Massachusetts General laws, Chapter 151B, Section 4.
Massachusetts General laws, Chapter 149, Section 24A.
Massachusetts General laws, Chapter 149, Section 24K.
Freedom of Political Activities.
Massachusetts General laws, Chapter 54, Section 33.

Jury Duty.
Massachusetts General laws, Chapter 234A, Section 61.
Military Leaves of Absence.
Massachusetts General laws, Chapter 149, Section 52A.
Employment Under False Pretense.
Massachusetts General laws, Chapter 149, Section 21.
Polygraph Restrictions.
Massachusetts General laws, Chapter 149, Section 19B.
Medical Examination Reports.
Massachusetts General laws, Chapter 149, Section 19A.
Arrest Records.
Massachusetts General laws, Chapter 276, Section 100A.
Hiring of Illegal Aliens.
Massachusetts General laws, Chapter 149, Section 19C.
Disclosure of Consumer Reports.
Massachusetts General laws, Chapter 93, Section 56.
Political Influence Relating to Public Utilities Employees.
Massachusetts General laws, Chapter 271, Section 40.
Plant Closures.
Massachusetts public laws, H.6120(1984).

Equal Pay.
Massachusetts General laws, Chapter 149, Section 105A.

{87.06} Wage and Hour Laws.
Child Labor.
Massachusetts General laws, Chapter 149, Section 56, et seq.

Voting Time.
Massachusetts General laws, Chapter 149, Section 178.
Payment of Wages.
Massachusetts General laws, Chapter 149, Section 148.
Payment Upon Termination.
Massachusetts General laws, Chapter 149, Section 148.
Medical Insurance Continuance.
Massachusetts General laws, Chapter 175, Section 110G.

{87.07} Safety and Health Laws.
General Provisions.
Massachusetts General laws, Chapter 149, Section 6-8.
Toxic Substances-Right to Know.
Massachusetts General laws, Chapter 111F, Section 1, et seq.

{87.08} Unemployment Compensation Laws.
Massachusetts General laws, Chapter 151A, Section 1, et seq.
Massachusetts General laws, Chapter 151A, Section 24, et seq.

{87.09} Workers Compensation Laws.
Massachusetts General laws, Chapter 152, Section 1, et seq.
Massachusetts General laws, Chapter 152, Section 25A, et seq.

{87.10} Employment-at-Will Development.
Public Policy Exception.
McKinney v. National Dairy Council, 491 F. Supp. 1108 (D. Mass., 1980).
Implied Contract Exception.
Garrity v. Valley View Nursing Home, Inc., 406 N.E.2d 423 (Mass. Ct. of App., 1980).
Covenant of Good Faith and Fair Dealing Exception.
Fortune v. National Cash Register Co., N.E.2d 1251 (Mass 1977).
Intentional Infliction of Emotional Distress Exception.
Agis v. Howard Johnson Co., 355 N.E.2d 315 (Mass., 1976).

Michigan

{88.01} Labor Relations Laws.
Public Employees Bargaining Rights.
Michigan Compiled laws, Section 423.209.
Michigan Compiled laws, Section 423.201.
Michigan Compiled laws, Section 423.201, et seq.
Michigan LMRA (Bonnie-Tripp Act).
Michigan Compiled laws, Annotated Section 423.8.
Michigan Compiled laws, Annotated Section 423.3.
Michigan Compiled laws, Annotated Section 423.1, et seq.
Police Officers and Firefighters Labor Disputes.
Michigan Compiled laws, Section 423.231, et seq.
Nonright-to-Work Policy.
Section 14(b) of the Labor Management Relations Act

{88.02} Strikes, Picketing, and Boycott Laws.
Michigan Compiled laws, Section 28-584.
Michigan Compiled laws, Section 423.17.
Sit-Down Strikes.
Michigan Compiled laws, Section 423.15
Unlawful Picketing.
Michigan Compiled laws, Section 423.9F.
Unlawful Assembly.
Michigan Compiled laws, Section 28-79.
Interference with Transportation.
Michigan Compiled laws, Section 28-655.
Strikebreakers and Replacements.
Michigan Statutes, Annotated Section 17-456(1).

{88.03} Mediation and Conciliation Laws.
Michigan Compiled laws, Section 423.9d

Michigan Compiled laws, Section 423.231, et seq.
Michigan public acts, No. 17 (1980), Section 1-16.

{88.04} Regulation of Union Activities.
Michigan Compiled laws, Section 750.265a
Michigan Compiled laws, Section 454.71, et seq.

{88.05} Regulation of Employment Practices.
Anti- Discrimination Laws.
Michigan Compiled laws, Section 37.2202.
Michigan Compiled laws, Section 37.1101, et seq.
Freedom of Political Activities.
Michigan Compiled laws, Annotated Section 6.1931.
Jury Duty.
Michigan Compiled laws, Section 600.1348.
Polygraph Restrictions.
Michigan Compiled laws, Section 37.201, et seq.
Whistle-Blowing Statute.
Michigan Compiled laws, Annotated Section 15.362.
Reemployment Rights of Military Personnel.
Michigan Compiled laws, Annotated Section 32.271, et seq.
Arrest Records.
Michigan Compiled laws, Section 37.2205a.
Access to Personnel Records.
Michigan Compiled laws, Annotated Section 423.501, et seq.
Michigan Compiled laws, Annotated Section 4423.501, et seq.
Wage Disclosure.
Michigan Compiled laws, Section 408.483.
Wage Evasion.
Michigan Compiled laws, Section 408.396.
Equal Pay.
Michigan Compiled laws, Section 408.397.

{88.06} Wage and Hour Laws.
Child Labor.
Michigan Compiled laws, Section 409.01 et seq.
Payment of Wages.
Michigan Compiled laws, Section 408.472.
Payment Upon Termination.
Michigan Compiled laws, Section 408.475.
Garnishment.
Michigan Compiled laws, Section 600.4015.

{88.07} Safety and Health Laws.
General Provisions.
Michigan Compiled laws, Section 408.1011.

Toxic Substances-Right to Know.
Michigan Compiled laws, Section 408.1001, et seq.

{88.08} Unemployment Compensation Laws.
Michigan Compiled laws, Section 421.1, et seq.

{88.09} Workers Compensation Laws.
Michigan Compiled laws, Section 411.1, et seq.

{88.10} Employment-at-Will Development.
Public Policy Exception.
Sventko v. Kroger Co., 245 N.W.2d 151(WD Mich., 1982).
Negligent Discharge Exception.
Chamberlain v. Bissell, Inc., 547 F. Supp. 1067 (WD Mich., 1982).
Intentional Infliction of Emotional Distress Exception.
See *Novesel v. Sears, Roebuck & Co.*, 495 F. Supp. 344 (ED Mich., 1980).
Implied Contract Exception.
See, for example, *Toussaint v. Blue Cross & Blue Shield of Michigan*, 292 N.W.2d 880 (Mich. S.Ct., 1980).

Minnesota

{89.01} Labor Relations Laws.
Minnesota Labor Relations Act.
Minnesota Statutes, Section 179.10.
Minnesota Statutes, Section 179.01, et seq.
Public Employees Bargaining Rights.
Minnesota Statutes, Section 179A.06.
Minnesota Statutes, Section 179A.01, et seq.
Nonright-to Work Policy.
Section 14(b) of the Labor Management Relations Act.
"Yellow-Dog" Contracts.
Minnesota Statutes, Section 179.60.

{89.02} Strikes, Picketing, and Boycott Laws.
Anti-Injunction Statute.
Minnesota Statutes, Section 185.02.
Unlawful Assembly.
Minnesota Statutes, Section 609.715.
Striker Replacements.
Minnesota Statutes, Section 184.38(10).
False Advertisement and Representations in Hiring.
Minnesota Statutes, Section 181.64.
Use of Police Office Limited During Strikes.
Minnesota Statutes, Section 299c.03.
Use of Licensed Private Detectives During Strikes.
Minnesota Statutes, Section 326.337.

Striker Employment Protection.
Minnesota Statutes, Section 181.52.
Secondary Boycotts.
Minnesota Statutes, Section 179.40, et seq.

{89.03} Mediation and Arbitration Laws.
Minnesota Statutes, Section 572.08, et seq.
Minnesota Statutes, Section 179.61, et seq.

{89.04} Regulation of Union Activities.
Minnesota Statutes, Section 179.11.

{89.05} Regulation of Employment Practices.
Anti-Discrimination Laws.
Minnesota Statutes, Section 363.01, et seq.
Protection of Political Freedom.
Minnesota Statutes, Section 3.083.
Backlisting.
Minnesota Statutes, Section 179.12.
Employment Under False Pretense.
Minnesota Statutes, Section 181.64.
Polygraph Restrictions.
Minnesota Statutes, Section 181.75.
National Guard Members.
Minnesota Statutes, Section 192.34.
Equal Pay.
Minnesota Statutes, Section 181.66-71.
Criminal and Arrest Records.
Minnesota Statutes, Section 364.04.
Jure Duty.
Minnesota Public Acts, Chapter 286 (L.1977).
Mandatory Insurance Participation Restricted.
Minnesota Statutes, Section 61A.091.
Pre-employment Medical Exams.
Minnesota Statutes, Section 363.02.
Maternity Leaves.
Minnesota Statutes, Section 181.92.

{89.06} Wage and Hour Laws.
Child Labor.
Minnesota Statutes, Section 181A.05, et seq.
Voting Time.
Minnesota Statutes, Section 204C.04.
Garnishment.
Minnesota Statutes, Section 571.61
Medical Coverage Continuance.
Minnesota Statutes, Section 62A.17.

{89.07} Safety and Health Laws.
General Provisions.
Minnesota Statutes, Section 182.653.
Toxic Substances-Right to Know.
Minnesota Statutes, Section 182.65, et seq.

{89.08} Unemployment Compensation.
Minnesota Statutes, Section 268.07-09.

{89.09} Workers Compensation Laws.
Minnesota Statutes, Section 176.021, et seq.
Minnesota Statutes, Section 176.66.

{89.10} Employment-at-Will Developments.
Promissory Estoppel Exception.
Grouse v. Group Health Plan, Inc., 306 N.W.2d 114 (Minn., 1981).
Implied Contract Exception.
Pine River State Bank v. Mettille, 333 N.W.2d 155 (Minn., 1972).
Independent Consideration Exception.
Bussard v. College of St. Thomas, Inc., 200 N.W.2d 155 (Minn., 1972).
Intentional Infliction of Emotional Distress Exception.
Hubbard v. United Press International, Inc., 330 N.W.2d 428 (Minn., 1983).
Covenant of Good Faith and Fair Dealing Exception.
Eklund v. Vincent Brass & Aluminum Co., 351 N.W.2d 371 (Minn., 1984).

Mississippi

{90.01} Labor Relations Laws.
Union Telegrapher's Rights.
Mississippi code, Section 77-9-725, et seq.
Right-to-Work Statute.
Mississippi Statutes, Section 71-1-47.
"Yellow-Dog" Contracts.
Mississippi Statutes, Section 6984.5(1)(b).

{90.02} Strikes, Picketing, and Boycott Laws.
Interference with Employment.
Mississippi Statutes, Section 97-1-1, et seq.
Unlawful Assembly.
Mississippi Statutes, Section 219-25-67
Interference with Business.
Mississippi Statutes, Section 97-23-83.
Interference with Transportation.
Mississippi Statutes, Section 97-25-43.
Restriction of Railroad Picketing.
Mississippi Statutes, Section 77-9-236.
Use of Highway Patrol During Strikes.
Mississippi Statutes, Section 45-3-21.

Use of Violence or Threats Forbidden.
Mississippi Statutes, Section 97-23-39.

{90.03} Mediation and Conciliation Laws.
Mississippi Statutes, Section 11-15-1.

{90.04} Regulation of Union Activities.
Mississippi Statutes, Section 97-21-53.
Mississippi Statutes, Section 71-1-49.
Mississippi Public Acts, H.B.123 (L.1962).

{90.05} Regulation of Employment Practices.
Anti-Discrimination Laws.
Mississippi Statutes, Section 25-9-103.
Jury Duty.
Mississippi Statutes, Section 13-5-23.
Protection of Military Personnel.
Mississippi Statutes, Section 33-1-19.
Unlawful Enticement.
Mississippi Statutes, Section 97-23-29.
Protection of Political Activities.
Mississippi Statutes, Section 23-3-29.

{90.06} Wage and Hour Laws.
Child Labor.
Mississippi Statutes, Section 71-1-1, et seq.
Payment of Wages.
Mississippi Statutes, Section 71-1-35.

{90.07} Safety and Health Laws.
Mississippi Statutes, Section 75-37-19.

{90.08} Unemployment Compensation.
Mississippi Statutes, Section 71-5-1, et seq.
Mississippi Statutes, Section 71-5-511.

{90.09} Workers Compensation Laws.
Mississippi Statutes, Section 71-3-1, et seq.

{90.10} Employment-at-Will Developments.
Implied Contract Exception
Conley v. Board of Trustees, 707 F.2d 175 (CA-5, 1983).
"Independent Consideration Exception"
McGlohn v. Gulf and S.I.R.R., 174 So. 250 (Miss., 1937); *Sartin v. City of Columbus Utilities Commission*, 421 F. Supp. 393 (N.D. Miss., 1976).

Intentional Tort Exception

Smith v. Atlas Off-Shore Boat Service, 653 F.2d 1057 (CA-5, 1981); See also: *Moeller v. Fuselier, Ott & McKee*, 115 LRRM 2600 (Miss., 1984).

Missouri

{91.01} Labor Relations Laws.

Public Employees Bargaining Rights.

Missouri Revised Statutes, Section 105.501, et seq.
Missouri Revised Statutes, Section 105.510.
Missouri Revised Statutes, Section 105.530.
Missouri Revised Statutes, Section 105.540.

Nonright-to-Work Policy.

Section 14(b) of the Labor Management Relations Act.

{91.02} Strikes, Picketing, and Boycott Laws.

Interference with Employment.

Missouri Revised Statutes, Section 559.460.

Unlawful Assembly.

Missouri Revised Statutes, Section 562.150.

Importation of Private Detectives During Strikes.

Missouri Revised Statutes, Section 562.200.

Restriction of Railroad Picketing.

Missouri Revised Statutes, Section 560.315.

Illegal Seizure of Property.

Missouri Revised Statutes, Section 560.435.

{91.03} Mediation and Arbitration Laws.

Missouri Revised Statutes, Section 295.010, et seq.
Missouri Revised Statutes, Section 295.180.

{91.04} Regulation of Union Activities.

Missouri Revised Statutes, Section 416.031.

{91.05} Regulation of Employment Practices.

Anti-Discrimination Laws.

Missouri Revised Statutes, Section 296.010, et seq.

Protection of Political Freedom.

Missouri Revised Statutes, Section 15.020(7).
Missouri Revised Statutes, Section 115.635(b).

Service Letter.

Missouri Revised Statutes, Section 290.140.

Employment Rights of Military.

Missouri Revised Statutes, Section 41.730.

Wage Reduction.

Missouri Revised Statutes, Section 296.100.

Equal Pay.

Missouri Revised Statutes, Section 290.400, et seq.

{91.06] Wage and Hour Laws.
Child Labor.
Missouri Revised Statutes, Section 294.400, et seq.
Voting Time.
Missouri Revised Statutes, Section 115.639
Payment of Wages.
Missouri Revised Statutes, Section 290.090.
Payment Upon Termination.
Missouri Revised Statutes, Section 115.639.
Garnishments.
Missouri Revised Statutes, Section 525.030.
Medical Coverage.
Missouri Revised Statutes, Section 376.397.

{91.07} Safety and Health Laws.
Missouri Revised Statutes, Section 291.101, et seq.
Missouri Revised Statutes, Section 291.130.

{91.08} Unemployment Compensation Laws.
Missouri Revised Statutes, Section 288.040.

{91.09} Workers Compensation Laws.
Missouri Revised Statutes, Section 287,780.

{91.10} Employment-at-Will Developments.
Public Policy Exception.
Arie v. Intertherm, Inc., 644 S.W.2d 142 (Mo. App., 1983).
Intentional Infliction of Emotional Distress Exception.
Bass v. Intertherm Co., 644 S.W.2d 765 (Mo. S.Ct., 1983).
Implied Contract Exception.
Arie v. Intertherm, Inc., 648 S.W.2d 142 (Mo. App., 1983); *Hinkeldey v. Cities Ins. Co.*, Case No. 36426 (Mo. Ct. of App., 1985).

Montana

{92.01} Labor Relations Laws.
Public Employees Bargaining Rights.
Montana Code Annotated, Section 39-31-201.
Nurses Bargaining Rights.
Montana Code Annotated, Section 39-32-101, et seq.
Nonright-to-Work Policy.
Section 14(b) of the Labor Management Relations Act.

{92.02} Strikes, Picketing, and Boycott Laws.
Anti-Injunction Statutes.
Montana Code Annotated, Section 27-19-103.
Unlawful Assembly.
Montana Code Annotated, Section 45-8-103.

Strikebreakers.
Montana Code Annotated, Section 39-33-201, et seq.
Striker Replacements.
Montana Code Annotated, Section 45-8-106.
Sit-Down Strikes.
Montana Code Annotated, Section 45-6-203.
Importation of Private Detectives During Strikes.
Montana Code Annotated, Section 47-5-102 and 47-5-103.

{92.04} Regulation of Union Activities.
Montana Code Annotated, Section 39-33-101.

{92.05} Regulation of Employment Practices.
Anti-Discrimination Laws.
Montana Code Annotated, Section 49-1-102, et seq.

Hiring of Illegal Aliens.
Montana Code Annotated, Section 39-2-305.
Polygraph Restrictions.
Montana Code Annotated, Section 39-2-304.
Blacklisting.
Montana Code Annotated, Section 39-2-802, et seq.
Service Letter.
Montana Code Annotated, Section 39-2-801.
Unlawful Enticement.
Montana Code Annotated, Section 39-2-303.
Comparable Pay.
Montana Code Annotated, Section 39-3-104.

{92.06} Wage and Hour Laws.
Child Labor.
Montana Code Annotated, Section 41-2-101.
Payment of Wages.
Montana Code Annotated, Section 39-3-204.
Payment Upon Termination.
Montana Code Annotated, Section 39-3-205.
Garnishments.
Montana Code Annotated, Section 39-2-302.

{92.07} Safety and Health Laws.
Montana Code Annotated, Section 50-70-113.

{92.08} Unemployment Compensation Laws.
Montana Code Annotated, Section 39-51-2103, et seq.
Montana Code Annotated, Section 39-51-2103.

{92.09} Works Compensation Laws.
Montana Code Annotated, Section 39-71-701, et seq.

{92.10} Employment-at-Will Developments.
Montana Code Annotated, Section 39-2-503.
Covenant of Good Faith and Fair Dealing Exception.
Gates v. Life of Montana Insurance Co., 668 P.2d 213 (Mont. S. Ct., 1983).
Public Policy Exception.
See Statement of Legal Principle in *Keneally v. Orgain*, 606 P2d 127 (Mont., 1980).

Nebraska

{93.01} Labor Relations Laws.
Teachers Bargaining Rights.
Revised Statutes of Nebraska, Section 79-1288.
Revised Statutes of Nebraska, Section 79-1290.
Revised Statutes of Nebraska, Section 79-1293.
Public Employees Bargaining Rights.
Revised Statutes of Nebraska, Section 48-801, et seq.
Revised Statutes of Nebraska, Section 48-838.
Right-to-Work Statute.
Revised Statutes of Nebraska, Section 48-217.

{93.02} Strikes, Picketing, and Boycott Laws.
Destruction of Property.
Revised Statutes of Nebraska, Section 25-580.
Unlawful Interference with Employment.
Nebraska Public Acts, L.B. 38 (L.1977), Section 301.
Mass Picketing.
Nebraska Public Acts, L.B. 38 (L.1977), Section 302.
Secondary Boycotts.
Revised Statutes of Nebraska, Section 48-901, et seq.

{92.03} Mediation and Conciliation Laws.
Revised Statutes of Nebraska, Section 25-2103.
Revised Statutes of Nebraska, Section 43-801.

{92.04} Regulation of Union Activities.
Revised Statutes of Nebraska, Section 28-548.
Revised Statutes of Nebraska, Section 4-106.
Nebraska Public Acts, L.B. 672 (L. 1963).

{93.05} Regulation of Employment Practices.
Anti-Discrimination Laws.
Revised Statutes of Nebraska, Section 48-1001, et seq.
Protection of Political Freedom.
Revised Statutes of Nebraska, Section 32-1223.

Medical Examinations and Coverage Continuance.
Revised Statutes of Nebraska, Section 44-1633.
Revised Statutes of Nebraska, Section 48-221.
Polygraph Restrictions.
Nebraska Public Acts, L.B. 485 (L. 1980).
Military Leaves of Absence.
Revised Statutes of Nebraska, Section 55-161.
Jury Duty.
Revised Statutes of Nebraska, Section 25-1640.
Service Letter.
Revised Statutes of Nebraska, Section 48-211.
Equal Pay.
Revised Statutes of Nebraska, Section 48-1221.

{93.06} Wage and Hour Laws.
Child Labor.
Revised Statutes of Nebraska, Section 48-304.
Revised Statutes of Nebraska, Section 48-302,303.
Voting Time.
Revised Statutes of Nebraska, Section 32-1046.
Payment of Wages.
Revised Statutes of Nebraska, Section 48-1230.
Payment Upon Termination.
Revised Statutes of Nebraska, Section 48-1231.
Garnishments.
Revised Statutes of Nebraska, Section 25-1558(b).

{93.07} Safety and Health Laws.
Revised Statutes of Nebraska, Section 48-401, et seq.
Revised Statutes of Nebraska, Section 48-404.

{93.08} Unemployment Compensation Laws.
Revised Statutes of Nebraska, Section 24-601, et seq.

{93.09} Workers Compensation Laws.
Revised Statutes of Nebraska, Section 48-101, et seq.
Revised Statutes of Nebraska, Section 48-101, et seq.

{93.10} Employment-at-Will Developments.
See, for example: *Mau v. Omaha Nat'l Bank*, 299 N.W.2d 147 (Neb., 1980).
Morris v. Lutheran Medical Center, 340 N.W.2d 388 (Neb., 1983); see also *Corso v. Creighton University*, 731 F.2d 529 (CA-8, 1984).

Nevada

{94.01} Labor Relations Laws.
General Right to Organize.
Nevada Revised Statutes, Section 614.100.

Municipal Employees Bargaining Right.
Nevada Revised Statutes, Section 288.140
Nevada Revised Statutes, Section 288.270.
Nevada Revised Statutes, Section 288.010, et seq.
Right-to-Work Statute.
Nevada Revised Statutes, Section 613.230, et seq.
"Yellow-Dog" Contracts.
Nevada Revised Statutes, Section 613.130.

{94.02} Strikes, Picketing, and Boycott Laws.
Interference with Employment.
Nevada Revised Statutes, Section 613.100.
Strike Notices.
Nevada Revised Statutes, Section 611.290.
Nevada Revised Statutes, Section 614.120.

{94.03} Mediation and Conciliation Laws.
Nevada Revised Statutes, Section 38.035, et seq.
Nevada Revised Statutes, Section 614.020.

{94.04} Regulation of Union Activities.
Nevada Revised Statutes, Section 205.205.

{94.05} Regulation of Employment Practices.
Anti-Discrimination Statutes.
Nevada Revised Statutes, Section 613.310, et seq.
Polygraph Restrictions.
Nevada Public Acts, chapter 676 (L.1981).
Protection of Political Freedom.
Nevada Revised Statutes, Section 613.040.
Employment Under False Pretense.
Nevada Revised Statutes, Section 613.010.
Jury and Witness Duty.
Nevada Public Acts, chapter 150 (L.1977).
Nevada Revised Statutes, Section 50.070.
Backlisting.
Nevada Revised Statutes, Section 613.210, et seq.
Service Letter.
Nevada Revised Statutes, Section 613.240.
Discharge of Volunteer Firefighters.
Nevada Public Acts, Chapter 381 (L.1983).
Restrictions on Detectives and "Spotters."
Nevada Revised Statutes, Section 613.160.
Equal Pay.
Nevada Revised Statutes, Section 608.017.

{94.06} Wage and Hour Laws.
Child Labor.
Nevada Revised Statutes, Section 609.220, et seq.
Voting Time.
Nevada Revised Statutes, Section 293.463.
Payment of Wages.
Nevada Revised Statutes, Section 608.060.
Payment Upon Termination.
Nevada Revised Statutes, Section 6608.020 and 608.030.

{94.07} Safety and Health Laws.
General Provisions.
Nevada Revised Statutes, Section 618.375, et seq.
Toxic Substances-Right to Know.
Nevada Revised Statutes, Section 668.380.
Nevada Revised Statutes, Section 618.370.

{94.08} Unemployment Compensation Laws.
Nevada Revised Statutes, Section 612.010, et seq.

{94.09} Workers Compensation Laws.
Nevada Revised Statutes, Section 616.570, et seq.

{94.10} Employment-at-Will Developments.
Public Police Exception
Hansen v. Harrah's, 675 P.2d 394 (Nev. S. Ct., 1984).
Implied Contract Exception.
Southwest Gas Crop. v. Ahmad, 668 P.2d 261 (Nev. S. Ct., 1983).

New Hampshire

{95.01} Labor Relations Laws.
State Employees Bargaining Rights.
New Hampshire Revised Statutes, Chapter 273-A:1, et seq.
Municipal Employees Bargaining Rights.
New Hampshire Public Acts, Chapter 255 (L.1955).
Nonright-to-Work Policy.
Section 14(b) of the Labor Management Relations Act.
"Yellow-Dog" Contracts.
New Hampshire Revised Statutes, Chapter 275:2.

{95.02} Strikes, Picketing, and Boycott Laws.
Unlawful Assembly.
New Hampshire Revised Statutes, Chapter 644:1
Striker Replacements.
New Hampshire Revised Statutes, Chapter 275-A:2.
Strikebreakers.
New Hampshire Revised Statutes, Chapter 275-A:2.

Restrictions on Use of National Guard in Strikes.
New Hampshire Revised Statutes, Chapter 111:1.

{95:03} Mediation and Conciliation Laws.
New Hampshire Revised Statutes, Chapter 210:12, et seq.
New Hampshire Revised Statutes, Chapter 542:1.

{95.04} Regulation of Union Activities.
New Hampshire Revised Statutes, Chapter 664:1, et seq.
New Hampshire Revised Statutes, Chapter 207:1-7.

{95.05} Regulation of Employment Practices.
Anti-Discrimination Laws.
New Hampshire Revised Statutes, Chapter 354-A:1, et seq.
Employment of Aliens.
New Hampshire Revised Statutes, Chapter 275-A:4-5.
Freedom of Expression (State Employees).
New Hampshire Revised Statutes, Chapter 98-E:2.
Protection of Military Personnel.
New Hampshire Revised Statutes, Chapter 110-B:65.
Unlawful Procurement of Employees.
New Hampshire Revised Statutes, Chapter 275:7.
Access to Personnel Records.
New Hampshire Revised Statutes, Chapter 275:55.
Consumer Reporting Agency Requirements.
New Hampshire Revised Statutes, Chapter 359-B:3.
Equal Pay.
New Hampshire Revised Statutes, Chapter 275:36-41.

{95.06} Wage and Hour Laws.
Child Labor.
New Hampshire Revised Statutes, Chapter 276A:4.
Payment of Wags.
New Hampshire Revised Statutes, Chapter 275:43.
Payment Upon Termination.
New Hampshire Revised Statutes, Chapter 277:44(I).
New Hampshire Revised Statutes, Chapter 277:44(II).
Medical Examinations and Insurance Coverage Continuance.
New Hampshire Revised Statutes, Chapter 275:3.
New Hampshire Revised Statutes, Chapter 500:3.

{95.07} Safety and Health Laws.
New Hampshire Revised Statutes, Chapter 277:11.
New Hampshire Revised Statutes, Chapter 277:12.
Toxic Substances-Right to Know.
New Hampshire Revised Statutes, Chapter 277-A:1, et seq.

{95.08} Unemployment Compensation Laws.
New Hampshire Revised Statutes, Chapter 282-A:1, et seq.

{95.09} Workers Compensation Laws.
New Hampshire Revised Statutes, Chapter 281-A:1, et seq.

{95.10} Employment-at-Will Developments.
Covenant of Good Faith and Fair Dealing Excepting.
Monge v. Beebe Rubber Co., 316 A.2d 549 (N.H., 1974).
Public Policy Exception.
Cloutier v. Great Atlantic and Pacific Tea Co., 436 A.2d 1140 (N.H., 1980).
Independent Consideration Exception.
Foley v. Community Oil Co., 64 F.R.D. 561 (D. N.H., 1974).

New Jersey

{96.01} Labor Relations Laws.
Public Utility Labor Disputes Act.
New Jersey Statutes Annotated, Section 34:13B-13.
New Jersey Statutes Annotated, Section 34:13B-1.
State Employees' Bargaining Rights.
New Jersey Statutes Annotated, Section 34:13A-1
New Jersey Statutes Annotated, Section 34:13A-5.4.
Nonright-to-Work Policy.
New Jersey Statutes Annotated, Section.14 (b).
"Yellow-Dog" Contracts.
New Jersey Statutes Annotated, Section 34: 12-2.

{96.02} Strikes, Picketing, and Boycott Laws.
Anti-Injunction Laws.
New Jersey Statutes Annotated, Section 2A: 15-51, et seq.
Unlawful Assembly.
New Jersey Statutes Annotated, Section 2A:126-4.
Restriction of Railroad Picketing.
New Jersey Statutes Annotated, Section 48:12-164, et seq.
Striker Replacements.
New Jersey Statutes Annotated, Section 30:8-40.
New Jersey Statutes Annotated, Section 34:8-25.

{96.03} Mediation and Conciliation Laws.
New Jersey Statutes Annotated, Section 34:13-1 et seq.

{96.04} Regulation of Union Activities.
New Jersey Statutes Annotated, Section 56:2-1, et seq.
New Jersey Public Acts, Chapter 246 (L. 1962, 1963).

{96.05} Regulation of Employment Practices.
Anti-Discrimination Laws.
New Jersey Statutes Annotated, Section 10:5-1, et seq.
Protection of Political Freedom.
New Jersey Statutes Annotated, Section 19:34-27.
Reemployment Rights of Military Personnel.
New Jersey Statutes Annotated, Section 10:5-5 (g); 10: 5-12.
Polygraph Restrictions.
New Jersey Statutes Annotated, Section 2C:40 A-1.
Jury Duty (Public Employees).
New Jersey Statutes Annotated, Section 2A:69-5.
Convictions.
New Jersey Statutes Annotated, Section 2C:52-1, et seq.
Medical Coverage Continuance.
New Jersey Statutes Annotated, Section 17B:27-51.12.
Authorization of State Employees' Union Dues Deduction.
New Jersey Statutes Annotated, Section 52:14-15.9C.
Equal Pay.
New Jersey Statutes Annotated, Section 34:11-56.2.

{96.06} Wage and Hour Laws.
Child Labor.
New Jersey Statutes Annotated, Section 34:2-3,8, 11 and 15.
Payment of Wages.
New Jersey Statutes Annotated, Section 34:11-4.2.
Payment Upon Termination.
New Jersey Statutes Annotated, Section 34:11-4.3.
Garnishment.
New Jersey Statutes Annotated, Section 2A:170-90.4

{96.07} Safety and health laws.
General Provisions.
New Jersey Statutes Annotated, Section 34:6A-3.
Toxic Substances-Right to Know.
New Jersey Statutes Annotated, Section 34:5A-1.

{96.08} Unemployment Compensation Laws.
New Jersey Statutes Annotated, Section 43:21-1, et seq.

{96.09} Worker's Compensation Laws.
New Jersey Statutes Annotated, Section 34:15-1, et seq.

{96.10} Employment-at-Will Development.
Public Police Exception.
Lalley v. Copygraphics, 428 A.2d 1317 (N.J., 1981).
Implied Contract Exception
Wooley v. Hoffman-LaRoche, Inc., Case No. A-98-82 (N.J. S. Ct., 1985).

New Mexico

{97.01} Labor Relations Laws.
State Employees' Bargaining Rights.
New Mexico Labor-Management Relations Regulations, Section 3.
New Mexico Labor-Management Relations Regulations, Section 16.
New Mexico Labor-Management Relations Regulations, Section 1, et seq.
Nonright-to-Work Policy.
New Mexico Labor-Management Relations Act, Section 14(b).
"Yellow-Dog" Contracts.
New Mexico Public Acts, S.B. 402 (L., 1967).

{97.02} Strikes, Picketing, and Boycott Laws.
Anti-Injunction Statute.
New Mexico Statutes, Section 59-2-1, et seq.
Unlawful Assembly.
New Mexico Statutes, Section 41-1210.
Interference with Ingress.
New Mexico Statutes, Section 50-2-2(b).

{97.03} Mediation and Arbitration Laws.
New Mexico Statutes, Section 44-7-1.

{97.04} Regulation of Union Activities.
New Mexico Statutes, Section 52-201, et seq.
New Mexico Statutes, Section 50-2-3.

{97.05} Regulation of Employment Practices.
Anti-Discrimination Laws.
New Mexico Statutes, Section 28-1-1, et seq.
Protection of Political Freedom.
New Mexico Statutes, Section 3-10-11.
Reemployment Rights of Military Personnel.
New Mexico Statutes, Section 28-15-1, et seq.
Jury Duty.
New Mexico Statutes, Section 38-5-18.
Blacklisting.
New Mexico Statutes, Section 30-13-3.
Arrest Records (Public Employees).
New Mexico Statutes, Section 28-2-3(B).
Credit Bureau Requirements.
New Mexico Statutes, Section 56-3-5.

{97.06} Wage and Hour Laws.
Child Labor.
New Mexico Statutes, Section 50-6-2.

Voting Time.
New Mexico Statutes, Section 1-12-42.
Payment of Wages.
New Mexico Statutes, Section 50-4-2.
Payment Upon Termination.
New Mexico Statutes, Section 50-4-4.
New Mexico Statutes, Section 50-4-5.

{97.07} Safety and Health Laws.
General Provisions.
New Mexico Statutes, Section 52-1-8.
Toxic Substance-Right to Know.
New Mexico Statutes, Section 50-9-1, et seq.

{97.08} Unemployment Compensation Laws.
New Mexico Statutes, Section 51-1-4B, et seq.

{97.09} Workers' Compensation Laws.
New Mexico Statutes, Section 52-1-8, et seq.
New Mexico Statutes, Section 52-1-9, et seq.

{97.10} Employment-at-Will Developments.
Public Policy Exception.
Vigil v. Arzola, 687 P.2d 1038. (N.M. S. Ct., 1984).
Implied Contract Exception.
Forrester v. Parker, 606 P.2d 191 (N.M. S. Ct., 1980).

New York

{98.01} Labor Relation Laws.
New York State Labor Relations Act.
Consolidated Laws of New York, Labor Laws, Section 700, et seq.
General Right to Organize.
New York Constitution, Section 17.
Port Authority Employees' Benefits.
New York Public Acts, Chapter 1203 (L. 1971).
Public Employees' Bargaining Rights.
Consolidated Laws of New York, Civil Service Laws, Section 202.
Consolidated Laws of New York, Civil Service Laws, Section 209.
Consolidated Laws of New York, Civil Service Laws, Section 200, et seq.
Labor and Management Improper Practices Act.
New York Labor and Management Improper Practices Act, Section 720, et seq.
Nonright-to-Work Policy.
New York Labor of Management Relations Acts, Section 14(b).
"Yellow-Dog" Contracts.
New York Public Acts, Chapter 11 (L. 1935).

{98.02} Strikes, Picketing, and Boycott Laws.
Anti-Injunction Statue.
New York Anti-Injunction Act, Section 807, et seq.
Interference with Employment.
Consolidated Laws of New York, Penal Code, Section 580.
Use of Detectives During Strikes:
Consolidated Laws of New York, General Business Law, Section 84.

{98.03} Mediation and Arbitration Laws.
State Civil Practice Law, Section 7501.
Consolidated Laws of New York, Labor Laws, Section 751, et seq.
General Municipal Law, Section 601, et seq.
Consolidated Laws of New York, Penal Code, Section 373.
Consolidated Laws of New York, Penal Code, Section 376 and 860.

{98.04} Regulation of Union Activities.
Consolidated Laws of New York, Labor Laws, Section 209.
Consolidated Laws of New York, Labor Laws, Section 209-a.
New York Labor and Management Improper Practices Act, Section 726.
New York Mitchell-Hollinger Law, Section 37, et seq.
Consolidated Laws of New York, Insurance and Banking Laws, Section 37, et seq.
Consolidated Laws of New York, Unconsolidated Laws, Section 9801, et seq.

{98.05} Regulation of Employment Practices.
Anti-Discrimination Laws.
Consolidated Laws of New York, Executive Laws, Section 291, et seq.
Employment Agency Requirements.
Consolidated Laws of New York, General Business Laws, Section 187.
Psychogalvanic Stress Evaluation Exams.
Consolidated Laws of New York, Labor Laws, Section 733, et seq.
Misrepresentation of Employment Opportunities.
Consolidated Laws of New York, General Business Laws, Section 396-1.
Arrest and Conviction Record.
New York Executive Laws, Section 296(15) and (16).
Fingerprinting Limitations.
Consolidated Laws of New York, Labor Laws, Section 201-a.
Disclosure of Consumer Reports.
New York General Business Laws, Section 380(a).

Jury and Witness Duty.
Consolidated Laws of New York, Judiciary Laws, Section 519.
Consolidated Laws of New York, Penal Code, Section 215.11.
Importation of Migratory Farm Labor.
Consolidated Laws of New York, Labor Laws, Section 212a.
Whistle-Blowing Statutes.
Consolidated Laws of New York, Labor Laws, Section 740.

Protection of Railroad Employees.
New York Railroad Laws, Section 54-b.
Protection of Military Personnel.
New York Military Laws, Section 318.
Blacklisting.
Consolidated Laws of New York, Labor Laws, Section 704(2).
Medical Examination Payment.
Consolidated Laws of New York, Labor Laws, Section 210b.
Eavesdropping on Union Activities.
Consolidated Laws of New York, Labor Laws, Section 704.
Equal Pay.
Consolidated Laws of New York, Labor Laws, Section 194.

{98.06} Wage and Hour Laws.
Child Labor.
Consolidated Laws of New York, Labor Laws, Section 130.
Consolidated Laws of New York, Labor Laws, Section 131(4).
Consolidated Laws of New York, Labor Laws, Section 3215(4).
Voting Time.
Consolidated Laws of New York, Election Laws, Section 3.110.
Payment Upon Termination.
Consolidated Laws of New York, Labor Laws, Section 191(3).
Garnishments.
New York Civil Practices Laws and Rules, Section 5252.
Medical Insurances Continuance.
Consolidated Laws of New York, Insurance Laws, Section 162(5)(a).

{98.07} Safety and Health Laws.
General Provision.
Consolidated Laws of New York, Labor Laws, Section 28(2)(a).
Toxic Substances-Right to Know.
New York Toxic Substances Laws, Section 875, et seq.

{98.08} Unemployment Compensation Laws.
Consolidated Laws of New York, Labor Laws, Section 591, et seq.

{98.09} Workers' Compensation.
Consolidated Laws of New York, Workers' Compensation Laws, Section 1, et seq.
Consolidated Laws of New York, Workers' Compensation Laws, Section 10 and 11.

{98.10} Employment-at-Will Developments.
Implied Contract Exception.
Weiner v. McGraw-Hill, 443 N.E.2d 441 (N.Y., 1982); as limited by certain requirement therein.

Promissory Estoppel Exception.
Myers v. Conradian Corp., 459 N.Y.S.2d 929 (1983).

North Carolina

{99.01} Labor Regulation Laws
Ban Against Joining Unions (Public Employees).
General Statutes of North Carolina, Section 95-98.
Right-to-Work Statutes.
General Statutes of North Carolina, Section 97-78, et seq.

{99.02} Strikes, Picketing, and Boycott Laws.
Unlawful Assembly.
General Statutes of North Carolina, Section 15-30.
Possession of Weapons on Picket Lines.
General Statutes of North Carolina, Section 14-277.2.
Public Employee Strikes Prohibited.
General Statutes of North Carolina, Section 95-98.1.

{99.03} Mediation and Arbitration Laws.
General Statutes of North Carolina, Section 95-32, et seq.
General Statutes of North Carolina, Section 95-36.1, et seq.

{99.04} Regulation of Union Activities.
General Statutes of North Carolina, Section 95-101.2.
General Statutes of North Carolina, Section 80-8 and 80-9.

{99.05} Regulation of Employment Practices.
Anti-Discrimination Laws.
General Statutes of North Carolina, Section 143-422.1.
General Statutes of North Carolina, Section 95-28.
Re-employment Right of Military Personnel.
General Statutes of North Carolina, Section 127A-201, et seq.
Blacklisting.
General Statutes of North Carolina, Section 14-355 and 14-356.
Unlawful Enticement.
General Statutes of North Carolina, Section 14-347.
General Statutes of North Carolina, Section 14-349.
Fraudulent Receipt of Wage Advancement.
General Statutes of North Carolina, Section 14-104.
Medical examination payment.
General Statutes of North Carolina, Section 14-357.

{99.06} Wage and Hour Laws.
Child Labor.
General Statutes of North Carolina, Section 95-25.
General Statutes of North Carolina, Section 95-25.5(a) and (d).

Payment of Wages.
General Statutes of North Carolina, Section 95-25.6.
General Statutes of North Carolina, Section 95-98.
Payment Upon.
General Statutes of North Carolina, Section 95.25.7.

{99.07} Safety and Hour Laws.
General Provisions.
General Statutes of North Carolina, Section 95-129(1).
General Statutes of North Carolina, Section 95-129(2).
Toxic Substance-Right to Know.
General Statutes of North Carolina, Section 95-143.

{99.08} Unemployment Compensation Laws.
General Statutes of North Carolina, Section 96-13, et seq.

{99.09} Workers' Compensation Laws.
General Statutes of North Carolina, Section 97-1 et seq.
General Statutes of North Carolina, Section 97-6.1.

{99.10} Employment-at-Will Developments.
Brooks v. Carolina Tel. and Co., 290 S.E.2d 370 (N.C., 1982).
Roberts v. Mays Mills, Inc., 114 S.E. 530 (N.C., 1922).
Implied Contract Exception.
Still v. Lance, 182 S.E.2d 403 (N.C. S. Ct., 1971); see also: *Sides v. Duke Hospital*, Case No. 83145C1308 (N.C. Ct. of App., 1985).
Public Policy Exception.
Sides v. Duke Hospital, supra.

North Dakota

{100.01} Labor Dakota Laws.
North Dakota Labor Management Relation.
North Dakota Century Code, Section 34-12-01, et seq.
Teachers' Bargaining Rights.
North Dakota Century Code, Section 15-38.01-07.
Court Enforcement of Collective Bargaining Contracts.
North Dakota Century Code, Section 34-09-08.
Public Policy Regarding Bargaining Rights.
North Dakota Century Code, Section 34-09-01.
Right-to-Work Statute.
North Dakota Century Code, Section 34-01-14.
"Yellow-Dog" Contracts.
North Dakota Century Code, Section 34-08-04.

{100.02} Strikes, Picketing, and Boycott Laws.
Anti-Injunction Statutes.
North Dakota Century Code, Section 34-08-03.

Stranger Picketing.
North Dakota Century Code, Section 34-09-12.
Secondary Boycotts.
North Dakota Century Code, Section 34-09-13.
Unlawful Assembly.
North Dakota Century Code, Section 34-19-06.
Striker Replacement Notices.
North Dakota Public Acts, Chapter 255, Section 10 (L. 1963).
Unlawful Interference with Employment.
North Dakota Century Code, Section 34-01-04, et seq.
Public Utility Strikes.
North Dakota Century Code, Section 37-01-06, et seq.

{100.03} Mediation and Arbitration Laws.
North Dakota Century Code, Section 34-11-01.
North Dakota Century Code, Section 34-11-01, et seq.
North Dakota Century Code, Section 34-11-03.
North Dakota Century Code, Section 34-29-01.

{100.04} Regulation of Union Activities.
North Dakota Century Code, Section 12-38-19.
North Dakota Century Code, Section 34-01-16.

{100.05} Regulation of Employment Practices.
Anti-Discrimination Laws.
North Dakota Century Code, Section 14-02.4-03.
Rights of Military Personnel (Public Employees).
North Dakota Century Code, Section 37-01-25.1.
Backlisting.
North Dakota Century Code, Section 34-12-03.
Jury Duty.
North Dakota Public Acts, S.B, 2320 (L.1971).
Fraudulent Receipt of Wage Advancement.
North Dakota Century Code, Section 34-01-10.
Protection of Political Activities (Public Employees).
North Dakota Century Code, Section 34-01-10.
Medical Examination Payment.
North Dakota Century Code, Section 34-01-15.
Polygraph Requirements.
North Dakota Century Code, Section 41-31-03.
Equal Pay.
North Dakota Century Code, Section 34-06.1-03.

{100.06} Wage and Hour Laws.
Child Labor.
North Dakota Century Code, Section 34-07-03, et seq.

Payment of Wage.
North Dakota Century Code, Section 34-14-02.
Payment Upon Termination.
North Dakota Century Code, Section 34-14-03(1).
North Dakota Century Code, Section 34-14-03(2).
Garnishments.
North Dakota Century Code, Section 32-09.1-18.

{100.07} Safety and Health Laws.
North Dakota Century Code, Section 65-03-01.

{100.08} Unemployment Compensation.
North Dakota Century Code, Section 52-06-01, et seq.

{100.09} Workers' Compensation Laws.
North Dakota Century Code, Section 56-01-01, et seq.
North Dakota Century Code, Section 56-05-01, et seq.

{100.10} Employment-at-Will Development.
Wood v. Buchanan, 5 N.W.2d 680 (N.D., 1942).
Aasmundstad v. Dickenson State College, 337 N.W.2d 792 (N.D., 1983).

Ohio

{101.01} Labor Relations Laws.
Public Employment Bargaining Right.
Ohio Statutes, Section 4117.03.
Ohio Statutes, Section 4117.11
Ohio Statutes, Section 4117.01, et seq.
Public Utility Bargaining.
Ohio Statutes, Section 717.03.
Successor Employers Bargaining.
Ohio Statutes, Section 4113.30.
Nonright-to-Work Policy.
Section 14(b) of the Labor Management Relations Act.
"Yellow-Dog" Contract.
Ohio Statutes, Section 4113.02.

{101.02} Strikes, Picketing, and Boycott Laws.
Unlawful Assembly.
Ohio Statutes, Section 3761.13.
Striker Replacements.
Ohio Statutes, Section 4143.12.
Injury to Person or Property.
Ohio Statutes, Section 2901.07.

{101.03} Mediation and Arbitration Laws.
Ohio Statutes, Section 2711.01, et seq.

Ohio Statutes, Section 4129.02.
Ohio Statutes, Section 4129.03, et seq.

{101.04} Regulation of Union Activities.
Ohio Statutes, Section 2911.27.

{101.05} Regulation of Employment Practices.
Anti-Discrimination Laws.
Ohio Statutes, Section 4112.02.
Arrest Records.
Ohio Statutes, Section 2953.42.
Conviction Records.
Ohio Statutes, Section 2953.32.
Access to Medical Records.
Ohio Statutes, Section 3113.23.
Jury and Witness Duty.
Ohio Statutes, Section 2313.18.
Ohio Statutes, Section 2313.18.
Military Leave of Absence.
Ohio Statutes, Section 5903.061.
Use of Railroad "Spotters."
Ohio Statutes, Section 4999.17.
Protection of Political Activities.
Ohio Statutes, Section 3599.06.
Blacklisting.
Ohio Statutes, Section 1331.03.
Medical Examination Payment.
Ohio Statutes, Section 4113.21.
Service Letter.
Ohio Statutes, Section 4973.03.
Equal Pay.
Ohio Statutes, Section 4111.17.

{101.06} Wage and Hour Laws.
Child Labor.
Ohio Statutes, Section 4109.01.
Ohio Statutes, Section 3331.01:4109.01.
Garnishments.
Ohio Statutes, Section 2716.05.
Notice of Medical Insurance Continuance.
Ohio Statutes, Section 1737.30.

{101.07} Safety and Health Laws.
Ohio Statutes, Section 4101.11.

{101.08} Unemployment Compensation Laws.
Ohio Statutes, Section 4141.01, et seq.

{101.09} Workers Compensation Laws.
Ohio Statutes, Section 4121.01, et seq.
Ohio Statutes, Section 4123.90.
Ohio Statutes, Section 4123.54.

{101.10} Employment-at-Will Developments.
Henkel v. Educational Research Council of America, 334 N.E.2d 118 (Ohio, 1976).
Implied Contract Exception.
See, for example: *Day v. Good Samaritan Hospital*, Case No. 8062(2nd App. Dist. Of Ohio, 1983).
Promissory Estoppel Exception.
Jones v. East Center for Community Mental Health, Case No. L-83-280 (6th App. Dist. Of Ohio, 1984).
Public Policy Exception.
Merkel v. Scovill, Inc., 570 F. Supp. 133 (S.D. Ohio, 1983).

Oklahoma

{102.01} Labor Relations Laws.
Collective Bargaining Laws.
Oklahoma Statutes Annotated 11:51-101.
Oklahoma Statutes Annotated 70:509.1, et seq.
Nonright-to-Work Police
Section 14(b) of the Labor Management Relations Act.

{102.02} Strikes, Picketing, and Boycott Laws.
Unlawful Assembly.
Oklahoma Statutes Annotated 21:1311, et seq.
Use of Prisoners During Strike.
Title 57 Section 543.
Use of Detectives During Strike.
Oklahoma Statutes Annotated 40:169.
Strikebreakers.
Title 40 Section 1991.1.
Strike Replacement.
Oklahoma Statutes Annotated 40:169.
Seizure of Property.
Title 21 Section 1351.
Unlawful Interference with Employment.
Oklahoma Statutes Annotated 21:837 and 21:838.

{102.03} Mediation and Arbitration Laws.
Oklahoma Statutes Annotated 59:94 (c).
Oklahoma Statutes Annotated 59:743 (c).

{102.04} Regulation of Union Activities.
Oklahoma Statutes Annotated 78:9, et seq.

{102.05} Regulation of Employment Practices.
Anti-Discrimination Laws.
Oklahoma Statutes Annotated 25:1302.
Employment Under False Pretense.
Oklahoma Statutes Annotated Title 21 Section 1351.
Protection of Political Freedom.
Oklahoma Statutes Annotated 26:440.
Leaves of Absence for Political Activities.
Oklahoma Statutes Annotated Title 40 Section 184, et seq.
Jury Service.
Oklahoma Statutes Annotated 38:34.
National Guard Members.
Oklahoma Statutes Annotated Title 44 Section 208.
Freedom of Expression.
See Oklahoma Public Acts, H.B.1128 (L.1981).
Service Letter.
Oklahoma Statutes Annotated Title 40 Section 171.
Backlisting.
Oklahoma Statutes Annotated 40:172.
Medical Examinations.
Oklahoma Statutes Annotated Title 40 Section 191, et seq.
Equal Pay.
Oklahoma Statutes Annotated 40:198.1.

{102.06} Wage and Hour Laws.
Child Labor.
Oklahoma Statutes Annotated 40:71.
Oklahoma Statutes Annotated 40:74 through 77.
Voting Time.
Oklahoma Statutes Annotated 26:7-101.
Payment of Wages.
Oklahoma Statutes Annotated Title 40 Section 165.
Payment Upon Termination.
Oklahoma Statutes Annotated 26:165.3.
Garnishments.
Oklahoma Statutes Annotated 14A:5-106.

{102.07} Safety and Health Laws.
Oklahoma Statutes Annotated Title 40 Section 403(1), et seq.

{102.08} Unemployment Compensation.
Oklahoma Statutes Annotated Title 40 Section 2-201, et seq.

{102.09} Workers Compensation Laws.
Oklahoma Statutes Annotated 85:1, et seq.

{102.10} Employment-at-Will Developments.
General Rule.
Foster v. Atlas Life Ins., Co., 6 P.2d 428 (CA-10, 1984); Langdon v. Sage Corp., 569 P.2d 524 (Okla. App., 1976).
Implied Contract Exception.
Vinyard v. King, 728 F.2d 428 (CA-10, 1984); *Langdon v. Sage Corp.*, 569 P.2d 524 (Okla. App., 1976).
Hall v. Farmers Ins. Exchange, Case no.59584 (Okla. S. Ct., 1985).

Oregon

{103.01} Labor Relations Laws.
Labor Peace Act.
Oregon Revised Statutes, Section 663.005, et seq.
Nurses Bargaining Rights.
Oregon Revised Statutes, Section 662.705.
Public Employees Bargaining Rights.
Oregon Revised Statutes, Section 243.662.
Oregon Revised Statutes, Section 243.672.
Oregon Revised Statutes, Section 243.650, et seq.
Nonright-to-Work Policy.
Section 14(b) of the Labor Management Relations Act.
"Yellow-Dog" Contracts.
Oregon Revised Statutes, Section 662.030.

{103.02} Strikes, Picketing, and Boycott Laws.
Anti-injunction laws.
Oregon Revised Statutes, Section 662.010 et seq.
Secondary Boycotts and Hot Cargo Clauses Forbidden.
Oregon Revised Statutes, Section 663.210, et seq.
Unlawful Interference with Employment.
Oregon Revised Statutes, Section 659.240.
Unlawful Assembly.
Oregon Revised Statutes, Section 23-801.
Picketing of Farms.
Oregon Public Acts, Chapter 543 (L.1963).
Striker Replacements.
Oregon Revised Statutes, Section 658.225.
Strikebreakers.
Oregon Public Acts, Chapter 645 (L.1975).

{103.03} Mediation and Arbitration Laws.
Oregon Revised Statutes, Section 662.405.

{103.04} Regulation of Union Activities.
Oregon Revised Statutes, Section 661.210, et seq.

{103.05} Regulation of Employment Practices.
Anti-Discrimination Laws.
Oregon Revised Statutes, Section 659.425.
Polygraph Restrictions.
Oregon Revised Statutes, Section 659.227.
Jury Services.
Oregon Public Acts, Chapter 160 (L.1975).
Access to Personnel Records.
Oregon Revised Statutes, Section 652.750.
Whistle-Blowing Statute.
Oregon Revised Statutes, Section 240.3165.
Employment Under False Pretense.
Oregon Revised Statutes, Section 659.260.
Fraudulent Receipt of Wage Advances.
Oregon Revised Statutes, Section 659.250.
Blacklisting.
Oregon Revised Statutes, Section 659.330.
National Guard Duty.
Oregon Revised Statutes, Section 408,210, et seq.
Anti-Nepotism Discrimination.
Oregon Revised Statutes, Section 659.340.
Medical Examinations.
Oregon Revised Statutes, Section 659.330.
Prior Testimony.
Oregon Revised Statutes, Section 659.270.
Juvenile Record.
Oregon Revised Statutes, Section 659.030 (a).

{103.06} Wage and Hour Laws.
Child Labor.
Oregon Wage and Hour Commission Rules, OAR 21-215(1).
Payment of Wages.
Oregon Revised Statutes, section 652.120.
Payment Upon Termination.
Oregon Revised Statutes, Section 652.140(1).
Garnishments.
Oregon Revised Statutes, Section 23.185(5).
Medical Insurance Conversion.
Oregon Revised Statutes, Section 743.850.

{103.07} Safety and Health Laws.
Oregon Revised Statutes, Section 654.010.

{103.08} Unemployment Compensation Laws.
Oregon Revised Statutes, Section 656.001, et seq.

{103.10} Employment-at-Will Developments.
Implied Contract Exception.
Fleming v. Kids and Kin Head Start, 693 P. 2d 1363 (Ore. S. Ct.,1985).
Public Policy Exception.
Nees v. Hocks, 536 p.2d 512 (Ore., 1975).
Intentional Infliction of Emotional Distress Exception.
Smithson v. Nordstrom, Inc., 664 P .2p 1119 (Ore Ct. of App.,1983).

Pennsylvania

{104.01} Labor Relations Laws.
General Right to Organize.
Pennsylvania Statutes, Title 43, Section 191.
Pennsylvania Labor Relations Act.
Pennsylvania Statutes, Title 43, Section 211.1, et seq.
Public Employees Bargaining Rights.
Pennsylvania Employee Relations Act, Section 401.
Pennsylvania Employee Relations Act, Section 101, et seq.
Public Transit Employees Bargaining Rights.
Pennsylvania Public Laws Act, Section 288 (L.1967).
Firefighters and Police Officers Bargaining Rights.
Pennsylvania Public Laws Act, Section 111 (L.1968).
Nonright-to-Work Policy
Section 14 (b) of the Labor Management Relations.
"Yellow-Dog" Contracts.
Pennsylvania Statutes, Title 18, Section 4669.

{104.02} Strikes, Picketing, and Boycott Laws.
Anti-Injunction Laws.
Pennsylvania Statutes, Title 43, Section 206a, et seq.
Strikebreakers Replacements.
Pennsylvania Statutes, Title 43, section 23.
Pennsylvania Public Laws Act, section 187 (L.1972).
Strikebreakers.
Pennsylvania Public Laws Act, section 111 (L.1968).
Right to Strike.
Pennsylvania Public Laws Act, section 187 (L.1972).
Interference with Railroads.
Pennsylvania Public Laws Act, 872 (L.1939).
Pennsylvania Statutes, Title 18, Section 4664, and Title 18, Section 4921.

{104.03} Mediation and Arbitration Laws.
Pennsylvania Statutes, Title 42, Section 7303.

{104.04} Regulation of Union Activities.
Pennsylvania Statutes, Title 73, Section 105.
Pennsylvania Statutes, Title 25, Section 3225.

{104.05} Anti-Discrimination Laws.
Pennsylvania Statutes, Title 43, Section 955.
Arrest Records.
Pennsylvania Public Laws, H.B.2095 (L.1978).
Polygraph Restrictions.
Pennsylvania Statutes, Title 18, Section 7321.
Voice Stress Analyzers.
Pennsylvania Statutes, Title 18, Section 7507.
Employment Under False Pretenses.
Pennsylvania Statutes, Title 18, Section 4856.
Interference with Employment.
Pennsylvania Statutes, Title 18, Section 4670.
Protection of Political Freedom.
Pennsylvania Statutes, Title 25, Section 3547.
Jury Service.
Pennsylvania Public Laws Act, 17 (L.1978).
Protection of Volunteer Firefighters.
Pennsylvania Public Laws Act, Section 83 (L.1977).
Access to Personnel Records.
Pennsylvania Public Laws Act, 286 (L.1978).
Medical Examination Payments.
Pennsylvania Statutes, Title 43, Section 1002.
Equal Pay.
Pennsylvania Public Laws Act, 694 (L.1968).

{104.06} Wage and Hour Laws.
Child Labor.
Pennsylvania Statutes, Title 24, Section 13-1391.
Payment of Wages.
Pennsylvania Public Laws Act, 329 (L.1961).
Payment on Termination.
Pennsylvania Public Laws Act, 329 (L.1961).
Medical Insurance Continuance.
Pennsylvania Statutes, Title 40, Section 756.2.

{104.07} Health and Safety Laws.
Pennsylvania Statutes, Title 43, Section 9, et seq.

{104.08} Unemployment Compensation Laws.
Pennsylvania Statutes, Title 43, Section 751, et seq.

{104.09} Workers Compensation.
Pennsylvania Statutes, Title 77, Section 1, et seq.

{104.10} Public Policy Exception.
McNulty v. Borden, Inc., 474 F. Supp. 1111 (E.D. Pa., 1979).

Independent Consideration Exception.
Cory v. SmithKline Beckman Corp., 116 LRRM 3361 (E.D. Pa., 1984).
Intentional Infliction of Emotional Distress Exception.
Shaffer v. National Can Corp., 565 F. Supp. 909 (E.D. Pa., 1983)

Rhode Island

{105.01} Labor Relations Laws.
Rhode Island State Labor Relations Act.
Section 28-7-1, et seq. of the Rhode Island General Laws.
Public Employees Bargaining Rights.
Rhode Island General Laws, Section 36-11-1, et seq.
Municipal Employees Bargaining Rights.
Rhode Island General Laws, Section 28-9.4-1, et seq.
Teachers Bargaining Rights.
Rhode Island General Laws, Section 28-9.3-9.
Firefighters and Police Officers Bargaining Rights.
Rhode Island General Laws, Section 28-9.1-2, et seq.
Rhode Island General Laws, Section 28-9.2-2 et seq.
State Police Officers Bargaining Rights.
Section 28-9. 5-1, et seq. of the Rhode Island General Laws.
Nonright-to-Work Police.
Section 14(b) of the Labor Management Relations Act.
"Yellow-Dog" Contracts.
Section 28-7-13, et seq. of the Rhode Island General Laws.

{105.02} Strikes, Picketing, and Boycott Laws.
Anti-Injunction Laws.
Rhode Island General Laws, Section 28-10-2 et seq.
Unlawful Assemble.
Rhode Island General Laws, Section 11-38-1.
Interference with Employment.
Rhode Island General Laws, Section 11-11-4.
Interference with Railroads.
Rhode Island General Laws, Section 11-36-3.
Use of Tear Gas Prohibited During Strikes.
Rhode Island General Laws, Section 28-10-7 and 28-10-8.
Strikebreakers and Replacement Notices.
Rhode Island General Laws, Section 28-10-10 et seq.
Rhode Island General Laws, Section 28-10-13.

{105.03} Mediation and Arbitration Laws.
Rhode Island General Laws, Section 28-9-1.

{105.04} Regulation of Union Activities.
Rhode Island General Laws, Section 11-14-5.

{105.05} Regulation of Employment Practices.
Anti-Discrimination Laws.
Rhode Island General Laws, Section 28-5-7 et seq.
Polygraph Restrictions.
Rhode Island General Laws, Section 28-6.1-1.
Arrest Records.
Rhode Island General Laws, Section 28-5-7.
Medical Examinations Payment.
Rhode Island General Laws, Section 28-6.2-1.
Protection of Political Freedom.
Rhode Island General Laws, Section 17-23-5.
Confidentiality of Medical Information.
Rhode Island General Laws, Section 5-37 .3-4(a).
Rhode Island Public laws. Chapter 119 (L.1968).
Reemployment Rights of Military Personnel.
Rhode Island General Laws, Section 30-11-2 et seq.
Jury Duty.
Rhode Island General Laws, Section 9-9-28.
Equal Pay.
Rhode Island General Laws, Section 28-6-17.

{105.06} Wage Labor Hour Laws.
Child Labor.
Rhode Island General Laws, Section 28-3-1 et seq.
Payment of Wages.
Rhode Island General Laws, Section 28-14-2.
Payment Upon Termination.
Rhode Island General Laws, Section 28-14-4.
Garnishments.
Rhode Island General Laws, Section 15-13-1.
Medical Insurance Continuance.
Rhode Island General Laws, Section 27-19.1-1.

{105.07} Health and Safety Laws.
General Provisions.
Rhode Island General Laws, Section 28-20-8(a).
Toxic Substances-Right to Know.
Rhode Island General Laws, Section 28-21-1 et seq.

{105.08} Unemployment Compensation Laws.
Rhode Island General Laws, Section 28-44-1, et seq.

{105.09} Worker Compensation Laws.
Rhode Island General Laws, Section 28-33-1 et seq.

{105.10} Employment-at-Will Developments.
Lamoureax v. Burrillville Racing Ass'n., 161 A.2d 213 (R.I., 1960).

South Carolina

{106.01} Labor Relations Laws.
Grievance Procedures.
South Carolina Public Laws. H.B. 2626 (L.1982).
Code of Laws of South Carolina. Section 8-17-110.
Right-to-Work Statute.
Code of Laws of South Carolina. Section 41-7-10.
"Yellow-Dog" Contracts.
Code of Laws of South Carolina. Section 41-7-30(2)

{106.02} Strikes, Picketing, and Boycott Laws.
Unlawful Assembly Against Political Beliefs.
Code of Laws of South Carolina. Section 16-101.
Striker Replacement.
Code of Laws of South Carolina. Section 41-25-50(c).
Use of Detectives During Strikes.
Article 8, Section 9, of the South Carolina Constitution.

{106.03} Mediation and Arbitration Laws.
Code of Laws of South Carolina. Section 41-17-10.
Article 6, Section 1, of the South Carolina Constitution.

{106.04} Regulation of Union Activities.
Code of Laws of South Carolina. Section 39-15-110.

{106.05} Regulation of Employment Practices.
Anti-Discrimination Laws.
Code of Laws of South Carolina. Section 1-13-80 and 43-33-550.
Protection of Political Freedom.
Code of Laws of South Carolina. Section 16-17-560.
Reemployment Right of Military Personnel.
Code of Laws of South Carolina. Section 25-1-2310.
Discrimination Against Union Members.
Code of Laws of South Carolina. Section 41-1-20.
Plant Closings.
Code of Laws of South Carolina. Section 41-1-40.
Voice Stress Analyzers.
Code of Laws of South Carolina. Section 40-53-40.

{106.06} Wage and Hours Laws.
Child Labor.
Code of Laws of South Carolina. Section 41-13-110, et seq.
Payment on Termination
Code of Laws of South Carolina. Section 41-11-170.
Garnishments
Code of Laws of South Carolina. Section 37-5-106.

Medical Insurance Conversion
Code of Laws of South Carolina. Section 38-35-946.

{106.07} Safety and Health Laws.
General Provisions
Code of Laws of South Carolina. Section 41-15-80.
Toxic Substances-Right to Know
Code of Laws of South Carolina. Section 41-15-210.

{106.08} Unemployment Compensation Laws.
Code of Laws and govern of South Carolina. Section 41-27-10.

{106.09} Workers Compensation Laws.
Code of Laws of South Carolina. Section 42-1-10, et seq.

{106.10} Employment-at-Will Developments.
Todd v. South Carolina Farm Bureau Mut. Ins. Co., 278 S.E.2d 607 (S.C., 1981).
Tyler v. Macks Stores of South Carolina, Inc., 272 S.E.2d 633 (S.C. S. Ct., 1980).
Ludwick v. This Minute of Carolina, Inc., Case no. 22408 (S.C. S. Ct., 1985).

South Dakota

{107.01} Labor Relation Laws.
South Dakota Labor Relations Act.
South Dakota Compiled Laws, 60-9A-1, et seq.
South Dakota Compiled Laws, 60-9A-7 and 12.
Public Employees Bargaining Rights.
South Dakota Compiled Laws, Section 3-18-1, et seq.
Right-to-Work Statute.
South Dakota Compiled Laws, Section 60-8-3, et seq.

{107.02} Strikes, Picketing, and Boycott Laws.
Unlawful Assembly.
South Dakota Compiled Laws, Section 22-10-1, et seq.
Unlawful Interference with Employment.
South Dakota Compiled Laws, Section 60-8-1.
Unlawful Picketing.
South Dakota Compiled Laws, Section 60-10-9.
Strike Notice.
South Dakota Compiled Laws, Section 60-6-19 and 20.

{107.03} Mediation and Arbitration Laws.
South Dakota Compiled Laws, Section 60-10-1.

{107.04} Regulation of Union Activities.
South Dakota Compiled Laws, Section 37-6-2.

South Dakota Compiled Laws, Section 60-9-2.
South Dakota Compiled Laws, Section 60-9-8.

{107.05} Regulation of Employment Practices.
Anti-Discrimination Laws.
South Dakota Compiled Laws, Section 20-13-1, et seq.
Protection of Political Freedom.
South Dakota Compiled Laws, Section 12-26-13.
Jury Duty.
South Dakota Compiled Laws, Section 16-13-41.1
Access to Personnel Records.
South Dakota Compiled Laws, Section 3-6A-31.
Military Duty.
South Dakota Compiled Laws, Section 33-17-15.
Medical Examination Payments.
South Dakota Compiled Laws, Section 60-11-2.
Equal Pay.
South Dakota Compiled Laws, Section 60-12-15.

{107.06} Wage and Hour Laws.
Child Labor.
South Dakota Compiled Laws, Section 60-12-2, et seq.
South Dakota Compiled Laws, Section 60-12-4.
Voting Time.
South Dakota Compiled Laws, Section 12-3-5.
Payment of Wages.
South Dakota Compiled Laws, Section 60-11-9.
Payment Upon Termination.
South Dakota Compiled Laws, Section 60-11-10.

{107.07} Safety and Health Laws.
South Dakota Compiled Laws, Section 60-12-7, et seq.

{107.08} Unemployment Compensation Laws.
South Dakota Compiled Laws, Section 61-6-20, et seq.

{107.09} Workers Compensation Laws.
South Dakota Compiled Laws, Section 62-5-1, et seq.

{107.10} Employment-at-Will Developments.
Implied Contract Exception.
Osterkamp v. Alkota Manufacturing, Inc., 332 N.W.2d 275 (S.D., 1983).
Presumption of Term Exception.
South Dakota Public Laws, S.B. 263 (L., 1985).

Tennessee

{108.01} Labor Relation Laws.
Professional Negotiations Act.
Tennessee Code Annotated Section 49-5-601, et seq.
Public Transit Employees Bargaining Rights.
Tennessee Code Annotated Section 7-56-102.
Collective Bargaining Sunshine Laws.
Tennessee Code Annotated Section 8-44-201.
Right-to-Work Statute.
Tennessee Code Annotated Section 50-1-201.

{108.02} Strikes, Picketing, and Boycott Laws.
Use of Armed Guards During Strikes.
Tennessee Code Annotated Section 50-1-102(d), (e).
"Sit-Down" Strikes.
Tennessee Code Annotated Section 50-1-303.
Strike Replacement Notices.
Tennessee Code Annotated Section 50-8-111.
Unlawful Inducement for Employment.
Tennessee Code Annotated Section 47-1706.

{108.03} Mediation and Arbitration Laws.
Tennessee Code Annotated Section 29-5-101.

{108.04} Regulation of Union Activities.
Tennessee Code Annotated Section 47-25-407, et seq.

{108.05} Regulation of Employment Practices.
Anti-Discrimination Laws.
Tennessee Code Annotated Section 4-21-105, et seq.
Tennessee Code Annotated Section 8-50-103.
Freedom of Political Activities.
Tennessee Code Annotated Section 2-19-34.
Jury Service.
Tennessee Code Annotated Section 22-4-108.
Unlawful Enticement.
Tennessee Code Annotated Section 50-1-101(a).
Selection of Physician.
Tennessee Code Annotated Section 50-1-302(a).
Access to Personnel Records.
Tennessee Code Annotated Section 8-50-108.
Tennessee Code Annotated Section 49-224.
Medical Examination Payments.
Tennessee Code Annotated Section 50-1-302.
Equal Pay.
Tennessee Code Annotated Section 50-2-202.

{108.06} Wage and Hour Laws.
Child Labor.
Tennessee Code Annotated Section 50-5-103, et seq.
Voting Time.
Tennessee Code Annotated Section 2-1-108, et seq.
Payment of Wages.
Tennessee Code Annotated Section 50-2-103(a).
Medical Coverage Conversion.
Tennessee Code Annotated Section 56-7-1501.
Wage Assignments for Child Support.
Tennessee Code Annotated Section 50-2-105

{108.07} Health and Safety Laws.
General Provisions.
Tennessee Code Annotated Section 50-3-105.
Toxic Substances-Right to Know.
Tennessee Public Laws, Ch. 417 (L. 1985).

{108.08} Unemployment Compensation Laws.
Tennessee Code Annotated, Section 50-7-301, et seq.

{108.09} Workers Compensation Laws.
Tennessee Code Annotated, Section 50-6-101, et seq.

{108.10} Employment-at-Will Developments.
General Rule.
Whittaker v. Care-More, Inc., 621 S.W.2d 395 (Tenn. App., 1981).
Public Policy Exception.
Clanton v. Cain-Sloan Co., 117 LRRM 2789 (Tenn., 1984).
Implied Contract Exception.
See, for example: *Gee v. Express Corp.*, 710 F.2d 1181 (CA-6, 1983), applying Tennessee law.
Hamby v. Genesco, 627 S.W.2d 373 (Tenn. App., 1981).

Texas

{109.01} Labor Relations Laws.
Fire and Police Employee Relations Act.
Texas Statutes, Article 5154(e)-1, Section 1, et seq.
General Right to Organize.
Texas Public Laws, Article 5152 (p.l.1899).
Right-to-Work Statute.
Vernon's Texas Statutes, Article 5207a, Section 2.

{109.02} Strikes, Picketing, and Boycott Laws.
Secondary Boycotts.
Section 1 of Article 5154F of the Texas Statutes.

Employment of Armed Detectives.
Texas Statutes, Article 5207.
Public Employment Strikes.
Section 3 of Article 5154c of the Texas Statutes.
Strikes Against Public Utilities.
Texas Statutes, Article 1446a, Section 3.
Mass Picketing.
Section 1 of Article 5154d of the Texas Statutes.
Interference with Employment.
Texas Statutes, Section 2 of Article 5154d, Section 2.

{109.03} Mediation and Arbitration Laws.
Texas Public Laws, Article 239 (L. 1895)

{109.04} Regulation of Union Activities.
Texas Public Laws, Article 1061 (L. 1895)
Texas Statues, Article 1690e.
Texas Statues, Article 5154a, Section 3.
Texas Statues, Article 5154a, Section 5.
Section 4b of Article 5154a of the Texas Statutes.
Texas Statues, Article 5154b.

{109.05} Regulation of Employment Practices.
Anti-Discrimination Laws.
Texas Statues, Article 5221K, Section 1, et seq.
Texas Statues, Article 5547-300, Section 9.
Protection of Military.
Texas Statues, Article 5765, Section 7A.
Jury Service.
Texas Statues, Article 5207b.
Blacklisting.
Texas Statues, Article 5196, et seq.
Whistle-Blowing Statute (Public Employees).
Section 16(a) of Article 6252 of the Texas Statue.
Voice Stress Analyzers.
Section 4 of Article 4413 of the Texas Statue.
Political Activities.
Texas Education Code, Article 13.34a.
Service Letters.
Article 5196 of the Texas Statues.
Equal Pay (Public Employees).
Texas Statues, Section 6825.

{109.06} Wage and Hour Laws.
Child Labor.
Texas Public Laws, Chapter 531 (L. 1981).

Voting Time.
Texas Election Code, Section 15.14 of Title 9.
Payment of Wages.
Texas Public Laws, H.B. 79 (L. 1983).
Payment Upon Termination.
Texas Statues, Article 5156, 5157, and 5158.
Texas Statues, Article 6431.
Wage Assignment.
West's Texas Family Code, Section 14.091(i).

{109.07} Safety Health Laws.
Texas Statues, Article 5182a, Section 10.23.

{109.08} Unemployment Compensation Laws.
Texas Statues, Article 5221b, Section 2, et seq.

{109.09} Workers' Compensation Laws.
Texas Statues, Article 8306, Section 1, et seq.
Texas Statues, Article 8307c.

{109.10} Employment-at-Will Developments.
Public Policy Exception.
Sabine Pilots, Inc. v. Hauck, 687 S.W.2d 733 (Tex. Ct. App., 1985).
Implied Contract Exception.
Reynolds Mfg. Co, Mendoza, 644 S.W.2d 536 (Tex. Ct. App., 1982).
Johnson v. Ford Motor Co., 690 S.W.2d 90 (Tex. Ct. App., 1985).

Utah

{110.01} Labor Relations Laws.
Utah Labor Relations Act.
Utah Code Annotated, Section 34-20-7.
Utah Code Annotated, Section 34-20-8.
Utah Code Annotated, Section 34-20-1, et seq.
Firefighters Bargaining Rights.
Utah Code Annotated, Section 34-20a-1.
Rights of Labor.
Utah Constitution, Article XVI, Section 1.
"Yellow-Dog" Contracts.
Utah Code Annotated, Section 34-1-24.

{110.02} Strikes, Picketing, and Boycott Laws.
Anti-Injunction Statutes.
Utah Code Annotated, Section 34-19-1.
Unlawful Assembly.
Utah Code Annotated, Section 76-9-101.
Interference with Employment.
Utah Code Annotated, Section 34-2-3.

Sabotage.
Utah Code Annotated, Section 76-8-802.
Employment of Armed Guards.
Utah Constitution, Article XII, Section 16.
Deputized Employees During Strikes.
Utah Code Annotated, Section 34-19-12.

{110.03} Mediation and Arbitration Laws.
Utah Code Annotated, Section 78-31-1, et seq.
Utah Public Laws, S.B. 133 (L.1977).

{110.04} Regulation of Union Activities.
Utah Code Annotated, Section 76-10-1002.

{110.05} Regulation of Employment Practices.
Utah Code Annotated, Section 35-35-6, et seq.
Protection of Political Freedom.
Utah Code Annotated, Section 20-13-6 and 20-13-7.
Polygraph Restrictions.
Utah Code Annotated, Section 34-37-16.
Blacklisting.
Utah Code Annotated, Section 34-24-1.
Access to Personnel Records.
Utah Code Annotated, Section 67-18-1, 63-2-66.
Military Duty.
Utah Code Annotated, Section 39-1-36.
Medical Examination Payments.
Utah Code Annotated, Section 34-33-1
Equal Pay.
Utah Code Annotated, Section 34-35-6.

{110.06} Wage and Hour Laws.
Child Labor.
Utah Code Annotated, Section 34-23-1, et seq.
Voting Time.
Utah Code Annotated, Section 20-13-18.
Payment of Wages.
Utah Code Annotated, Section 34-28-3.
Payment Upon Termination.
Utah Code Annotated, Section 34-28-5(1).
Utah Code Annotated, Section 34-28-5(2).
Garnishments.
Utah Code Annotated, Section 70B-5-106.

{110.07} Safety and Health Laws.
Utah Code Annotated, Section 35-9-5

{110.08} Unemployment Compensation.
Utah Code Annotated, Section 35-4-1, et seq.

{110.09} Workers' Compensation Laws.
Utah Code Annotated, Section 35-1-1 et seq.

{110.10} Employment-at-Will Developments.
Bihlmaier v. Carson, 630 P.2d 790 (Utah, 1979).

Vermont

{111.01} Labor Relations Laws.
Vermont Labor Relations Law.
Vermont Statutes Annotated, Title 21, Section 1501, et seq.
State Employees Bargaining Rights.
Vermont Statutes Annotated, Title 27, Section 901, et seq.
Vermont Statutes Annotated, Title 27, Section 961 and 962.
Teachers Bargaining Rights.
Vermont Public Laws, Chapter 57 (L.1969).
Vermont Municipal Labor Relations Act.
Vermont Statutes Annotated, Title 21, Section 172, et seq.
Nonright-to-Work Policy.

{111.02} Strikes, Picketing, and Boycott Laws.
Unlawful Assembly.
Vermont Statutes Annotated, Title 21, Section 521, et seq.
Interference with Employment.
Vermont Statutes Annotated, Title 13, Section 931, et seq.
"Sit-Down" Strikes.
Vermont Statutes Annotated, Title 13, Section 933.

{111.03} Mediation and Arbitration Laws.
Vermont Statutes Annotated, Title 21, Section 521, et seq.

{111.04} Regulation of Union.
Vermont Statutes, Ch. 337, Section 7759.

{111.05} Regulation of Employment.
Anti-Discrimination Laws.
Vermont Statutes Annotated, Title 21, Section 495.
Employment of Aliens.
Vermont Public Laws, Act 99 (L.1977).
Jury or Witness Duty.
Vermont Public Laws, S.B. 98 (L.1969).
Military Duty.
Vermont Statutes Annotated, Title 21, Section 491, et seq.
Access to Personnel Records.
Vermont Statutes Annotated, Title 1, Section 317 (b)(7).

Medical Examination Payments.
Vermont Statutes Annotated, Title 21, Section 301.
Political Leaves of Absence.
Vermont Statutes Annotated, Title 21, Section 496.
Equal Pay.
Vermont Statutes Annotated, Title 21, Section 495.

{111.06} Wage and Hour Laws.
Child Labor.
Vermont Statutes Annotated, Title 21, Section 431, et seq.
Payment Upon Termination.
Vermont Statutes Annotated, Title 21, Section 342 (c) (2).
Vermont Statutes Annotated, Title 21, Section 342 (c) (1).
Garnishments.
Vermont Statutes Annotated, Title 12, Section 3172.

{111.07} Safety and Health Laws.
Vermont Statutes Annotated, Title 21, Section 1301, et seq.

{111.08} Unemployment Compensation Laws.
Vermont Statutes Annotated, Title 21, Section 1301, et seq.

{111.09} Workers Compensation Laws.
Vermont Statutes Annotated, Title 21, Section 601, et seq.
Vermont Statutes Annotated, Title 21, Section 618, et seq.

{111.10} Employment-at-Will-Developments.
Brower v. Holmes Transp., Inc., 435 A.2d 952 (Vermont, 1981); *Jones v. Keough*, 409 A.2d 581 (Vermont, 1979).

Virginia

{112.01} Labor Relations Laws.
Right-to-Work Statute.
Code of Virginia, Section 40.1-60.
"Yellow-Dog" Contract.
Code of Virginia, Section 40.1-61.

{112.02} Strikes, Picketing, and Boycott Laws.
Unlawful Assembly.
Code of Virginia, Section 18.1-254.
Interference with Employment.
Code of Virginia, Section 40.1-53.
Interference with Ingress or Egress.
Code of Virginia, Section 40.1-53.
Picketing of Residence or Dwelling.
Virginia Public Laws, Chapter 711 (L.1970).

Public Employee Strikes.
Code of Virginia, Section 40.1-55.
Hospital Strikes.
Code of Virginia, Section 40.1-54(1).
See also Code of Virginia, Section 40.1-54(2). (However, this provision may well be in contravention of the Federal Labor-Management Relation Act and thereby pre-empted by it.)
Public Utilities Seizure Act.
Code of Virginia, Section 56.1-509, et seq.
Coal Industry Seizure Act.
Code of Virginia, Section 45.1-145, et seq.

{112.03} Mediation and Arbitration Laws.
Code of Virginia, Section 8.01-577, et seq.
Code of Virginia, Section 22.1-306, et seq.
Code of Virginia, Section 2.1-114.5:1

{112.04} Regulation of Union Activities.
Code of Virginia, Section 40.1-76.
Code of Virginia, Section 18.1-410.
Code of Virginia, Section 40.1-54.

{112.05} Regulation of Employment Practice.
Anti-Discrimination Laws (State Employee).
Code of Virginia, Section 2.1-376.
Employment of Aliens.
Code of Virginia, Section 40.1-11.1.
Polygraph Restrictions.
Code of Virginia, Section 40.1-51.4:3.
Blacklisting.
Code of Virginia, Section 40.1-27.
Medical Examination Payments.
Code of Virginia, Section 40.1-28.
Arrest Inquiries.
Code of Virginia, Section 19.2-392.
Jury Duty.
Virginia Public Laws (L. 1985, C. 436).
Military Duty.
Code of Virginia, Section 44-98.
Code of Virginia, Section 44-93
Equal Pay.
Code of Virginia, Section 40.1-28.

{112.06} Wage and Hour Laws.
Child Labor.
Code of Virginia, Section 40.1-100, et seq.

Code of Virginia, Section 40.1-105, et seq.
Code of Virginia, Section 40.1-78, et seq.

Payment of Wages.
Code of Virginia, Section 40.1-29.

Payment Upon Termination.
Code of Virginia, Section 40.1-29(a).

Garnishments.
Code of Virginia, Section 34-29(f).

Medical Insurance Conversion.
Code of Virginia, Section 38.1-348.11.

{112.07} Safety and Health Laws.

General Provision.
Code of Virginia, Section 40.1-51.

Toxic Substances-Compensation Laws.
Code of Virginia, Section 40.1-51.1(c).

{112.08} Unemployment Compensation Laws.
Code of Virginia, Section 60.1-1, et seq.

{112.09} Workers' Compensation Laws.
Code of Virginia, Section 65.1-1, et seq.
Code of Virginia, Section 65.1-40, et seq.

{112.10} Employment-at-Will Development.

Public Policy Exception.
Bowman v. State Bank of Keysville, 331 SE2d 799 (Va. Ct. App. 1985).

Implied Contract Exception.
Frazier v. Colonial Williamsburg Foundation, 574 F. Supp. 318 (E.D. Va., 1983).

Presumption of Term Exception.
Hoffman Specialty Co. v. Pelouze, 164 S.E. 397 (Va., 1932)

Washington

{113.01} Labor Relations Laws.

General Right to Organize.
Revised Code of Washington, Section 49.36.010.

Public Employees' Bargaining Right.
Revised Code of Washington, Section 41.56.010.
Revised Code of Washington, Section 41.56.010, et seq.

Educational Employment Relation Act.
Revised Code of Washington, Section 41.59.010, et seq.

Academic Employees' Bargaining Rights.
Revised Code of Washington, Section 28.B.52.010, et seq.

Higher Education Collective Bargaining.
Revised Code of Washington, Section 28B.16.100, et seq.

Healthcare Collective Bargaining.
Revised Code of Washington, Section 49.66.010, et seq.
Marine Employees' Bargaining Rights.
Revised Code of Washington, Section 47.64.010, et seq.
Port District Employees' Bargaining Rights.
Revised Code of Washington, Section 53.18.010, et seq.
Public Utility Employees' Bargaining Rights.
Revised Code of Washington, Section 54.04.170, et seq.
Nonright-to-Work Policy.
Labor Management Relation, Section 14(b).
"Yellow-Dog" Contract.
Revised Code of Washington, Section 49.32.030.

{113.02} Strikes, Picketing, and Boycott Laws.
Anti-Injunction Statue.
Revised Code of Washington, Section 49.32.011.
Injury to Property.
Revised Code of Washington, Section 9.05.060.
Interference with Employment.
Revised Code of Washington, Section 9.22.010.
Unlawful Assembly.
Revised Code of Washington, Section 9.27.040.
Unlawful Breach of Contract.
Revised Code of Washington, Section 49.44.080.
"Sit-Down" Strikes.
Revised Code of Washington, Section 9.05.070.
Strikebreakers.
Revised Code of Washington, Section 49.44.100.

{113.03} Mediation and Arbitration Laws.
Revised Code of Washington, Section 7.04.010.
Revised Code of Washington, Section 49.08.010, et seq.

{113.04} Regulation of Union Activities.
Revised Code of Washington, Section 9.16.030.
Revised Code of Washington, Section 49.44.030.

{113.05} Regulation of Employment Practices.
Anti-Discrimination Laws.
Revised Code of Washington, Section 49.60.010.
Military Duty.
Revised Code of Washington, Section 73.16.033.
Polygraph Restrictions.
Revised Code of Washington, Section 49.44.120.
Employment Under False Pretenses.
Revised Code of Washington, Section 49.44.040.

Whistle-Blowing Statute.
Washington Public Laws, Chapter 208 (L. 1982).
Service Letters.
WAC, Section 296-126-050.
Blacklisting.
Revised Code of Washington, Section 49.44.010.
Equal Pay.
Revised Code of Washington, Section 49.12.175.

{113.06} Wage and Hour Laws.
Child Labor.
Revised Code of Washington, Section 26.28.060, et seq.
Revised Code of Washington, Section 26.28.060 and 49.12.12.
Payment Upon Termination.
Revised Code of Washington, Section 49.48.010.
Garnishments.
Revised Code of Washington, Section 7.33.160.

{113.07} Safety and Health Laws.
General Provisions.
Washington Public Laws, S.B. 2386 (L. 1973).
Toxic Substances-Right to Know.
Washington Public Laws, S.B. 4831, Section 15 (L. 1984).

{113.08} Unemployment Compensation Laws.
Revised Code of Washington, Section 50.20.001, et seq.

{113.09} Workers Compensation Laws.
Revised Code of Washington, Section 50.04.010, et seq.

{113.10} Employment-at-Will Developments.
Public Policy Exception.
Thompson v. St. Regis Paper Co., 685 P.2d 1081 (Wash., 1984).
Implied Contract Exception.
Roberts v. Atlantic Richfield Co., 568 P.2d 764 (Wash., 1977).
Intentional Infliction of Emotional Distress Exception.
Contreras v. Crown Zellerbach Corp., 565 P.2d 1173 (Wash., 1977).

West Virginia

{114.01} Labor Relation Laws.
West Virginia Labor Management Relation.
West Virginia Code, Section 21-1A-1, et seq.
Nonright-to-Work Policy.
Labor Management Relation Act, Section 14(b).

{114.02} Strikes, Picketing, and Boycotts Laws.
Unlawful Assembly.
West Virginia Code, Section 61-6-6.
Use of Out-of-State Police.
West Virginia Code, Section 61-6-11.
Interference with Employment at Mines.
West Virginia Code, Section 22-2-77.

{114.03} Mediation and Arbitration Laws.
West Virginia Code, Section 55-10-1, et seq.
West Virginia Code, Section 21-1A-1.

{114.04} Regulation of Union Activities.
West Virginia Code, Section 47-2-3.

{114.05} Regulation of Employment Practices.
Anti-Discrimination Laws.
West Virginia Code, Section 5-11-19, et seq.
Protection of Political Freedom.
West Virginia Code, Section 3-9-15.
Employment Under False Pretense.
West Virginia Code, Section 21-2-6.
Jury Duty.
West Virginia Code, Section 52-3-1.
Military Duty.
West Virginia Code, Section 15-1F-8
Polygraph Restrictions.
West Virginia Code, Section 21-5-5b
Medical Examination Payment.
West Virginia Code, Section 21-3-17.
Equal Pay.
West Virginia Code, Section 21-5B-3

{114.06} Wage and Hour Laws.
Child Labor.
West Virginia Code, Section 21-6-1, et seq.
Voting Time.
West Virginia Code, Section 3-1-42.
Payment of Wages.
West Virginia Code, Section 21-5-3.
Payment Upon Termination.
West Virginia Code, Section 21-5-4(b).
West Virginia Code, Section 21-5-4(c).
Garnishments.
West Virginia Code, Section 46A-2-131.
Medical Insurance Conversion.
West Virginia Code, Section 33-16A-1.

{114.07} Safety and Health Laws.
General Provisions.
West Virginia Code, Section 21-3-1.
Toxic Substances-Right to Know.
West Virginia Code, Section 21-3-18.

{114.08} Unemployment Compensation Laws.
West Virginia Code, Section 21A-1-1, et seq.

{114.09} Workers Compensation Laws.
West Virginia Code, Section 23-1-1, et seq.

{114.10} Employment-at-Will Developments.
Public Police Exception.
Harless v. First Nat'l Bank in Fairmont, 246 S.E.2d 270 (W.Va., 1978).
Implied Contract Exception
McMillion v. Appalachian Power Co., 701 F.2d 166 (CA-4, 1983).
Intentional Infliction of Emotional Distress Exception
Harless v. First Nat'l Bank in Fairmont, supra.

Wisconsin

{115.01} Labor Relations Laws.
Employment Peace Act.
Wisconsin Statutes, Section 111.01, et, seq.
Municipal Employees Bargaining Rights.
Wisconsin Statutes, Section 111.70, et, seq.
State Employees Bargaining Rights.
Wisconsin Statutes, Section 111.80, et seq.
Police Officers and Firefighters Bargaining.
Wisconsin Statutes, Section 111.77, et seq.
Nonright-to-Work Police.
Section 14(b) of the Labor Management Relations Act.
"Yellow-Dog" Contracts.
Wisconsin Statutes, Section 101.52.

{115.02] Strikes, Picketing, and Boycott Laws.
Anti-Injunction Statues.
Wisconsin Statutes, Section 103.56.
Unlawful Assembly.
Wisconsin Statutes, Section 347.02.
Interference with Employment.
Wisconsin Statutes, Section 343.683.
Disorderly Conduct.
Wisconsin Statutes, Section 348.5.
Mass Picketing.
Wisconsin Statutes, Section 111.06(2)(e).

"Sit-Down" Strikes.
Wisconsin Statutes, Section 111.06(2)(b).
Strikebreakers.
Wisconsin Statutes, Section 348.472.
Picketing of Resident or Dwelling.
Wisconsin Statutes, Section 111.06(2)(a).
Secondary Boycotts.
Wisconsin Statutes, Section 111.06(2)(g).

{115.03} Mediation and Arbitration Laws.
Wisconsin Statutes, Section 298.01.
Wisconsin Statutes, Section 111.50, et seq.

{115.04} Regulation of Union Activities.
Wisconsin Statutes, Section 132.19.
Wisconsin Statutes, Section 211.01, et seq.
Wisconsin Statutes, Section 111.08.

{115.05} Regulation of Employment Practices.
Anti-Discrimination Laws.
Wisconsin Statutes, Section 111.321, et seq.
Polygraph Restrictions.
Wisconsin Statutes, Section 111.37.
Protection of Political Freedom.
Wisconsin Statutes, Section 103.18.
Jury Duty.
Wisconsin Statutes, Section 756.25.
Arrest Records.
Wisconsin Statutes, Section 111.31, 111.335.
Access to Personnel Records.
Wisconsin Statutes, Section 103.13. et seq.
Backlisting.
Wisconsin Statutes, Section 343.682.
Plant Closure.
Wisconsin Statutes, Section 109.07.
Medical Examination Payments.
Wisconsin Statutes, Section 103.37.
Whistle-Blowing Statute.
Wisconsin Statutes, Section 230.80, et seq.
Equal Pay.
Wisconsin Statutes, Section 111.36.

{115.06} Wage and Hour Laws.
Child Labor.
Wisconsin Statutes, Section 103.66, et seq.
Wisconsin Statutes, Section 103.78.

Voting Time.
Wisconsin Statutes, Section 6.76.
Payment Upon Termination.
Wisconsin Statutes, Section 109.03.
Garnishments.
Wisconsin Statutes, Section 812.235.
Medical Insurance Conversion.
Wisconsin Statutes, Section 632.897.

{115.07} Safety and Health Laws.
General Provision.
Wisconsin Statutes, Section 101.11, et seq.
Toxic Substances-Right to Know.
Wisconsin Statutes, Section 101,58, et seq.

{115.08} Unemployment Compensation Laws.
Wisconsin Statutes, Section 108.01, et seq.

{115.09} Workers Compensation Laws.
Wisconsin Statutes, Section 102.01, et seq.
Wisconsin Statutes, Section 102.01, et seq.

{115.10} Employment-at-Will Developments.
Implied Contract Exception.
Ferrero v. Voelsch, 350 N.W.2d 735 (Wis. S. Ct., 1985),
Public Policy Exception
Ward v. Frito-Lay Inc., 290 N.W.2d 356 (Wis. Ct. App., 1980).

Wyoming

{116.01} Labor Relations Laws.
General Right to Organize.
Wyoming Statutes, Section 27-7-101.
Firefighters Bargaining Rights.
Wyoming Statutes, Section 27-10-101, et seq.
Right-to-Work Statute.
Wyoming Statutes, Section 27-7-108.
"Yellow-Dog" Contracts.
Wyoming Statutes, Section 27-245.3.

{116.02} Strikes, Picketing, and Boycott Laws.
Anti-Injunction Statute.
Wyoming Statutes, Section 27-7-103.
Unlawful Assembly.
Wyoming Statutes, Section 6-10-108.
Use of Police During Strikes.
Wyoming Constitution, Article 19, Section 6.

{116.03} Mediation and Arbitration Laws.
Wyoming Statutes, Section 1-36-103.

{116.04} Regulation of Union Activities.
Wyoming Statutes, Section 27-7-245.

{116.05} Regulating of Employment Practices.
Anti-Discrimination Laws.
Wyoming Statutes, Section 27-9-105, et seq.
Protection of Political of Political Freedom.
Wyoming Statutes, Section 22-341, et seq.
Jury Service.
Wyoming Statutes, Section 1-11-401(a).
Military Duty.
Wyoming Statutes, Section 19-3-105€.
Access to Medical Records.
Wyoming Statutes, Section 27-11-113.
Equal Pay.
Wyoming Statutes, Section 27-4-301.

{116.06} Wage and Hour Laws.
Child Labor.
Wyoming Statutes, Section 27-6-107, et seq.
Voting Time.
Wyoming Statutes, Section 22-2-111.
Payment on Termination.
Wyoming Statutes, Section 27-7-104.
Garnishments.
Wyoming Statutes, Section 40-14-506.

{116.07} Safety and Health Laws.
Wyoming Statutes, Section 27-11-105(a).

{116.08} Unemployment Compensation Laws.
Wyoming Statutes, Section 27-3-101, et seq.

{116.09} Workers Compensation Laws.
Wyoming Statutes, Section 27-12-101, et seq.
Wyoming Statutes, Section 27-12-401, et seq.

NOTES

1. Equal Employment Opportunity Commission website located at www.eeoc.gov.
2. US Department of Labor website located at www.dol.gov.

7 OSHA and State Plan Programs

***

> "To acquire knowledge, one must study: but to acquire wisdom, one must observe."
>
> **Marilyn Vos Savant**

> "The ability to think straight, some knowledge of the past, some vision of the future, some skill to do useful service, some urge to fit that service into the well-being of the community – these are the most vital things education must try to produce."
>
> **Virginia Gildersleeve**

***

For many safety professionals, managing the safety and health function in multiple locations throughout the United States, it is important to recognize the jurisdictional boundaries between the federal Occupational Safety and Health Administration (OSHA) and individual state plan programs. For example, if the safety and health professional possesses responsibilities for operations in Ohio, Kentucky, and California, the safety professional would be required to ensure compliance with federal OSHA standards (Ohio) as well as individual state plan standards and regulations in the Commonwealth of Kentucky (KYOSH) and California (Cal OSHA). The jurisdictional boundaries among and between federal OSHA and state plan programs are state boundary lines.

Safety and health professionals should recognize that there are currently 26 states and Puerto Rico and the US Virgin Islands that operate a federal OSHA-approved state plan program. In 21 of these states, the state plan program covers compliance in both private sector companies as well as local government operations. Six states cover only state and local government operations. Although each of these state plan programs operate independently and under the laws of the individual state, federal OSHA approval is required and federal OSHA provides oversight and funding. State plan programs must adopt federally approved standards within a specified time period after promulgation and approval may modify the federal standard as long as the proposed standard is as stringent or more stringent than the federal standard. Safety professionals should be aware that the standard identification numbering system may be different; however, virtually all states possess a website permitting access to each specific standard within the individual state plan program. Additionally, safety and health professionals should be aware that individual state plan programs possess an appeals process which culminates in legal challenges within the individual state court system.

State plan programs usually mirror the federal OSHA system in providing consultation, education, and training services free of charge to companies within their jurisdiction. Additionally, state plan programs usually provide, paralleling the federal OSHA system, an opportunity for variance petitions to address compliance issues as well as other educational and research resources.

KENTUCKY

Overview

- Initial Approval: July 31, 1973 (38 FR 20324)
- State Plan Certification: February 8, 1980 (45 FR 8596, 8598)
- 18(e) Final Approval: June 13, 1985 (50 FR 24896)

Kentucky Occupational Safety and Health (Kentucky OSH) is part of the Kentucky Labor Cabinet's Department of Workplace Standards. The main office is located in Frankfort.

Coverage

The Kentucky State Plan applies to all private sector workplaces in the state with the exception of:

1. Maritime employment, including shipyard employment, marine terminals, and longshoring;
2. Employment at Tennessee Valley Authority facilities, military bases, and other properties ceded to the federal government;
3. Contract workers and contractor-operated facilities engaged in United States Postal Service (USPS) mail operations;
4. The enforcement of the field sanitation standard (29 CFR 1928.110) and the enforcement of the temporary labor camps standard (29 CFR 1910.142) with respect to any agricultural establishment where workers are engaged in "agricultural employment" within the meaning of the Migrant and Seasonal Agricultural Worker Protection Act, 29 U.S.C. 1802(3)—regardless of the number of workers—including workers engaged in hand packing of produce into containers, whether done on the ground, on a moving machine, or in a temporary packing shed, except that Kentucky retains enforcement responsibility over agricultural temporary labor camps for workers engaged in egg, poultry, or red meat production, or the post-harvest processing of agricultural or horticultural commodities;
5. All working conditions of aircraft cabin crewmembers onboard aircraft in operation; and
6. Any hazard, industry, geographical area, operation or facility over which the state is unable to effectively exercise jurisdiction for reasons not related to the required performance or structure of the State Plan.

The Kentucky State Plan also applies to state and local government employers. It does not apply to federal government employers including USPS. Federal OSHA covers the issues not covered by the Kentucky State Plan, except for the enforcement of the field sanitation and temporary labor camp standards, as described above, which is the responsibility of the Wage and Hour Division of the US Department of Labor. In addition, federal OSHA retains enforcement of the anti-retaliation provision of the Occupational Safety and Health Act of 1970, Section 11(c), 29 U.S.C. 660(c), with respect to the private sector. Kentucky OSH also investigates private and state and local government workplace retaliation cases under a provision analogous to Section 11(c).

A brief summary of the Kentucky State Plan is included in the Code of Federal Regulations at 29 CFR 1952.11. Federal OSHA retains the authority to promulgate, modify, or revoke occupational safety and health standards under Section 6 of the OSH Act. In the event that federal OSHA resumes enforcement, those federal standards will be enforced. Federal OSHA also retains the authority to monitor the State Plan under Section 18(f) of the OSH Act.

State Plan Standards

The Kentucky OSH Standards Board is a 13-member body empowered to adopt, modify, or repeal OSH standards in the Commonwealth of Kentucky. The Board is chaired by the Secretary of the Labor Cabinet. The remaining 12 members are appointed by the governor to equally represent agriculture, labor, management, and the safety and health profession. Kentucky OSH standards consist of state-specific standards unique to Kentucky, OSHA standards incorporated with state-specific provisions, and OSHA standards incorporated without change.

General Industry

- Access to Employee Exposure and Medical Records
- Batteries
- Bloodborne Pathogens
- Contest of Citations
- Control of Hazardous Energy (Lockout)
- Discrimination
- Electrical Testing
- Employers' Responsibility Where Employees are Exposed to Toxic Substances
- Fire Apparatus and Fire Department Facilities
- Flammable and Combustible Liquids
- Guarding Floor and Wall Openings and Holes
- Laboratory Activities
- Machinery and Machine Guarding
- Materials Handling and Storage
- Medical Services and First Aid
- Occupational Noise
- Off-Highway Motor Vehicles and Equipment
- Powered Industrial Trucks

- Receiving and Unloading Bulk Hazardous Liquids
- Recordkeeping; Reporting; Statistics
- Refuse Collection and Compaction Equipment
- Rollover Protective Structures; Overhead Equipment
- Safety Belts, Lanyards, and Life Lines
- Sanitation
- Selection and Use of Work Practices (Electrical)
- Toxic and Hazardous Substances (including 4,4 Methylene bis (2-Chloroaniline))

Construction

- Blasting and Use of Explosives
- Bloodborne Pathogens
- Contest of Citations
- Discrimination
- Employers Responsibilities Where Employees Are Exposed to Toxic Substances
- Fall Protection, Cranes and Derricks in Construction
- Fall Protection, Residential
- Fall Protection, Steel Erection
- Material Hoists, Personnel Hoists, and Elevators
- Power Transmission and Distribution
- Receiving and Unloading Bulk Hazardous Liquids
- Recordkeeping; Reporting; Statistics
- Refuse Collection and Compaction Equipment
- Safety and Testing of Supply Lines in Excess of 600 Volts

Enforcement Programs

Kentucky OSH Compliance is responsible for the enforcement of safety and health standards. Compliance officers inspect workplaces for hazardous conditions and issue citations where violations of standards are found. Inspections may be the result of regular scheduling, imminent danger reports, fatalities, and worker complaints or referrals. More information on enforcement can be found on the Kentucky State Plan website.

Voluntary and Cooperative Programs

Kentucky OSH offers voluntary and cooperative programs focused on reducing injuries, illnesses, and fatalities. It also offers on-site consultation services which help employers—both state and local government and private—comply with Kentucky OSH or OSHA standards and identify and correct potential safety and health hazards. For more information on these programs, please visit the Kentucky State Plan website.

Informal Conferences and Appeals

Employers, workers, and worker representatives have the right to contest citations issued by Kentucky OSH. When a citation is contested, the case is reviewed by the

Review Commission. For more information on these proceedings, please visit the Kentucky State Plan website.[1]

SEC. 18.

State Jurisdiction and State Plans

(a)
29 U.S.C. 667
Nothing in this Act shall prevent any State agency or court from asserting jurisdiction under State law over any occupational safety or health issue with respect to which no standard is in effect under section 6.

(b)
Any State which, at any time, desires to assume responsibility for development and enforcement therein of occupational safety and health standards relating to any occupational safety or health issue with respect to which a Federal standard has been promulgated under section 6 shall submit a State plan for the development of such standards and their enforcement.

(c)
The Secretary shall approve the plan submitted by a State under subsection (b), or any modification thereof, if such plan in his judgement --

(1)
designates a State agency or agencies as the agency or agencies responsible for administering the plan throughout the State,

(2)
provides for the development and enforcement of safety and health standards relating to one or more safety or health issues, which standards (and the enforcement of which standards) are or will be at least as effective in providing safe and healthful employment and places of employment as the standards promulgated under section 6 which relate to the same issues, and which standards, when applicable to products which are distributed or used in interstate commerce, are required by compelling local conditions and do not unduly burden interstate commerce,

(3)
provides for a right of entry and inspection of all workplaces subject to the Act which is at least as effective as that provided in section 8, and includes a prohibition on advance notice of inspections,

(4)
contains satisfactory assurances that such agency or agencies have or will have the legal authority and qualified personnel necessary for the enforcement of such standards,

(5)
gives satisfactory assurances that such State will devote adequate funds to the administration and enforcement of such standards,

(6)

contains satisfactory assurances that such State will, to the extent permitted by its law, establish and maintain an effective and comprehensive occupational safety and health program applicable to all employees of public agencies of the State and its political subdivisions, which program is as effective as the standards contained in an approved plan,

(7)

requires employers in the State to make reports to the Secretary in the same manner and to the same extent as if the plan were not in effect, and

(8)

provides that the State agency will make such reports to the Secretary in such form and containing such information, as the Secretary shall from time to time require.

(d)

If the Secretary rejects a plan submitted under subsection (b), he shall afford the State submitting the plan due notice and opportunity for a hearing before so doing.

(e)

After the Secretary approves a State plan submitted under subsection (b), he may, but shall not be required to, exercise his authority under sections 8, 9, 10, 13, and 17 with respect to comparable standards promulgated under section 6, for the period specified in the next sentence. The Secretary may exercise the authority referred to above until he determines, on the basis of actual operations under the State plan, that the criteria set forth in subsection (c) are being applied, but he shall not make such determination for at least three years after the plan's approval under subsection (c). Upon making the determination referred to in the preceding sentence, the provisions of sections 5(a)(2), 8 (except for the purpose of carrying out subsection (f) of this section), 9, 10, 13, and 17, and standards promulgated under section 6 of this Act, shall not apply with respect to any occupational safety or health issues covered under the plan, but the Secretary may retain jurisdiction under the above provisions in any proceeding commenced under section 9 or 10 before the date of determination.

(f)

The Secretary shall, on the basis of reports submitted by the State agency and his own inspections make a continuing evaluation of the manner in which each State having a plan approved under this section is carrying out such plan. Whenever the Secretary finds, after affording due notice and opportunity for a hearing, that in the administration of the State plan there is a failure to comply substantially with any provision of the State plan (or any assurance contained therein), he shall notify the State agency of his withdrawal of approval of such plan and upon receipt of such notice such plan shall cease to be in effect, but the State may retain jurisdiction in any case commenced before the withdrawal of the plan in order to enforce standards under the plan whenever the issues involved do not relate to the reasons for the withdrawal of the plan.

(g)

The State may obtain a review of a decision of the Secretary withdrawing approval of or rejecting its plan by the United States court of appeals for the circuit in which the State is located by filing in such court within thirty days following receipt of notice of such decision a petition to modify or set aside in whole or in part the action of the Secretary. A copy of such petition shall forthwith be served upon the Secretary, and thereupon the Secretary shall certify and file in the court the record upon which the decision complained of was issued as provided in section 2112 of title 28, United States Code. Unless the court finds that the Secretary's decision in rejecting a proposed State plan or withdrawing his approval of such a plan is not supported by substantial evidence the court shall affirm the Secretary's decision. The judgment of the court shall be subject to review by the Supreme Court of the United States upon certiorari or certification as provided in section 1254 of title 28, United States Code.

(h)

The Secretary may enter into an agreement with a State under which the State will be permitted to continue to enforce one or more occupational health and safety standards in effect in such State until final action is taken by the Secretary with respect to a plan submitted by a State under subsection (b) of this section, or two years from the date of enactment of this Act, whichever is earlier.[2]

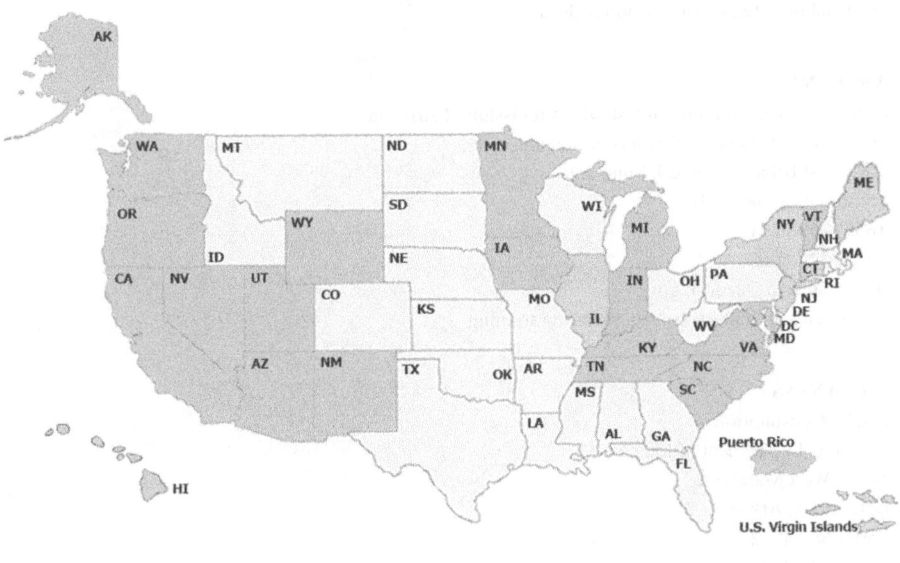

Note: 26 states, Puerto Rico, and the Virgin Islands have OSHA-approved State Plans; 22 State Plans (21 states and one US territory) cover both private and state and local government workplaces. The remaining six State Plans (five states and one US territory) cover state and local government workers only.[3]

As noted above, federal OSHA as well as state plan programs provides consultation, education, and training services free of charge to entities within their specific jurisdiction. Please find the contact information for federal as well as state plan consultation services.

Consultation Services and Contact Information

ALABAMA
Safe State Program
University of Alabama
432 Martha Parham West, Rm 432, Box 870388
Tuscaloosa, Alabama 35487
(800) 452-5928
(205) 348-9286
jsikes@ccs.ua.edu
http://alabamasafestate.ua.edu/safety-consultation/index.php

ALASKA
Consultation and Training
Alaska Occupational Safety and Health
1251 Muldoon Road, Suite 109
Anchorage, Alaska 99504
(800) 656-4972 or (907) 269-4957
(907) 269-4950
Anchorage.LSS-OSH@alaska.gov
www.labor.state.ak.us/lss/oshhome.htm

ARIZONA
Consultation & Training, Industrial Commission of Arizona
Division of Occupational Safety & Health
2675 East Broadway Road, Suite 239
Tucson, Arizona 85716
(855) 268-5251
(520) 322-8008
Jessie.atencio@azdosh.gov
www.azica.gov/adosh-consultation-and-training

ARKANSAS
OSHA Consultation
Arkansas Department of Labor
10421 West Markham
Little Rock, Arkansas 72205
(501) 682-4522
(501) 682-4532
clark.thomas@arkansas.gov
www.labor.arkansas.gov/osha-consultation

CALIFORNIA
CAL/OSHA Consultation Services Branch
Department of Industrial Relations
1901 Gateway Boulevard, Suite 102
Fresno, California 23727
(800) 963-9424
(559) 454-0858
InfoCons@hq.dir.ca.gov
www.dir.ca.gov/DOSH/consultation.html

COLORADO
Colorado State University
Health & Safety Consultation Department of Environmental and Radiological Health Services
1681 Campus Delivery
Fort Collins, Colorado 80523
(970) 491-6151
(970) 491-7778
OHSS@lamar.colostate.edu
http://csu-cvmbs.colostate.edu/academics/erhs/osha/Pages/default.aspx

CONNECTICUT
Connecticut Department of Labor
Division of Occupational Safety & Health
38 Wolcott Hill Road
Wethersfield, Connecticut 06109
(860) 263-6900
(860) 263-6940
rosa.john@dol.gov
www.ctdol.state.ct.us/osha/consulti.htm

DELAWARE
Office of Safety and Health Consultation
Delaware Department of Labor—Division of Industrial Affairs
4425 North Market Street, 3rd Floor
Wilmington, Delaware 19802
(302) 761-8219
(302) 762-3590
DOL_OSH_Consultation@state.de.us
https://dia.delawareworks.com/osha-consultation/

DISTRICT OF COLUMBIA
Office of Occupational Safety & Health
D.C. Department of Employment Services
4058 Minnesota Ave., N.E.—Room 3600
Washington, District of Columbia 20019
(202) 724-3690
(202) 673-6411
does@dc.gov
https://does.dc.gov/service/occupational-safety-and-health

FLORIDA
Safety Florida Consultation Program
University of South Florida
13201 Bruce B. Downs Boulevard, MDC 56
Tampa, Florida 33612
(866) 273-1105
(813) 974-8270
usfsafetyflorida@health.usf.edu
www.usfsafetyflorida.com/

GEORGIA
Occupational Safety and Health Consultation Program
Georgia Tech, Enterprise Innovation Institute
171 17th Street NW, Suite 1250
Atlanta, Georgia 30363
(404) 407-7431
(404) 407-8275
Sarah.Cooper@gtri.gatech.edu
www.oshainfo.gatech.edu/about.html

GUAM
OSHA On-site Consultation Division
Guam Department of Labor
GCIC Building, Suite 400, 414 West Soledad Avenue
Hagatna, Guam 96910
(671) 300-4618
(671) 475-7126
anderson.anthony@dol.gov
http://dol.guam.gov/safety-and-health/osha/

HAWAII
Consultation & Training Branch
Department of Labor & Industrial Relations
830 Punchbowl Street, Room 425
Honolulu, Hawaii 96813
(808) 586-9100
(808) 586-9104
dlir.hiosh@hawaii.gov
http://labor.hawaii.gov/hiosh/home/for-employers/consultation/

IDAHO
Idaho OSHA Consultation Program
Boise State University
1910 University Drive, M.S. 1825
Boise, Idaho 83725-1825
(208) 426-3283
(208) 426-4411
consultation@boisestate.edu
https://oshcon.boisestate.edu/

ILLINOIS
Illinois On-site Consultation
Illinois Department of Labor
160 N. LaSalle Street, Suite C-1300
Chicago, Illinois 60601
(800) 972-4216
(312) 793-2081
dol.consultation@illinois.gov
www.illinois.gov/idol/Laws-Rules/safety/Consultation/Pages/default.aspx

INDIANA
INSafe
Indiana Department of Labor
402 West Washington, W195
Indianapolis, Indiana 46204
(317) 232-2688
(317) 233-3790
insafe@dol.in.gov
www.in.gov/dol/insafe.htm

IOWA
OSHA Consultation and Education
Iowa Workforce Development, Division of Labor Services
1000 East Grand Avenue
Des Moines, Iowa 50319
(515) 281-7629
(515) 281-5522
oshaconsultation@iwd.iowa.gov
www.iowadivisionoflabor.gov/iowa-osha-site-consultation

KANSAS
Kansas Department of Labor
Division of Industrial Safety & Health
417 SW Jackson St.
Topeka, Kansas 66603-3327
(785) 296-4386
(785) 296-1775
indsafetyhealth@dol.ks.gov
www.dol.ks.gov/Safety/assistance.aspx

KENTUCKY
OSH Education and Training
Kentucky Labor Cabinet—Kentucky OSH Program
1047 US Highway 127, South, Suite 4
Frankfort, Kentucky 40601
(502) 564-4087
(502) 564-4769
kim.perry@ky.gov
https://kysafe.ky.gov/services/Pages/consult.aspx

LOUISIANA
21(d) Consultation Program
Louisiana Department of Labor—Louisiana Workforce Commission
1001 North 23rd Street, Room 402 Annex
Baton Rouge, Louisiana 70804-9094
(800) 201-2495 or (225) 342-8441
(225) 342-5158
workplacesafety@lwc.la.gov
www.laworks.net/FAQs/FAQ_WorkComp_OSHA.asp

MAINE
Maine Bureau of Labor Standards
Workplace Safety & Health Division
State House Station #45
Augusta, Maine 04333-0045
(877) 723-3345 or (207) 623-7900
(207) 623-7934
general.bls-safetyworks@maine.gov
www.safetyworksmaine.gov/consultations/

MARYLAND
MOSH Consultation Services
Department of Labor, Licensing and Regulation
10946 Golden West Drive, Suite 160
Hunt Valley, Maryland 21031
(410) 527-4499
(410) 527-4481
dldlimosh-dllr@maryland.gov
www.dllr.state.md.us/labor/mosh/volc.shtml

MASSACHUSETTS
OSHA Consultation Program
Massachusetts Department of Labor Standards
37 Shattuck Street
Lawrence, Massachusetts 01843
(617) 626-6504
(978) 687-0013
masscon@state.ma.us
www.mass.gov/on-site-consultation-program

MICHIGAN
Consultation Education & Training Division
Michigan Occupational Safety & Health Administration, Licensing & Regulatory Affairs
525 W. Allegan Street, PO Box 30643
Lansing, Michigan 48909-8413
(800) 866-4674 or (517) 284-7720
(517) 284-7725
cetrca@michigan.gov
www.michigan.gov/lara/0,4601,7-154-11407_15317---,00.html

MINNESOTA
Workplace Safety Consultation Unit
Minnesota Department of Labor & Industry
443 LaFayette Road North
Saint Paul, Minnesota 55155
(800) 657-3776
(651) 284-5739
osha.consultation@state.mn.us
www.doli.state.mn.us/Wsc.asp

MISSISSIPPI
Industrial Health and Safety Program
Mississippi State University Extension Service—Center for Continuing Education
2151 Hwy 18, Suite B
Brandon, Mississippi 39042
(601) 825-0783
(601) 825-6609
j.dale@msstate.edu
http://ce.extension.msstate.edu/programs/health-safety-consultation-services

MISSOURI
Onsite Consultation Program, Division of Labor Standards
Dept. of Labor & Industrial Relations
3315 West Truman Boulevard, Post Office Box 449
Jefferson City, Missouri 65109
(573) 751-3403
(573) 751-3721
laborstandards@labor.mo.gov
https://labor.mo.gov/onsite

MONTANA
Montana Consultation Program
Montana Department of Labor & Industry—Safety & Health Bureau
2201 White Blvd
Butte, Montana 59701
(406) 494-0324
(406) 494-5481
tott@mt.gov
http://erd.dli.mt.gov/safety-health/onsite-consultation

NEBRASKA
Onsite Safety & Health Consultation Program
Nebraska Department of Labor
PO Box 94600, 550 South 16th Street
Lincoln, Nebraska 68509-4600
(402) 471-4717
(402) 471-4728
ndol.safety@nebraska.gov
https://dol.nebraska.gov/Safety/OnsiteConsultationProgram/Overview

NEVADA
Safety, Consultation & Training Section
Department of Business & Industry—Division of Industrial Relations
1301 North Green Valley Parkway #200
Henderson, Nevada 89074
(877) 472-3368
(702) 990-0362
tschultz@business.nv.gov
http://4safenv.state.nv.us/consultation

NEW HAMPSHIRE
WorkWISE NH
Keene State College
229 Main Street, Elliot Hall M-3925
Keene, New Hampshire 03435
(603) 358-2943
(603) 358-2399
whartz@keene.edu
www.keene.edu/academics/conted/safety/workwise/

NEW JERSEY
Workforce Development
New Jersey Department of Labor
1 John Fitch Plaza, PO Box 953
Trenton, New Jersey 08625-0953
(609) 984-0785
(609) 292-4409
safetyhealth_onsiteconsult@dol.nj.gov
http://lwd.dol.state.nj.us/labor/lsse/employer/Occupational_Safety_and_Health_Onsite_Consultation_Program.html

NEW MEXICO
Occupational Health & Safety Bureau
New Mexico Environment Department
525 Camino de Los Marquez Suite 3
Santa Fe, New Mexico 87502
(877) 610-6742 or (505) 476-8702
(505) 476-8734
Debbie.Odell@state.nm.us
www.env.nm.gov/Ohsb_Website/Consultation/

NEW YORK
New York State Department of Labor
On-site Consultation
State Campus Office Room 168
Albany, New York 12240
(518) 457-2238
(518) 457-1639
onsite@labor.ny.gov
www.labor.state.ny.us/workerprotection/safetyhealth/DOSH_ONSITE_CONSULTATION.shtm

NORTH CAROLINA
Bureau of Consultative Services
North Carolina Department of Labor
1101 Mail Service Center
Raleigh, North Carolina 27699-1101
(919) 807-2905
(919) 807-2902
dol.consultationmail@labor.nc.gov
www.labor.nc.gov/safety-and-health/occupational-safety-and-health/safety-and-health-consultation-program

NORTH DAKOTA
Continuing Education, Training and Innovation
North Dakota Occupational Safety and Health
Bismarck State College, 1700 Schafer Street
Bismarck, North Dakota 58506-5587
(877) 846-9387 or (701) 224-5539
(701) 224-5763
albert.koch@bismarckstate.edu
https://bismarckstate.edu/continuingeducation/ndosh/

NORTHERN MARIANA ISLANDS
CNMI Department of Labor
CNMI-OSHA Onsite Consultation Program
Post Office Box 10007, 2nd Floor-Afetnas Building, San Antonio
Saipan, Northern Mariana Islands 96950
(670) 664-3157
(670) 664-3158
guerrero.ray@dol.gov
http://marianaslabor.net/sec.asp?secID=25

OHIO
OSHA On-site Consultation Program
Bureau of Workers' Compensation
13430 Yarmouth Drive
Pickerington, Ohio 43147
(800) 282-1425
(614) 621-1245
greg.collins@bwc.state.oh.us
www.bwc.ohio.gov/employer/programs/safety/SandHOSHAOnsiteDetails.asp

OKLAHOMA
Oklahoma Department of Labor
OSHA Consultation Division
3017 N. Stiles, Suite 100
Oklahoma City, Oklahoma 73105-5212
(405) 521-6100
(405) 521-6020
Jason.Hudson@labor.ok.gov
www.ok.gov/odol/About_ODOL/Divisions/Workplace_Saftey_and_Health/Consultation_Services.html

OREGON
Oregon OSHA Consultation and Services
Department of Consumer & Business Services
350 Winter Street, N.E., Room 430
Salem, Oregon 97301-3882
(800) 922-2689 or (503) 947-7434
(503) 947-7462
consult.web@oregon.gov
http://osha.oregon.gov/consult/Pages/index.aspx

PENNSYLVANIA
Pennsylvania OSHA Consultation Office
Indiana University of Pennsylvania
57 South 9th Street, Suite 305
Indiana, Pennsylvania 15701
(800) 382-1241
(724) 357-2385
samuel.gualardo@iup.edu
www.iup.edu/pa-oshaconsultation/

PUERTO RICO
PR Occupational Safety and Health Administration
Department of Labor & Human Resources
505 Munoz Rivera Avenue, 21st Floor
Hato Rey, Puerto Rico 00918
(787) 754-2172
(787) 767-6051
Cruz.Judith@dol.gov
www.trabajo.pr.gov/

RHODE ISLAND
OSHA Consultation Program, Office of Healthy Housing and Environment
Rhode Island Department of Health
3 Capitol Hill, Cannon Building, Room 206
Providence, Rhode Island 02908
(401) 222-7745
(401) 222-2456
Gina.Friedman@health.ri.gov
http://health.ri.gov/programs/detail.php?pgm_id=24/

SOUTH CAROLINA
Office of OSHA Voluntary Programs
South Carolina Department of Labor—Licensing and Regulation
121 Executive Center Drive, Suite 230
Columbia, South Carolina 29210
(803) 896-7744
(803) 896-7750
harvey.jessup@llr.sc.gov
www.scosha.llronline.com/index.asp?file=scovp/consult.htm

SOUTH DAKOTA
South Dakota State University
Engineering Extension—OSHA Consultation
Box 2219, Crothers Engineering Hall 316
Brookings, South Dakota 57007
(605) 688-4101
(605) 688-6290
sdsu.engineeringextension@sdstate.edu
www.sdstate.edu/engineering-extension

TENNESSEE
Tennessee Department of Labor and Workforce Development
TOSHA/Consultation
220 French Landing Drive
Nashville, Tennessee 37243
(800) 325-9901
(615) 532-2997
garrett.rea@tn.gov
www.tn.gov/workforce/employees/safety-health/tosha-redirect/consultative-services.html

TEXAS
Workers' Health & Safety Division- MS 22
Texas Department of Insurance-Division of Workers Compensation
7551 Metro Center Drive, Suite 100
Austin, Texas 78744-1609
(800) 252-7031 or (512) 804-4640
(512) 804-4619
OSHCON@tdi.texas.gov
www.tdi.texas.gov/oshcon/index.html

UTAH
UOSH Consultation and Education Services
UOSH Consultation Program
160 East 300 South, Third Floor
Salt Lake City, Utah 84114-6650
(801) 530-6855
(801) 530-6992
consultation_program@utah.gov
https://laborcommission.utah.gov/divisions/UOSH/index.html

VERMONT
Project WorkSAFE
Vermont Department of Labor
PO Box 488, 5 Green Mountain Drive
Montpelier, Vermont 05601-0488
(888) 723-3973 or (802) 888-0620
(802) 888-0730
labor.WPSafety@vermont.gov
http://labor.vermont.gov/project-worksafe/

VIRGINIA
Occupational Safety and Health—Training and Consultation
Virginia Department of Labor & Industry
Main Street Centre Building 600 East Main Street Suite 207
Richmond, Virginia 23219-4101
(804) 786-8707
(804) 786-6359
Dennis.Edwards@doli.virginia.gov
www.doli.virginia.gov/vosh_coop/vosh_consultation_p1.html

VIRGIN ISLANDS
Safety in Paradise, University of the Virgin Islands
Community Engagement and Lifelong Learning Center
#2 John Brewer's Bay
St. Thomas, Virgin Islands 00803
(340) 693-1146
(340) 693-1189
safetyinparadise@uvi.edu
https://safety.uvi.edu/default.aspx

WASHINGTON
Division of Occupational Safety and Health
Washington Department of Labor & Industries
7273 Linderson Way SW
Tumwater, Washington 98501-5414
(360) 902-5554
(360) 902-5438
Flos235@lni.wa.gov
www.lni.wa.gov/SAFETY/CONSULTATION/DEFAULT.ASP

WEST VIRGINIA
OSHA Safety Consultation Program
West Virginia Division of Labor
1900 Kanawha Boulevard East, State Capitol Complex, Building 3- Room 200
Charleston, West Virginia 25305
(304) 558-7890
(304) 558-2415
carla.j.campbell@wv.gov
http://labor.wv.gov/Safety/Occupational_Safety/WV_OSHA/Pages/default.aspx

WISCONSIN
OSHA Onsite Consultation (WisCon) Program
University of Wisconsin—Wisconsin State Laboratory of Hygiene
2601 Agriculture Drive
Madison, Wisconsin 53718
(800) 947-0553 or (608) 226-5240
(608) 226-5249
steve.strebel@slh.wisc.edu
www.slh.wisc.edu/occupational/wiscon/

WYOMING
OSHA Division
Wyoming Department of Workforce Services
1510 East Pershing Blvd., West Wing
Cheyenne, Wyoming 82002
(307) 777-7786
(307) 777-3646
daniel.bulkley@wyo.gov
www.wyomingworkforce.org/businesses/osha/consultation/

NOTES

1. Kentucky OSHA located at www.osha.gov/dcsp/osp/stateprogs/kentucky.html.
2. www.osha.gov/dcsp/osp/stateprogs/kentucky.html.
3. OSHA website located at www.osha.gov.

8 Ethics, Professionalism, and Discipline in Safety

***

"An error is always the more dangerous in proportion to the degree of truth which it contains."

Henri Frederic Amiel

"The journey of a thousand miles begins with a single step."

Laozi

***

Is safety and health a profession or a job? Is the management of safety and health an art or a science? Is there a code of professional ethics which governs the safety and health profession? Is there a minimum level of education required to be considered a safety and health professional? Is there any type of proficiency testing or licensure necessary to be a safety and health professional? Are there any continuing education requirements for safety and health professionals? If a safety and health professional performs unethically or unprofessionally, is there any mandatory system to sanction or penalize the safety and health professional? These and many other questions and related issues challenge the individuals who are tasked with safeguarding the safety and health of others within the American workplace. Within the confines of what we identify as safety and health in today's workplace, there is a myriad of voluntary associations and affiliations that provide guidance; however, there is no federal, state, or professional agency or association which possesses the responsibility and power to develop professional or ethical standards, require a level of proficiency, and enforce sanctions for disciplinary violations. In essence, literally any individual with any level of education and expertise can hold themselves out as being a safety and health professional. For companies and organizations who depend upon individuals to safeguard their employees and company assets as well as depend upon the individual's knowledge, experience, and expertise, it is "Buyer Beware"!

Is the management of the safety and health function a profession or a job? When comparing the safety and health function with other recognized "professions" such as medicine and law, individuals serving in managerial capacities with responsibilities for the safety and health function would correlate with a job function. In law and medicine, there is a major educational requirement. In safety and health, anyone with any degree, or no degree, can serve as the responsible person for the safety and health function. In the law and medical professions, professionals must successfully pass a mandatory competency test and maintain their competency through continuing education requirements. In safety and health, there are voluntary associations and organizations with voluntary certifications and continuing education requirements; however, these are voluntary in nature, i.e., not required. In the law and medical professions, there is a prescribed code of professional conduct and ethics, which is strictly enforced through sanction or removal of the license to practice. In safety and health, there is no mandatory code of professional conduct or ethics, no licensure, and no method of enforcement.

Individuals employed by companies or organizations to manage the safety and health function have the lives and livelihood of each and every employee in their sphere of responsibility. From a company or organization's perspective, safety and health is a job function—a job function that does not produce a profit. However, many companies or organizations value employee safety and health, as well as the environment and sustainability, and have embedded this function within their corporate structure. Other companies have reluctantly added the safety and health job position to their corporate hierarchy as the result of compliance issues and penalties, workers compensation cost, or insurance issues and/or major workplace incidents. The knowledge, skill, experience, and expertise of the individual hired to perform this important function will have a direct impact on the success of the function and thus the safety and health of the employees. In essence, since there are no mandatory licensure or professional standards, it is solely up to the company or organization to do their due diligence in screening and selecting the appropriate candidate to manage this important and life-impacting function.

Is the safety and health function an "art" or a "science"? In most circumstances, to perform within the safety and health realm requires knowledge, skills, and abilities that encompass both the skills of an artist and the knowledge of a scientist. Safety and health professionals must possess a keen knowledge level of not only the OSHA standards but also the science involved in adapting those standards to the workplace. Knowledge of scientific principles, chemistry, biology, and related areas permit the safety and health professional to achieve and maintain a safe and healthful workplace. However, simply achieving compliance does not create a safe and healthful workplace. Safety and health professionals should go far beyond simple compliance and this often requires skills and artistry to acquire funding, effective verbal and written communication, and other components to elevate the safety and health function to greater heights. This combination of art and science is unique to the safety and health profession.

In many companies and organizations, safety and health policies are intertwined with policies applicable to the human resource function. Company and organizational policies, ranging from attendance through discipline, can and do impact the safety and health function. In many companies and organizations, the safety and health function possesses direct or indirect reporting to the human resource function. On a day-to-day basis, the safety and health function often works with the human resource function to achieve the goals and objectives of each unit.

Disciplinary policies as related to the failure to follow safety and health rules and requirements are often an area of intersection. Although disciplinary policies and processes can vary, in general, most disciplinary policies possess a form of "stair step," starting with verbal warnings and escalating through a more formal or written level in order to change behavior or performance. If this formal disciplinary action is not effective, most policies move to more direct monetary penalties through suspension without pay. The ultimate "workplace death penalty" is the involuntary termination of employment.

However, within the safety and health function, the methodology through which the disciplinary action is issued can have a major impact on the effectiveness of the safety and health function. If the safety and health manager is responsible for the actual issuance of disciplinary action, he/she can often be labeled as the "safety cop" and employees often only exhibit safety and health behavior and maintain compliance when being observed. Empowerment of employees and acceptance of responsibility for the safety and health function is usually not feasible. Enforcement of disciplinary policies, which includes violation of safety and health rules, should be within the function of the supervisory ranks or the human resource function. Fair and consistent enforcement of safety and health rules is imperative as well as the necessary documentation and recordkeeping required within our litigious society.

In summation, the safety and health function is responsible for safeguarding the safety and health of all employees each and every day. This responsibility requires a unique and exceptional skill set within the art as well as science arenas as well as the fortitude and professionalism to champion the safety and health function within the organization or company. Safety and health is far more than a job; however, it has not been recognized as such. With the lives and health of workers in their hands every day, it is imperative that those working within the field ensure a level of competency as well as continuing education and professionalism.

9 Human Resource Errors by Safety and Health Professionals

***

"Hold yourself responsible for a higher standard than anybody else expects of you. Never excuse yourself. Never Pity yourself. Be a hard master to yourself – and be lenient to everybody else."

Henry Ward Beecher

"Only the man who can impose discipline on himself is fit to discipline others or can impose discipline on others."

William Feather

***

To err is human; however, safety and health professionals should be aware that he/she is often held to a higher standard as a member of the management team, as well as having a high level of trust of employees and staff. Conversely, a safety and health professional should also be aware that he/she is also an "at-will employee" of the company or organization and thus is subject to all human resource policies and procedures and subsequent disciplinary actions, up to and including termination for any violations. The safety and health professional must be competent and effective at all levels from the grassroots employee level to the boardroom. However, despite often "juggling" variables beyond the safety and health professional's control, he/she is often held accountable for the performance and outcomes of the function.

The safety and health function does not work in a vacuum. Although safety and health professionals often focus their time and attention on achieving and maintaining compliance with OSHA standards and working with employees with workers compensation claims among their other responsibilities, safety and health professionals should also be cognizant of the number of other federal and state laws which also impact function. Safety and health professionals should acquire at least an awareness level of the requirements of these laws in order to avoid complications that can impact their employment status.

Most safety and health professionals are salaried management employees, which categorizes them as "at-will" employees. The "at-will" employment doctrine is a term used for the relationship between the employee and employer where the employee can be terminated from employment for any reason and without warning without having to establish just cause for the termination and as long as the termination is not for an illegal purpose or in violation of a protected category. Conversely, the employee can voluntarily quit at any time and for any reason (usually providing notice to the employer). Protections afforded through collective bargaining agreements and individual employment contracts are not applicable. State courts and legislatures have also identified additional exceptions to the "at-will" doctrine specific to the individual state.

In general, safety and health professionals are usually not terminated in the same manner as hourly employees. So not to damage further career opportunities, safety and health professionals, as with other members of management, are often provided an opportunity to voluntarily terminate rather than being dismissed. And, in many circumstances, the reason for the termination, whether voluntary or involuntary, involves circumstances involving other laws or policies which are not directly related to the individual's safety and health work performance.

Most companies or organizations possess policies prohibiting the use of alcohol and controlled substances during employment. Many companies also test for alcohol and controlled substances and a positive test can be used as grounds for termination. Safety and health professionals are human and, as with all professions, can acquire a substance use disorder. With most professions, the safety and health professional with the substance use disorder can continue to perform the functions of the job despite the impacts of the disorder. The most common disorders include binge drinking, marijuana, and cocaine.[1] According to the National Council on Alcoholism and Drug Dependence, "Drug abuse and addiction costs American companies $81 billion each year."[2]

Safety and health professionals often possess the means as well as the access to alcohol and illegal controlled substances as well as legal pharmaceuticals. Some of the signs of a substance use disorder can include talking about financial problems, a decline in personal appearance or hygiene, failing relationships at home, and taking time off for family issues or vague illnesses. Safety and health professionals encountering difficulties with alcohol or controlled substances can often acquire the necessary assistance through employee assistance programs (EAP). However, safety and health professionals should be aware that the violation of company policies regarding the use of alcohol and controlled substance while at work (or at company functions) can often result in investigation and appropriate disciplinary action, up to and including involuntary termination.

Safety and health professionals should be aware of your company's policies regarding dating and relationships among and between employees and between management and nonexempt personnel. As a result of the potential for legal liability for the company (and potential violations of the company's sexual harassment policy) resulting from relationships which have gone badly resulting in EEOC claims, litigation, and other issues and costs, companies have developed a wide variety of policies ranging from no dating among employees to personalized contracts wherein employees who wish to become involved romantically are required to sign a waiver of liability. With most workplaces being a microcosm of society, there is a relatively high probability of interaction and relationships being formed among and between employees. Prudent safety and health professionals should be aware of their company's policy, as well as culture, if considering a relationship with a fellow employee.

Safety and health professionals should be aware that often when the relationship ends and accusations fly, violation of this policy can result in disciplinary action, up to and including termination, as well as the long-term impact on career and family.

Safety and health professionals should be knowledgeable regarding company policies as well as federal and state laws prohibiting sexual harassment or harassment of any kind in the workplace. In most companies, there is "zero tolerance" for harassment of any kind and enforcement is rigorous. Safety and health professionals who engage in any type of harassment in the workplace should be aware that disciplinary action, up to and including termination, usually results after allegations are raised and an appropriate investigation is completed. Prudent safety and health professionals should always be professional in the workplace as well as at company-sponsored events and be cognizant to maintain not only a safe and healthful workplace but also a workplace free of any form of harassment.

SEXUAL HARASSMENT

It is unlawful to harass a person (an applicant or employee) because of that person's sex. Harassment can include "sexual harassment" or unwelcome sexual advances, requests for sexual favors, and other verbal or physical harassment of a sexual nature.

Harassment does not have to be of a sexual nature, however, and can include offensive remarks about a person's sex. For example, it is illegal to harass a woman by making offensive comments about women in general.

Both victim and the harasser can be either a woman or a man, and the victim and harasser can be the same sex.

Although the law doesn't prohibit simple teasing, offhand comments, or isolated incidents that are not very serious, harassment is illegal when it is so frequent or severe that it creates a hostile or offensive work environment or when it results in an adverse employment decision (such as the victim being fired or demoted).

The harasser can be the victim's supervisor, a supervisor in another area, a co-worker, or someone who is not an employee of the employer, such as a client or customer.[3]

Title VII of the Civil Rights Act[4] (known as "Title VII") forbids discrimination based on race, sex, color, religion, and national origin. The Age Discrimination in Employment Act[5] (known as "ADEA") prohibits arbitrary age (between ages 40 and 70) discrimination in employment. The Pregnancy Discrimination Act,[6] an amendment to Title II, prohibits discrimination against pregnant employees in the workplace. For safety and health professionals, it is important to know the protected classes and ensure there is no discrimination of any type, direct or implied, in any aspect of the safety and health program. Additionally, safety and health professionals, as the "eyes and ears" of management, possess a unique opportunity to be in every area of operations and can identify and address areas of potential discrimination within the workplace.

Most employers expect a higher level of professionalism and honesty from their safety and health professional. Areas in which safety and health professionals can encounter situations where a company's disciplinary policy and possible sanctions, up to and including termination from employment, can occur include unexcused absences from work, lying or falsifying information, stealing of company property and inappropriate language, unpermitted relationships, and other unprofessional or prohibited activities. Safety and health professionals should hold themselves to the highest level of ethical behavior and professional conduct at all times.

NOTES

1. National Council on Alcoholism and Drug Dependence (NCADD) located at www.ncadd.org.
2. Ibid.
3. Equal Employment Opportunity Commission located at www.eeoc.gov.
4. 42 USCA Section 2000E through 2000E-7.
5. 29 USCA Section 621 et seq.
6. 42 USCA Section 2000e(k).

10 Setting a Foundation for Cultural Change

***

"Low morale plus low loyalty equals high turnover. That can hobble a company."

Karen Carlisle

***

Human resources and safety have been getting progressively more integrated over the past several years. Whether it is because of the cost of the bottom line or the thought that safety culture can only get so far without the help of human resources and vice versa. There is no right or wrong way to combine or separate safety and human resource departments. It is dependent on the company, the culture, and processes that are in place. Many aspects of the two departments overlap, which is why you are seeing companies combine the two areas. From employee discipline, drug and alcohol usage, record keeping, onboarding, training, and termination, these two departments have more in common than they think. This chapter is going to help you set a foundation for cultural change to build a strong safety program.

NO. 1: DO NOT RELY ON ONE PERSON TO FULFILL THE DUTIES OF HUMAN RESOURCES AND SAFETY

When this happens, the responsibilities of safety usually get put on the back burner due to the day-to-day involvement of HR from employee issues, benefits questions, recruiting, hiring, onboarding, and so on. Depending on the size of the facility, juggling both duties can be difficult for one person and will adversely affect the facility, potentially causing higher incident rates, workers compensation claims, and poor employee morale.

It is important to take an in-depth look at your facility's safety and human resource needs before staffing the position or positions. Both areas have a large impact on the functionality of your facility and needs to be staffed appropriately with either two separate managers or one depending on the needs and scope of responsibility to ensure neither safety nor human resources are given less priority or attention. If you are confident that a single person can assume these responsibilities depending on the size of your facility, you may be able to consolidate the position. If your facility has a larger number of employees, safety issues, or human resource weaknesses, you may need to decide on splitting these responsibilities to ensure both duties get the attention they need.

The role of human resources should be responsible for the following:

- Serve as a link between management and employees when handling questions, interpreting procedures, and assisting in resolving work-related problems.
- Management of employee documentation such as employee personnel files and the employee handbook.
- Advice and training management on the organization's policies and procedures (equal employment opportunity, sexual harassment, conduct and ethics, etc.).
- Administer disciplinary actions (dismissal of employees, layoffs, etc.).
- Conduct new hire orientations.
- Identify staff vacancies, triggering recruitment, interviewing, and selection of new employees.
- Coordinate, maintain, and track employee performance evaluations.
- Maintain the work structure and all supporting documents such as org charts and job descriptions.
- Preparing and managing employee wage budgets through annual wage surveys, job evaluations, and pay structure revisions.
- Manage employee benefit elections through needs assessments, benefit trends, and obtaining and evaluating benefit contract bids. Conduct annual benefits rollout and the open enrollment process.

The role of safety should be responsible for the following:

- Provide guidance for regulatory requirements in the area of OSHA.
- Coordinate accident investigation and return-to-work programs.
- Maintain OSHA and MSDS documentation.
- Ensure employees are trained and are aware of all regulatory standards such as OSHA, EPA, and so on.
- Conduct safety audits on personnel, equipment, and materials.
- Develop and maintain organization safety and health programs and procedures.
- Track incident metrics and use findings to reduce or eliminate risks.
- Assess and alter safety procedures and programs as needed to benefit employees.
- Receive all OSHA citations and determine the appropriate responses and actions under the direction of the general counsel and chief administer officer.
- Receive and review all safety audits, risk assessments, and job hazard analysis to measure and report on the safety of the plant.
- Focus on eliminating accidents and injuries that could occur at the facility.

Both human resources and safety should be responsible for the following:

- Ensuring compliance with policies and procedures.
- Analyzing training needs whether it is lockout tagout for safety or sexual harassment for human resources.

- Conduct a Gap analysis to identify gaps in processes, procedures, or training and use resources to improve on the identified gap.
- Ensuring employees work in a safe environment away from physical and mental health and safety hazards (workplace violence, hazardous substances, mechanical safety hazards, etc.).

NO. 2: DO NOT WAIT FOR THE ANNUAL REVIEW TO TAKE CORRECTIVE ACTION

Waiting for the annual performance review to correct an action will not only allow potentially hazardous behavior to continue but could cause employee issues that will be harder to correct in the future. Instead, provide constant feedback, both positive and negative, in order to create a positive impact on employee engagement. Take a step back and think if you were an employee who believed they were doing the right thing over the course of several months, your manager has seen you do this action a number of times and did not say anything. Then your annual performance review rolls around, they sit you down and tell you the action that you had been doing for several months, and that they had seen you performing for several months, was not the correct practice. These types of situations can cause negative employee emotions and can bleed into the entirety of your employee population. This is why it is important to correct negative behaviors when you see them and not to wait. It is always a good time to protect someone from a potentially dangerous action.

NO. 3: DOCUMENT. DOCUMENT. DOCUMENT

Unfortunately, we live in the age of pointless lawsuits. Imagine getting sued for an employee who got hurt while standing on a rolling chair. You had told this employee before not to stand on a rolling chair to perform overhead work because that was not the intention of the chair and it is a safety hazard. The employee swears you never said anything to him before. If you had documented the first conversation, you would have written proof that the conversation did take place and the employee knew not to do that.

As a safety professional, you encounter some heated situations when trying to correct unsafe acts or rolling out a new safety program that changes the way employees conduct day-to-day work. Whether it is a small or large conversation, make sure to document and have the employee sign-off that the conversation took place and the main points of the conversation. If the employee refuses to sign-off, make a note of that as well. Simple documentation of your conversations, corrective actions, and disciplinary ventures could save you and your company countless dollars in legal fees.

NO. 4: SELL SAFETY

Instead of telling employees this is what you do, and this is how you do it, explain why this is how they should do it and the benefits short- and long-term. Encouraging employees to act safely will have a larger impact and greater return compared with instructing the employees to do something a certain way and threaten disciplinary action if not done correctly.

The idea and the importance of safety should begin to be instilled in every employee's head on day one, before they hit the manufacturing floor or start their normal day-to-day work. When onboarding new employees, go over all safety precautions, personal protective equipment, safety procedures, and so on. During this training, be sure to explain why each procedure is in place and why they are to wear every piece of PPE.

As a child, you were told "no" a lot, but it was not until you knew why you were told no that you actually stopped doing something. Some learn this the hard way, OSHA has daily reports of lost limbs or digits due to employees knowing they should not do something but they did it anyway, sadly learning the consequences of the "no" the hard way. This is why it is important for employees to know the equipment and the potential hazards associated with it to prevent safety incidents.

NO. 5: TAKE THE TIME TO CREATE A SAFETY PROGRAM

Time is money but when you rush through creating a program, I can guarantee it will not be effective and will not be received well by employees. If you have to spend a few more dollars to create a state-of-the-art safety program for your facility that is tailored to your employee's needs and wants, it will be worth the time and money spent in the long run. Creating the best safety program for your facility is about taking risks. Go outside the normal box of safety programs that have been implemented by your facility and find what drives your employees, what will catch their attention and cause them to think twice before any unsafe behavior.

Doing different things with your safety programs will keep your employees engaged and excited for safety. With recent OSHA changes, safety programs have to be careful about rewarding employees in the wrong way that would promote hiding safety incidents. Instead, do small things such as promote attendance for your safety meetings or rewards for corrective action behaviors. The takeaway with this error is to stay away from falling into a consecutive and boring cycle with your safety programs. Change it up and keep employees engaged in safety, even though it may take more time, money, and resources.

NO. 6: FOSTER IDEAS

As a manager, team lead, supervisor, and so on, you will have the opportunity to either develop ideas yourself and delegate responsibilities to others or you can share ideas with your reports and foster the ideas they may have to reach the end goal. This is the concept of "more heads are better than one" which will be discussed more in the next chapter. The term "human resources" is often overlooked by safety professionals, but using your humans as resources is the best move you can make in order to get buy-in for new programs, knowledge on a procedure you may not be familiar with, and communicating a concept or idea throughout the workforce.

NO. 7: BE A LEADER AND HIRE LEADERS

As a supervisor, manager, team lead, and so on, you must be able to draw that fine line between friend and boss with your employees. Your employees should be able to trust you enough to come with you for issues but should respect you enough to take direction from you. Some managers make the mistake of becoming "buddy buddy" with a few of their employees, making it harder to discipline or correct behaviors that they may see them doing. This can also cause some animosity among your other reports that you have not developed this type of friendly relationship with. As a safety professional, try to do just that, keep things professional between you and your employees.

Hiring leaders within your organization, even if they do not hold manager roles, is important for organizational culture and change. Leaders help move change forward through leading those employees who follow. Bringing leaders into an organization will double as change agents. These change agents will be able to help make the connections of the best way to act through change management to get the best buy-in, engagement, and following.

NO. 8: BUILD TRUST WITH YOUR EMPLOYEES

Expanding on No. 7, not building trust with your employees can also be a huge downfall. As mentioned before, there is a fine line between friend and boss. You need to make an effort to get to know every one of your employees on a personal level and find out what "makes them tick." Having simple conversations that do not involve work can help build these relationships. When gaining the trust of your employees, you want to make sure your actions match your words. Do as you say and say as you do so to speak.

A key to building trust with employees is walking in their shoes, literally. Make an effort to walk around and either help or offer assistance on a task they may be working on. If you work in a manufacturing facility, change out of your dress clothes and put on the uniform that your employees wear. Go out on the manufacturing floor and visit with your employees. This simple act can help gain trust and respect with those employees.

NO. 9: BE CONSISTENT WITH YOUR DISCIPLINE

Consistency is key! In recent years, employers are being faced with employment discrimination and retaliation cases due to inconsistent disciplinary actions. In *Carmelita Vazquez v. Caesar's Paradise Stream Resort*,[1] the court awarded an African American housekeeper $25,000 for discrimination because she was fired for refusing to take her braids out of her hair. Their policy stated that cornrows were prohibited, but braids could be worn. She won her case by proving that non-African American housekeepers wore braids and were never disciplined, but when she wore braids, she was fired for refusing to remove them.

This is where safety and human resources need to work together to ensure consistent disciplinary actions are taken throughout the organization. Human resources should be knowledgeable of all procedures and policies and what disciplinary action has taken place for each employee. This helps prevent inconsistent discipline throughout departments, employees, and teams.

NO. 10: BE A LEADER IN TOUGH SITUATIONS

As a safety professional, you can be put into some heated discussions or situations that can easily get out of hand with emotions. From a screaming feud between employees that you may need to cease, to an employee who may refuse to take any constructive feedback from you, causing you to get frustrated and wanting to raise your voice, or an employee that is frustrated and raising their voice at you. There are countless situations that can cause your emotions to fly and act in a way you may not want to. This is why it is important to stay calm in high stress or tension situations and work toward a calm solution.

NO. 11: ENSURE SAFETY IS PART OF THE ONBOARDING PROCESS

New hire orientation and the onboarding process set the stage for the entire employment life of those individuals. When safety is not involved in this process, it encourages those employees to think that safety is not a primary focus of this employer. Whether it is a corporate or manufacturing setting, safety needs to be part of the onboarding process. The amount of safety touched on is also directly related to how safety is perceived in the company, outside the controlled environment of orientation.

As a safety professional and the key safety contact for your facility, you want to work closely with HR to implement a safety training session during the new hire orientation and onboarding process. Whether it is for an hour or for a week, you need to take responsibility for instilling a strong safety culture into your new employees before they hit the floor. If you are in a manufacturing facility, use the new hire orientation to make the new employees familiar with all safety procedures, PPE, and equipment and their hazards. By doing this before they are overwhelmed with "real work," they are able to process and remember it for when they are actually working with or close to that piece of equipment.

NO. 12: MAINTAIN SAFETY TRAINING DOCUMENTATION

Did safety training happen if there are no records to prove it? As an employer, it is your responsibility to provide a safe and healthful work environment. Improper record keeping of safety training can not only lead to confusion and lack of refreshers but can lead to citations and fines from OSHA. Not all training is required to maintain documentation, but it is suggested to keep records of all safety training that occurs. Some of the OSHA general industry standards[2] that require documentation of training include the following:

- Asbestos (1910.1001)
- Carcinogens (1910.1003)
- Confined Spaces (1910.146)

- Fire Brigades (1910.156)
- Fire Extinguishers (1910.157)
- Lockout/Tagout (1910.147)
- Powered Industrial Trucks (1910.178)
- Respiratory Protection (1910.134)

Benefits of maintaining safety training documentation include tracking employee progression, the ability to determine when annual or required refresher training is needed, showing proof of training, and determining whether any safety performance issues were caused by a lack of training. OSHA also requires that employers have training certification documentation. This provides proof to OSHA that you as an employer have successfully given said training to your employees. OSHA standard section 17(g) states that it is a crime to knowingly create falsified certifications or required training documents. This is why it is important to stress the importance of the documents to the management staff that will be completing and filing these documents.

When documenting the safety training ensure that the following items are listed:

- Employee Name
- Date of Training
- Name of Trainer
- Signature of Trainer
- Training Subject
- Summary of Training Session
- Proof of Competency/Awareness

This information must then be filed into employee records, along with all other health and safety records. This can be done either via print or web archive. Whichever record keeping platforms you prefer, ensure that you can produce the training records at any time. This will be helpful for auditing and retrieval purposes. The length of required record retention varies but it would be suggested to keep them for the length of employment and plus x years at the discretion of your company's record retention policy. Having the documents for a time past employment will allow an easy retrieval for any re-employment case.

All new hires should go through the relevant and required safety training for their job duties and responsibilities fairly early in their onboarding process. This will allow a safety culture to be instilled into their day-to-day activities and will provide them safety guidance when performing their duties on the job. For both seasoned and new employees, safety training can provide an employer with an exponential return on investment by causing fewer injuries and illnesses, creating better employee morale, lower insurance premiums, and so on.

In addition to training documentation, OSHA 29 CFR 1904 recordkeeping regulates that all employers must keep and maintain an OSHA 300 log that identifies all serious occupational illnesses, injuries, and fatalities. If an incident were to occur at your site, not only would the OSHA log be looked at, but training records will be requested. An OSHA log should be kept at for every physical location where said operations are performed.

Government agencies such as OSHA and EPA that require documentation to be kept should have training records to compliment. Several federal laws require employers to retain several forms of employment records but are dependent on the size of the company (ex. 11–14 employees: OSHA records, 15–19 employees: Americans with Disabilities Act of 1990, 20–49: Age Discrimination in Employment Act of 1967). These records should be kept forever whether they are hard copy, cloud based, internal server, and so on.

NO. 13: DEVELOP AN EMERGENCY RESPONSE PLAN

Emergencies can happen at any time, this is why it is important to determine the emergencies that are relevant or have a potential to occur at your site and develop an emergency response plan. Emergency response plans help mitigate the resource and financial loss if said emergency were to occur. Emergency response plans should be developed no matter the size, location, or industry of the company.

In order to decide the type of emergency response plan needed for your facility, a risk assessment must be performed. This will identify potential emergency situations for your facility so you can determine the resources, plans, and procedures needed to prepare. Severe weather is an example of an event that can be foreseeable in most cases. This will allow valuable time in preparing and protecting your people and facility. For events that are unforeseeable, it is important to have a prepared and practiced plan that can quickly be activated.

To begin creating an emergency response plan for situations that would require evacuation, find all potential evacuation roots and emergency exits and choose the safest route to an exit. Designate a primary and secondary route and ensure they are clearly marked and lit in case of a power outage. For events that involve shelter in place, an interior room inside your facility that has no windows should be designated as a "Shelter-in-Place" safe area.

In the event of an evacuation, an emergency exit route must comply with 29 CFR 1910 36, this includes having an adequate number of exit routes. Two or more exit routes are preferable to prevent panic and clutter. You want to ensure that the exit routes are available to all employees and are located as far away as practical from one another to ensure that if one exit is blocked due to fire or smoke that the employees have an alternative exit.

The exit must lead directly outside or to an open space with access to outside. These areas must be able to accommodate all occupants of the building in which you are exiting. Ensure that the exit route is permanent, with an exit door that is separated by fire resistant materials, connected via side-hinges and swing outward.

When your exit route has been established, work with local emergency management authorities to coordinate plans. Ensure to designate "evacuation wardens" who will assist others during the evacuation process and will be identified via fluorescent vests. Once everyone has evacuated, they will then account for all employees. This can be split up into departments, floors, teams, and so on, as long as there are sufficient wardens to assist with the evacuation. The duty and appointment of evacuation warden can be written into the company's emergency response plan.

Accounting for all employees once evacuated can be simplified by designating an area for each department to meet. From there, make managers responsible for the accounting of their direct reports and then reporting their attendance to the evacuation warden. This will simplify the responsibility of the evacuation warden and will provide a better knowledge of missing employees. Evacuation wardens should be equipped with a contact and emergency contact list for all employees in case of an absence during roll-call.

Employee training should be created in order to delegate individual roles, responsibilities (i.e., department heads, evacuation wardens), evacuation routes, potential threats, notification systems, emergency response procedures, accountability procedures, department evacuation locations, and hazards and protective actions for all emergencies. Emergency drills should be held at least once a year to ensure all employees have knowledge of what to do in an emergency and to test the efficiency of the exit routes, assembly areas, and accountability of departments. Training and refreshers should take place at least once a year, similar to drills.

NO. 14: DEVELOP A CONSISTENT EMPLOYEE COMPENSATION PROGRAM

With an increasingly competitive work environment, you as a company must have a well-designed compensation program that will help motivate employees, control costs, and are equitable both internally and externally. To do this, you must access market data and conduct a compensation analysis to ensure you have placed your positions in the correct pay grade for your company. Every wage decision should be based on market data collected and the overall company strategy. Compensation programs are not a shoot from the hip type of decision.

Market data can consist of salary surveys, compensation consultants, government labor databases, and so on. A salary survey is the most highly used compensation source by organizations. These surveys allow companies to find the low, average, and high wages given for certain positions, along with benefits provided. When collecting such data, it is important to compare apples to apples, this is done by ensuring the surveys were taken in the same or similar geographic area, job descriptions/position, education level needed, operating budget, type of organization, number of participants, and year of data collected.

Wage surveys can be conducted either internally or externally by a third party. If you are creating a wage survey internally, here are some quantifiable and nonquantifiable areas that can be measured:

Quantifiable

- Base Salaries
- Percentage Increases
- Frequency of Increases
- Merit Increases
- Salary Ranges

- Benefits
- Working Hours
- Number of Employees

Nonquantifiable

- Company Information
- Educational Requirements
- Geographic Location
- Hiring Source (Internal/External)
- Union Presence (Yes/No)

The overall goals of the company will help decide whether to set wages either below, above, or equal to the market data collected. Some companies decide to place their wages above the market rate for certain positions because they want to attract and retain high performing employees. For other positions, companies can choose to set wages below market data because the position does not require substantial education or experience. Either way, as an employer, you want to drive the right behaviors from your employees and wages drive a small but important piece of the big picture of profitability.

When the compensation program has been developed, it should be effectively communicated to all employees. The communication should include the bonus target with the understanding that this can only be paid if both the employee and company meet all set objectives, an in-depth understanding of the objectives that must be achieved by both employee and company, and the dates and deadlines of all set objectives.

It is important to have management communicate how the company and employee are performing throughout the year and how the performance is aligning with the set objectives. Keeping an open flow of communication about performance and goals provides higher employee morale, which is linked to company performance.

NO. 15: HAVE AN EMPLOYEE HANDBOOK

Policies and procedures are the framework for consistent and fair treatment of employees. The policies help to define standards of management and procedures assist with the consistency of decisions. The creation of a policy and procedure manual for the company can not only help it run but can save both time and resources in the long term.

Written policies and procedures provide a basis of communication, providing guidance to employees through rules and guidelines of the organization. Policies and procedures also help save management's time that would have otherwise been wasted on countless training and explanations. Policies help organize and communicate management's plans for the future and help explain the investment the organization has in its employees through the explanation of benefits provided to them.

A policy and procedure manual should be developed and distributed in order to assist both management and staff in day-to-day business activities. Policies are created with the mission and vision of the organization in mind, to serve as rules, and as a business process. Procedures help save hours worked on figuring out the way of doing something. Policy and procedure manuals should be one of the first things given to onboarding employees and be able to be accessed by all other employees in some fashion (hardcopy, internet portal, etc.). This manual should stay up-to-date and revised whenever a change is made and communicated to all employees.

From a legal standpoint, policies and procedures are used to protect the company. Written policies and procedures define the obligations, rights, and rules that govern employees. These written documents protect the company but must be fair to both the employee and the company. This must be shown via the policies and procedures and must be clear, concise, and coherent.

Organization/Introduction

This should include an introduction and/or history of the organization, mission and vision statement, and the rights of management and employees. This section should be brief and reflect the entirety of the policy and procedure manual.

Employment

This section should include everything a new employee should know about their rights, including all required local, state, and federal laws that correlate with company policies (Title VII, Affirmative Action, EEOC Guidelines, Anti-Discrimination Law, etc.).

This will include the orientation process, probationary period, equal employment opportunity (EEO) statement, reporting structure, employment practices, job classification, Fair Labor Standards Act, hours of work, hour reporting, overtime, conflict of interest policy, outside employment policy, employee conduct, performance management, progressive discipline, grievance and appeals, separation of employment, and so on.

Employee Responsibilities

This section needs to include all general responsibilities your employees will have while being employed with your company. This will include personnel record information, absenteeism, harassment, drug free workplace, and garnishments. It is important that this section is written to be easily understood and includes a sign-off for the employee to sign and be placed in their personnel file.

Compensation and Benefits

This section needs to include all compensation and benefit information your company provides to employees. It shall include everything from the process of getting paid, any tuition assistance, holidays, vacation, sick leave, unpaid leave, military leave, bereavement leave, jury duty, FMLA, ADA, workers compensation, and any resources for living (social development, stress management, anxiety and depression help, legal and financial matters assistance, work life balance assistance, education planning, substance abuse, etc.)

Example Table of Contents

1. **Introduction**
   a. Mission Statement
   b. Vision Statement
   c. Policy Changes
2. **Employment**
   a. Affirmative Action
   b. EEO
   c. Employment Procedure
   d. New Employee Orientation
   e. Rehire Orientation
   f. Dress Code
   g. Confidentiality
   h. Employee Conduct
      i. Harassment
      ii. Termination
3. **Policies and Procedures**
   a. Attendance
      i. Absence
      ii. Tardiness
   b. Lunch and Breaks
   c. Discrimination
   d. Injury Reporting
   e. Conflict of Interest
   f. Written Communications
   g. Personnel Records
   h. Smoking Policy
   i. Drug Policy
   j. Visitors and Vendors
4. **Wages**
   a. Employee Classification
   b. Independent Contractors
   c. Hours of Work
   d. Overtime
   e. Paid Time Off (PTO)
   f. Shift Differential
   g. Performance Reviews
   h. Pay Periods
   i. Garnishments and Wage Attachments
5. **Benefits**
   a. Vacation
   b. Holidays

c. Sick Leave
d. Insurance
e. Retirement
f. Leave of Absence
  i. FMLA
  ii. Military Leave
  iii. Personal Leave
  iv. Bereavement Leave
  v. Jury Duty

After you have completed the manual, it is important to proofread and edit and get all the proper approvals (management, legal, compliance, etc.) before publishing and distributing it to your employees. You want to ensure that the policies and procedures are able to be read and understood by all levels of employees. Before distribution, whether it is via print or web, you want to make sure all formatting and page styles allow a smooth flow throughout the manual. You want to have a good balance between text and visuals if you decide to add visuals. Create an appealing and professional cover to the manual.

Once the manual has been completed, have it reviewed and approved via the appropriate chain of command. Allow for editing, feedback, and comments to be made with a deadline set and communicated. Once the feedback has been received, make the appropriate changes and edits before getting final approval and publishing it for employees.

NO. 16: ENSURE FRONT LINE LEADERS ARE AWARE OF POLICIES/PROCEDURES FOR TERMINATION

Managers and supervisors are often involved in the voluntary termination (i.e., "I quit") as well as involuntary termination (i.e., "Your fired") of their employees. Whether it is the manager or supervisor who informs the employee or the human resource department or other entity within the company communicating with the employee, it is important for managers and supervisors to know the process and identify the potential risks with regards to this severing of the employment relationship.

A voluntary termination is when an employee quits your employment to pursue other ventures or activities. This type of termination is directly generated by the employee and the manager or supervisor is often one of the first representatives of the company to be notified. Managers and supervisors should be aware of company policy or procedures through which this termination can be addressed. An employee who is voluntarily terminating himself/herself from the employment relationship often has a number of questions including when will I receive my last check? How to I withdraw my money from the 401K plan? Why do I need to provide two weeks' notice? and other questions. Additionally, managers and supervisors should be aware that the employee may possess equipment, company credit cards, security clearance cards, employer identification, and other equipment that is required to be returned by the employee. Managers and supervisors should be aware of the company's processes and ensure all aspects and communications with the employee are properly documented. As we have seen from the unemployment section, the reason for termination provided by the employee at the time of termination may change by the time he/she reaches the unemployment office.

Involuntary termination, or the last stage in a progressive disciplinary system, is when the employee has been provided numerous chances to rectify previously identified actions, behavior, and so on. Managers and supervisors should be aware that there is a high likelihood of protracted litigation, from unemployment through the courts, resulting from this action.

Additionally, managers and supervisors should be aware that the act of involuntary termination often leads to emotions, ranging from fear to aggression, by the employee which can result in harm to personnel and company property. Although most employees are aware of the impending involuntary termination as a result of the earlier progressive discipline, the act of actual termination can create risk which the manager or supervisor should anticipate and mitigate.

Before conducting any disciplinary action, it is imperative that the manager or supervisor analyzes the process and considers the various potential risks of the process. It is important that human resources is also involved in the process. Some of the considerations in this analysis include the following:

- Where do you want to terminate the employee?

Managers and supervisors should consider the location where the employee will be notified of his/her termination. Considering the safety of the manager or supervisor, can the manager or supervisor escape from the room if the employee becomes violent? Should security be available during the termination? Should the door be open or closed? Should there be a witness in the room? What objects are on the desk or in the room which can be used as weapons?

- Should the employee be permitted to return to the operations AFTER the termination?

Managers and supervisors should consider how the employee will acquire his/her personal items from the locker, workstation, or other location within the operations. Do you want an individual who has just been terminated returning to the workplace? Who is responsible if the individual injures himself/herself while in the operations? Can the individual sabotage the operations? Should the individual be escorted from the company property?

- What should I say to the employee?

Keep it brief and to the point! Managers and supervisors should be prepared, always remaining calm, and maintaining an even tone of voice. Given the progressive discipline the employee has previously received, the termination should not be a surprise. Managers and supervisors should be unambiguous in stating to the employee that his/her employment with the company is terminated effective immediately. Managers and supervisors should ask for company identification, tools, computers, or other items of company property and know the method through which you are going to acquire the items (such as escorting the employee to his/her locker, mailing items within 30 days, etc.).

- What questions should I be prepared to answer?

Managers and supervisors should be prepared for questions by the employee after informing him/her of the involuntary termination. Usual questions include their paycheck, termination of benefits, unused sick leave, unused vacation, references, and related questions. Where possible, the manager and supervisors should address these questions before the termination to be prepared respond to the inquiries. If a question arises which is out of the norm, the manager should simply respond that he/she will check and get back to the employee in writing.

- What date/time should I conduct the termination?

When applicable, the date and time of the termination can be planned in order to acquire the final paycheck, vacation pay, and related items in order to be able to provide to the employee at the termination session. Additionally, managers and supervisors may consider a date/time which would not disrupt the operations. Terminated individuals should be removed from company property immediately after the termination, if possible, and terminated employees should not be permitted to return to the operation.

- What should be documented?

Managers and supervisors should document all aspects of the termination including, but not limited to, all conversations, all documents, and all aspects of the termination. As noted previously in our unemployment section, the perception of the truth sometimes achieves a different perspective as the terminated individual applies for unemployment benefits or pursues legal action. Every aspect of the termination should be documented!

Concluding, termination of an employee is never a fun event. It is the manager or supervisor's duty, as a representative of the company, to protect the company through properly conducting the termination process. You, as the manager or supervisor, are not terminating the employee. The employee is terminating himself/herself due to their failure to follow company policies and procedures despite several warnings through the progressive disciplinary system. Managers and supervisors should handle this difficult duty in an effective, efficient, and professional manner while safeguarding yourself and the company through appropriate documentation to be utilized in the future.

NO. 17: CONDUCT ADEQUATE COMPLIANCE TRAINING

Detecting and preventing improper behavior of employees can be eliminated with effective human resources compliance training. This training can include, but not limited to, sexual harassment, sexual violence, workplace violence, discrimination, and so on. While there is no one-size-fits-all training program, there are actions you can take to facilitate an effective training that can positively impact your organization.

Pre-Training

Prior to any training, ensure the policy, procedures, code of conduct, and any other training relevant material is up-to-date with any relevant changes in regulations, case law, and company published materials. Ensure this material is posted and employees are informed of where to find or reference such information.

Next, assign leadership oversight. This will help distinguish lines of communication to the operating areas of your facility about the compliance effort and easily get feedback and answer any questions from employees post-training. Although not all leadership will have direct oversight, all leadership must understand the company policies in order to act as a governing authority and engage in actions consistent with the program.

When creating the training you will want to do the following:

- Pair your information with a visual presentation in order to educate both your auditory and visual learners.
- Only use enough of the information to have a clear message of purpose.
- Know your information, this will allow the training to deliver in a confident, knowledgeable manner.
- Support your facts with examples, data, surveys, personal anecdotes, etc.
- Have a clear point of contact for any questions that may arise after the presentation to continue to have a clear and consistent message.

Deciding when to facilitate the training is just as important as the creation of the training. Whether your facility is a 9–5 M–F or 24/7 you must coordinate the training at the convenience of your employees. For example, in a 24/7 plant with four different teams that work 12-hour shifts, you will need to conduct 4 to 5 training sessions. An example schedule is seen below:

Monday	6:00 am	A Team
Monday	11:00 am	5 Day (Day Shift)
Monday	6:00 pm	B Team
Tuesday	6:00 am	C Team
Tuesday	6:00 pm	D Team

Training

Compliance training should be required for all employees and attendance should be taken. This can be done via a sign-in sheet and should be kept forever. These records can become a principle artifact if a case were to occur involving an employee who is in the disciplinary system for noncompliance of said policy/procedure that was discussed in attended training.

During the training, ask your employees for day-to-day examples of discussed criteria in their everyday work environment. This will allow you as a facilitator to expand on the "gray areas" of the policy. Such a discussion will help your employees better understand the compliance topic being discussed and be able to implement positive actions into their day-to-day activities.

Post-Training

When the training has been completed, you can conduct an acknowledgment of training and policy adherence quiz or a signed statement that will allow employees to acknowledge they understand both the training and policy. The training and correlating policies should be continuously audited to ensure proper actions and disciplinary procedures are taking place. The audits will also help identify any knowledge or training gaps that need to be addressed.

In conclusion, it is easier to prepare for change if you know where you are. Ensuring you have a solid operating practice from onboarding, compliance, termination, and day-to-day operations will safeguard against legal liability, employee inconsistencies, and preferential treatment. Having a strong organizational structure in place will allow you as a safety professional to focus on what matters. If these things are not in place, they will be identified in the Gap analysis you conduct during the change management process and will give you a baseline/starting point to move toward a consistent and compliant organization.

NOTES

1. *Carmelita Vazquez v. Caesar's Paradise Stream Resort*, United States District Court for the Middle District of Pennsylvania (http://hr.cch.com/ELD/VazquezCaesars2.pdf).
2. 29 CFR Parts 1910 and 1926 Safety Standards for General Industry and Construction, Occupational Safety and Health Administration (www.osha.gov/laws-regs/federalregister/1993-06-30).

11 Why Care about Safe Change?

***

"It's not necessary to be extraordinary to get extraordinary results."

Warren Buffett

***

Why change? Why care about changing? Why care about safely changing? Change is essential for all companies and industries in order to stay relevant, successful, and profitable. Think of all the companies that have shut their doors in the last 5, 10, 20 years. Why did that happen? What could they have done to prevent it? Change is important for organizations, but it is also important to ensure the changes are done in a safe manner.

WHY CHANGE?

Nintendo was founded in 1889 as a company that made handmade playing cards. By the late 1960s, the company's stock was at an all-time low, where it remained until the 1970s, when an assembly line maintenance engineer designed an extendable robotic arm.[1] Nintendo made its first venture into the toy market with a handheld gaming device called the Ultra Hand. A few years later in 1974, Nintendo decided to enter the video game industry when Donkey Kong was born and their brand became widely known. Without these changes to Nintendo, the brand and company would have become obsolete in the 1960s.

Blockbuster opened its doors in 1985 and had early success providing films for rent to customers via individual storefronts. Blockbuster was able to customize their store video selection to best fit their local visitors. In 1997, Netflix emerged to provide a DVD mail subscription service for one flat fee per month. By 2010, Blockbuster had closed 1,061 domestic storefronts.[2] Where could they have changed to be more competitive?

To stay relevant, change is needed, the companies and industries that were booming 5, 10, 20 years ago that did not change are no longer relevant in 2018. Your site, company, or organization must change in order to stay industry leading, profitable, and a consumer want.

WHY CARE ABOUT CHANGING?

Change management is the act of transitioning, re-directing, and allocating resources to reshape your site, company, or organization. Think of anything you have ever purposely changed, like a paint color in your dining room. The process normally starts out with identifying what you want to change the color to and preparing for the change by moving furniture, covering the floor, and taping off the section of wall. Then, you create or pick out the color you want, followed by the execution of the painting of the dining room. You sustain the color by preventing nicks, tears, or a child size wall coloring scenario. You monitor the color over the years to ensure there isn't any touch-up spots needed. These same steps are applied to any change no matter the caliber, as you will see in Chapter 12.

Caring about change allows you to see a problem as an opportunity instead of trying to fix something to its original state. Identifying the root cause and changing the process will bring your site, company, or organization to the next level because no longer are you working through the original problem time after time.

Caring about change allows you to think long term instead of short term. This allows you to see the impact of the changes being made. Change takes time and by changing your mindset to look beyond the cost of a change today and see the benefits in the future, you will be able to see the investments today make a return on investment for the future.

Caring about an effective change will allow you to stray from the map. Instead of sticking to the same direction, vision, goal, and so on, open your mind to a new direction, a new road that may lead to continuous improvement opportunities and long-term cost savings.

Effective change takes buy-in from all levels. There is not one person at a business, facility, or operation that knows everything about everything. When trying to make a change, it is important to get input, buy-in, and assistance from a wide variety of individuals that will be affected. When you care about the changes being made, it is easier to get the buy-in, respect, and resources you need to effectively implement the changes needed to move the business forward.

WHY CARE ABOUT SAFE CHANGE?

Change can impact the safety of an organization due to the lack of knowledge of new procedures, processes, and equipment. A reliable and consistent change management system is essential for every company and facility. Effective change management can reduce safety incidents and process errors by employees.

Think of it like this, if you were to walk into your house at night and someone had forgotten to tell you that the light switches had been relocated, you will stumble around, stub a toe, trip over some shoes, and cuss a little bit until you finally find the light switches. This scenario could have been avoided if the individual, whether it is a contractor or your significant other, had contacted you before you went home in order to tell you that the light switches had been moved. It would be even better if they sent you a layout of the house showing where they had been moved to or even involved you in the moving decision of the light switches to get your opinion.

Part of change management is providing your employees and those who would need to know about changes being made in order to avoid confusion, stress, and injury. If a change impacts the productivity, delivery, quality, cost, safety, or the business, it should be managed through a change management process. During the change management process, the following things should be identified:

1. What is the problem that identifies the need for the change?
2. Has this problem been analyzed?
3. Is the desired end condition identified?
4. Have the appropriate people given approval for the change?
5. Have all affected individuals been identified?
6. Are all affected individuals knowledgeable of the change and why the change is happening?
7. Has proper training been developed for the change?
8. Will the training be implemented prior to the change being completed?
9. Is there a timeline of the change process and all actions taking place around the change?
10. If it is a behavior change, have positive reinforcement activities been developed and communicated to all employees affected?
11. Have follow-up actions been scheduled to ensure the change is successful?

If your facility does not have a change management system in place, begin by designating an individual that will manage the change management system. This ensures that change management stays at the forefront of all projects, system changes, and safety implementation.

In conclusion, caring about safe change will allow your site, company, or organization to make the changes without negatively impacting safety through change management. Having a standardized change management process for your facility/organization ensures standardized methods and procedures are used for efficient and effective handling of changes. This will prevent safety, quality, environmental, service, or any other type of adverse impact due to change. The change management process should include a representative from every department that a change request touches in order to identify potential safety hazards, process gaps, training needs, and so on.

NOTES

1. Kim, Lamont, Ogasawara, Park, and Takaoka (2011). https://mitsloan.mit.edu/LearningEdge/CaseDocs/11-124.Nintendo_Revolution_lamont.FINAL.pdf.
2. Davis and Higgins (2013). "A Blockbuster Failure: How an Outdated Business Model Destroyed a Giant." Chapter 11 Bankruptcy Case Studies. http://trace.tennessee.edu/utk_studlawbankruptcy/11.

12 Changing the Safety Culture

***

"New beginnings are often disguised as painful endings."

Lao Tzu

"Any change, even a change for the better, is always accompanied by drawbacks and discomforts."

Arnold Bennett

***

Changing a culture is never an easy thing to do. It takes time to create a culture and it takes time to change it. Your company culture is made up of the attitudes, values, beliefs, assumptions, goals, vision, and behavior of the organization. It is the overall feeling when you go into work every day. The culture impacts the leadership and employee commitment, customer satisfaction, innovation, and ultimately reflects your success. Changing a safety culture embodies more than just safety, and involves more than just the employees, frontline leaders, and management. It embodies everyone from the top down.

What is culture? Culture is the personality of the organization. Just like an individual's personality impacts on first impressions and the ability to succeed, motivate, and engage others. Culture is driven by all levels of the organization. The founder, CEO, or president sets the strategic direction and objectives showing where the resources will be allocated. Senior management helps create the roadmap for their reports to achieve such objectives. Frontline leaders are the level that ensures company and employee goals align to achieve a mutual trust and customer satisfaction.

You can pick up on a company culture as early as the first interview. What is important to the company and how do they go about achieving what is important to them? The more time you spend within an organization, the more you will understand the culture; just like a person, the more time you spend with them, the more you get to know them.

Picture scenario one, it is your first day at a new company and your hiring manager greets you with everything you need from a computer, office, and the outline of your onboarding plan for the next 30-60-90 days. He/she lays out objectives and goals for you during your onboarding period. You are assigned a buddy that does similar work or knows the job that you can rely on and ask questions. You get introduced to the team and all the resources you may need during your time with the company. Your manager sets up weekly one-to-one meetings to go over your progress and assist with anything you may need help with.

Then picture scenario number two. It is your first day at a new company and you meet your hiring manager who shows you to your office. Your manager shows you where the restrooms, copy machine, and break room are. Your manager then tells you to work with HR to get onboarded and they are there if you have any questions. You only talk to your manager during monthly staff meetings but never have individual time with them until mid-year and year-end reviews.

Which scenario do you feel like people are a priority for the company? This scenario can be played out in different scenarios throughout your career. An injury occurs and the employee is immediately fired for bypassing LOTO. Did anyone ask the employee why they did it? Did anyone ask the employee what they believe could have been done to prevent the injury? How an organization or site handles day-to-day activities reflect the culture. Do you say anything when you see someone not wearing safety glasses? Is there an open reporting culture? All these things help you picture the type of culture your organization may have.

Changing a safety culture involves six steps that all take time and cannot be rushed. Each step must get employee buy-in, leadership involvement, and resource allocation. If these steps are rushed, the change will fail because the proper buy-in, trust, and resources were not there. If these steps take too long, you lose employee trust that change will happen and resources may be pulled in other directions. Proper timing is determined by the size of the organization, the number of culture changes you are wishing to implement, and the current company culture.

Managing the change will take place in six steps: Identify, Prepare, Create, Execute, Sustain, and Monitor. Change management is shown in a cycle because it is a never-ending process that you will be continuously be monitoring, sustaining, executing, creating, identifying, and preparing. You may sit in one function longer than others due to the state of the business, but it will always be a continuous process.

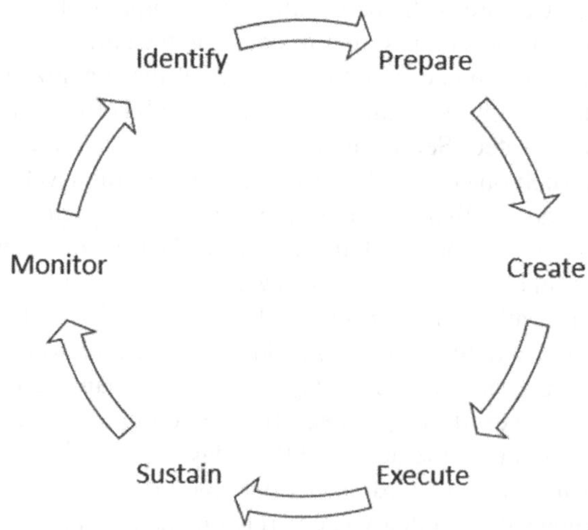

Identify: Here you will identify the current strengths, weakness, opportunities, and threats (SWOT analysis) of your current safety culture. During this stage, conducting a SWOT analysis, employee roundtables, manager roundtables, executive one-to-one's, and reviewing current and historic OSHA data (near-misses, recordable, etc.) will help identify the current safety culture. Next, conduct a Gap analysis where you take the information you have gathered and compare it against the company's current vision, goals, and objectives for safety.

During this phase, it is important to build a rapport and trust with the employees and managers you interface with. This will allow information to be free flowing allowing you to get a better understanding of what is really going on and where the gaps are located.

Prepare: Taking the information you gathered while identifying the current state of your safety culture and the gaps that will be addressed during the changing of the safety culture. During the preparation phase, you and your leadership team need to be fully involved. Leadership needs to be part of the culture shift at every phase to ensure that there is consistency, a clear vision, and unified coaching to create one message. During this phase, you prepare to be actionable on any resources you may need whether that is monetary, headcount, time off the floor, and so on.

Design: What is it about the current culture you are wishing to change? How are you going to do it? Designing the communication, training, or roll-out of the execution is vital. For example, during your Gap analysis and employee roundtables, you were made aware that supervisor A was allowing employees to conduct servicing work alone and supervisor B was mandating that service work must be completed in pairs as per the company policy. Supervisor A was new to the company and had not been properly trained or made aware of the company policy because his predecessor had "always done it this way." Supervisor B's employees where frustrated because they had to work in pairs rather than solo, making the work quicker and creating tension between employees and in their relationship with their manager.

During the design phase, you are designing your program, communication, training, and so on, that will correct the gaps in which you identified during the Identify Phase. For this example, supervisor training would be essential to ensure consistency between all your supervisors from discipline, day-to-day work, and employee relations. This training will capture your current supervisor population, but the gap began with a new supervisor entering the field with proper training. Identifying the root cause of the issue will help capture and eliminate the gap in the future. For this example, creating an onboarding training for all new supervisors will help ensure that supervisors are made aware of policies and procedures going forward.

Execute: Now that you have identified and designed your plan, it is time to execute it. Execution must be done timely and properly. The execution must be taken seriously and build trust with employees. During the execution phase, communication is important to build a reputation of transparency. For this example, communication is vital for all levels. Employees may need to be aware of the policy and procedures to ensure everyone is acting in a safe manner, those employees of Supervisor B need to be aware that their team has been following proper safety protocol for having to work in pairs, although it adds time, it is there for their safety. Supervisors will now be aware of all safety policies and procedures moving forward and executive or top line leaders will be supporting these training efforts and communicating the importance of working in pairs.

Sustain: You have executed the actions that help shift your culture to better align with your safety vision. Now you must sustain those actions. For the example, this would be continuing to have onboarding for all new supervisors prior to entering the field, promising that their knowledge of policies, procedures, and practices are aligned with all other supervisors. This will ensure consistent practices across the field. To ensure the feasibility of sustaining the action, how will it be assured that this onboarding will happen to every new supervisor? Will HR add it to the onboarding checklist? Will the hiring manager ensure it is completed? This is something as an organization you must decide to ensure that every new supervisor is consistently getting the proper training they need.

Monitor: Monitoring ensures that the organization is consensually looking at the actions taken, how they are impacting the culture, and are the actions moving the safety culture in the right direction to align with the organization's vision and objectives? Monitoring also comes in various forms. Quarterly employee roundtables, one-to-one meetings, and audits can help monitor the behavior, day-to-day activities, OSHA statistics (near-misses, recordable, etc.), and employee morale. This is once again where it is important to have transparency and trust with employees and leadership to gather as much information as possible to see where there are potential challenges with implementing a change. While monitoring, it is important to document as much as possible to use during the identify, prepare, and create phases.

Identify: The human factors of change ensure that tweaking will be needed to ensure the changes resonate with all populations, locations, and employees. While monitoring, you may realize that the changes are being reflected in half of your sites but in the other half, there has been little to no change. Use the documentation you took during the monitor phase to help identify the gaps and the potential cause of the gap. Identify what needs to be changed to capture the employees who did not resonate with the first roll-out of change initiatives.

From here, you begin back at the Prepare phase. Change is every evolving; this system will never end and continuous improvement can only happen if the process stays continuous. During this change management process, it is important to build trust with employees at all levels and create an atmosphere of transparency showing the end goal, objective, or vision, this will allow unity from all levels to achieve the changes needed to reach that goal.

WHILE SHIFTING YOUR SAFETY CULTURE THERE ARE SOME KEY PHRASES TO BE AWARE OF

"We have always done it this way"—Change does not happen without the buy-in from employees. If you have an employee population that insists change does not need to happen because they have a "this is how they have always done it" mentality, communication and transparency is even more vital. You must show this employee population that the change that is being made is for the better and will improve their day-to-day lives. Bring into perspective the bigger picture of a strong safety culture, sending people home the same way they arrived.

"This is how we do it around here"—Ask the questions and learn the process they have always done. Ask how you can better the system or make their lives easier.

"Doesn't matter what we say, nothing will change"—This is especially worrying during a time of cultural change. Employees who say this have no trust in management. In these circumstances, it is important to listen to the employee and understand why they feel this way. Usually, it is because they have brought up a safety concern or hazard and nothing was done about it, reflecting poorly upon management. In these circumstances, it is important to "do as you say" to build trust back into management and the importance of employee safety.

"It's not my problem"—Safety is everyone's problem, and that is the mindset shift that will need to happen for these types of employees. Why do they feel like it is not their problem? What is the root cause of this thinking? Did something happen to cause this mindset? These employees feel as if the safety of someone else is not their problem or a safety concern is not their problem when in fact it is everyone's problem. The mindset shift that needs to happen is instead of thinking about themselves, they need to think of the organization as a team that everyone is responsible for.

"If it's not broken, don't fix it"—It may not be broken, but we can better the system. Understand what is working well and what is not working well for them. If they claim it is a perfect system, ask them to walk them through their day, what are the largest time consumers? What do they not enjoy about their day-to-day work that you could potentially assist with? It is important to show these employees that change can make their lives easier, better, or safer.

Finding the root cause for a lot of the employee frustrations will allow the organization to address, correct, and proceed. Building trust with the employees during this shift is vital to the organization, the culture, and employee involvement. While developing your safety culture, it is important for leaders to lead the change, reflect the change in day-to-day activities, and find what motivates employees to do the same.

In conclusion, changing a safety culture is a time-consuming, continuous process but, when done correctly, can re-align your organization through identifying the direction, preparing the message, creating a plan, executing the actions, sustaining the change, monitoring the change, and identifying continuous improvements in the process. It is important to get buy-in from all levels and be transparent while gaining the trust of employees. A good safety culture is something that embodies the employees, their actions, and their attitude.

Appendix 1: Americans with Disabilities Act

AN ACT

To establish a clear and comprehensive prohibition of discrimination on the basis of disability.

Be it enacted by the Senate and House of Representatives of the United States of America in Congress assembled,

SECTION 1. SHORT TITLE; TABLE OF CONTENTS.

(a) SHORT TITLE—This Act may be cited as the "Americans with Disabilities Act of 1990".

(b) TABLE OF CONTENTS—The table of contents is as follows:

Sec. 1. Short title; table of contents.

Sec. 2. Findings and purposes.

Sec. 3. Definitions.

TITLE I—EMPLOYMENT

Sec. 101. Definitions.

Sec. 102. Discrimination.

Sec. 103. Defenses.

Sec. 104. Illegal use of drugs and alcohol.

Sec. 105. Posting notices.

Sec. 106. Regulations.

Sec. 107. Enforcement.

Sec. 108. Effective date.

TITLE II—PUBLIC SERVICES

Subtitle A—Prohibition Against Discrimination and Other Generally Applicable Provisions

Subtitle B—Actions Applicable to Public Transportation Provided by Public Entities Considered Discriminatory

Part I—Public Transportation Other Than by Aircraft or Certain Rail Operations

Part II—Public Transportation by Intercity and Commuter Rail

Sec. 243. Conformance of accessibility standards.

Sec. 244. Regulations.

Sec. 245. Interim accessibility requirements.

Sec. 246. Effective date.

TITLE III—PUBLIC ACCOMMODATIONS AND SERVICES OPERATED BY PRIVATE ENTITIES

Sec. 301. Definitions.

Sec. 302. Prohibition of discrimination by public accommodations.

Sec. 303. New construction and alterations in public accommodations and commercial facilities.

Sec. 304. Prohibition of discrimination in specified public transportation services provided by private entities.

Sec. 305. Study.

Sec. 306. Regulations.

Sec. 307. Exemptions for private clubs and religious organizations.

Sec. 308. Enforcement.

Sec. 309. Examinations and courses.

Sec. 310. Effective date.

TITLE IV—TELECOMMUNICATIONS

Sec. 401. Telecommunications relay services for hearing-impaired and speech-impaired individuals.

Sec. 402. Closed-captioning of public service announcements.

TITLE V—MISCELLANEOUS PROVISIONS

Sec. 501. Construction.

Sec. 502. State immunity.

Sec. 503. Prohibition against retaliation and coercion.

Sec. 504. Regulations by the Architectural and Transportation Barriers Compliance Board.

Sec. 505. Attorney's fees.

Sec. 506. Technical assistance.

Sec. 507. Federal wilderness areas.

Sec. 508. Transvestites.

Sec. 509. Coverage of Congress and the agencies of the legislative branch.

Sec. 510. Illegal use of drugs.

Sec. 511. Definitions.

Sec. 512. Amendments to the Rehabilitation Act.

Sec. 513. Alternative means of dispute resolution.

Sec. 514. Severability.

SEC. 2. FINDINGS AND PURPOSES

(a) FINDINGS- The Congress finds that—

(1) some 43,000,000 Americans have one or more physical or mental disabilities, and this number is increasing as the population as a whole is growing older;

(2) historically, society has tended to isolate and segregate individuals with disabilities, and, despite some improvements, such forms of discrimination against individuals with disabilities continue to be a serious and pervasive social problem;

(3) discrimination against individuals with disabilities persists in such critical areas as employment, housing, public accommodations, education, transportation, communication, recreation, institutionalization, health services, voting, and access to public services;

(4) unlike individuals who have experienced discrimination on the basis of race, color, sex, national origin, religion, or age, individuals who have experienced discrimination on the basis of disability have often had no legal recourse to redress such discrimination;

(5) individuals with disabilities continually encounter various forms of discrimination, including outright intentional exclusion, the discriminatory effects of architectural, transportation, and communication barriers, overprotective rules and policies, failure to make modifications to existing facilities and practices, exclusionary qualification standards and criteria, segregation, and relegation to lesser services, programs, activities, benefits, jobs, or other opportunities;

(6) census data, national polls, and other studies have documented that people with disabilities, as a group, occupy an inferior status in our society, and are severely disadvantaged socially, vocationally, economically, and educationally;

(7) individuals with disabilities are a discrete and insular minority who have been faced with restrictions and limitations, subjected to a history of purposeful unequal treatment, and relegated to a position of political powerlessness in our society, based on characteristics that are beyond the control of such individuals and resulting from stereotypic assumptions not truly indicative of the individual ability of such individuals to participate in, and contribute to, society;

(8) the Nation's proper goals regarding individuals with disabilities are to assure equality of opportunity, full participation, independent living, and economic self-sufficiency for such individuals; and

(9) the continuing existence of unfair and unnecessary discrimination and prejudice denies people with disabilities the opportunity to compete on an equal basis and to pursue those opportunities for which our free society is justifiably famous, and costs the United States billions of dollars in unnecessary expenses resulting from dependency and nonproductivity.

(b) PURPOSE- It is the purpose of this Act—

(1) to provide a clear and comprehensive national mandate for the elimination of discrimination against individuals with disabilities;

(2) to provide clear, strong, consistent, enforceable standards addressing discrimination against individuals with disabilities;

(3) to ensure that the Federal Government plays a central role in enforcing the standards established in this Act on behalf of individuals with disabilities; and

(4) to invoke the sweep of congressional authority, including the power to enforce the fourteenth amendment and to regulate commerce, in order to address the major areas of discrimination faced day-to-day by people with disabilities.

SEC. 3. DEFINITIONS

As used in this Act:

(1) AUXILIARY AIDS AND SERVICES- The term "auxiliary aids and services" includes—

(A) qualified interpreters or other effective methods of making aurally delivered materials available to individuals with hearing impairments;

(B) qualified readers, taped texts, or other effective methods of making visually delivered materials available to individuals with visual impairments;

(C) acquisition or modification of equipment or devices; and

(D) other similar services and actions.

(2) DISABILITY- The term "disability" means, with respect to an individual—

(A) a physical or mental impairment that substantially limits one or more of the major life activities of such individual;

(B) a record of such an impairment; or

(C) being regarded as having such an impairment.

(3) STATE- The term "State" means each of the several States, the District of Columbia, the Commonwealth of Puerto Rico, Guam, American Samoa, the Virgin Islands, the Trust Territory of the Pacific Islands, and the Commonwealth of the Northern Mariana Islands.

TITLE I—EMPLOYMENT

SEC. 101. DEFINITIONS

As used in this title:

(1) COMMISSION- The term "Commission" means the Equal Employment Opportunity Commission established by section 705 of the Civil Rights Act of 1964 (42 U.S.C. 2000e-4).

(2) COVERED ENTITY- The term "covered entity" means an employer, employment agency, labor organization, or joint labor-management committee.

(3) DIRECT THREAT- The term "direct threat" means a significant risk to the health or safety of others that cannot be eliminated by reasonable accommodation.

(4) EMPLOYEE- The term "employee" means an individual employed by an employer.

(5) EMPLOYER-

(A) IN GENERAL- The term "employer" means a person engaged in an industry affecting commerce who has 15 or more employees for each working day in each of 20 or more calendar weeks in the current or preceding calendar year, and any agent of such person, except that, for two years following the effective date of this title, an employer means a person engaged in an industry affecting commerce who has 25 or more employees for each working day in each of 20 or more calendar weeks in the current or preceding year, and any agent of such person.

(B) EXCEPTIONS- The term "employer" does not include—

(i) the United States, a corporation wholly owned by the government of the United States, or an Indian tribe; or

(ii) a bona fide private membership club (other than a labor organization) that is exempt from taxation under section 501(c) of the Internal Revenue Code of 1986.

(6) ILLEGAL USE OF DRUGS-

(A) IN GENERAL- The term "illegal use of drugs" means the use of drugs, the possession or distribution of which is unlawful under the Controlled Substances Act (21 U.S.C. 812). Such term does not include the use of a drug taken under supervision by a licensed health care professional, or other uses authorized by the Controlled Substances Act or other provisions of Federal law.

(B) DRUGS- The term "drug" means a controlled substance, as defined in schedules I through V of section 202 of the Controlled Substances Act.

(7) PERSON, ETC- The terms "person", "labor organization", "employment agency", "commerce", and "industry affecting commerce", shall have the same meaning given such terms in section 701 of the Civil Rights Act of 1964 (42 U.S.C. 2000e).

(8) QUALIFIED INDIVIDUAL WITH A DISABILITY- The term "qualified individual with a disability" means an individual with a disability who, with or without reasonable accommodation, can perform the essential functions of the employment position that such individual holds or desires. For the purposes of this title, consideration shall be given to the employer's judgment as to what functions of a job are essential, and if an employer has prepared a written description before advertising or interviewing applicants for the job, this description shall be considered evidence of the essential functions of the job.

(9) REASONABLE ACCOMMODATION- The term "reasonable accommodation" may include—

(A) making existing facilities used by employees readily accessible to and usable by individuals with disabilities; and

(B) job restructuring, part-time or modified work schedules, reassignment to a vacant position, acquisition or modification of equipment or devices, appropriate adjustment or modifications of examinations, training materials or policies, the provision of qualified readers or interpreters, and other similar accommodations for individuals with disabilities.

(10) UNDUE HARDSHIP-

(A) IN GENERAL- The term "undue hardship" means an action requiring significant difficulty or expense, when considered in light of the factors set forth in subparagraph (B).

(B) FACTORS TO BE CONSIDERED- In determining whether an accommodation would impose an undue hardship on a covered entity, factors to be considered include—

(i) the nature and cost of the accommodation needed under this Act;

(ii) the overall financial resources of the facility or facilities involved in the provision of the reasonable accommodation; the number of persons employed at such facility; the effect on expenses and resources, or the impact otherwise of such accommodation upon the operation of the facility;

(iii) the overall financial resources of the covered entity; the overall size of the business of a covered entity with respect to the number of its employees; the number, type, and location of its facilities; and

(iv) the type of operation or operations of the covered entity, including the composition, structure, and functions of the workforce of such entity; the geographic separateness, administrative, or fiscal relationship of the facility or facilities in question to the covered entity.

SEC. 102. DISCRIMINATION

(a) GENERAL RULE- No covered entity shall discriminate against a qualified individual with a disability because of the disability of such individual in regard to job application procedures, the hiring, advancement, or discharge of employees, employee compensation, job training, and other terms, conditions, and privileges of employment.

(b) CONSTRUCTION- As used in subsection (a), the term "discriminate" includes—

(1) limiting, segregating, or classifying a job applicant or employee in a way that adversely affects the opportunities or status of such applicant or employee because of the disability of such applicant or employee;

(2) participating in a contractual or other arrangement or relationship that has the effect of subjecting a covered entity's qualified applicant or employee with a disability to the discrimination prohibited by this title (such relationship includes a relationship with an employment or referral agency, labor union, an organization providing fringe benefits to an employee of the covered entity, or an organization providing training and apprenticeship programs);

(3) utilizing standards, criteria, or methods of administration—

(A) that have the effect of discrimination on the basis of disability; or

(B) that perpetuate the discrimination of others who are subject to common administrative control;

(4) excluding or otherwise denying equal jobs or benefits to a qualified individual because of the known disability of an individual with whom the qualified individual is known to have a relationship or association;

(5)(A) not making reasonable accommodations to the known physical or mental limitations of an otherwise qualified individual with a disability who is an applicant or employee, unless such covered entity can demonstrate that the accommodation would impose an undue hardship on the operation of the business of such covered entity; or

(B) denying employment opportunities to a job applicant or employee who is an otherwise qualified individual with a disability, if such denial is based on the need of such covered entity to make reasonable accommodation to the physical or mental impairments of the employee or applicant;

(6) using qualification standards, employment tests or other selection criteria that screen out or tend to screen out an individual with a disability or a class of individuals with disabilities unless the standard, test or other selection criteria, as used by the covered entity, is shown to be job-related for the position in question and is consistent with business necessity; and

(7) failing to select and administer tests concerning employment in the most effective manner to ensure that, when such test is administered to a job applicant or employee who has a disability that impairs sensory, manual, or speaking skills, such test results accurately reflect the skills, aptitude, or whatever other factor of such applicant or employee that such test purports to measure, rather than reflecting the impaired sensory, manual, or speaking skills of such employee or applicant (except where such skills are the factors that the test purports to measure).

(c) MEDICAL EXAMINATIONS AND INQUIRIES-

(1) IN GENERAL- The prohibition against discrimination as referred to in subsection (a) shall include medical examinations and inquiries.

(2) PRE-EMPLOYMENT-

(A) PROHIBITED EXAMINATION OR INQUIRY- Except as provided in paragraph (3), a covered entity shall not conduct a medical examination or make inquiries of a job applicant as to whether such applicant is an individual with a disability or as to the nature or severity of such disability.

(B) ACCEPTABLE INQUIRY- A covered entity may make pre-employment inquiries into the ability of an applicant to perform job-related functions.

(3) EMPLOYMENT ENTRANCE EXAMINATION- A covered entity may require a medical examination after an offer of employment has been made to a job applicant and prior to the commencement of the employment duties of such applicant, and may condition an offer of employment on the results of such examination, if—

(A) all entering employees are subjected to such an examination regardless of disability;

(B) information obtained regarding the medical condition or history of the applicant is collected and maintained on separate forms and in separate medical files and is treated as a confidential medical record, except that—

(i) supervisors and managers may be informed regarding necessary restrictions on the work or duties of the employee and necessary accommodations;

(ii) first aid and safety personnel may be informed, when appropriate, if the disability might require emergency treatment; and

(iii) government officials investigating compliance with this Act shall be provided relevant information on request; and

(C) the results of such examination are used only in accordance with this title.

(4) EXAMINATION AND INQUIRY-

(A) PROHIBITED EXAMINATIONS AND INQUIRIES- A covered entity shall not require a medical examination and shall not make inquiries of an employee as to whether such employee is an individual with a disability or as to the nature or severity of the disability, unless such examination or inquiry is shown to be job related and consistent with business necessity.

(B) ACCEPTABLE EXAMINATIONS AND INQUIRIES- A covered entity may conduct voluntary medical examinations, including voluntary medical histories, which are part of an employee health program available to employees at that work site. A covered entity may make inquiries into the ability of an employee to perform job-related functions.

(C) REQUIREMENT- Information obtained under subparagraph (B) regarding the medical condition or history of any employee are subject to the requirements of subparagraphs (B) and (C) of paragraph (3).

SEC. 103. DEFENSES

(a) IN GENERAL- It may be a defense to a charge of discrimination under this Act that an alleged application of qualification standards, tests, or selection criteria that screen out or tend to screen out or otherwise deny a job or benefit to an individual with a disability has been shown to be job related and consistent with business necessity, and such performance cannot be accomplished by reasonable accommodation, as required under this title.

(b) QUALIFICATION STANDARDS- The term “qualification standards” may include a requirement that an individual shall not pose a direct threat to the health or safety of other individuals in the workplace.

(c) RELIGIOUS ENTITIES-

(1) IN GENERAL- This title shall not prohibit a religious corporation, association, educational institution, or society from giving preference in employment to

individuals of a particular religion to perform work connected with the carrying on by such corporation, association, educational institution, or society of its activities.

(2) RELIGIOUS TENETS REQUIREMENT- Under this title, a religious organization may require that all applicants and employees conform to the religious tenets of such organization.

(d) List of Infectious and Communicable Diseases-

(1) IN GENERAL- The Secretary of Health and Human Services, not later than 6 months after the date of enactment of this Act, shall—

(A) review all infectious and communicable diseases which may be transmitted through handling the food supply;

(B) publish a list of infectious and communicable diseases which are transmitted through handling the food supply;

(C) publish the methods by which such diseases are transmitted; and

(D) widely disseminate such information regarding the list of diseases and their modes of transmissibility to the general public.

Such list shall be updated annually.

(2) APPLICATIONS- In any case in which an individual has an infectious or communicable disease that is transmitted to others through the handling of food, that is included on the list developed by the Secretary of Health and Human Services under paragraph (1), and which cannot be eliminated by reasonable accommodation, a covered entity may refuse to assign or continue to assign such individual to a job involving food handling.

(3) CONSTRUCTION- Nothing in this Act shall be construed to preempt, modify, or amend any State, county, or local law, ordinance, or regulation applicable to food handling which is designed to protect the public health from individuals who pose a significant risk to the health or safety of others, which cannot be eliminated by reasonable accommodation, pursuant to the list of infectious or communicable diseases and the modes of transmissibility published by the Secretary of Health and Human Services.

SEC. 104. ILLEGAL USE OF DRUGS AND ALCOHOL

(a) QUALIFIED INDIVIDUAL WITH A DISABILITY- For purposes of this title, the term "qualified individual with a disability" shall not include any employee or applicant who is currently engaging in the illegal use of drugs, when the covered entity acts on the basis of such use.

(b) RULES OF CONSTRUCTION- Nothing in subsection (a) shall be construed to exclude as a qualified individual with a disability an individual who—

(1) has successfully completed a supervised drug rehabilitation program and is no longer engaging in the illegal use of drugs, or has otherwise been rehabilitated successfully and is no longer engaging in such use;

(2) is participating in a supervised rehabilitation program and is no longer engaging in such use; or

(3) is erroneously regarded as engaging in such use, but is not engaging in such use;

except that it shall not be a violation of this Act for a covered entity to adopt or administer reasonable policies or procedures, including but not limited to drug testing, designed to ensure that an individual described in paragraph (1) or (2) is no longer engaging in the illegal use of drugs.

(c) AUTHORITY OF COVERED ENTITY- A covered entity—

(1) may prohibit the illegal use of drugs and the use of alcohol at the workplace by all employees;

(2) may require that employees shall not be under the influence of alcohol or be engaging in the illegal use of drugs at the workplace;

(3) may require that employees behave in conformance with the requirements established under the Drug-Free Workplace Act of 1988 (41 U.S.C. 701 et seq.);

(4) may hold an employee who engages in the illegal use of drugs or who is an alcoholic to the same qualification standards for employment or job performance and behavior that such entity holds other employees, even if any unsatisfactory performance or behavior is related to the drug use or alcoholism of such employee; and

(5) may, with respect to Federal regulations regarding alcohol and the illegal use of drugs, require that—

(A) employees comply with the standards established in such regulations of the Department of Defense, if the employees of the covered entity are employed in an industry subject to such regulations, including complying with regulations (if any) that apply to employment in sensitive positions in such an industry, in the case of employees of the covered entity who are employed in such positions (as defined in the regulations of the Department of Defense);

(B) employees comply with the standards established in such regulations of the Nuclear Regulatory Commission, if the employees of the covered entity are employed in an industry subject to such regulations, including complying with regulations (if any) that apply to employment in sensitive positions in such an industry, in the case of employees of the covered entity who are employed in such positions (as defined in the regulations of the Nuclear Regulatory Commission); and

(C) employees comply with the standards established in such regulations of the Department of Transportation, if the employees of the covered entity are employed in a transportation industry subject to such regulations, including complying with such regulations (if any) that apply to employment in sensitive positions in such an industry, in the case of employees of the covered entity who are employed in such positions (as defined in the regulations of the Department of Transportation).

(d) DRUG TESTING-

(1) IN GENERAL- For purposes of this title, a test to determine the illegal use of drugs shall not be considered a medical examination.

(2) CONSTRUCTION- Nothing in this title shall be construed to encourage, prohibit, or authorize the conducting of drug testing for the illegal use of drugs by job applicants or employees or making employment decisions based on such test results.

(e) TRANSPORTATION EMPLOYEES- Nothing in this title shall be construed to encourage, prohibit, restrict, or authorize the otherwise lawful exercise by entities subject to the jurisdiction of the Department of Transportation of authority to—

(1) test employees of such entities in, and applicants for, positions involving safety-sensitive duties for the illegal use of drugs and for on-duty impairment by alcohol; and

(2) remove such persons who test positive for illegal use of drugs and on-duty impairment by alcohol pursuant to paragraph (1) from safety-sensitive duties in implementing subsection (c).

SEC. 105. POSTING NOTICES

Every employer, employment agency, labor organization, or joint labor-management committee covered under this title shall post notices in an accessible format to applicants, employees, and members describing the applicable provisions of this Act, in the manner prescribed by section 711 of the Civil Rights Act of 1964 (42 U.S.C. 2000e-10).

SEC. 106. REGULATIONS

Not later than 1 year after the date of enactment of this Act, the Commission shall issue regulations in an accessible format to carry out this title in accordance with subchapter II of chapter 5 of title 5, United States Code.

SEC. 107. ENFORCEMENT

(a) POWERS, REMEDIES, AND PROCEDURES- The powers, remedies, and procedures set forth in sections 705, 706, 707, 709, and 710 of the Civil Rights Act of 1964 (42 U.S.C. 2000e-4, 2000e-5, 2000e-6, 2000e-8, and 2000e-9) shall be the powers, remedies, and procedures this title provides to the Commission, to the Attorney General, or to any person alleging discrimination on the basis of disability in violation of any provision of this Act, or regulations promulgated under section 106, concerning employment.

(b) COORDINATION- The agencies with enforcement authority for actions which allege employment discrimination under this title and under the Rehabilitation Act of 1973 shall develop procedures to ensure that administrative complaints filed under this title and under the Rehabilitation Act of 1973 are dealt with in a manner that avoids duplication of effort and prevents imposition of inconsistent or conflicting standards for the same requirements under this title and the Rehabilitation Act of 1973. The Commission, the Attorney General, and the Office of Federal Contract Compliance Programs shall establish such coordinating mechanisms (similar to provisions contained in the joint regulations promulgated by the Commission and the Attorney General at part 42 of title 28 and part 1691 of title 29, Code of Federal Regulations, and the Memorandum of Understanding between the Commission and the Office of Federal Contract Compliance Programs dated January 16, 1981 (46 Fed. Reg. 7435, January 23, 1981)) in regulations implementing this title and Rehabilitation Act of 1973 not later than 18 months after the date of enactment of this Act.

SEC. 108. EFFECTIVE DATE

This title shall become effective 24 months after the date of enactment.

TITLE II—PUBLIC SERVICES

Subtitle A—Prohibition Against Discrimination and Other Generally Applicable Provisions

SEC. 201. DEFINITION

As used in this title:

(1) PUBLIC ENTITY- The term “public entity” means—

(A) any State or local government;

(B) any department, agency, special purpose district, or other instrumentality of a State or States or local government; and

(C) the National Railroad Passenger Corporation, and any commuter authority (as defined in section 103(8) of the Rail Passenger Service Act).

(2) QUALIFIED INDIVIDUAL WITH A DISABILITY- The term “qualified individual with a disability” means an individual with a disability who, with or without reasonable modifications to rules, policies, or practices, the removal of architectural,

communication, or transportation barriers, or the provision of auxiliary aids and services, meets the essential eligibility requirements for the receipt of services or the participation in programs or activities provided by a public entity.

SEC. 202. DISCRIMINATION

Subject to the provisions of this title, no qualified individual with a disability shall, by reason of such disability, be excluded from participation in or be denied the benefits of the services, programs, or activities of a public entity, or be subjected to discrimination by any such entity.

SEC. 203. ENFORCEMENT

The remedies, procedures, and rights set forth in section 505 of the Rehabilitation Act of 1973 (29 U.S.C. 794a) shall be the remedies, procedures, and rights this title provides to any person alleging discrimination on the basis of disability in violation of section 202.

SEC. 204. REGULATIONS

(a) IN GENERAL- Not later than 1 year after the date of enactment of this Act, the Attorney General shall promulgate regulations in an accessible format that implement this subtitle. Such regulations shall not include any matter within the scope of the authority of the Secretary of Transportation under section 223, 229, or 244.

(b) RELATIONSHIP TO OTHER REGULATIONS- Except for "program accessibility, existing facilities", and "communications", regulations under subsection (a) shall be consistent with this Act and with the coordination regulations under part 41 of title 28, Code of Federal Regulations (as promulgated by the Department of Health, Education, and Welfare on January 13, 1978), applicable to recipients of Federal financial assistance under section 504 of the Rehabilitation Act of 1973 (29 U.S.C. 794). With respect to "program accessibility, existing facilities", and "communications", such regulations shall be consistent with regulations and analysis as in part 39 of title 28 of the Code of Federal Regulations, applicable to federally conducted activities under such section 504.

(c) STANDARDS- Regulations under subsection (a) shall include standards applicable to facilities and vehicles covered by this subtitle, other than facilities, stations, rail passenger cars, and vehicles covered by subtitle B. Such standards shall be consistent with the minimum guidelines and requirements issued by the Architectural and Transportation Barriers Compliance Board in accordance with section 504(a) of this Act.

SEC. 205. EFFECTIVE DATE

(a) GENERAL RULE- Except as provided in subsection (b), this subtitle shall become effective 18 months after the date of enactment of this Act.

(b) EXCEPTION- Section 204 shall become effective on the date of enactment of this Act.

SUBTITLE B—ACTIONS APPLICABLE TO PUBLIC TRANSPORTATION PROVIDED BY PUBLIC ENTITIES CONSIDERED DISCRIMINATORY

PART I—PUBLIC TRANSPORTATION OTHER THAN BY AIRCRAFT OR CERTAIN RAIL OPERATIONS

SEC. 221. DEFINITIONS

As used in this part:

(1) DEMAND RESPONSIVE SYSTEM- The term "demand responsive system" means any system of providing designated public transportation which is not a fixed route system.

(2) DESIGNATED PUBLIC TRANSPORTATION- The term "designated public transportation" means transportation (other than public school transportation) by bus, rail, or any other conveyance (other than transportation by aircraft or intercity or commuter rail transportation (as defined in section 241)) that provides the general public with general or special service (including charter service) on a regular and continuing basis.

(3) FIXED ROUTE SYSTEM- The term "fixed route system" means a system of providing designated public transportation on which a vehicle is operated along a prescribed route according to a fixed schedule.

(4) OPERATES- The term "operates", as used with respect to a fixed route system or demand responsive system, includes operation of such system by a person under a contractual or other arrangement or relationship with a public entity.

(5) PUBLIC SCHOOL TRANSPORTATION- The term "public school transportation" means transportation by schoolbus vehicles of schoolchildren, personnel, and equipment to and from a public elementary or secondary school and school-related activities.

(6) SECRETARY- The term "Secretary" means the Secretary of Transportation.

SEC. 222. PUBLIC ENTITIES OPERATING FIXED ROUTE SYSTEMS

(a) PURCHASE AND LEASE OF NEW VEHICLES- It shall be considered discrimination for purposes of section 202 of this Act and section 504 of the Rehabilitation Act of 1973 (29 U.S.C. 794) for a public entity which operates a fixed route system to purchase or lease a new bus, a new rapid rail vehicle, a new light rail vehicle, or any other new vehicle to be used on such system, if the solicitation for such purchase or lease is made after the 30th day following the effective date of this subsection and if such bus, rail vehicle, or other vehicle is not readily accessible to and usable by individuals with disabilities, including individuals who use wheelchairs.

(b) PURCHASE AND LEASE OF USED VEHICLES- Subject to subsection (c)(1), it shall be considered discrimination for purposes of section 202 of this Act and section 504 of the Rehabilitation Act of 1973 (29 U.S.C. 794) for a public entity which operates a fixed route system to purchase or lease, after the 30th day following the effective date of this subsection, a used vehicle for use on such system unless such entity makes demonstrated good faith efforts to purchase or lease a used vehicle for use on such system that is readily accessible to and usable by individuals with disabilities, including individuals who use wheelchairs.

(c) REMANUFACTURED VEHICLES-

(1) GENERAL RULE- Except as provided in paragraph (2), it shall be considered discrimination for purposes of section 202 of this Act and section 504 of the Rehabilitation Act of 1973 (29 U.S.C. 794) for a public entity which operates a fixed route system—

(A) to remanufacture a vehicle for use on such system so as to extend its usable life for 5 years or more, which remanufacture begins (or for which the solicitation is made) after the 30th day following the effective date of this subsection; or

(B) to purchase or lease for use on such system a remanufactured vehicle which has been remanufactured so as to extend its usable life for 5 years or more, which purchase or lease occurs after such 30th day and during the period in which the usable life is extended; unless, after remanufacture, the vehicle is, to the maximum extent feasible, readily accessible to and usable by individuals with disabilities, including individuals who use wheelchairs.

(2) EXCEPTION FOR HISTORIC VEHICLES-

(A) GENERAL RULE- If a public entity operates a fixed route system any segment of which is included on the National Register of Historic Places and if making a vehicle of historic character to be used solely on such segment readily accessible to and usable by individuals with disabilities would significantly alter the historic character of such vehicle, the public entity only has to make (or to purchase or lease a remanufactured vehicle with) those modifications which are necessary to meet the requirements of paragraph (1) and which do not significantly alter the historic character of such vehicle.

(B) VEHICLES OF HISTORIC CHARACTER DEFINED BY REGULATIONS- For purposes of this paragraph and section 228(b), a vehicle of historic character shall be defined by the regulations issued by the Secretary to carry out this subsection.

SEC. 223. PARATRANSIT AS A COMPLEMENT TO FIXED ROUTE SERVICE

(a) GENERAL RULE- It shall be considered discrimination for purposes of section 202 of this Act and section 504 of the Rehabilitation Act of 1973 (29 U.S.C. 794) for a public entity which operates a fixed route system (other than a system which provides solely commuter bus service) to fail to provide with respect to the operations of its fixed route system, in accordance with this section, paratransit and other special transportation services to individuals with disabilities, including individuals who use wheelchairs, that are sufficient to provide to such individuals a level of service (1) which is comparable to the level of designated public transportation services provided to individuals without disabilities using such system; or (2) in the case of response time, which is comparable, to the extent practicable, to the level of designated public transportation services provided to individuals without disabilities using such system.

(b) ISSUANCE OF REGULATIONS- Not later than 1 year after the effective date of this subsection, the Secretary shall issue final regulations to carry out this section.

(c) REQUIRED CONTENTS OF REGULATIONS-

(1) ELIGIBLE RECIPIENTS OF SERVICE- The regulations issued under this section shall require each public entity which operates a fixed route system to provide the paratransit and other special transportation services required under this section—

(A)(i) to any individual with a disability who is unable, as a result of a physical or mental impairment (including a vision impairment) and without the assistance of another individual (except an operator of a wheelchair lift or other boarding assistance device), to board, ride, or disembark from any vehicle on the system which is readily accessible to and usable by individuals with disabilities;

(ii) to any individual with a disability who needs the assistance of a wheelchair lift or other boarding assistance device (and is able with such assistance) to board, ride, and disembark from any vehicle which is readily accessible to and usable by individuals with disabilities if the individual wants to travel on a route on the system during the hours of operation of the system at a time (or within a reasonable period of such time) when such a vehicle is not being used to provide designated public transportation on the route; and

(iii) to any individual with a disability who has a specific impairment-related condition which prevents such individual from traveling to a boarding location or from a disembarking location on such system;

(B) to one other individual accompanying the individual with the disability; and

(C) to other individuals, in addition to the one individual described in subparagraph (B), accompanying the individual with a disability provided that space for these additional individuals is available on the paratransit vehicle carrying the individual with a disability and that the transportation of such additional individuals will not result in a denial of service to individuals with disabilities.

For purposes of clauses (i) and (ii) of subparagraph (A), boarding or disembarking from a vehicle does not include travel to the boarding location or from the disembarking location.

(2) SERVICE AREA- The regulations issued under this section shall require the provision of paratransit and special transportation services required under this section in the service area of each public entity which operates a fixed route system, other than any portion of the service area in which the public entity solely provides commuter bus service.

(3) SERVICE CRITERIA- Subject to paragraphs (1) and (2), the regulations issued under this section shall establish minimum service criteria for determining the level of services to be required under this section.

(4) UNDUE FINANCIAL BURDEN LIMITATION- The regulations issued under this section shall provide that, if the public entity is able to demonstrate to the satisfaction of the Secretary that the provision of paratransit and other special transportation services otherwise required under this section would impose an undue financial burden on the public entity, the public entity, notwithstanding any other provision of this section (other than paragraph (5)), shall only be required to provide such services to the extent that providing such services would not impose such a burden.

(5) ADDITIONAL SERVICES- The regulations issued under this section shall establish circumstances under which the Secretary may require a public entity to provide, notwithstanding paragraph (4), paratransit and other special transportation services under this section beyond the level of paratransit and other special transportation services which would otherwise be required under paragraph (4).

(6) PUBLIC PARTICIPATION- The regulations issued under this section shall require that each public entity which operates a fixed route system hold a public hearing, provide an opportunity for public comment, and consult with individuals with disabilities in preparing its plan under paragraph (7).

(7) PLANS- The regulations issued under this section shall require that each public entity which operates a fixed route system—

(A) within 18 months after the effective date of this subsection, submit to the Secretary, and commence implementation of, a plan for providing paratransit and other special transportation services which meets the requirements of this section; and

(B) on an annual basis thereafter, submit to the Secretary, and commence implementation of, a plan for providing such services.

(8) PROVISION OF SERVICES BY OTHERS- The regulations issued under this section shall—

(A) require that a public entity submitting a plan to the Secretary under this section identify in the plan any person or other public entity which is providing a paratransit or other special transportation service for individuals with disabilities in the service area to which the plan applies; and

(B) provide that the public entity submitting the plan does not have to provide under the plan such service for individuals with disabilities.

(9) OTHER PROVISIONS- The regulations issued under this section shall include such other provisions and requirements as the Secretary determines are necessary to carry out the objectives of this section.

(d) REVIEW OF PLAN-

(1) GENERAL RULE- The Secretary shall review a plan submitted under this section for the purpose of determining whether or not such plan meets the requirements of this section, including the regulations issued under this section.

(2) DISAPPROVAL- If the Secretary determines that a plan reviewed under this subsection fails to meet the requirements of this section, the Secretary shall disapprove the plan and notify the public entity which submitted the plan of such disapproval and the reasons therefor.

(3) MODIFICATION OF DISAPPROVED PLAN- Not later than 90 days after the date of disapproval of a plan under this subsection, the public entity which submitted the plan shall modify the plan to meet the requirements of this section and shall submit to the Secretary, and commence implementation of, such modified plan.

(e) DISCRIMINATION DEFINED- As used in subsection (a), the term "discrimination" includes—

(1) a failure of a public entity to which the regulations issued under this section apply to submit, or commence implementation of, a plan in accordance with subsections (c)(6) and (c)(7);

(2) a failure of such entity to submit, or commence implementation of, a modified plan in accordance with subsection (d)(3);

(3) submission to the Secretary of a modified plan under subsection (d)(3) which does not meet the requirements of this section; or

(4) a failure of such entity to provide paratransit or other special transportation services in accordance with the plan or modified plan the public entity submitted to the Secretary under this section.

(f) STATUTORY CONSTRUCTION- Nothing in this section shall be construed as preventing a public entity—

(1) from providing paratransit or other special transportation services at a level which is greater than the level of such services which are required by this section,

(2) from providing paratransit or other special transportation services in addition to those paratransit and special transportation services required by this section, or

(3) from providing such services to individuals in addition to those individuals to whom such services are required to be provided by this section.

SEC. 224. PUBLIC ENTITY OPERATING A DEMAND RESPONSIVE SYSTEM

If a public entity operates a demand responsive system, it shall be considered discrimination, for purposes of section 202 of this Act and section 504 of the Rehabilitation Act of 1973 (29 U.S.C. 794), for such entity to purchase or lease a new vehicle for use on such system, for which a solicitation is made after the 30th day following the effective date of this section, that is not readily accessible to and usable by individuals with disabilities, including individuals who use wheelchairs, unless such system, when viewed in its entirety, provides a level of service to such individuals equivalent to the level of service such system provides to individuals without disabilities.

SEC. 225. TEMPORARY RELIEF WHERE LIFTS ARE UNAVAILABLE

(a) GRANTING- With respect to the purchase of new buses, a public entity may apply for, and the Secretary may temporarily relieve such public entity from the obligation under section 222(a) or 224 to purchase new buses that are readily accessible to and usable by individuals with disabilities if such public entity demonstrates to the satisfaction of the Secretary—

(1) that the initial solicitation for new buses made by the public entity specified that all new buses were to be lift-equipped and were to be otherwise accessible to and usable by individuals with disabilities;

(2) the unavailability from any qualified manufacturer of hydraulic, electromechanical, or other lifts for such new buses;

(3) that the public entity seeking temporary relief has made good faith efforts to locate a qualified manufacturer to supply the lifts to the manufacturer of such buses in sufficient time to comply with such solicitation; and

(4) that any further delay in purchasing new buses necessary to obtain such lifts would significantly impair transportation services in the community served by the public entity.

(b) DURATION AND NOTICE TO CONGRESS- Any relief granted under subsection (a) shall be limited in duration by a specified date, and the appropriate committees of Congress shall be notified of any such relief granted.

(c) FRAUDULENT APPLICATION- If, at any time, the Secretary has reasonable cause to believe that any relief granted under subsection (a) was fraudulently applied for, the Secretary shall—

(1) cancel such relief if such relief is still in effect; and

(2) take such other action as the Secretary considers appropriate.

SEC. 226. NEW FACILITIES

For purposes of section 202 of this Act and section 504 of the Rehabilitation Act of 1973 (29 U.S.C. 794), it shall be considered discrimination for a public entity to construct a new facility to be used in the provision of designated public transportation services unless such facility is readily accessible to and usable by individuals with disabilities, including individuals who use wheelchairs.

SEC. 227. ALTERATIONS OF EXISTING FACILITIES

(a) GENERAL RULE- With respect to alterations of an existing facility or part thereof used in the provision of designated public transportation services that affect or could affect the usability of the facility or part thereof, it shall be considered discrimination, for purposes of section 202 of this Act and section 504 of the Rehabilitation Act of 1973 (29 U.S.C. 794), for a public entity to fail to make such alterations (or to ensure that the alterations are made) in such a manner that, to the maximum extent feasible, the altered portions of the facility are readily accessible to and usable by individuals with disabilities, including individuals who use wheelchairs, upon the completion of such alterations. Where the public entity is undertaking an alteration that affects or could affect usability of or access to an area of the facility containing a primary function, the entity shall also make the alterations in such a manner that, to the maximum extent feasible, the path of travel to the altered area and the bathrooms, telephones, and drinking fountains serving the altered area, are readily accessible to and usable by individuals with disabilities, including individuals who use wheelchairs, upon completion of such alterations, where such alterations to the path of travel or the bathrooms, telephones, and drinking fountains serving the altered area are not disproportionate to the overall alterations in terms of cost and scope (as determined under criteria established by the Attorney General).

(b) Special Rule for Stations-

(1) GENERAL RULE- For purposes of section 202 of this Act and section 504 of the Rehabilitation Act of 1973 (29 U.S.C. 794), it shall be considered discrimination for a public entity that provides designated public transportation to fail, in accordance with the provisions of this subsection, to make key stations (as determined under criteria established by the Secretary by regulation) in rapid rail and light rail systems readily accessible to and usable by individuals with disabilities, including individuals who use wheelchairs.

(2) Rapid rail and light rail key stations-

(A) ACCESSIBILITY- Except as otherwise provided in this paragraph, all key stations (as determined under criteria established by the Secretary by regulation) in rapid rail and light rail systems shall be made readily accessible to and usable by individuals with disabilities, including individuals who use wheelchairs, as soon as practicable but in no event later than the last day of the 3-year period beginning on the effective date of this paragraph.

(B) EXTENSION FOR EXTRAORDINARILY EXPENSIVE STRUCTURAL CHANGES- The Secretary may extend the 3-year period under subparagraph (A) up to a 30-year period for key stations in a rapid rail or light rail system which stations need extraordinarily expensive structural changes to, or replacement of, existing facilities; except that by the last day of the 20th year following the date of the enactment of this Act at least 2/3 of such key stations must be readily accessible to and usable by individuals with disabilities.

(3) PLANS AND MILESTONES- The Secretary shall require the appropriate public entity to develop and submit to the Secretary a plan for compliance with this subsection—

(A) that reflects consultation with individuals with disabilities affected by such plan and the results of a public hearing and public comments on such plan, and

(B) that establishes milestones for achievement of the requirements of this subsection.

SEC. 228. PUBLIC TRANSPORTATION PROGRAMS AND ACTIVITIES IN EXISTING FACILITIES AND ONE CAR PER TRAIN RULE

(a) PUBLIC TRANSPORTATION PROGRAMS AND ACTIVITIES IN EXISTING FACILITIES-

(1) IN GENERAL- With respect to existing facilities used in the provision of designated public transportation services, it shall be considered discrimination, for purposes of section 202 of this Act and section 504 of the Rehabilitation Act of 1973 (29 U.S.C. 794), for a public entity to fail to operate a designated public transportation program or activity conducted in such facilities so that, when viewed in the entirety, the program or activity is readily accessible to and usable by individuals with disabilities.

(2) EXCEPTION- Paragraph (1) shall not require a public entity to make structural changes to existing facilities in order to make such facilities accessible to individuals who use wheelchairs, unless and to the extent required by section 227(a) (relating to alterations) or section 227(b) (relating to key stations).

(3) UTILIZATION- Paragraph (1) shall not require a public entity to which paragraph (2) applies, to provide to individuals who use wheelchairs services made available to the general public at such facilities when such individuals could not utilize or benefit from such services provided at such facilities.

(b) ONE CAR PER TRAIN RULE-

(1) GENERAL RULE- Subject to paragraph (2), with respect to 2 or more vehicles operated as a train by a light or rapid rail system, for purposes of section 202 of this Act and section 504 of the Rehabilitation Act of 1973 (29 U.S.C. 794), it shall be considered discrimination for a public entity to fail to have at least 1 vehicle per

train that is accessible to individuals with disabilities, including individuals who use wheelchairs, as soon as practicable but in no event later than the last day of the 5-year period beginning on the effective date of this section.

(2) HISTORIC TRAINS- In order to comply with paragraph (1) with respect to the remanufacture of a vehicle of historic character which is to be used on a segment of a light or rapid rail system which is included on the National Register of Historic Places, if making such vehicle readily accessible to and usable by individuals with disabilities would significantly alter the historic character of such vehicle, the public entity which operates such system only has to make (or to purchase or lease a remanufactured vehicle with) those modifications which are necessary to meet the requirements of section 222(c)(1) and which do not significantly alter the historic character of such vehicle.

SEC. 229. REGULATIONS

(a) IN GENERAL- Not later than 1 year after the date of enactment of this Act, the Secretary of Transportation shall issue regulations, in an accessible format, necessary for carrying out this part (other than section 223).

(b) STANDARDS- The regulations issued under this section and section 223 shall include standards applicable to facilities and vehicles covered by this subtitle. The standards shall be consistent with the minimum guidelines and requirements issued by the Architectural and Transportation Barriers Compliance Board in accordance with section 504 of this Act.

SEC. 230. INTERIM ACCESSIBILITY REQUIREMENTS

If final regulations have not been issued pursuant to section 229, for new construction or alterations for which a valid and appropriate State or local building permit is obtained prior to the issuance of final regulations under such section, and for which the construction or alteration authorized by such permit begins within one year of the receipt of such permit and is completed under the terms of such permit, compliance with the Uniform Federal Accessibility Standards in effect at the time the building permit is issued shall suffice to satisfy the requirement that facilities be readily accessible to and usable by persons with disabilities as required under sections 226 and 227, except that, if such final regulations have not been issued one year after the Architectural and Transportation Barriers Compliance Board has issued the supplemental minimum guidelines required under section 504(a) of this Act, compliance with such supplemental minimum guidelines shall be necessary to satisfy the requirement that facilities be readily accessible to and usable by persons with disabilities prior to issuance of the final regulations.

SEC. 231. EFFECTIVE DATE

(a) GENERAL RULE- Except as provided in subsection (b), this part shall become effective 18 months after the date of enactment of this Act.

(b) EXCEPTION- Sections 222, 223 (other than subsection (a)), 224, 225, 227(b), 228(b), and 229 shall become effective on the date of enactment of this Act.

PART II—PUBLIC TRANSPORTATION BY INTERCITY AND COMMUTER RAIL

SEC. 241. DEFINITIONS

As used in this part:

(1) COMMUTER AUTHORITY- The term "commuter authority" has the meaning given such term in section 103(8) of the Rail Passenger Service Act (45 U.S.C. 502(8)).

(2) COMMUTER RAIL TRANSPORTATION- The term "commuter rail transportation" has the meaning given the term "commuter service" in section 103(9) of the Rail Passenger Service Act (45 U.S.C. 502(9)).

(3) INTERCITY RAIL TRANSPORTATION- The term "intercity rail transportation" means transportation provided by the National Railroad Passenger Corporation.

(4) RAIL PASSENGER CAR- The term "rail passenger car" means, with respect to intercity rail transportation, single-level and bi-level coach cars, single-level and bi-level dining cars, single-level and bi-level sleeping cars, single-level and bi-level lounge cars, and food service cars.

(5) RESPONSIBLE PERSON- The term "responsible person" means—

(A) in the case of a station more than 50 percent of which is owned by a public entity, such public entity;

(B) in the case of a station more than 50 percent of which is owned by a private party, the persons providing intercity or commuter rail transportation to such station, as allocated on an equitable basis by regulation by the Secretary of Transportation; and

(C) in a case where no party owns more than 50 percent of a station, the persons providing intercity or commuter rail transportation to such station and the owners of the station, other than private party owners, as allocated on an equitable basis by regulation by the Secretary of Transportation.

(6) STATION- The term "station" means the portion of a property located appurtenant to a right-of-way on which intercity or commuter rail transportation is operated, where such portion is used by the general public and is related to the provision of such transportation, including passenger platforms, designated waiting areas, ticketing areas, restrooms, and, where a public entity providing rail transportation owns the property, concession areas, to the extent that such public entity exercises control over the selection, design, construction, or alteration of the property, but such term does not include flag stops.

SEC. 242. INTERCITY AND COMMUTER RAIL ACTIONS CONSIDERED DISCRIMINATORY

(a) INTERCITY RAIL TRANSPORTATION-

(1) ONE CAR PER TRAIN RULE- It shall be considered discrimination for purposes of section 202 of this Act and section 504 of the Rehabilitation Act of 1973 (29 U.S.C. 794) for a person who provides intercity rail transportation to fail to have at least one passenger car per train that is readily accessible to and usable by individuals with disabilities, including individuals who use wheelchairs, in accordance with regulations issued under section 244, as soon as practicable, but in no event later than 5 years after the date of enactment of this Act.

(2) NEW INTERCITY CARS-

(A) GENERAL RULE- Except as otherwise provided in this subsection with respect to individuals who use wheelchairs, it shall be considered discrimination for purposes of section 202 of this Act and section 504 of the Rehabilitation Act of 1973 (29 U.S.C. 794) for a person to purchase or lease any new rail passenger cars for use in intercity rail transportation, and for which a solicitation is made later than 30 days after the effective date of this section, unless all such rail cars are readily accessible to and usable by individuals with disabilities, including individuals who use wheelchairs, as prescribed by the Secretary of Transportation in regulations issued under section 244.

(B) SPECIAL RULE FOR SINGLE-LEVEL PASSENGER COACHES FOR INDIVIDUALS WHO USE WHEELCHAIRS- Single-level passenger coaches shall be required to—

(i) be able to be entered by an individual who uses a wheelchair;

(ii) have space to park and secure a wheelchair;

(iii) have a seat to which a passenger in a wheelchair can transfer, and a space to fold and store such passenger's wheelchair; and

(iv) have a restroom usable by an individual who uses a wheelchair, only to the extent provided in paragraph (3).

(C) SPECIAL RULE FOR SINGLE-LEVEL DINING CARS FOR INDIVIDUALS WHO USE WHEELCHAIRS- Single-level dining cars shall not be required to—

(i) be able to be entered from the station platform by an individual who uses a wheelchair; or

(ii) have a restroom usable by an individual who uses a wheelchair if no restroom is provided in such car for any passenger.

(D) SPECIAL RULE FOR BI-LEVEL DINING CARS FOR INDIVIDUALS WHO USE WHEELCHAIRS- Bi-level dining cars shall not be required to—

(i) be able to be entered by an individual who uses a wheelchair;

(ii) have space to park and secure a wheelchair;

(iii) have a seat to which a passenger in a wheelchair can transfer, or a space to fold and store such passenger's wheelchair; or

(iv) have a restroom usable by an individual who uses a wheelchair.

(3) ACCESSIBILITY OF SINGLE-LEVEL COACHES-

(A) GENERAL RULE- It shall be considered discrimination for purposes of section 202 of this Act and section 504 of the Rehabilitation Act of 1973 (29 U.S.C. 794) for a person who provides intercity rail transportation to fail to have on each train which includes one or more single-level rail passenger coaches—

(i) a number of spaces—

(I) to park and secure wheelchairs (to accommodate individuals who wish to remain in their wheelchairs) equal to not less than one-half of the number of single-level rail passenger coaches in such train; and

(II) to fold and store wheelchairs (to accommodate individuals who wish to transfer to coach seats) equal to not less than one-half of the number of single-level rail passenger coaches in such train, as soon as practicable, but in no event later than 5 years after the date of enactment of this Act; and

(ii) a number of spaces—

(I) to park and secure wheelchairs (to accommodate individuals who wish to remain in their wheelchairs) equal to not less than the total number of single-level rail passenger coaches in such train; and

(II) to fold and store wheelchairs (to accommodate individuals who wish to transfer to coach seats) equal to not less than the total number of single-level rail passenger coaches in such train, as soon as practicable, but in no event later than 10 years after the date of enactment of this Act.

(B) LOCATION- Spaces required by subparagraph (A) shall be located in single-level rail passenger coaches or food service cars.

(C) LIMITATION- Of the number of spaces required on a train by subparagraph (A), not more than two spaces to park and secure wheelchairs nor more than two spaces to fold and store wheelchairs shall be located in any one coach or food service car.

(D) OTHER ACCESSIBILITY FEATURES- Single-level rail passenger coaches and food service cars on which the spaces required by subparagraph (A) are located shall have a restroom usable by an individual who uses a wheelchair and shall be able to be entered from the station platform by an individual who uses a wheelchair.

(4) FOOD SERVICE-

(A) SINGLE-LEVEL DINING CARS- On any train in which a single-level dining car is used to provide food service—

(i) if such single-level dining car was purchased after the date of enactment of this Act, table service in such car shall be provided to a passenger who uses a wheelchair if—

(I) the car adjacent to the end of the dining car through which a wheelchair may enter is itself accessible to a wheelchair;

(II) such passenger can exit to the platform from the car such passenger occupies, move down the platform, and enter the adjacent accessible car described in subclause (I) without the necessity of the train being moved within the station; and

(III) space to park and secure a wheelchair is available in the dining car at the time such passenger wishes to eat (if such passenger wishes to remain in a wheelchair), or space to store and fold a wheelchair is available in the dining car at the time such passenger wishes to eat (if such passenger wishes to transfer to a dining car seat); and

(ii) appropriate auxiliary aids and services, including a hard surface on which to eat, shall be provided to ensure that other equivalent food service is available to individuals with disabilities, including individuals who use wheelchairs, and to passengers traveling with such individuals.

Unless not practicable, a person providing intercity rail transportation shall place an accessible car adjacent to the end of a dining car described in clause (i) through which an individual who uses a wheelchair may enter.

(B) BI-LEVEL DINING CARS- On any train in which a bi-level dining car is used to provide food service—

(i) if such train includes a bi-level lounge car purchased after the date of enactment of this Act, table service in such lounge car shall be provided to individuals who use wheelchairs and to other passengers; and

(ii) appropriate auxiliary aids and services, including a hard surface on which to eat, shall be provided to ensure that other equivalent food service is available to individuals with disabilities, including individuals who use wheelchairs, and to passengers traveling with such individuals.

(b) COMMUTER RAIL TRANSPORTATION-

(1) ONE CAR PER TRAIN RULE- It shall be considered discrimination for purposes of section 202 of this Act and section 504 of the Rehabilitation Act of 1973

(29 U.S.C. 794) for a person who provides commuter rail transportation to fail to have at least one passenger car per train that is readily accessible to and usable by individuals with disabilities, including individuals who use wheelchairs, in accordance with regulations issued under section 244, as soon as practicable, but in no event later than 5 years after the date of enactment of this Act.

(2) NEW COMMUTER RAIL CARS-

(A) GENERAL RULE- It shall be considered discrimination for purposes of section 202 of this Act and section 504 of the Rehabilitation Act of 1973 (29 U.S.C. 794) for a person to purchase or lease any new rail passenger cars for use in commuter rail transportation, and for which a solicitation is made later than 30 days after the effective date of this section, unless all such rail cars are readily accessible to and usable by individuals with disabilities, including individuals who use wheelchairs, as prescribed by the Secretary of Transportation in regulations issued under section 244.

(B) ACCESSIBILITY- For purposes of section 202 of this Act and section 504 of the Rehabilitation Act of 1973 (29 U.S.C. 794), a requirement that a rail passenger car used in commuter rail transportation be accessible to or readily accessible to and usable by individuals with disabilities, including individuals who use wheelchairs, shall not be construed to require—

(i) a restroom usable by an individual who uses a wheelchair if no restroom is provided in such car for any passenger;

(ii) space to fold and store a wheelchair; or

(iii) a seat to which a passenger who uses a wheelchair can transfer.

(c) USED RAIL CARS- It shall be considered discrimination for purposes of section 202 of this Act and section 504 of the Rehabilitation Act of 1973 (29 U.S.C. 794) for a person to purchase or lease a used rail passenger car for use in intercity or commuter rail transportation, unless such person makes demonstrated good faith efforts to purchase or lease a used rail car that is readily accessible to and usable by individuals with disabilities, including individuals who use wheelchairs, as prescribed by the Secretary of Transportation in regulations issued under section 244.

(d) REMANUFACTURED RAIL CARS-

(1) REMANUFACTURING- It shall be considered discrimination for purposes of section 202 of this Act and section 504 of the Rehabilitation Act of 1973 (29 U.S.C. 794) for a person to remanufacture a rail passenger car for use in intercity or commuter rail transportation so as to extend its usable life for 10 years or more, unless the rail car, to the maximum extent feasible, is made readily accessible to and usable by individuals with disabilities, including individuals who use wheelchairs, as prescribed by the Secretary of Transportation in regulations issued under section 244.

(2) PURCHASE OR LEASE- It shall be considered discrimination for purposes of section 202 of this Act and section 504 of the Rehabilitation Act of 1973 (29 U.S.C. 794)

for a person to purchase or lease a remanufactured rail passenger car for use in intercity or commuter rail transportation unless such car was remanufactured in accordance with paragraph (1).

(e) STATIONS-

(1) NEW STATIONS- It shall be considered discrimination for purposes of section 202 of this Act and section 504 of the Rehabilitation Act of 1973 (29 U.S.C. 794) for a person to build a new station for use in intercity or commuter rail transportation that is not readily accessible to and usable by individuals with disabilities, including individuals who use wheelchairs, as prescribed by the Secretary of Transportation in regulations issued under section 244.

(2) EXISTING STATIONS-

(A) FAILURE TO MAKE READILY ACCESSIBLE-

(i) GENERAL RULE- It shall be considered discrimination for purposes of section 202 of this Act and section 504 of the Rehabilitation Act of 1973 (29 U.S.C. 794) for a responsible person to fail to make existing stations in the intercity rail transportation system, and existing key stations in commuter rail transportation systems, readily accessible to and usable by individuals with disabilities, including individuals who use wheelchairs, as prescribed by the Secretary of Transportation in regulations issued under section 244.

(ii) PERIOD FOR COMPLIANCE-

(I) INTERCITY RAIL- All stations in the intercity rail transportation system shall be made readily accessible to and usable by individuals with disabilities, including individuals who use wheelchairs, as soon as practicable, but in no event later than 20 years after the date of enactment of this Act.

(II) COMMUTER RAIL- Key stations in commuter rail transportation systems shall be made readily accessible to and usable by individuals with disabilities, including individuals who use wheelchairs, as soon as practicable but in no event later than 3 years after the date of enactment of this Act, except that the time limit may be extended by the Secretary of Transportation up to 20 years after the date of enactment of this Act in a case where the raising of the entire passenger platform is the only means available of attaining accessibility or where other extraordinarily expensive structural changes are necessary to attain accessibility.

(iii) DESIGNATION OF KEY STATIONS- Each commuter authority shall designate the key stations in its commuter rail transportation system, in consultation with individuals with disabilities and organizations representing such individuals, taking into consideration such factors as high ridership and whether such station serves as a transfer or feeder station. Before the final designation of key stations under this clause, a commuter authority shall hold a public hearing.

(iv) PLANS AND MILESTONES- The Secretary of Transportation shall require the appropriate person to develop a plan for carrying out this subparagraph that reflects consultation with individuals with disabilities affected by such plan and that establishes milestones for achievement of the requirements of this subparagraph.

(B) REQUIREMENT WHEN MAKING ALTERATIONS-

(i) GENERAL RULE- It shall be considered discrimination, for purposes of section 202 of this Act and section 504 of the Rehabilitation Act of 1973 (29 U.S.C. 794), with respect to alterations of an existing station or part thereof in the intercity or commuter rail transportation systems that affect or could affect the usability of the station or part thereof, for the responsible person, owner, or person in control of the station to fail to make the alterations in such a manner that, to the maximum extent feasible, the altered portions of the station are readily accessible to and usable by individuals with disabilities, including individuals who use wheelchairs, upon completion of such alterations.

(ii) ALTERATIONS TO A PRIMARY FUNCTION AREA- It shall be considered discrimination, for purposes of section 202 of this Act and section 504 of the Rehabilitation Act of 1973 (29 U.S.C. 794), with respect to alterations that affect or could affect the usability of or access to an area of the station containing a primary function, for the responsible person, owner, or person in control of the station to fail to make the alterations in such a manner that, to the maximum extent feasible, the path of travel to the altered area, and the bathrooms, telephones, and drinking fountains serving the altered area, are readily accessible to and usable by individuals with disabilities, including individuals who use wheelchairs, upon completion of such alterations, where such alterations to the path of travel or the bathrooms, telephones, and drinking fountains serving the altered area are not disproportionate to the overall alterations in terms of cost and scope (as determined under criteria established by the Attorney General).

(C) REQUIRED COOPERATION- It shall be considered discrimination for purposes of section 202 of this Act and section 504 of the Rehabilitation Act of 1973 (29 U.S.C. 794) for an owner, or person in control, of a station governed by subparagraph (A) or (B) to fail to provide reasonable cooperation to a responsible person with respect to such station in that responsible person's efforts to comply with such subparagraph. An owner, or person in control, of a station shall be liable to a responsible person for any failure to provide reasonable cooperation as required by this subparagraph. Failure to receive reasonable cooperation required by this subparagraph shall not be a defense to a claim of discrimination under this Act.

SEC. 243. CONFORMANCE OF ACCESSIBILITY STANDARDS

Accessibility standards included in regulations issued under this part shall be consistent with the minimum guidelines issued by the Architectural and Transportation Barriers Compliance Board under section 504(a) of this Act.

SEC. 244. REGULATIONS

Not later than 1 year after the date of enactment of this Act, the Secretary of Transportation shall issue regulations, in an accessible format, necessary for carrying out this part.

SEC. 245. INTERIM ACCESSIBILITY REQUIREMENTS

(a) STATIONS- If final regulations have not been issued pursuant to section 244, for new construction or alterations for which a valid and appropriate State or local building permit is obtained prior to the issuance of final regulations under such section, and for which the construction or alteration authorized by such permit begins within one year of the receipt of such permit and is completed under the terms of such permit, compliance with the Uniform Federal Accessibility Standards in effect at the time the building permit is issued shall suffice to satisfy the requirement that stations be readily accessible to and usable by persons with disabilities as required under section 242(e), except that, if such final regulations have not been issued one year after the Architectural and Transportation Barriers Compliance Board has issued the supplemental minimum guidelines required under section 504(a) of this Act, compliance with such supplemental minimum guidelines shall be necessary to satisfy the requirement that stations be readily accessible to and usable by persons with disabilities prior to issuance of the final regulations.

(b) RAIL PASSENGER CARS- If final regulations have not been issued pursuant to section 244, a person shall be considered to have complied with the requirements of section 242 (a) through (d) that a rail passenger car be readily accessible to and usable by individuals with disabilities, if the design for such car complies with the laws and regulations (including the Minimum Guidelines and Requirements for Accessible Design and such supplemental minimum guidelines as are issued under section 504(a) of this Act) governing accessibility of such cars, to the extent that such laws and regulations are not inconsistent with this part and are in effect at the time such design is substantially completed.

SEC. 246. EFFECTIVE DATE

(a) GENERAL RULE- Except as provided in subsection (b), this part shall become effective 18 months after the date of enactment of this Act.

(b) EXCEPTION- Sections 242 and 244 shall become effective on the date of enactment of this Act.

TITLE III—PUBLIC ACCOMMODATIONS AND SERVICES OPERATED BY PRIVATE ENTITIES

SEC. 301. DEFINITIONS

As used in this title:

(1) COMMERCE- The term “commerce” means travel, trade, traffic, commerce, transportation, or communication—

(A) among the several States;

(B) between any foreign country or any territory or possession and any State; or

(C) between points in the same State but through another State or foreign country.

(2) COMMERCIAL FACILITIES- The term "commercial facilities" means facilities—

(A) that are intended for nonresidential use; and

(B) whose operations will affect commerce.

Such term shall not include railroad locomotives, railroad freight cars, railroad cabooses, railroad cars described in section 242 or covered under this title, railroad rights-of-way, or facilities that are covered or expressly exempted from coverage under the Fair Housing Act of 1968 (42 U.S.C. 3601 et seq.).

(3) DEMAND RESPONSIVE SYSTEM- The term "demand responsive system" means any system of providing transportation of individuals by a vehicle, other than a system which is a fixed route system.

(4) FIXED ROUTE SYSTEM- The term "fixed route system" means a system of providing transportation of individuals (other than by aircraft) on which a vehicle is operated along a prescribed route according to a fixed schedule.

(5) OVER-THE-ROAD BUS- The term "over-the-road bus" means a bus characterized by an elevated passenger deck located over a baggage compartment.

(6) PRIVATE ENTITY- The term "private entity" means any entity other than a public entity (as defined in section 201(1)).

(7) PUBLIC ACCOMMODATION- The following private entities are considered public accommodations for purposes of this title, if the operations of such entities affect commerce—

(A) an inn, hotel, motel, or other place of lodging, except for an establishment located within a building that contains not more than five rooms for rent or hire and that is actually occupied by the proprietor of such establishment as the residence of such proprietor;

(B) a restaurant, bar, or other establishment serving food or drink;

(C) a motion picture house, theater, concert hall, stadium, or other place of exhibition or entertainment;

(D) an auditorium, convention center, lecture hall, or other place of public gathering;

(E) a bakery, grocery store, clothing store, hardware store, shopping center, or other sales or rental establishment;

(F) a laundromat, dry-cleaner, bank, barber shop, beauty shop, travel service, shoe repair service, funeral parlor, gas station, office of an accountant or lawyer, pharmacy, insurance office, professional office of a health care provider, hospital, or other service establishment;

(G) a terminal, depot, or other station used for specified public transportation;

(H) a museum, library, gallery, or other place of public display or collection;

(I) a park, zoo, amusement park, or other place of recreation;

(J) a nursery, elementary, secondary, undergraduate, or postgraduate private school, or other place of education;

(K) a day care center, senior citizen center, homeless shelter, food bank, adoption agency, or other social service center establishment; and

(L) a gymnasium, health spa, bowling alley, golf course, or other place of exercise or recreation.

(8) RAIL AND RAILROAD- The terms “rail” and “railroad” have the meaning given the term “railroad” in section 202(e) of the Federal Railroad Safety Act of 1970 (45 U.S.C. 431(e)).

(9) READILY ACHIEVABLE- The term “readily achievable” means easily accomplishable and able to be carried out without much difficulty or expense. In determining whether an action is readily achievable, factors to be considered include—

(A) the nature and cost of the action needed under this Act;

(B) the overall financial resources of the facility or facilities involved in the action; the number of persons employed at such facility; the effect on expenses and resources, or the impact otherwise of such action upon the operation of the facility;

(C) the overall financial resources of the covered entity; the overall size of the business of a covered entity with respect to the number of its employees; the number, type, and location of its facilities; and

(D) the type of operation or operations of the covered entity, including the composition, structure, and functions of the workforce of such entity; the geographic separateness, administrative or fiscal relationship of the facility or facilities in question to the covered entity.

(10) SPECIFIED PUBLIC TRANSPORTATION- The term “specified public transportation” means transportation by bus, rail, or any other conveyance (other than by aircraft) that provides the general public with general or special service (including charter service) on a regular and continuing basis.

(11) VEHICLE- The term “vehicle” does not include a rail passenger car, railroad locomotive, railroad freight car, railroad caboose, or a railroad car described in section 242 or covered under this title.

SEC. 302. PROHIBITION OF DISCRIMINATION BY PUBLIC ACCOMMODATIONS

(a) GENERAL RULE- No individual shall be discriminated against on the basis of disability in the full and equal enjoyment of the goods, services, facilities, privileges, advantages, or accommodations of any place of public accommodation by any person who owns, leases (or leases to), or operates a place of public accommodation.

(b) CONSTRUCTION-

(1) GENERAL PROHIBITION-

(A) ACTIVITIES-

(i) DENIAL OF PARTICIPATION- It shall be discriminatory to subject an individual or class of individuals on the basis of a disability or disabilities of such individual or class, directly, or through contractual, licensing, or other arrangements, to a denial of the opportunity of the individual or class to participate in or benefit from the goods, services, facilities, privileges, advantages, or accommodations of an entity.

(ii) PARTICIPATION IN UNEQUAL BENEFIT- It shall be discriminatory to afford an individual or class of individuals, on the basis of a disability or disabilities of such individual or class, directly, or through contractual, licensing, or other arrangements with the opportunity to participate in or benefit from a good, service, facility, privilege, advantage, or accommodation that is not equal to that afforded to other individuals.

(iii) SEPARATE BENEFIT- It shall be discriminatory to provide an individual or class of individuals, on the basis of a disability or disabilities of such individual or class, directly, or through contractual, licensing, or other arrangements with a good, service, facility, privilege, advantage, or accommodation that is different or separate from that provided to other individuals, unless such action is necessary to provide the individual or class of individuals with a good, service, facility, privilege, advantage, or accommodation, or other opportunity that is as effective as that provided to others.

(iv) INDIVIDUAL OR CLASS OF INDIVIDUALS- For purposes of clauses (i) through (iii) of this subparagraph, the term "individual or class of individuals" refers to the clients or customers of the covered public accommodation that enters into the contractual, licensing or other arrangement.

(B) INTEGRATED SETTINGS- Goods, services, facilities, privileges, advantages, and accommodations shall be afforded to an individual with a disability in the most integrated setting appropriate to the needs of the individual.

(C) OPPORTUNITY TO PARTICIPATE- Notwithstanding the existence of separate or different programs or activities provided in accordance with this section, an individual with a disability shall not be denied the opportunity to participate in such programs or activities that are not separate or different.

(D) ADMINISTRATIVE METHODS- An individual or entity shall not, directly or through contractual or other arrangements, utilize standards or criteria or methods of administration—

(i) that have the effect of discriminating on the basis of disability; or

(ii) that perpetuate the discrimination of others who are subject to common administrative control.

(E) ASSOCIATION- It shall be discriminatory to exclude or otherwise deny equal goods, services, facilities, privileges, advantages, accommodations, or other opportunities to an individual or entity because of the known disability of an individual with whom the individual or entity is known to have a relationship or association.

(2) SPECIFIC PROHIBITIONS-

(A) DISCRIMINATION- For purposes of subsection (a), discrimination includes—

(i) the imposition or application of eligibility criteria that screen out or tend to screen out an individual with a disability or any class of individuals with disabilities from fully and equally enjoying any goods, services, facilities, privileges, advantages, or accommodations, unless such criteria can be shown to be necessary for the provision of the goods, services, facilities, privileges, advantages, or accommodations being offered;

(ii) a failure to make reasonable modifications in policies, practices, or procedures, when such modifications are necessary to afford such goods, services, facilities, privileges, advantages, or accommodations to individuals with disabilities, unless the entity can demonstrate that making such modifications would fundamentally alter the nature of such goods, services, facilities, privileges, advantages, or accommodations;

(iii) a failure to take such steps as may be necessary to ensure that no individual with a disability is excluded, denied services, segregated or otherwise treated differently than other individuals because of the absence of auxiliary aids and services, unless the entity can demonstrate that taking such steps would fundamentally alter the nature of the good, service, facility, privilege, advantage, or accommodation being offered or would result in an undue burden;

(iv) a failure to remove architectural barriers, and communication barriers that are structural in nature, in existing facilities, and transportation barriers in existing vehicles and rail passenger cars used by an establishment for transporting individuals (not including barriers that can only be removed through the retrofitting of vehicles or rail passenger cars by the installation of a hydraulic or other lift), where such removal is readily achievable; and

(v) where an entity can demonstrate that the removal of a barrier under clause (iv) is not readily achievable, a failure to make such goods, services, facilities, privileges, advantages, or accommodations available through alternative methods if such methods are readily achievable.

(B) FIXED ROUTE SYSTEM-

(i) ACCESSIBILITY- It shall be considered discrimination for a private entity which operates a fixed route system and which is not subject to section 304 to purchase or lease a vehicle with a seating capacity in excess of 16 passengers (including the driver) for use on such system, for which a solicitation is made after the 30th day following the effective date of this subparagraph, that is not readily accessible to and usable by individuals with disabilities, including individuals who use wheelchairs.

(ii) EQUIVALENT SERVICE- If a private entity which operates a fixed route system and which is not subject to section 304 purchases or leases a vehicle with a seating capacity of 16 passengers or less (including the driver) for use on such system after the effective date of this subparagraph that is not readily accessible to or usable by individuals with disabilities, it shall be considered discrimination for such entity to fail to operate such system so that, when viewed in its entirety, such system ensures a level of service to individuals with disabilities, including individuals who use wheelchairs, equivalent to the level of service provided to individuals without disabilities.

(C) DEMAND RESPONSIVE SYSTEM- For purposes of subsection (a), discrimination includes—

(i) a failure of a private entity which operates a demand responsive system and which is not subject to section 304 to operate such system so that, when viewed in its entirety, such system ensures a level of service to individuals with disabilities, including individuals who use wheelchairs, equivalent to the level of service provided to individuals without disabilities; and

(ii) the purchase or lease by such entity for use on such system of a vehicle with a seating capacity in excess of 16 passengers (including the driver), for which solicitations are made after the 30th day following the effective date of this subparagraph, that is not readily accessible to and usable by individuals with disabilities (including individuals who use wheelchairs) unless such entity can demonstrate that such system, when viewed in its entirety, provides a level of service to individuals with disabilities equivalent to that provided to individuals without disabilities.

(D) OVER-THE-ROAD BUSES-

(i) LIMITATION ON APPLICABILITY- Subparagraphs (B) and (C) do not apply to over-the-road buses.

(ii) ACCESSIBILITY REQUIREMENTS- For purposes of subsection (a), discrimination includes (I) the purchase or lease of an over-the-road bus which does not comply with the regulations issued under section 306(a)(2) by a private entity which provides transportation of individuals and which is not primarily engaged in the business of transporting people, and (II) any other failure of such entity to comply with such regulations.

(3) SPECIFIC CONSTRUCTION- Nothing in this title shall require an entity to permit an individual to participate in or benefit from the goods, services, facilities, privileges, advantages and accommodations of such entity where such individual poses a direct threat to the health or safety of others. The term "direct threat" means a significant risk to the health or safety of others that cannot be eliminated by a

modification of policies, practices, or procedures or by the provision of auxiliary aids or services.

SEC. 303. NEW CONSTRUCTION AND ALTERATIONS IN PUBLIC ACCOMMODATIONS AND COMMERCIAL FACILITIES

(a) APPLICATION OF TERM- Except as provided in subsection (b), as applied to public accommodations and commercial facilities, discrimination for purposes of section 302(a) includes—

(1) a failure to design and construct facilities for first occupancy later than 30 months after the date of enactment of this Act that are readily accessible to and usable by individuals with disabilities, except where an entity can demonstrate that it is structurally impracticable to meet the requirements of such subsection in accordance with standards set forth or incorporated by reference in regulations issued under this title; and

(2) with respect to a facility or part thereof that is altered by, on behalf of, or for the use of an establishment in a manner that affects or could affect the usability of the facility or part thereof, a failure to make alterations in such a manner that, to the maximum extent feasible, the altered portions of the facility are readily accessible to and usable by individuals with disabilities, including individuals who use wheelchairs. Where the entity is undertaking an alteration that affects or could affect usability of or access to an area of the facility containing a primary function, the entity shall also make the alterations in such a manner that, to the maximum extent feasible, the path of travel to the altered area and the bathrooms, telephones, and drinking fountains serving the altered area, are readily accessible to and usable by individuals with disabilities where such alterations to the path of travel or the bathrooms, telephones, and drinking fountains serving the altered area are not disproportionate to the overall alterations in terms of cost and scope (as determined under criteria established by the Attorney General).

(b) ELEVATOR- Subsection (a) shall not be construed to require the installation of an elevator for facilities that are less than three stories or have less than 3,000 square feet per story unless the building is a shopping center, a shopping mall, or the professional office of a health care provider or unless the Attorney General determines that a particular category of such facilities requires the installation of elevators based on the usage of such facilities.

SEC. 304. PROHIBITION OF DISCRIMINATION IN SPECIFIED PUBLIC TRANSPORTATION SERVICES PROVIDED BY PRIVATE ENTITIES

(a) GENERAL RULE- No individual shall be discriminated against on the basis of disability in the full and equal enjoyment of specified public transportation services provided by a private entity that is primarily engaged in the business of transporting people and whose operations affect commerce.

(b) CONSTRUCTION- For purposes of subsection (a), discrimination includes—

(1) the imposition or application by a entity described in subsection (a) of eligibility criteria that screen out or tend to screen out an individual with a disability or any class of individuals with disabilities from fully enjoying the specified public transportation services provided by the entity, unless such criteria can be shown to be necessary for the provision of the services being offered;

(2) the failure of such entity to—

(A) make reasonable modifications consistent with those required under section 302(b)(2)(A)(ii);

(B) provide auxiliary aids and services consistent with the requirements of section 302(b)(2)(A)(iii); and

(C) remove barriers consistent with the requirements of section 302(b)(2)(A) and with the requirements of section 303(a)(2);

(3) the purchase or lease by such entity of a new vehicle (other than an automobile, a van with a seating capacity of less than 8 passengers, including the driver, or an over-the-road bus) which is to be used to provide specified public transportation and for which a solicitation is made after the 30th day following the effective date of this section, that is not readily accessible to and usable by individuals with disabilities, including individuals who use wheelchairs; except that the new vehicle need not be readily accessible to and usable by such individuals if the new vehicle is to be used solely in a demand responsive system and if the entity can demonstrate that such system, when viewed in its entirety, provides a level of service to such individuals equivalent to the level of service provided to the general public;

(4)(A) the purchase or lease by such entity of an over-the-road bus which does not comply with the regulations issued under section 306(a)(2); and

(B) any other failure of such entity to comply with such regulations; and

(5) the purchase or lease by such entity of a new van with a seating capacity of less than 8 passengers, including the driver, which is to be used to provide specified public transportation and for which a solicitation is made after the 30th day following the effective date of this section that is not readily accessible to or usable by individuals with disabilities, including individuals who use wheelchairs; except that the new van need not be readily accessible to and usable by such individuals if the entity can demonstrate that the system for which the van is being purchased or leased, when viewed in its entirety, provides a level of service to such individuals equivalent to the level of service provided to the general public;

(6) the purchase or lease by such entity of a new rail passenger car that is to be used to provide specified public transportation, and for which a solicitation is made later than 30 days after the effective date of this paragraph, that is not readily accessible

to and usable by individuals with disabilities, including individuals who use wheelchairs; and

(7) the remanufacture by such entity of a rail passenger car that is to be used to provide specified public transportation so as to extend its usable life for 10 years or more, or the purchase or lease by such entity of such a rail car, unless the rail car, to the maximum extent feasible, is made readily accessible to and usable by individuals with disabilities, including individuals who use wheelchairs.

(c) HISTORICAL OR ANTIQUATED CARS-

(1) EXCEPTION- To the extent that compliance with subsection (b)(2)(C) or (b)(7) would significantly alter the historic or antiquated character of a historical or antiquated rail passenger car, or a rail station served exclusively by such cars, or would result in violation of any rule, regulation, standard, or order issued by the Secretary of Transportation under the Federal Railroad Safety Act of 1970, such compliance shall not be required.

(2) DEFINITION- As used in this subsection, the term "historical or antiquated rail passenger car" means a rail passenger car—

(A) which is not less than 30 years old at the time of its use for transporting individuals;

(B) the manufacturer of which is no longer in the business of manufacturing rail passenger cars; and

(C) which—

(i) has a consequential association with events or persons significant to the past; or

(ii) embodies, or is being restored to embody, the distinctive characteristics of a type of rail passenger car used in the past, or to represent a time period which has passed.

SEC. 305. STUDY

(a) PURPOSES- The Office of Technology Assessment shall undertake a study to determine—

(1) the access needs of individuals with disabilities to over-the-road buses and over-the-road bus service; and

(2) the most cost-effective methods for providing access to over-the-road buses and over-the-road bus service to individuals with disabilities, particularly individuals who use wheelchairs, through all forms of boarding options.

(b) CONTENTS- The study shall include, at a minimum, an analysis of the following:

(1) The anticipated demand by individuals with disabilities for accessible over-the-road buses and over-the-road bus service.

(2) The degree to which such buses and service, including any service required under sections 304(b)(4) and 306(a)(2), are readily accessible to and usable by individuals with disabilities.

(3) The effectiveness of various methods of providing accessibility to such buses and service to individuals with disabilities.

(4) The cost of providing accessible over-the-road buses and bus service to individuals with disabilities, including consideration of recent technological and cost saving developments in equipment and devices.

(5) Possible design changes in over-the-road buses that could enhance accessibility, including the installation of accessible restrooms which do not result in a loss of seating capacity.

(6) The impact of accessibility requirements on the continuation of over-the-road bus service, with particular consideration of the impact of such requirements on such service to rural communities.

(c) ADVISORY COMMITTEE- In conducting the study required by subsection (a), the Office of Technology Assessment shall establish an advisory committee, which shall consist of—

(1) members selected from among private operators and manufacturers of over-the-road buses;

(2) members selected from among individuals with disabilities, particularly individuals who use wheelchairs, who are potential riders of such buses; and

(3) members selected for their technical expertise on issues included in the study, including manufacturers of boarding assistance equipment and devices.

The number of members selected under each of paragraphs (1) and (2) shall be equal, and the total number of members selected under paragraphs (1) and (2) shall exceed the number of members selected under paragraph (3).

(d) DEADLINE- The study required by subsection (a), along with recommendations by the Office of Technology Assessment, including any policy options for legislative action, shall be submitted to the President and Congress within 36 months after the date of the enactment of this Act. If the President determines that compliance with the regulations issued pursuant to section 306(a)(2)(B) on or before the applicable deadlines specified in section 306(a)(2)(B) will result in a significant reduction in intercity over-the-road bus service, the President shall extend each such deadline by 1 year.

(e) REVIEW- In developing the study required by subsection (a), the Office of Technology Assessment shall provide a preliminary draft of such study to the Architectural and Transportation Barriers Compliance Board established under section 502 of the Rehabilitation Act of 1973 (29 U.S.C. 792). The Board shall have an opportunity to comment on such draft study, and any such comments by the Board made in writing within 120 days after the Board's receipt of the draft study shall be incorporated as part of the final study required to be submitted under subsection (d).

SEC. 306. REGULATIONS

(a) TRANSPORTATION PROVISIONS-

(1) GENERAL RULE- Not later than 1 year after the date of the enactment of this Act, the Secretary of Transportation shall issue regulations in an accessible format to carry out sections 302(b)(2) (B) and (C) and to carry out section 304 (other than subsection (b)(4)).

(2) SPECIAL RULES FOR PROVIDING ACCESS TO OVER-THE-ROAD BUSES-

(A) INTERIM REQUIREMENTS-

(i) ISSUANCE- Not later than 1 year after the date of the enactment of this Act, the Secretary of Transportation shall issue regulations in an accessible format to carry out sections 304(b)(4) and 302(b)(2)(D)(ii) that require each private entity which uses an over-the-road bus to provide transportation of individuals to provide accessibility to such bus; except that such regulations shall not require any structural changes in over-the-road buses in order to provide access to individuals who use wheelchairs during the effective period of such regulations and shall not require the purchase of boarding assistance devices to provide access to such individuals.

(ii) EFFECTIVE PERIOD- The regulations issued pursuant to this subparagraph shall be effective until the effective date of the regulations issued under subparagraph (B).

(B) FINAL REQUIREMENT-

(i) REVIEW OF STUDY AND INTERIM REQUIREMENTS- The Secretary shall review the study submitted under section 305 and the regulations issued pursuant to subparagraph (A).

(ii) ISSUANCE- Not later than 1 year after the date of the submission of the study under section 305, the Secretary shall issue in an accessible format new regulations to carry out sections 304(b)(4) and 302(b)(2)(D)(ii) that require, taking into account the purposes of the study under section 305 and any recommendations resulting from such study, each private entity which uses an over-the-road bus to provide transportation to individuals to provide accessibility to such bus to individuals with disabilities, including individuals who use wheelchairs.

(iii) EFFECTIVE PERIOD- Subject to section 305(d), the regulations issued pursuant to this subparagraph shall take effect—

(I) with respect to small providers of transportation (as defined by the Secretary), 7 years after the date of the enactment of this Act; and

(II) with respect to other providers of transportation, 6 years after such date of enactment.

(C) LIMITATION ON REQUIRING INSTALLATION OF ACCESSIBLE RESTROOMS- The regulations issued pursuant to this paragraph shall not require the installation of accessible restrooms in over-the-road buses if such installation would result in a loss of seating capacity.

(3) STANDARDS- The regulations issued pursuant to this subsection shall include standards applicable to facilities and vehicles covered by sections 302(b)(2) and 304.

(b) OTHER PROVISIONS- Not later than 1 year after the date of the enactment of this Act, the Attorney General shall issue regulations in an accessible format to carry out the provisions of this title not referred to in subsection (a) that include standards applicable to facilities and vehicles covered under section 302.

(c) CONSISTENCY WITH ATBCB GUIDELINES- Standards included in regulations issued under subsections (a) and (b) shall be consistent with the minimum guidelines and requirements issued by the Architectural and Transportation Barriers Compliance Board in accordance with section 504 of this Act.

(d) INTERIM ACCESSIBILITY STANDARDS-

(1) FACILITIES- If final regulations have not been issued pursuant to this section, for new construction or alterations for which a valid and appropriate State or local building permit is obtained prior to the issuance of final regulations under this section, and for which the construction or alteration authorized by such permit begins within one year of the receipt of such permit and is completed under the terms of such permit, compliance with the Uniform Federal Accessibility Standards in effect at the time the building permit is issued shall suffice to satisfy the requirement that facilities be readily accessible to and usable by persons with disabilities as required under section 303, except that, if such final regulations have not been issued one year after the Architectural and Transportation Barriers Compliance Board has issued the supplemental minimum guidelines required under section 504(a) of this Act, compliance with such supplemental minimum guidelines shall be necessary to satisfy the requirement that facilities be readily accessible to and usable by persons with disabilities prior to issuance of the final regulations.

(2) VEHICLES AND RAIL PASSENGER CARS- If final regulations have not been issued pursuant to this section, a private entity shall be considered to have complied with the requirements of this title, if any, that a vehicle or rail passenger car be readily accessible to and usable by individuals with disabilities, if the design for such vehicle or car complies with the laws and regulations (including the Minimum Guidelines and Requirements for Accessible Design and such supplemental minimum guidelines as are issued under section 504(a) of this Act) governing accessibility of such vehicles or cars, to the extent that such laws and regulations are not inconsistent with this title and are in effect at the time such design is substantially completed.

SEC. 307. EXEMPTIONS FOR PRIVATE CLUBS AND RELIGIOUS ORGANIZATIONS

The provisions of this title shall not apply to private clubs or establishments exempted from coverage under title II of the Civil Rights Act of 1964 (42 U.S.C. 2000-a(e)) or to religious organizations or entities controlled by religious organizations, including places of worship.

SEC. 308. ENFORCEMENT

(a) IN GENERAL-

(1) AVAILABILITY OF REMEDIES AND PROCEDURES- The remedies and procedures set forth in section 204(a) of the Civil Rights Act of 1964 (42 U.S.C. 2000a-3(a)) are the remedies and procedures this title provides to any person who is being subjected to discrimination on the basis of disability in violation of this title or who has reasonable grounds for believing that such person is about to be subjected to discrimination in violation of section 303. Nothing in this section shall require a person with a disability to engage in a futile gesture if such person has actual notice that a person or organization covered by this title does not intend to comply with its provisions.

(2) INJUNCTIVE RELIEF- In the case of violations of sections 302(b)(2)(A)(iv) and section 303(a), injunctive relief shall include an order to alter facilities to make such facilities readily accessible to and usable by individuals with disabilities to the extent required by this title. Where appropriate, injunctive relief shall also include requiring the provision of an auxiliary aid or service, modification of a policy, or provision of alternative methods, to the extent required by this title.

(b) ENFORCEMENT BY THE ATTORNEY GENERAL-

(1) DENIAL OF RIGHTS-

(A) DUTY TO INVESTIGATE-

(i) IN GENERAL- The Attorney General shall investigate alleged violations of this title, and shall undertake periodic reviews of compliance of covered entities under this title.

(ii) ATTORNEY GENERAL CERTIFICATION- On the application of a State or local government, the Attorney General may, in consultation with the Architectural and Transportation Barriers Compliance Board, and after prior notice and a public hearing at which persons, including individuals with disabilities, are provided an opportunity to testify against such certification, certify that a State law or local building code or similar ordinance that establishes accessibility requirements meets or exceeds the minimum requirements of this Act for the accessibility and usability of covered facilities under this title. At any enforcement proceeding under this section, such certification by the Attorney General shall be rebuttable evidence that such State law or local ordinance does meet or exceed the minimum requirements of this Act.

(B) POTENTIAL VIOLATION- If the Attorney General has reasonable cause to believe that—

(i) any person or group of persons is engaged in a pattern or practice of discrimination under this title; or

(ii) any person or group of persons has been discriminated against under this title and such discrimination raises an issue of general public importance, the Attorney General may commence a civil action in any appropriate United States district court.

(2) AUTHORITY OF COURT- In a civil action under paragraph (1)(B), the court—

(A) may grant any equitable relief that such court considers to be appropriate, including, to the extent required by this title—

(i) granting temporary, preliminary, or permanent relief;

(ii) providing an auxiliary aid or service, modification of policy, practice, or procedure, or alternative method; and

(iii) making facilities readily accessible to and usable by individuals with disabilities;

(B) may award such other relief as the court considers to be appropriate, including monetary damages to persons aggrieved when requested by the Attorney General; and

(C) may, to vindicate the public interest, assess a civil penalty against the entity in an amount—

(i) not exceeding $50,000 for a first violation; and

(ii) not exceeding $100,000 for any subsequent violation.

(3) SINGLE VIOLATION- For purposes of paragraph (2)(C), in determining whether a first or subsequent violation has occurred, a determination in a single action, by judgment or settlement, that the covered entity has engaged in more than one discriminatory act shall be counted as a single violation.

(4) PUNITIVE DAMAGES- For purposes of subsection (b)(2)(B), the term "monetary damages" and "such other relief" does not include punitive damages.

(5) JUDICIAL CONSIDERATION- In a civil action under paragraph (1)(B), the court, when considering what amount of civil penalty, if any, is appropriate, shall give consideration to any good faith effort or attempt to comply with this Act by the entity. In evaluating good faith, the court shall consider, among other factors it

deems relevant, whether the entity could have reasonably anticipated the need for an appropriate type of auxiliary aid needed to accommodate the unique needs of a particular individual with a disability.

SEC. 309. EXAMINATIONS AND COURSES

Any person that offers examinations or courses related to applications, licensing, certification, or credentialing for secondary or postsecondary education, professional, or trade purposes shall offer such examinations or courses in a place and manner accessible to persons with disabilities or offer alternative accessible arrangements for such individuals.

SEC. 310. EFFECTIVE DATE

(a) GENERAL RULE- Except as provided in subsections (b) and (c), this title shall become effective 18 months after the date of the enactment of this Act.

(b) CIVIL ACTIONS- Except for any civil action brought for a violation of section 303, no civil action shall be brought for any act or omission described in section 302 which occurs—

(1) during the first 6 months after the effective date, against businesses that employ 25 or fewer employees and have gross receipts of $1,000,000 or less; and

(2) during the first year after the effective date, against businesses that employ 10 or fewer employees and have gross receipts of $500,000 or less.

(c) EXCEPTION- Sections 302(a) for purposes of section 302(b)(2) (B) and (C) only, 304(a) for purposes of section 304(b)(3) only, 304(b)(3), 305, and 306 shall take effect on the date of the enactment of this Act.

TITLE IV—TELECOMMUNICATIONS

SEC. 401. TELECOMMUNICATIONS RELAY SERVICES FOR HEARING-IMPAIRED AND SPEECH-IMPAIRED INDIVIDUALS

(a) TELECOMMUNICATIONS- Title II of the Communications Act of 1934 (47 U.S.C. 201 et seq.) is amended by adding at the end thereof the following new section:

SEC. 225. TELECOMMUNICATIONS SERVICES FOR HEARING-IMPAIRED AND SPEECH-IMPAIRED INDIVIDUALS.

(a) DEFINITIONS- As used in this section—

(1) COMMON CARRIER OR CARRIER- The term “common carrier” or “carrier” includes any common carrier engaged in interstate communication by wire or radio

as defined in section 3(h) and any common carrier engaged in intrastate communication by wire or radio, notwithstanding sections 2(b) and 221(b).

(2) TDD- The term “TDD” means a Telecommunications Device for the Deaf, which is a machine that employs graphic communication in the transmission of coded signals through a wire or radio communication system.

(3) TELECOMMUNICATIONS RELAY SERVICES- The term “telecommunications relay services” means telephone transmission services that provide the ability for an individual who has a hearing impairment or speech impairment to engage in communication by wire or radio with a hearing individual in a manner that is functionally equivalent to the ability of an individual who does not have a hearing impairment or speech impairment to communicate using voice communication services by wire or radio. Such term includes services that enable two-way communication between an individual who uses a TDD or other nonvoice terminal device and an individual who does not use such a device.

(b) AVAILABILITY OF TELECOMMUNICATIONS RELAY SERVICES-

(1) IN GENERAL- In order to carry out the purposes established under section 1, to make available to all individuals in the United States a rapid, efficient nationwide communication service, and to increase the utility of the telephone system of the Nation, the Commission shall ensure that interstate and intrastate telecommunications relay services are available, to the extent possible and in the most efficient manner, to hearing-impaired and speech-impaired individuals in the United States.

(2) USE OF GENERAL AUTHORITY AND REMEDIES- For the purposes of administering and enforcing the provisions of this section and the regulations prescribed thereunder, the Commission shall have the same authority, power, and functions with respect to common carriers engaged in intrastate communication as the Commission has in administering and enforcing the provisions of this title with respect to any common carrier engaged in interstate communication. Any violation of this section by any common carrier engaged in intrastate communication shall be subject to the same remedies, penalties, and procedures as are applicable to a violation of this Act by a common carrier engaged in interstate communication.

(c) PROVISION OF SERVICES- Each common carrier providing telephone voice transmission services shall, not later than 3 years after the date of enactment of this section, provide in compliance with the regulations prescribed under this section, throughout the area in which it offers service, telecommunications relay services, individually, through designees, through a competitively selected vendor, or in concert with other carriers. A common carrier shall be considered to be in compliance with such regulations—

(1) with respect to intrastate telecommunications relay services in any State that does not have a certified program under subsection (f) and with respect to interstate telecommunications relay services, if such common carrier (or other entity

through which the carrier is providing such relay services) is in compliance with the Commission's regulations under subsection (d); or

(2) with respect to intrastate telecommunications relay services in any State that has a certified program under subsection (f) for such State, if such common carrier (or other entity through which the carrier is providing such relay services) is in compliance with the program certified under subsection (f) for such State.

(d) REGULATIONS-

(1) IN GENERAL- The Commission shall, not later than 1 year after the date of enactment of this section, prescribe regulations to implement this section, including regulations that—

(A) establish functional requirements, guidelines, and operations procedures for telecommunications relay services;

(B) establish minimum standards that shall be met in carrying out subsection (c);

(C) require that telecommunications relay services operate every day for 24 hours per day;

(D) require that users of telecommunications relay services pay rates no greater than the rates paid for functionally equivalent voice communication services with respect to such factors as the duration of the call, the time of day, and the distance from point of origination to point of termination;

(E) prohibit relay operators from failing to fulfill the obligations of common carriers by refusing calls or limiting the length of calls that use telecommunications relay services;

(F) prohibit relay operators from disclosing the content of any relayed conversation and from keeping records of the content of any such conversation beyond the duration of the call; and

(G) prohibit relay operators from intentionally altering a relayed conversation.

(2) TECHNOLOGY- The Commission shall ensure that regulations prescribed to implement this section encourage, consistent with section 7(a) of this Act, the use of existing technology and do not discourage or impair the development of improved technology.

(3) JURISDICTIONAL SEPARATION OF COSTS-

(A) IN GENERAL- Consistent with the provisions of section 410 of this Act, the Commission shall prescribe regulations governing the jurisdictional separation of costs for the services provided pursuant to this section.

(B) RECOVERING COSTS- Such regulations shall generally provide that costs caused by interstate telecommunications relay services shall be recovered from all subscribers for every interstate service and costs caused by intrastate telecommunications relay services shall be recovered from the intrastate jurisdiction. In a State that has a certified program under subsection (f), a State commission shall permit a common carrier to recover the costs incurred in providing intrastate telecommunications relay services by a method consistent with the requirements of this section.

(e) ENFORCEMENT-

(1) IN GENERAL- Subject to subsections (f) and (g), the Commission shall enforce this section.

(2) COMPLAINT- The Commission shall resolve, by final order, a complaint alleging a violation of this section within 180 days after the date such complaint is filed.

(f) CERTIFICATION-

(1) STATE DOCUMENTATION- Any State desiring to establish a State program under this section shall submit documentation to the Commission that describes the program of such State for implementing intrastate telecommunications relay services and the procedures and remedies available for enforcing any requirements imposed by the State program.

(2) REQUIREMENTS FOR CERTIFICATION- After review of such documentation, the Commission shall certify the State program if the Commission determines that—

(A) the program makes available to hearing-impaired and speech-impaired individuals, either directly, through designees, through a competitively selected vendor, or through regulation of intrastate common carriers, intrastate telecommunications relay services in such State in a manner that meets or exceeds the requirements of regulations prescribed by the Commission under subsection (d); and

(B) the program makes available adequate procedures and remedies for enforcing the requirements of the State program.

(3) METHOD OF FUNDING- Except as provided in subsection (d), the Commission shall not refuse to certify a State program based solely on the method such State will implement for funding intrastate telecommunication relay services.

(4) SUSPENSION OR REVOCATION OF CERTIFICATION- The Commission may suspend or revoke such certification if, after notice and opportunity for hearing, the Commission determines that such certification is no longer warranted. In a State whose program has been suspended or revoked, the Commission shall take such steps as may be necessary, consistent with this section, to ensure continuity of telecommunications relay services.

(g) COMPLAINT-

(1) REFERRAL OF COMPLAINT- If a complaint to the Commission alleges a violation of this section with respect to intrastate telecommunications relay services

within a State and certification of the program of such State under subsection (f) is in effect, the Commission shall refer such complaint to such State.

(2) JURISDICTION OF COMMISSION- After referring a complaint to a State under paragraph (1), the Commission shall exercise jurisdiction over such complaint only if—

(A) final action under such State program has not been taken on such complaint by such State—

(i) within 180 days after the complaint is filed with such State; or

(ii) within a shorter period as prescribed by the regulations of such State; or

(B) the Commission determines that such State program is no longer qualified for certification under subsection (f).

(b) CONFORMING AMENDMENTS- The Communications Act of 1934 (47 U.S.C. 151 et seq.) is amended—

(1) in section 2(b) (47 U.S.C. 152(b)), by striking "section 224" and inserting "sections 224 and 225"; and

(2) in section 221(b) (47 U.S.C. 221(b)), by striking "section 301" and inserting "sections 225 and 301".

SEC. 402. CLOSED-CAPTIONING OF PUBLIC SERVICE ANNOUNCEMENTS

Section 711 of the Communications Act of 1934 is amended to read as follows:

SEC. 711. CLOSED-CAPTIONING OF PUBLIC SERVICE ANNOUNCEMENTS

Any television public service announcement that is produced or funded in whole or in part by any agency or instrumentality of Federal Government shall include closed captioning of the verbal content of such announcement. A television broadcast station licensee—

(1) shall not be required to supply closed captioning for any such announcement that fails to include it; and

(2) shall not be liable for broadcasting any such announcement without transmitting a closed caption unless the licensee intentionally fails to transmit the closed caption that was included with the announcement.

TITLE V—MISCELLANEOUS PROVISIONS

SEC. 501. CONSTRUCTION

(a) IN GENERAL- Except as otherwise provided in this Act, nothing in this Act shall be construed to apply a lesser standard than the standards applied under title V of the Rehabilitation Act of 1973 (29 U.S.C. 790 et seq.) or the regulations issued by Federal agencies pursuant to such title.

(b) RELATIONSHIP TO OTHER LAWS- Nothing in this Act shall be construed to invalidate or limit the remedies, rights, and procedures of any Federal law or law of any State or political subdivision of any State or jurisdiction that provides greater or equal protection for the rights of individuals with disabilities than are afforded by this Act. Nothing in this Act shall be construed to preclude the prohibition of, or the imposition of restrictions on, smoking in places of employment covered by title I, in transportation covered by title II or III, or in places of public accommodation covered by title III.

(c) INSURANCE- Titles I through IV of this Act shall not be construed to prohibit or restrict—

(1) an insurer, hospital or medical service company, health maintenance organization, or any agent, or entity that administers benefit plans, or similar organizations from underwriting risks, classifying risks, or administering such risks that are based on or not inconsistent with State law; or

(2) a person or organization covered by this Act from establishing, sponsoring, observing or administering the terms of a bona fide benefit plan that are based on underwriting risks, classifying risks, or administering such risks that are based on or not inconsistent with State law; or

(3) a person or organization covered by this Act from establishing, sponsoring, observing or administering the terms of a bona fide benefit plan that is not subject to State laws that regulate insurance.

Paragraphs (1), (2), and (3) shall not be used as a subterfuge to evade the purposes of title I and III.

(d) ACCOMMODATIONS AND SERVICES- Nothing in this Act shall be construed to require an individual with a disability to accept an accommodation, aid, service, opportunity, or benefit which such individual chooses not to accept.

SEC. 502. STATE IMMUNITY

A State shall not be immune under the eleventh amendment to the Constitution of the United States from an action in Federal or State court of competent jurisdiction for a violation of this Act. In any action against a State for a violation of the requirements of this Act, remedies (including remedies both at law and in equity) are available for such a violation to the same extent as such remedies are available for such a violation in an action against any public or private entity other than a State.

SEC. 503. PROHIBITION AGAINST RETALIATION AND COERCION

(a) RETALIATION- No person shall discriminate against any individual because such individual has opposed any act or practice made unlawful by this Act or because such individual made a charge, testified, assisted, or participated in any manner in an investigation, proceeding, or hearing under this Act.

(b) INTERFERENCE, COERCION, OR INTIMIDATION- It shall be unlawful to coerce, intimidate, threaten, or interfere with any individual in the exercise or enjoyment of, or on account of his or her having exercised or enjoyed, or on account of his or her having aided or encouraged any other individual in the exercise or enjoyment of, any right granted or protected by this Act.

(c) REMEDIES AND PROCEDURES- The remedies and procedures available under sections 107, 203, and 308 of this Act shall be available to aggrieved persons for violations of subsections (a) and (b), with respect to title I, title II and title III, respectively.

SEC. 504. REGULATIONS BY THE ARCHITECTURAL AND TRANSPORTATION BARRIERS COMPLIANCE BOARD

(a) ISSUANCE OF GUIDELINES- Not later than 9 months after the date of enactment of this Act, the Architectural and Transportation Barriers Compliance Board shall issue minimum guidelines that shall supplement the existing Minimum Guidelines and Requirements for Accessible Design for purposes of titles II and III of this Act.

(b) CONTENTS OF GUIDELINES- The supplemental guidelines issued under subsection (a) shall establish additional requirements, consistent with this Act, to ensure that buildings, facilities, rail passenger cars, and vehicles are accessible, in terms of architecture and design, transportation, and communication, to individuals with disabilities.

(c) QUALIFIED HISTORIC PROPERTIES-

(1) IN GENERAL- The supplemental guidelines issued under subsection (a) shall include procedures and requirements for alterations that will threaten or destroy the historic significance of qualified historic buildings and facilities as defined in 4.1.7(1)(a) of the Uniform Federal Accessibility Standards.

(2) SITES ELIGIBLE FOR LISTING IN NATIONAL REGISTER- With respect to alterations of buildings or facilities that are eligible for listing in the National Register of Historic Places under the National Historic Preservation Act (16 U.S.C. 470 et seq.), the guidelines described in paragraph (1) shall, at a minimum, maintain the procedures and requirements established in 4.1.7 (1) and (2) of the Uniform Federal Accessibility Standards.

(3) OTHER SITES- With respect to alterations of buildings or facilities designated as historic under State or local law, the guidelines described in paragraph (1) shall establish procedures equivalent to those established by 4.1.7(1) (b) and (c) of the Uniform Federal Accessibility Standards, and shall require, at a minimum, compliance with the requirements established in 4.1.7(2) of such standards.

SEC. 505. ATTORNEY'S FEES

In any action or administrative proceeding commenced pursuant to this Act, the court or agency, in its discretion, may allow the prevailing party, other than the United States, a reasonable attorney's fee, including litigation expenses, and costs, and the United States shall be liable for the foregoing the same as a private individual.

SEC. 506. TECHNICAL ASSISTANCE

(a) PLAN FOR ASSISTANCE-

(1) IN GENERAL- Not later than 180 days after the date of enactment of this Act, the Attorney General, in consultation with the Chair of the Equal Employment Opportunity Commission, the Secretary of Transportation, the Chair of the Architectural and Transportation Barriers Compliance Board, and the Chairman of the Federal Communications Commission, shall develop a plan to assist entities covered under this Act, and other Federal agencies, in understanding the responsibility of such entities and agencies under this Act.

(2) PUBLICATION OF PLAN- The Attorney General shall publish the plan referred to in paragraph (1) for public comment in accordance with subchapter II of chapter 5 of title 5, United States Code (commonly known as the Administrative Procedure Act).

(b) AGENCY AND PUBLIC ASSISTANCE- The Attorney General may obtain the assistance of other Federal agencies in carrying out subsection (a), including the National Council on Disability, the President's Committee on Employment of People with Disabilities, the Small Business Administration, and the Department of Commerce.

(c) IMPLEMENTATION-

(1) RENDERING ASSISTANCE- Each Federal agency that has responsibility under paragraph (2) for implementing this Act may render technical assistance to individuals and institutions that have rights or duties under the respective title or titles for which such agency has responsibility.

(2) IMPLEMENTATION OF TITLES-

(A) TITLE I- The Equal Employment Opportunity Commission and the Attorney General shall implement the plan for assistance developed under subsection (a), for title I.

(B) TITLE II-

(i) SUBTITLE A- The Attorney General shall implement such plan for assistance for subtitle A of title II.

(ii) SUBTITLE B- The Secretary of Transportation shall implement such plan for assistance for subtitle B of title II.

(C) TITLE III- The Attorney General, in coordination with the Secretary of Transportation and the Chair of the Architectural Transportation Barriers Compliance Board, shall implement such plan for assistance for title III, except for section 304, the plan for assistance for which shall be implemented by the Secretary of Transportation.

(D) TITLE IV- The Chairman of the Federal Communications Commission, in coordination with the Attorney General, shall implement such plan for assistance for title IV.

(3) TECHNICAL ASSISTANCE MANUALS- Each Federal agency that has responsibility under paragraph (2) for implementing this Act shall, as part of its implementation responsibilities, ensure the availability and provision of appropriate technical assistance manuals to individuals or entities with rights or duties under this Act no later than six months after applicable final regulations are published under titles I, II, III, and IV.

(d) GRANTS AND CONTRACTS-

(1) IN GENERAL- Each Federal agency that has responsibility under subsection (c)(2) for implementing this Act may make grants or award contracts to effectuate the purposes of this section, subject to the availability of appropriations. Such grants and contracts may be awarded to individuals, institutions not organized for profit and no part of the net earnings of which inures to the benefit of any private shareholder or individual (including educational institutions), and associations representing individuals who have rights or duties under this Act. Contracts may be awarded to entities organized for profit, but such entities may not be the recipients or grants described in this paragraph.

(2) DISSEMINATION OF INFORMATION- Such grants and contracts, among other uses, may be designed to ensure wide dissemination of information about the rights and duties established by this Act and to provide information and technical assistance about techniques for effective compliance with this Act.

(e) FAILURE TO RECEIVE ASSISTANCE- An employer, public accommodation, or other entity covered under this Act shall not be excused from compliance with the requirements of this Act because of any failure to receive technical assistance under this section, including any failure in the development or dissemination of any technical assistance manual authorized by this section.

SEC. 507. FEDERAL WILDERNESS AREAS

(a) STUDY- The National Council on Disability shall conduct a study and report on the effect that wilderness designations and wilderness land management practices have on the ability of individuals with disabilities to use and enjoy the National Wilderness Preservation System as established under the Wilderness Act (16 U.S.C. 1131 et seq.).

(b) SUBMISSION OF REPORT- Not later than 1 year after the enactment of this Act, the National Council on Disability shall submit the report required under subsection (a) to Congress.

(c) SPECIFIC WILDERNESS ACCESS-

(1) IN GENERAL- Congress reaffirms that nothing in the Wilderness Act is to be construed as prohibiting the use of a wheelchair in a wilderness area by an individual whose disability requires use of a wheelchair, and consistent with the Wilderness Act no agency is required to provide any form of special treatment or accommodation, or to construct any facilities or modify any conditions of lands within a wilderness area in order to facilitate such use.

(2) DEFINITION- For purposes of paragraph (1), the term "wheelchair" means a device designed solely for use by a mobility-impaired person for locomotion, that is suitable for use in an indoor pedestrian area.

SEC. 508. TRANSVESTITES

For the purposes of this Act, the term "disabled" or "disability" shall not apply to an individual solely because that individual is a transvestite.

SEC. 509. COVERAGE OF CONGRESS AND THE AGENCIES OF THE LEGISLATIVE BRANCH

(a) COVERAGE OF THE SENATE-

(1) COMMITMENT TO RULE XLII- The Senate reaffirms its commitment to Rule XLII of the Standing Rules of the Senate which provides as follows:

No member, officer, or employee of the Senate shall, with respect to employment by the Senate or any office thereof—

(a) fail or refuse to hire an individual;

(b) discharge an individual; or

(c) otherwise discriminate against an individual with respect to promotion, compensation, or terms, conditions, or privileges of employment on the basis of such individual's race, color, religion, sex, national origin, age, or state of physical handicap.

(2) APPLICATION TO SENATE EMPLOYMENT- The rights and protections provided pursuant to this Act, the Civil Rights Act of 1990 (S. 2104, 101st Congress), the Civil Rights Act of 1964, the Age Discrimination in Employment Act of 1967, and the Rehabilitation Act of 1973 shall apply with respect to employment by the United States Senate.

(3) INVESTIGATION AND ADJUDICATION OF CLAIMS- All claims raised by any individual with respect to Senate employment, pursuant to the Acts referred to

in paragraph (2), shall be investigated and adjudicated by the Select Committee on Ethics, pursuant to S. Res. 338, 88th Congress, as amended, or such other entity as the Senate may designate.

(4) RIGHTS OF EMPLOYEES- The Committee on Rules and Administration shall ensure that Senate employees are informed of their rights under the Acts referred to in paragraph (2).

(5) APPLICABLE REMEDIES- When assigning remedies to individuals found to have a valid claim under the Acts referred to in paragraph (2), the Select Committee on Ethics, or such other entity as the Senate may designate, should to the extent practicable apply the same remedies applicable to all other employees covered by the Acts referred to in paragraph (2). Such remedies shall apply exclusively.

(6) MATTERS OTHER THAN EMPLOYMENT-

(A) IN GENERAL- The rights and protections under this Act shall, subject to subparagraph (B), apply with respect to the conduct of the Senate regarding matters other than employment.

(B) REMEDIES- The Architect of the Capitol shall establish remedies and procedures to be utilized with respect to the rights and protections provided pursuant to subparagraph (A). Such remedies and procedures shall apply exclusively, after approval in accordance with subparagraph (C).

(C) PROPOSED REMEDIES AND PROCEDURES- For purposes of subparagraph (B), the Architect of the Capitol shall submit proposed remedies and procedures to the Senate Committee on Rules and Administration. The remedies and procedures shall be effective upon the approval of the Committee on Rules and Administration.

(7) EXERCISE OF RULEMAKING POWER- Notwithstanding any other provision of law, enforcement and adjudication of the rights and protections referred to in paragraph (2) and (6)(A) shall be within the exclusive jurisdiction of the United States Senate. The provisions of paragraph (1), (3), (4), (5), (6)(B), and (6)(C) are enacted by the Senate as an exercise of the rulemaking power of the Senate, with full recognition of the right of the Senate to change its rules, in the same manner, and to the same extent, as in the case of any other rule of the Senate.

(b) COVERAGE OF THE HOUSE OF REPRESENTATIVES-

(1) IN GENERAL- Notwithstanding any other provision of this Act or of law, the purposes of this Act shall, subject to paragraphs (2) and (3), apply in their entirety to the House of Representatives.

(2) EMPLOYMENT IN THE HOUSE-

(A) APPLICATION- The rights and protections under this Act shall, subject to subparagraph (B), apply with respect to any employee in an employment position in the House of Representatives and any employing authority of the House of Representatives.

(B) ADMINISTRATION-

(i) IN GENERAL- In the administration of this paragraph, the remedies and procedures made applicable pursuant to the resolution described in clause (ii) shall apply exclusively.

(ii) RESOLUTION- The resolution referred to in clause (i) is House Resolution 15 of the One Hundred First Congress, as agreed to January 3, 1989, or any other provision that continues in effect the provisions of, or is a successor to, the Fair Employment Practices Resolution (House Resolution 558 of the One Hundredth Congress, as agreed to October 4, 1988).

(C) EXERCISE OF RULEMAKING POWER- The provisions of subparagraph (B) are enacted by the House of Representatives as an exercise of the rulemaking power of the House of Representatives, with full recognition of the right of the House to change its rules, in the same manner, and to the same extent as in the case of any other rule of the House.

(3) MATTERS OTHER THAN EMPLOYMENT-

(A) IN GENERAL- The rights and protections under this Act shall, subject to subparagraph (B), apply with respect to the conduct of the House of Representatives regarding matters other than employment.

(B) REMEDIES- The Architect of the Capitol shall establish remedies and procedures to be utilized with respect to the rights and protections provided pursuant to subparagraph (A). Such remedies and procedures shall apply exclusively, after approval in accordance with subparagraph (C).

(C) APPROVAL- For purposes of subparagraph (B), the Architect of the Capitol shall submit proposed remedies and procedures to the Speaker of the House of Representatives. The remedies and procedures shall be effective upon the approval of the Speaker, after consultation with the House Office Building Commission.

(c) INSTRUMENTALITIES OF CONGRESS-

(1) IN GENERAL- The rights and protections under this Act shall, subject to paragraph (2), apply with respect to the conduct of each instrumentality of the Congress.

(2) ESTABLISHMENT OF REMEDIES AND PROCEDURES BY INSTRUMENTALITIES- The chief official of each instrumentality of the Congress shall establish remedies and procedures to be utilized with respect to the rights and protections provided pursuant to paragraph (1). Such remedies and procedures shall apply exclusively.

(3) REPORT TO CONGRESS- The chief official of each instrumentality of the Congress shall, after establishing remedies and procedures for purposes of paragraph (2), submit to the Congress a report describing the remedies and procedures.

(4) DEFINITION OF INSTRUMENTALITIES- For purposes of this section, instrumentalities of the Congress include the following: the Architect of the Capitol, the Congressional Budget Office, the General Accounting Office, the Government Printing Office, the Library of Congress, the Office of Technology Assessment, and the United States Botanic Garden.

(5) CONSTRUCTION- Nothing in this section shall alter the enforcement procedures for individuals with disabilities provided in the General Accounting Office Personnel Act of 1980 and regulations promulgated pursuant to that Act.

SEC. 510. ILLEGAL USE OF DRUGS

(a) IN GENERAL- For purposes of this Act, the term "individual with a disability" does not include an individual who is currently engaging in the illegal use of drugs, when the covered entity acts on the basis of such use.

(b) RULES OF CONSTRUCTION- Nothing in subsection (a) shall be construed to exclude as an individual with a disability an individual who—

(1) has successfully completed a supervised drug rehabilitation program and is no longer engaging in the illegal use of drugs, or has otherwise been rehabilitated successfully and is no longer engaging in such use;

(2) is participating in a supervised rehabilitation program and is no longer engaging in such use; or

(3) is erroneously regarded as engaging in such use, but is not engaging in such use; except that it shall not be a violation of this Act for a covered entity to adopt or administer reasonable policies or procedures, including but not limited to drug testing, designed to ensure that an individual described in paragraph (1) or (2) is no longer engaging in the illegal use of drugs; however, nothing in this section shall be construed to encourage, prohibit, restrict, or authorize the conducting of testing for the illegal use of drugs.

(c) HEALTH AND OTHER SERVICES- Notwithstanding subsection (a) and section 511(b)(3), an individual shall not be denied health services, or services provided in connection with drug rehabilitation, on the basis of the current illegal use of drugs if the individual is otherwise entitled to such services.

(d) DEFINITION OF ILLEGAL USE OF DRUGS-

(1) IN GENERAL- The term "illegal use of drugs" means the use of drugs, the possession or distribution of which is unlawful under the Controlled Substances Act (21 U.S.C. 812). Such term does not include the use of a drug taken under supervision by a licensed health care professional, or other uses authorized by the Controlled Substances Act or other provisions of Federal law.

(2) DRUGS- The term "drug" means a controlled substance, as defined in schedules I through V of section 202 of the Controlled Substances Act.

SEC. 511. DEFINITIONS

(a) HOMOSEXUALITY AND BISEXUALITY- For purposes of the definition of "disability" in section 3(2), homosexuality and bisexuality are not impairments and as such are not disabilities under this Act.

(b) CERTAIN CONDITIONS- Under this Act, the term "disability" shall not include—

(1) transvestism, transsexualism, pedophilia, exhibitionism, voyeurism, gender identity disorders not resulting from physical impairments, or other sexual behavior disorders;

(2) compulsive gambling, kleptomania, or pyromania; or

(3) psychoactive substance use disorders resulting from current illegal use of drugs.

SEC. 512. AMENDMENTS TO THE REHABILITATION ACT

(a) DEFINITION OF HANDICAPPED INDIVIDUAL- Section 7(8) of the Rehabilitation Act of 1973 (29 U.S.C. 706(8)) is amended by redesignating subparagraph (C) as subparagraph (D), and by inserting after subparagraph (B) the following subparagraph:

(C)(i) For purposes of title V, the term "individual with handicaps" does not include an individual who is currently engaging in the illegal use of drugs, when a covered entity acts on the basis of such use.

(ii) Nothing in clause (i) shall be construed to exclude as an individual with handicaps an individual who—

(I) has successfully completed a supervised drug rehabilitation program and is no longer engaging in the illegal use of drugs, or has otherwise been rehabilitated successfully and is no longer engaging in such use;

(II) is participating in a supervised rehabilitation program and is no longer engaging in such use; or

(III) is erroneously regarded as engaging in such use, but is not engaging in such use; except that it shall not be a violation of this Act for a covered entity to adopt or administer reasonable policies or procedures, including but not limited to drug testing, designed to ensure that an individual described in subclause (I) or (II) is no longer engaging in the illegal use of drugs.

(iii) Notwithstanding clause (i), for purposes of programs and activities providing health services and services provided under titles I, II and III, an individual shall not be excluded from the benefits of such programs or activities on the basis of his or her current illegal use of drugs if he or she is otherwise entitled to such services.

(iv) For purposes of programs and activities providing educational services, local educational agencies may take disciplinary action pertaining to the use or possession of illegal drugs or alcohol against any handicapped student who currently is engaging in the illegal use of drugs or in the use of alcohol to the same extent that such disciplinary action is taken against nonhandicapped students. Furthermore, the due process procedures at 34 CFR 104.36 shall not apply to such disciplinary actions.

(v) For purposes of sections 503 and 504 as such sections relate to employment, the term "individual with handicaps" does not include any individual who is an alcoholic whose current use of alcohol prevents such individual from performing the duties of the job in question or whose employment, by reason of such current alcohol abuse, would constitute a direct threat to property or the safety of others.

(b) DEFINITION OF ILLEGAL DRUGS- Section 7 of the Rehabilitation Act of 1973 (29 U.S.C. 706) is amended by adding at the end the following new paragraph:

(22)(A) The term "drug" means a controlled substance, as defined in schedules I through V of section 202 of the Controlled Substances Act (21 U.S.C. 812).

(B) The term "illegal use of drugs" means the use of drugs, the possession or distribution of which is unlawful under the Controlled Substances Act. Such term does not include the use of a drug taken under supervision by a licensed health care professional, or other uses authorized by the Controlled Substances Act or other provisions of Federal law.

(c) CONFORMING AMENDMENTS- Section 7(8)(B) of the Rehabilitation Act of 1973 (29 U.S.C. 706(8)(B)) is amended—

(1) in the first sentence, by striking "Subject to the second sentence of this subparagraph," and inserting "Subject to subparagraphs (C) and (D)"; and

(2) by striking the second sentence.

SEC. 513. ALTERNATIVE MEANS OF DISPUTE RESOLUTION

Where appropriate and to the extent authorized by law, the use of alternative means of dispute resolution, including settlement negotiations, conciliation, facilitation, mediation, factfinding, minitrials, and arbitration, is encouraged to resolve disputes arising under this Act.

SEC. 514. SEVERABILITY

Should any provision in this Act be found to be unconstitutional by a court of law, such provision shall be severed from the remainder of the Act, and such action shall not affect the enforceability of the remaining provisions of the Act.

Appendix 2: Title VII of the Civil Rights Act of 1964

AN ACT

To enforce the constitutional right to vote, to confer jurisdiction upon the district courts of the United States to provide injunctive relief against discrimination in public accommodations, to authorize the attorney General to institute suits to protect constitutional rights in public facilities and public education, to extend the Commission on Civil Rights, to prevent discrimination in federally assisted programs, to establish a Commission on Equal Employment Opportunity, and for other purposes.

Be it enacted by the Senate and House of Representatives of the United States of America in Congress assembled, That this Act may be cited as the "Civil Rights Act of 1964".

* * *

DEFINITIONS

SEC. 2000e. [*Section 701*]

For the purposes of this subchapter-

(a) The term "person" includes one or more individuals, governments, governmental agencies, political subdivisions, labor unions, partnerships, associations, corporations, legal representatives, mutual companies, joint-stock companies, trusts, unincorporated organizations, trustees, trustees in cases under Title 11 [*originally, bankruptcy*], or receivers.

(b) The term "employer" means a person engaged in an industry affecting commerce who has fifteen or more employees for each working day in each of twenty or more calendar weeks in the current or preceding calendar year, and any agent of such a person, but such term does not include (1) the United States, a corporation wholly owned by the Government of the United States, an Indian tribe, or any department or agency of the District of Columbia subject by statute to procedures of the competitive service (as defined in section 2102 of Title 5 [*United States Code*]), or

(2) a bona fide private membership club (other than a labor organization) which is exempt from taxation under section 501(c) of Title 26 [*the Internal Revenue Code of 1986*], except that during the first year after March 24, 1972 [*the date of enactment of the Equal Employment Opportunity Act of 1972*], persons having fewer than twenty-five employees (and their agents) shall not be considered employers.

(c) The term "employment agency" means any person regularly undertaking with or without compensation to procure employees for an employer or to procure for employees opportunities to work for an employer and includes an agent of such a person.

(d) The term "labor organization" means a labor organization engaged in an industry affecting commerce, and any agent of such an organization, and includes any organization of any kind, any agency, or employee representation committee, group, association, or plan so engaged in which employees participate and which exists for the purpose, in whole or in part, of dealing with employers concerning grievances, labor disputes, wages, rates of pay, hours, or other terms or conditions of employment, and any conference, general committee, joint or system board, or joint council so engaged which is subordinate to a national or international labor organization.

(e) A labor organization shall be deemed to be engaged in an industry affecting commerce if (1) it maintains or operates a hiring hall or hiring office which procures employees for an employer or procures for employees opportunities to work for an employer, or (2) the number of its members (or, where it is a labor organization composed of other labor organizations or their representatives, if the aggregate number of the members of such other labor organization) is (A) twenty-five or more during the first year after March 24, 1972 [*the date of enactment of the Equal Employment Opportunity Act of 1972*], or (B) fifteen or more thereafter, and such labor organization-

> (1) is the certified representative of employees under the provisions of the National Labor Relations Act, as amended [*29 U.S.C. 151 et seq.*], or the Railway Labor Act, as amended [*45 U.S.C. 151 et seq.*];
>
> (2) although not certified, is a national or international labor organization or a local labor organization recognized or acting as the representative of employees of an employer or employers engaged in an industry affecting commerce; or
>
> (3) has chartered a local labor organization or subsidiary body which is representing or actively seeking to represent employees of employers within the meaning of paragraph (1) or (2); or
>
> (4) has been chartered by a labor organization representing or actively seeking to represent employees within the meaning of paragraph (1) or (2) as the local or subordinate body through which such employees may enjoy membership or become affiliated with such labor organization; or
>
> (5) is a conference, general committee, joint or system board, or joint council subordinate to a national or international labor organization, which includes a labor organization engaged in an industry affecting commerce within the meaning of any of the preceding paragraphs of this subsection.

(f) The term "employee" means an individual employed by an employer, except that the term "employee" shall not include any person elected to public office in any State or political subdivision of any State by the qualified voters thereof, or any person chosen by such officer to be on such officer's personal staff, or an appointee on the policy making level or an immediate adviser with respect to the exercise of the constitutional or legal powers of the office. The exemption set forth in the preceding sentence shall not include employees subject to the civil service laws of a State government, governmental agency or political subdivision. With respect to employment in a foreign country, such term includes an individual who is a citizen of the United States.

(g) The term "commerce" means trade, traffic, commerce, transportation, transmission, or communication among the several States; or between a State and any place outside thereof; or within the District of Columbia, or a possession of the United States; or between points in the same State but through a point outside thereof.

(h) The term "industry affecting commerce" means any activity, business, or industry in commerce or in which a labor dispute would hinder or obstruct commerce or the free flow of commerce and includes any activity or industry "affecting commerce" within the meaning of the Labor-Management Reporting and Disclosure Act of 1959 [*29 U.S.C. 401 et seq.*], and further includes any governmental industry, business, or activity.

(i) The term "State" includes a State of the United States, the District of Columbia, Puerto Rico, the Virgin Islands, American Samoa, Guam, Wake Island, the Canal Zone, and Outer Continental Shelf lands defined in the Outer Continental Shelf Lands Act [*43 U.S.C. 1331 et seq.*].

(j) The term "religion" includes all aspects of religious observance and practice, as well as belief, unless an employer demonstrates that he is unable to reasonably accommodate to an employee's or prospective employee's religious observance or practice without undue hardship on the conduct of the employer's business.

(k) The terms "because of sex" or "on the basis of sex" include, but are not limited to, because of or on the basis of pregnancy, childbirth, or related medical conditions; and women affected by pregnancy, childbirth, or related medical conditions shall be treated the same for all employment-related purposes, including receipt of benefits under fringe benefit programs, as other persons not so affected but similar in their ability or inability to work, and nothing in section 2000e-2(h) of this title [*section 703(h)*] shall be interpreted to permit otherwise. This subsection shall not require an employer to pay for health insurance benefits for abortion, except where the life of the mother would be endangered if the fetus were carried to term, or except where medical complications have arisen from an abortion: Provided, That nothing herein shall preclude an employer from providing abortion benefits or otherwise affect bargaining agreements in regard to abortion.

(l) The term "complaining party" means the Commission, the Attorney General, or a person who may bring an action or proceeding under this subchapter.

(m) The term "demonstrates" means meets the burdens of production and persuasion.

(n) The term "respondent" means an employer, employment agency, labor organization, joint labor management committee controlling apprenticeship or other training or retraining program, including an on-the-job training program, or Federal entity subject to section 2000e-16 of this title.

APPLICABILITY TO FOREIGN AND RELIGIOUS EMPLOYMENT

SEC. 2000e-1. [*Section 702*]

(a) Inapplicability of subchapter to certain aliens and employees of religious entities

This subchapter shall not apply to an employer with respect to the employment of aliens outside any State, or to a religious corporation, association, educational institution, or society with respect to the employment of individuals of a particular religion to perform work connected with the carrying on by such corporation, association, educational institution, or society of its activities.

(b) Compliance with statute as violative of foreign law

It shall not be unlawful under section 2000e-2 or 2000e-3 of this title [*section 703 or 704*] for an employer (or a corporation controlled by an employer), labor organization, employment agency, or joint labor-management committee controlling apprenticeship or other training or retraining (including on-the-job training programs) to take any action otherwise prohibited by such section, with respect to an employee in a workplace in a foreign country if compliance with such section would cause such employer (or such corporation), such organization, such agency, or such committee to violate the law of the foreign country in which such workplace is located.

(c) Control of corporation incorporated in foreign country

(1) If an employer controls a corporation whose place of incorporation is a foreign country, any practice prohibited by section 2000e-2 or 2000e-3 of this title [*section 703 or 704*] engaged in by such corporation shall be presumed to be engaged in by such employer.

(2) Sections 2000e-2 and 2000e-3 of this title [*sections 703 and 704*] shall not apply with respect to the foreign operations of an employer that is a foreign person not controlled by an American employer.

(3) For purposes of this subsection, the determination of whether an employer controls a corporation shall be based on-

(A) the interrelation of operations;

(B) the common management;

(C) the centralized control of labor relations; and

(D) the common ownership or financial control, of the employer and the corporation.

UNLAWFUL EMPLOYMENT PRACTICES

SEC. 2000e-2. [*Section 703*]

(a) Employer practices

It shall be an unlawful employment practice for an employer -

(1) to fail or refuse to hire or to discharge any individual, or otherwise to discriminate against any individual with respect to his compensation, terms, conditions, or privileges of employment, because of such individual's race, color, religion, sex, or national origin; or

(2) to limit, segregate, or classify his employees or applicants for employment in any way which would deprive or tend to deprive any individual of employment opportunities or otherwise adversely affect his status as an employee, because of such individual's race, color, religion, sex, or national origin.

(b) Employment agency practices

It shall be an unlawful employment practice for an employment agency to fail or refuse to refer for employment, or otherwise to discriminate against, any individual because of his race, color, religion, sex, or national origin, or to classify or refer for employment any individual on the basis of his race, color, religion, sex, or national origin.

(c) Labor organization practices

It shall be an unlawful employment practice for a labor organization-

(1) to exclude or to expel from its membership, or otherwise to discriminate against, any individual because of his race, color, religion, sex, or national origin;

(2) to limit, segregate, or classify its membership or applicants for membership, or to classify or fail or refuse to refer for employment any individual, in any way which would deprive or tend to deprive any individual of employment opportunities, or would limit such employment opportunities or otherwise adversely affect his status as an employee or as an applicant for employment, because of such individual's race, color, religion, sex, or national origin; or

(3) to cause or attempt to cause an employer to discriminate against an individual in violation of this section.

(d) Training programs

It shall be an unlawful employment practice for any employer, labor organization, or joint labor-management committee controlling apprenticeship or other training or retraining, including on-the-job training programs to discriminate against any individual because of his race, color, religion, sex, or national origin in admission to, or employment in, any program established to provide apprenticeship or other training.

(e) Businesses or enterprises with personnel qualified on basis of religion, sex, or national origin; educational institutions with personnel of particular religion

Notwithstanding any other provision of this subchapter, (1) it shall not be an unlawful employment practice for an employer to hire and employ employees, for an employment agency to classify, or refer for employment any individual, for a labor organization to classify its membership or to classify or refer for employment any individual, or for an employer, labor organization, or joint labor management committee controlling apprenticeship or other training or retraining programs to admit or employ any individual in any such program, on the basis of his religion, sex, or national origin in those certain instances where religion, sex, or national origin is a bona fide occupational qualification reasonably necessary to the normal operation of that particular business or enterprise, and (2) it shall not be an unlawful employment practice for a school, college, university, or other educational institution or institution of learning to hire and employ employees of a particular religion if such school, college, university, or other educational institution or institution of learning is, in whole or in substantial part, owned, supported, controlled, or managed by a particular religion or by a particular religious corporation, association, or society, or if the curriculum of such school, college, university, or other educational institution or institution of learning is directed toward the propagation of a particular religion.

(f) Members of Communist Party or Communist-action or Communist-front organizations

As used in this subchapter, the phrase "unlawful employment practice" shall not be deemed to include any action or measure taken by an employer, labor organization, joint labor management committee, or employment agency with respect to an individual who is a member of the Communist Party of the United States or of any other organization required to register as a Communist-action or Communist-front organization by final order of the Subversive Activities Control Board pursuant to the Subversive Activities Control Act of 1950 [*50 U.S.C. 781 et seq.*].

(g) National security

Notwithstanding any other provision of this subchapter, it shall not be an unlawful employment practice for an employer to fail or refuse to hire and employ any individual for any position, for an employer to discharge any individual from any position, or for an employment agency to fail or refuse to refer any individual for employment in any position, or for a labor organization to fail or refuse to refer any individual for employment in any position, if-

(1) the occupancy of such position, or access to the premises in or upon which any part of the duties of such position is performed or is to be performed, is subject to any requirement imposed in the interest of the national security of the United States under any security program in effect pursuant to or administered under any statute of the United States or any Executive order of the President; and

(2) such individual has not fulfilled or has ceased to fulfill that requirement.

(h) Seniority or merit system; quantity or quality of production; ability tests; compensation based on sex and authorized by minimum wage provisions

Notwithstanding any other provision of this subchapter, it shall not be an unlawful employment practice for an employer to apply different standards of compensation, or different terms, conditions, or privileges of employment pursuant to a bona fide seniority or merit system, or a system which measures earnings by quantity or quality of production or to employees who work in different locations, provided that such differences are not the result of an intention to discriminate because of race, color, religion, sex, or national origin, nor shall it be an unlawful employment practice for an employer to give and to act upon the results of any professionally developed ability test provided that such test, its administration or action upon the results is not designed, intended or used to discriminate because of race, color, religion, sex or national origin. It shall not be an unlawful employment practice under this subchapter for any employer to differentiate upon the basis of sex in determining the amount of the wages or compensation paid or to be paid to employees of such employer if such differentiation is authorized by the provisions of section 206(d) of Title 29 [*section 6(d) of the Labor Standards Act of 1938, as amended*].

(i) Businesses or enterprises extending preferential treatment to Indians

Nothing contained in this subchapter shall apply to any business or enterprise on or near an Indian reservation with respect to any publicly announced employment practice of such business or enterprise under which a preferential treatment is given to any individual because he is an Indian living on or near a reservation.

(j) Preferential treatment not to be granted on account of existing number or percentage imbalance

Nothing contained in this subchapter shall be interpreted to require any employer, employment agency, labor organization, or joint labor-management committee subject to this subchapter to grant preferential treatment to any individual or to any group because of the race, color, religion, sex, or national origin of such individual or group on account of an imbalance which may exist with respect to the total number or percentage of persons of any race, color, religion, sex, or national origin employed by any employer, referred or classified for employment by any employment agency or labor organization, admitted to membership or classified by any labor organization, or admitted to, or employed in, any apprenticeship or other training program, in comparison with the total number or percentage of persons of such race, color, religion, sex, or national origin in any community, State, section, or other area, or in the available work force in any community, State, section, or other area.

(k) Burden of proof in disparate impact cases

(1) (A) An unlawful employment practice based on disparate impact is established under this subchapter only if-

(i) a complaining party demonstrates that a respondent uses a particular employment practice that causes a disparate impact on the basis of race, color, religion, sex, or national origin and the respondent fails to demonstrate that the challenged practice is job related for the position in question and consistent with business necessity; or

(ii) the complaining party makes the demonstration described in subparagraph (C) with respect to an alternative employment practice and the respondent refuses to adopt such alternative employment practice.

(B) (i) With respect to demonstrating that a particular employment practice causes a disparate impact as described in subparagraph (A)(i), the complaining party shall demonstrate that each particular challenged employment practice causes a disparate impact, except that if the complaining party can demonstrate to the court that the elements of a respondent's decisionmaking process are not capable of separation for analysis, the decisionmaking process may be analyzed as one employment practice.

(ii) If the respondent demonstrates that a specific employment practice does not cause the disparate impact, the respondent shall not be required to demonstrate that such practice is required by business necessity.

(C) The demonstration referred to by subparagraph (A)(ii) shall be in accordance with the law as it existed on June 4, 1989, with respect to the concept of "alternative employment practice".

(2) A demonstration that an employment practice is required by business necessity may not be used as a defense against a claim of intentional discrimination under this subchapter.

(3) Notwithstanding any other provision of this subchapter, a rule barring the employment of an individual who currently and knowingly uses or possesses a controlled substance, as defined in schedules I and II of section 102(6) of the Controlled Substances Act (21 U.S.C. 802(6)), other than the use or possession of a drug taken under the supervision of a licensed health care professional, or any other use or possession authorized by the Controlled Substances Act [*21 U.S.C. 801 et seq.*] or any other provision of Federal law, shall be considered an unlawful employment practice under this subchapter only if such rule is adopted or applied with an intent to discriminate because of race, color, religion, sex, or national origin.

(l) Prohibition of discriminatory use of test scores

It shall be an unlawful employment practice for a respondent, in connection with the selection or referral of applicants or candidates for employment or promotion, to adjust the scores of, use different cutoff scores for, or otherwise alter the results of, employment related tests on the basis of race, color, religion, sex, or national origin.

(m) Impermissible consideration of race, color, religion, sex, or national origin in employment practices

Except as otherwise provided in this subchapter, an unlawful employment practice is established when the complaining party demonstrates that race, color, religion, sex, or national origin was a motivating factor for any employment practice, even though other factors also motivated the practice.

(n) Resolution of challenges to employment practices implementing litigated or consent judgments or orders

(1) (A) Notwithstanding any other provision of law, and except as provided in paragraph (2), an employment practice that implements and is within the scope of a litigated or consent judgment or order that resolves a claim of employment discrimination under the Constitution or Federal civil rights laws may not be challenged under the circumstances described in subparagraph (B).

(B) A practice described in subparagraph (A) may not be challenged in a claim under the Constitution or Federal civil rights laws-

(i) by a person who, prior to the entry of the judgment or order described in subparagraph (A), had-

(I) actual notice of the proposed judgment or order sufficient to apprise such person that such judgment or order might adversely affect the interests and legal rights of such person and that an opportunity was available to present objections to such judgment or order by a future date certain; and

(II) a reasonable opportunity to present objections to such judgment or order; or

(ii) by a person whose interests were adequately represented by another person who had previously challenged the judgment or order on the same legal grounds and with a similar factual situation, unless there has been an intervening change in law or fact.

(2) Nothing in this subsection shall be construed to-

(A) alter the standards for intervention under rule 24 of the Federal Rules of Civil Procedure or apply to the rights of parties who have successfully intervened pursuant to such rule in the proceeding in which the parties intervened;

(B) apply to the rights of parties to the action in which a litigated or consent judgment or order was entered, or of members of a class represented or sought to be represented in such action, or of members of a group on whose behalf relief was sought in such action by the Federal Government;

(C) prevent challenges to a litigated or consent judgment or order on the ground that such judgment or order was obtained through collusion or fraud, or is transparently invalid or was entered by a court lacking subject matter jurisdiction; or

(D) authorize or permit the denial to any person of the due process of law required by the Constitution.

(3) Any action not precluded under this subsection that challenges an employment consent judgment or order described in paragraph (1) shall be brought in the court, and if possible before the judge, that entered such judgment or order. Nothing in this subsection shall preclude a transfer of such action pursuant to section 1404 of Title 28 [*United States Code*].

OTHER UNLAWFUL EMPLOYMENT PRACTICES

SEC. 2000e-3. [*Section 704*]

(a) Discrimination for making charges, testifying, assisting, or participating in enforcement proceedings

It shall be an unlawful employment practice for an employer to discriminate against any of his employees or applicants for employment, for an employment agency, or joint labor-management committee controlling apprenticeship or other training or retraining, including on-the-job training programs, to discriminate against any individual, or for a labor organization to discriminate against any member thereof or applicant for membership, because he has opposed any practice made an unlawful employment practice by this subchapter, or because he has made a charge, testified, assisted, or participated in any manner in an investigation, proceeding, or hearing under this subchapter.

(b) Printing or publication of notices or advertisements indicating prohibited preference, limitation, specification, or discrimination; occupational qualification exception

It shall be an unlawful employment practice for an employer, labor organization, employment agency, or joint labor-management committee controlling apprenticeship or other training or retraining, including on-the-job training programs, to print or publish or cause to be printed or published any notice or advertisement relating to employment by such an employer or membership in or any classification or referral for employment by such a labor organization, or relating to any classification or referral for employment by such an employment agency, or relating to admission to, or employment in, any program established to provide apprenticeship or other training by such a joint labor-management committee, indicating any preference, limitation, specification, or discrimination, based on race, color, religion, sex, or national origin, except that such a notice or advertisement may indicate a preference, limitation, specification, or discrimination based on religion, sex, or national origin when religion, sex, or national origin is a bona fide occupational qualification for employment.

EQUAL EMPLOYMENT OPPORTUNITY COMMISSION

SEC. 2000e-4. [*Section 705*]

(a) Creation; composition; political representation; appointment; term; vacancies; Chairman and Vice Chairman; duties of Chairman; appointment of personnel; compensation of personnel

There is hereby created a Commission to be known as the Equal Employment Opportunity Commission, which shall be composed of five members, not more than three of whom shall be members of the same political party. Members of the Commission shall be appointed by the President by and with the advice and consent of the Senate for a term of five years. Any individual chosen to fill a vacancy shall be appointed only for the unexpired term of the member whom he shall succeed, and all members of the Commission shall continue to serve until their successors are appointed and qualified, except that no such member of the Commission shall continue to serve (1) for more than sixty days when the Congress is in session unless a nomination to fill such vacancy shall have been submitted to the Senate, or (2) after the adjournment sine die of the session of the Senate in which such nomination was submitted. The President shall designate one member to serve as Chairman of the Commission, and one member to serve as Vice Chairman. The Chairman shall be responsible on behalf of the Commission for the administrative operations of the Commission, and, except as provided in subsection (b) of this section, shall appoint, in accordance with the provisions of Title 5 [*United States Code*] governing appointments in the competitive service, such officers, agents, attorneys, administrative law judges [*originally, hearing examiners*], and employees as he deems necessary to assist it in the performance of its functions and to fix their compensation in accordance with the provisions of chapter 51 and subchapter III of chapter 53 of Title 5 [*United States Code*], relating to classification and General Schedule pay rates: Provided, That assignment, removal, and compensation of administrative law judges [*originally, hearing examiners*] shall be in accordance with sections 3105, 3344, 5372, and 7521 of Title 5 [*United States Code*].

(b) General Counsel; appointment; term; duties; representation by attorneys and Attorney General

(1) There shall be a General Counsel of the Commission appointed by the President, by and with the advice and consent of the Senate, for a term of four years. The General Counsel shall have responsibility for the conduct of litigation as provided in sections 2000e-5 and 2000e-6 of this title [*sections 706 and 707*]. The General Counsel shall have such other duties as the Commission may prescribe or as may be provided by law and shall concur with the Chairman of the Commission on the appointment and supervision of regional attorneys. The General Counsel of the Commission on the effective date of this Act shall continue in such position and perform the functions specified in this subsection until a successor is appointed and qualified.

(2) Attorneys appointed under this section may, at the direction of the Commission, appear for and represent the Commission in any case in court, provided that the Attorney General shall conduct all litigation to which the Commission is a party in the Supreme Court pursuant to this subchapter.

(c) Exercise of powers during vacancy; quorum

A vacancy in the Commission shall not impair the right of the remaining members to exercise all the powers of the Commission and three members thereof shall constitute a quorum.

(d) Seal; judicial notice

The Commission shall have an official seal which shall be judicially noticed.

(e) Reports to Congress and the President

The Commission shall at the close of each fiscal year report to the Congress and to the President concerning the action it has taken [*originally, the names, salaries, and duties of all individuals in its employ*] and the moneys it has disbursed. It shall make such further reports on the cause of and means of eliminating discrimination and such recommendations for further legislation as may appear desirable.

(f) Principal and other offices

The principal office of the Commission shall be in or near the District of Columbia, but it may meet or exercise any or all its powers at any other place. The Commission may establish such regional or State offices as it deems necessary to accomplish the purpose of this subchapter.

(g) Powers of Commission

The Commission shall have power-

(1) to cooperate with and, with their consent, utilize regional, State, local, and other agencies, both public and private, and individuals;

(2) to pay to witnesses whose depositions are taken or who are summoned before the Commission or any of its agents the same witness and mileage fees as are paid to witnesses in the courts of the United States;

(3) to furnish to persons subject to this subchapter such technical assistance as they may request to further their compliance with this subchapter or an order issued thereunder;

(4) upon the request of (i) any employer, whose employees or some of them, or (ii) any labor organization, whose members or some of them, refuse or threaten to refuse to cooperate in effectuating the provisions of this subchapter, to assist

in such effectuation by conciliation or such other remedial action as is provided by this subchapter;

(5) to make such technical studies as are appropriate to effectuate the purposes and policies of this subchapter and to make the results of such studies available to the public;

(6) to intervene in a civil action brought under section 2000e-5 of this title [*section 706*] by an aggrieved party against a respondent other than a government, governmental agency or political subdivision.

(h) Cooperation with other departments and agencies in performance of educational or promotional activities; outreach activities

(1) The Commission shall, in any of its educational or promotional activities, cooperate with other departments and agencies in the performance of such educational and promotional activities.

(2) In exercising its powers under this subchapter, the Commission shall carry out educational and outreach activities (including dissemination of information in languages other than English) targeted to-

(A) individuals who historically have been victims of employment discrimination and have not been equitably served by the Commission; and

(B) individuals on whose behalf the Commission has authority to enforce any other law prohibiting employment discrimination, concerning rights and obligations under this subchapter or such law, as the case may be.

(i) Personnel subject to political activity restrictions

All officers, agents, attorneys, and employees of the Commission shall be subject to the provisions of section 7324 of Title 5 [*originally, section 9 of the Act of August 2, 1939, as amended (the Hatch Act)*], notwithstanding any exemption contained in such section.

(j) Technical Assistance Training Institute

(1) The Commission shall establish a Technical Assistance Training Institute, through which the Commission shall provide technical assistance and training regarding the laws and regulations enforced by the Commission.

(2) An employer or other entity covered under this subchapter shall not be excused from compliance with the requirements of this subchapter because of any failure to receive technical assistance under this subsection.

(3) There are authorized to be appropriated to carry out this subsection such sums as may be necessary for fiscal year 1992.

(k) EEOC Education, Technical Assistance, and Training Revolving Fund

(1) There is hereby established in the Treasury of the United States a revolving fund to be known as the "EEOC Education, Technical Assistance, and Training Revolving Fund" (hereinafter in this subsection referred to as the "Fund") and to pay the cost (including administrative and personnel expenses) of providing education, technical assistance, and training relating to laws administered by the Commission. Monies in the Fund shall be available without fiscal year limitation to the Commission for such purposes.

(2)(A) The Commission shall charge fees in accordance with the provisions of this paragraph to offset the costs of education, technical assistance, and training provided with monies in the Fund. Such fees for any education, technical assistance, or training—

(i) shall be imposed on a uniform basis on persons and entities receiving such education, assistance, or training,

(ii) shall not exceed the cost of providing such education, assistance, and training, and

(iii) with respect to each person or entity receiving such education, assistance, or training, shall bear a reasonable relationship to the cost of providing such education, assistance, or training to such person or entity.

(B) Fees received under subparagraph (A) shall be deposited in the Fund by the Commission.

(C) The Commission shall include in each report made under subsection (e) of this section information with respect to the operation of the Fund, including information, presented in the aggregate, relating to—

(i) the number of persons and entities to which the Commission provided education, technical assistance, or training with monies in the Fund, in the fiscal year for which such report is prepared,

(ii) the cost to the Commission to provide such education, technical assistance, or training to such persons and entities, and

(iii) the amount of any fees received by the Commission from such persons and entities for such education, technical assistance, or training.

(3) The Secretary of the Treasury shall invest the portion of the Fund not required to satisfy current expenditures from the Fund, as determined by the Commission, in obligations of the United States or obligations guaranteed as to principal by the United States. Investment proceeds shall be deposited in the Fund.

(4) There is hereby transferred to the Fund $1,000,000 from the Salaries and Expenses appropriation of the Commission.

ENFORCEMENT PROVISIONS

SEC. 2000e-5. [*Section 706*]

(a) Power of Commission to prevent unlawful employment practices

The Commission is empowered, as hereinafter provided, to prevent any person from engaging in any unlawful employment practice as set forth in section 2000e-2 or 2000e-3 of this title [*section 703 or 704*].

(b) Charges by persons aggrieved or member of Commission of unlawful employment practices by employers, etc.; filing; allegations; notice to respondent; contents of notice; investigation by Commission; contents of charges; prohibition on disclosure of charges; determination of reasonable cause; conference, conciliation, and persuasion for elimination of unlawful practices; prohibition on disclosure of informal endeavors to end unlawful practices; use of evidence in subsequent proceedings; penalties for disclosure of information; time for determination of reasonable cause

Whenever a charge is filed by or on behalf of a person claiming to be aggrieved, or by a member of the Commission, alleging that an employer, employment agency, labor organization, or joint labor management committee controlling apprenticeship or other training or retraining, including on-the-job training programs, has engaged in an unlawful employment practice, the Commission shall serve a notice of the charge (including the date, place and circumstances of the alleged unlawful employment practice) on such employer, employment agency, labor organization, or joint labor-management committee (hereinafter referred to as the "respondent") within ten days, and shall make an investigation thereof. Charges shall be in writing under oath or affirmation and shall contain such information and be in such form as the Commission requires. Charges shall not be made public by the Commission. If the Commission determines after such investigation that there is not reasonable cause to believe that the charge is true, it shall dismiss the charge and promptly notify the person claiming to be aggrieved and the respondent of its action. In determining whether reasonable cause exists, the Commission shall accord substantial weight to final findings and orders made by State or local authorities in proceedings commenced under State or local law pursuant to the requirements of subsections (c) and (d) of this section. If the Commission determines after such investigation that there is reasonable cause to believe that the charge is true, the Commission shall endeavor to eliminate any such alleged unlawful employment practice by informal methods of conference, conciliation, and persuasion. Nothing said or done during and as a part of such informal endeavors may be made public by the Commission, its officers or employees, or used as evidence in a subsequent proceeding without the written consent of the persons concerned. Any person who makes public information in violation of this subsection shall be fined not more than $1,000 or imprisoned for not more than one year, or both. The Commission shall make its determination on reasonable cause as promptly as possible and, so far as practicable, not later than one hundred and twenty days from the filing of the charge or, where applicable under subsection

(c) or (d) of this section, from the date upon which the Commission is authorized to take action with respect to the charge.

(c) State or local enforcement proceedings; notification of State or local authority; time for filing charges with Commission; commencement of proceedings

In the case of an alleged unlawful employment practice occurring in a State, or political subdivision of a State, which has a State or local law prohibiting the unlawful employment practice alleged and establishing or authorizing a State or local authority to grant or seek relief from such practice or to institute criminal proceedings with respect thereto upon receiving notice thereof, no charge may be filed under subsection (a) of this section by the person aggrieved before the expiration of sixty days after proceedings have been commenced under the State or local law, unless such proceedings have been earlier terminated, provided that such sixty- day period shall be extended to one hundred and twenty days during the first year after the effective date of such State or local law. If any requirement for the commencement of such proceedings is imposed by a State or local authority other than a requirement of the filing of a written and signed statement of the facts upon which the proceeding is based, the proceeding shall be deemed to have been commenced for the purposes of this subsection at the time such statement is sent by registered mail to the appropriate State or local authority.

(d) State or local enforcement proceedings; notification of State or local authority; time for action on charges by Commission

In the case of any charge filed by a member of the Commission alleging an unlawful employment practice occurring in a State or political subdivision of a State which has a State or local law prohibiting the practice alleged and establishing or authorizing a State or local authority to grant or seek relief from such practice or to institute criminal proceedings with respect thereto upon receiving notice thereof, the Commission shall, before taking any action with respect to such charge, notify the appropriate State or local officials and, upon request, afford them a reasonable time, but not less than sixty days (provided that such sixty-day period shall be extended to one hundred and twenty days during the first year after the effective day of such State or local law), unless a shorter period is requested, to act under such State or local law to remedy the practice alleged.

(e) Time for filing charges; time for service of notice of charge on respondent; filing of charge by Commission with State or local agency; seniority system

(1) A charge under this section shall be filed within one hundred and eighty days after the alleged unlawful employment practice occurred and notice of the charge (including the date, place and circumstances of the alleged unlawful employment practice) shall be served upon the person against whom such charge is made within ten days thereafter, except that in a case of an unlawful employment practice with respect to which the person aggrieved has initially instituted proceedings with a State or local agency with authority to grant or seek relief from such practice or to

institute criminal proceedings with respect thereto upon receiving notice thereof, such charge shall be filed by or on behalf of the person aggrieved within three hundred days after the alleged unlawful employment practice occurred, or within thirty days after receiving notice that the State or local agency has terminated the proceedings under the State or local law, whichever is earlier, and a copy of such charge shall be filed by the Commission with the State or local agency.

(2) For purposes of this section, an unlawful employment practice occurs, with respect to a seniority system that has been adopted for an intentionally discriminatory purpose in violation of this subchapter (whether or not that discriminatory purpose is apparent on the face of the seniority provision), when the seniority system is adopted, when an individual becomes subject to the seniority system, or when a person aggrieved is injured by the application of the seniority system or provision of the system.

(3)(A) For purposes of this section, an unlawful employment practice occurs, with respect to discrimination in compensation in violation of this title, when a discriminatory compensation decision or other practice is adopted, when an individual becomes subject to a discriminatory compensation decision or other practice, or when an individual is affected by application of a discriminatory compensation decision or other practice, including each time wages, benefits, or other compensation is paid, resulting in whole or in part from such a decision or other practice.

(B) In addition to any relief authorized by section 1977A of the Revised Statutes (42 U.S.C. 1981a), liability may accrue and an aggrieved person may obtain relief as provided in subsection (g)(1), including recovery of back pay for up to two years preceding the filing of the charge, where the unlawful employment practices that have occurred during the charge filing period are similar or related to unlawful employment practices with regard to discrimination in compensation that occurred outside the time for filing a charge.

(f) Civil action by Commission, Attorney General, or person aggrieved; preconditions; procedure; appointment of attorney; payment of fees, costs, or security; intervention; stay of Federal proceedings; action for appropriate temporary or preliminary relief pending final disposition of charge; jurisdiction and venue of United States courts; designation of judge to hear and determine case; assignment of case for hearing; expedition of case; appointment of master

(1) If within thirty days after a charge is filed with the Commission or within thirty days after expiration of any period of reference under subsection (c) or (d) of this section, the Commission has been unable to secure from the respondent a conciliation agreement acceptable to the Commission, the Commission may bring a civil action against any respondent not a government, governmental agency, or political subdivision named in the charge. In the case of a respondent which is a government, governmental agency, or political subdivision, if the Commission has been unable to secure from the respondent a conciliation agreement acceptable to the Commission, the Commission shall take no further action and shall refer the

case to the Attorney General who may bring a civil action against such respondent in the appropriate United States district court. The person or persons aggrieved shall have the right to intervene in a civil action brought by the Commission or the Attorney General in a case involving a government, governmental agency, or political subdivision. If a charge filed with the Commission pursuant to subsection (b) of this section is dismissed by the Commission, or if within one hundred and eighty days from the filing of such charge or the expiration of any period of reference under subsection (c) or (d) of this section, whichever is later, the Commission has not filed a civil action under this section or the Attorney General has not filed a civil action in a case involving a government, governmental agency, or political subdivision, or the Commission has not entered into a conciliation agreement to which the person aggrieved is a party, the Commission, or the Attorney General in a case involving a government, governmental agency, or political subdivision, shall so notify the person aggrieved and within ninety days after the giving of such notice a civil action may be brought against the respondent named in the charge (A) by the person claiming to be aggrieved or (B) if such charge was filed by a member of the Commission, by any person whom the charge alleges was aggrieved by the alleged unlawful employment practice. Upon application by the complainant and in such circumstances as the court may deem just, the court may appoint an attorney for such complainant and may authorize the commencement of the action without the payment of fees, costs, or security. Upon timely application, the court may, in its discretion, permit the Commission, or the Attorney General in a case involving a government, governmental agency, or political subdivision, to intervene in such civil action upon certification that the case is of general public importance. Upon request, the court may, in its discretion, stay further proceedings for not more than sixty days pending the termination of State or local proceedings described in subsection (c) or (d) of this section or further efforts of the Commission to obtain voluntary compliance.

(2) Whenever a charge is filed with the Commission and the Commission concludes on the basis of a preliminary investigation that prompt judicial action is necessary to carry out the purposes of this Act, the Commission, or the Attorney General in a case involving a government, governmental agency, or political subdivision, may bring an action for appropriate temporary or preliminary relief pending final disposition of such charge. Any temporary restraining order or other order granting preliminary or temporary relief shall be issued in accordance with rule 65 of the Federal Rules of Civil Procedure. It shall be the duty of a court having jurisdiction over proceedings under this section to assign cases for hearing at the earliest practicable date and to cause such cases to be in every way expedited.

(3) Each United States district court and each United States court of a place subject to the jurisdiction of the United States shall have jurisdiction of actions brought under this subchapter. Such an action may be brought in any judicial district in the State in which the unlawful employment practice is alleged to have been committed, in the judicial district in which the employment records relevant to such practice are maintained and administered, or in the judicial district in which the

aggrieved person would have worked but for the alleged unlawful employment practice, but if the respondent is not found within any such district, such an action may be brought within the judicial district in which the respondent has his principal office. For purposes of sections 1404 and 1406 of Title 28 [*United States Code*], the judicial district in which the respondent has his principal office shall in all cases be considered a district in which the action might have been brought.

(4) It shall be the duty of the chief judge of the district (or in his absence, the acting chief judge) in which the case is pending immediately to designate a judge in such district to hear and determine the case. In the event that no judge in the district is available to hear and determine the case, the chief judge of the district, or the acting chief judge, as the case may be, shall certify this fact to the chief judge of the circuit (or in his absence, the acting chief judge) who shall then designate a district or circuit judge of the circuit to hear and determine the case.

(5) It shall be the duty of the judge designated pursuant to this subsection to assign the case for hearing at the earliest practicable date and to cause the case to be in every way expedited. If such judge has not scheduled the case for trial within one hundred and twenty days after issue has been joined, that judge may appoint a master pursuant to rule 53 of the Federal Rules of Civil Procedure.

(g) Injunctions; appropriate affirmative action; equitable relief; accrual of back pay; reduction of back pay; limitations on judicial orders

(1) If the court finds that the respondent has intentionally engaged in or is intentionally engaging in an unlawful employment practice charged in the complaint, the court may enjoin the respondent from engaging in such unlawful employment practice, and order such affirmative action as may be appropriate, which may include, but is not limited to, reinstatement or hiring of employees, with or without back pay (payable by the employer, employment agency, or labor organization, as the case may be, responsible for the unlawful employment practice), or any other equitable relief as the court deems appropriate. Back pay liability shall not accrue from a date more than two years prior to the filing of a charge with the Commission. Interim earnings or amounts earnable with reasonable diligence by the person or persons discriminated against shall operate to reduce the back pay otherwise allowable.

(2) (A) No order of the court shall require the admission or reinstatement of an individual as a member of a union, or the hiring, reinstatement, or promotion of an individual as an employee, or the payment to him of any back pay, if such individual was refused admission, suspended, or expelled, or was refused employment or advancement or was suspended or discharged for any reason other than discrimination on account of race, color, religion, sex, or national origin or in violation of section 2000e-3(a) of this Title [*section 704(a)*].

(B) On a claim in which an individual proves a violation under section 2000e-2(m) of this title [*section 703(m)*] and a respondent demonstrates that the respondent would have taken the same action in the absence of the impermissible motivating factor, the court-

(i) may grant declaratory relief, injunctive relief (except as provided in clause (ii)), and attorney's fees and costs demonstrated to be directly attributable only to the pursuit of a claim under section 2000e-2(m) of this title [*section 703(m)*]; and

(ii) shall not award damages or issue an order requiring any admission, reinstatement, hiring, promotion, or payment, described in subparagraph (A).

(h) Provisions of chapter 6 of Title 29 not applicable to civil actions for prevention of unlawful practices

The provisions of chapter 6 of title 29 [*the Act entitled "An Act to amend the Judicial Code and to define and limit the jurisdiction of courts sitting in equity, and for other purposes," approved March 23, 1932 (29 U.S.C. 105-115)*] shall not apply with respect to civil actions brought under this section.

(i) Proceedings by Commission to compel compliance with judicial orders In any case in which an employer, employment agency, or labor organization fails to comply with an order of a court issued in a civil action brought under this section, the Commission may commence proceedings to compel compliance with such order.

(j) Appeals

Any civil action brought under this section and any proceedings brought under subsection (i) of this section shall be subject to appeal as provided in sections 1291 and 1292, Title 28 [*United States Code*].

(k) Attorney's fee; liability of Commission and United States for costs

In any action or proceeding under this subchapter the court, in its discretion, may allow the prevailing party, other than the Commission or the United States, a reasonable attorney's fee (including expert fees) as part of the costs, and the Commission and the United States shall be liable for costs the same as a private person.

CIVIL ACTIONS BY THE ATTORNEY GENERAL

SEC. 2000e-6. [*Section 707*]

(a) Complaint

Whenever the Attorney General has reasonable cause to believe that any person or group of persons is engaged in a pattern or practice of resistance to the full enjoyment of any of the rights secured by this subchapter, and that the pattern or practice is of such a nature and is intended to deny the full exercise of the rights herein described, the Attorney General may bring a civil action in the appropriate district court of the United States by filing with it a complaint (1) signed by him (or in his absence the Acting Attorney General), (2) setting forth facts pertaining to such pattern or practice, and (3) requesting such relief, including an application for a permanent or temporary injunction, restraining order or other order against the person or persons responsible for such pattern or practice, as he deems necessary to insure the full enjoyment of the rights herein described.

(b) Jurisdiction; three-judge district court for cases of general public importance: hearing, determination, expedition of action, review by Supreme Court; single judge district court: hearing, determination, expedition of action

The district courts of the United States shall have and shall exercise jurisdiction of proceedings instituted pursuant to this section, and in any such proceeding the Attorney General may file with the clerk of such court a request that a court of three judges be convened to hear and determine the case. Such request by the Attorney General shall be accompanied by a certificate that, in his opinion, the case is of general public importance. A copy of the certificate and request for a three-judge court shall be immediately furnished by such clerk to the chief judge of the circuit (or in his absence, the presiding circuit judge of the circuit) in which the case is pending. Upon receipt of such request it shall be the duty of the chief judge of the circuit or the presiding circuit judge, as the case may be, to designate immediately three judges in such circuit, of whom at least one shall be a circuit judge and another of whom shall be a district judge of the court in which the proceeding was instituted, to hear and determine such case, and it shall be the duty of the judges so designated to assign the case for hearing at the earliest practicable date, to participate in the hearing and determination thereof, and to cause the case to be in every way expedited. An appeal from the final judgment of such court will lie to the Supreme Court.

In the event the Attorney General fails to file such a request in any such proceeding, it shall be the duty of the chief judge of the district (or in his absence, the acting chief judge) in which the case is pending immediately to designate a judge in such district to hear and determine the case. In the event that no judge in the district is available to hear and determine the case, the chief judge of the district, or the acting chief judge, as the case may be, shall certify this fact to the chief judge of the circuit (or in his absence, the acting chief judge) who shall then designate a district or circuit judge of the circuit to hear and determine the case.

It shall be the duty of the judge designated pursuant to this section to assign the case for hearing at the earliest practicable date and to cause the case to be in every way expedited.

(c) Transfer of functions, etc., to Commission; effective date; prerequisite to transfer; execution of functions by Commission

Effective two years after March 24, 1972 [*the date of enactment of the Equal Employment Opportunity Act of 1972*], the functions of the Attorney General under this section shall be transferred to the Commission, together with such personnel, property, records, and unexpended balances of appropriations, allocations, and other funds employed, used, held, available, or to be made available in connection with such functions unless the President submits, and neither House of Congress vetoes, a reorganization plan pursuant to chapter 9 of Title 5 [*United States Code*], inconsistent with the provisions of this subsection. The Commission shall carry out such functions in accordance with subsections (d) and (e) of this section.

(d) Transfer of functions, etc., not to affect suits commenced pursuant to this section prior to date of transfer

Upon the transfer of functions provided for in subsection (c) of this section, in all suits commenced pursuant to this section prior to the date of such transfer, proceedings shall continue without abatement, all court orders and decrees shall remain in effect, and the Commission shall be substituted as a party for the United States of America, the Attorney General, or the Acting Attorney General, as appropriate.

(e) Investigation and action by Commission pursuant to filing of charge of discrimination; procedure

Subsequent to March 24, 1972 [*the date of enactment of the Equal Employment Opportunity Act of 1972*], the Commission shall have authority to investigate and act on a charge of a pattern or practice of discrimination, whether filed by or on behalf of a person claiming to be aggrieved or by a member of the Commission. All such actions shall be conducted in accordance with the procedures set forth in section 2000e-5 of this title [*section 706*].

EFFECT ON STATE LAWS

SEC. 2000e-7. [*Section 708*]

Nothing in this subchapter shall be deemed to exempt or relieve any person from any liability, duty, penalty, or punishment provided by any present or future law of any State or political subdivision of a State, other than any such law which purports to require or permit the doing of any act which would be an unlawful employment practice under this subchapter.

INVESTIGATIONS

SEC. 2000e-8. [*Section 709*]

(a) Examination and copying of evidence related to unlawful employment practices

In connection with any investigation of a charge filed under section 2000e-5 of this title [*section 706*], the Commission or its designated representative shall at all reasonable times have access to, for the purposes of examination, and the right to copy any evidence of any person being investigated or proceeded against that relates to unlawful employment practices covered by this subchapter and is relevant to the charge under investigation.

(b) Cooperation with State and local agencies administering State fair employment practices laws; participation in and contribution to research and other projects; utilization of services; payment in advance or reimbursement; agreements and rescission of agreements

The Commission may cooperate with State and local agencies charged with the administration of State fair employment practices laws and, with the consent of such agencies, may, for the purpose of carrying out its functions and duties under this subchapter and within the limitation of funds appropriated specifically for such purpose, engage in and contribute to the cost of research and other projects of mutual interest undertaken by such agencies, and utilize the services of such agencies and their employees, and, notwithstanding any other provision of law, pay by advance or reimbursement such agencies and their employees for services rendered to assist the Commission in carrying out this subchapter. In furtherance of such cooperative efforts, the Commission may enter into written agreements with such State or local agencies and such agreements may include provisions under which the Commission shall refrain from processing a charge in any cases or class of cases specified in such agreements or under which the Commission shall relieve any person or class of persons in such State or locality from requirements imposed under this section. The Commission shall rescind any such agreement whenever it determines that the agreement no longer serves the interest of effective enforcement of this subchapter.

(c) Execution, retention, and preservation of records; reports to Commission; training program records; appropriate relief from regulation or order for undue hardship; procedure for exemption; judicial action to compel compliance

Every employer, employment agency, and labor organization subject to this subchapter shall (1) make and keep such records relevant to the determinations of whether unlawful employment practices have been or are being committed, (2) preserve such records for such periods, and (3) make such reports therefrom as the Commission shall prescribe by regulation or order, after public hearing, as reasonable, necessary, or appropriate for the enforcement of this subchapter or the regulations or orders thereunder. The Commission shall, by regulation, require each employer, labor organization, and joint labor-management committee subject to this subchapter which controls an apprenticeship or other training program to maintain such records as are reasonably necessary to carry out the purposes of this subchapter, including, but not limited to, a list of applicants who wish to participate in such program, including the chronological order in which applications were received, and to furnish to the Commission upon request, a detailed description of the manner in which persons are selected to participate in the apprenticeship or other training program. Any employer, employment agency, labor organization, or joint labor-management committee which believes that the application to it of any regulation or order issued under this section would result in undue hardship may apply to the Commission for an exemption from the application of such regulation or order, and, if such application for an exemption is denied, bring a civil action in the United States district court for the district where such records are kept. If the Commission or the court, as the case may be, finds that the application of the regulation or order to the employer, employment agency, or labor organization in question would impose an undue hardship, the Commission or the court, as the case may be, may grant appropriate relief. If any person required to comply with the provisions of this subsection fails or refuses to do so, the United States district court for the district in which such person is found, resides, or transacts business, shall, upon application of the Commission, or the Attorney General in a case involving a government, governmental agency or political subdivision, have jurisdiction to issue to such person an order requiring him to comply.

(d) Consultation and coordination between Commission and interested State and Federal agencies in prescribing recordkeeping and reporting requirements; availability of information furnished pursuant to recordkeeping and reporting requirements; conditions on availability

In prescribing requirements pursuant to subsection (c) of this section, the Commission shall consult with other interested State and Federal agencies and shall endeavor to coordinate its requirements with those adopted by such agencies. The Commission shall furnish upon request and without cost to any State or local agency charged with the administration of a fair employment practice law information obtained pursuant to subsection (c) of this section from any employer, employment agency, labor organization, or joint labor-management committee subject to the jurisdiction of such agency. Such information shall be furnished on condition that it not be made public by the recipient agency prior to the institution of a proceeding under State or local law involving such information. If this condition is violated by a recipient agency, the Commission may decline to honor subsequent requests pursuant to this subsection.

(e) Prohibited disclosures; penalties

It shall be unlawful for any officer or employee of the Commission to make public in any manner whatever any information obtained by the Commission pursuant to its authority under this section prior to the institution of any proceeding under this subchapter involving such information. Any officer or employee of the Commission who shall make public in any manner whatever any information in violation of this subsection shall be guilty of a misdemeanor and upon conviction thereof, shall be fined not more than $1,000, or imprisoned not more than one year.

CONDUCT OF HEARINGS AND INVESTIGATIONS PURSUANT TO SECTION 161 OF TITLE 29

SEC. 2000e-9. [*Section 710*]

For the purpose of all hearings and investigations conducted by the Commission or its duly authorized agents or agencies, section 161 of Title 29 [*section 11 of the National Labor Relations Act*] shall apply.

POSTING OF NOTICES; PENALTIES

SEC. 2000e-10. [*Section 711*]

(a) Every employer, employment agency, and labor organization, as the case may be, shall post and keep posted in conspicuous places upon its premises where notices to employees, applicants for employment, and members are customarily posted a notice to be prepared or approved by the Commission setting forth excerpts from or, summaries of, the pertinent provisions of this subchapter and information pertinent to the filing of a complaint.

(b) A willful violation of this section shall be punishable by a fine of not more than $100 for each separate offense.

VETERANS' SPECIAL RIGHTS OR PREFERENCE

SEC. 2000e-11. [*Section 712*]

Nothing contained in this subchapter shall be construed to repeal or modify any Federal, State, territorial, or local law creating special rights or preference for veterans.

REGULATIONS; CONFORMITY OF REGULATIONS WITH ADMINISTRATIVE PROCEDURE PROVISIONS; RELIANCE ON INTERPRETATIONS AND INSTRUCTIONS OF COMMISSION

SEC. 2000e-12. [*Section 713*]

(a) The Commission shall have authority from time to time to issue, amend, or rescind suitable procedural regulations to carry out the provisions of this subchapter. Regulations issued under this section shall be in conformity with the standards and limitations of subchapter II of chapter 5 of Title 5 [*originally, the Administrative Procedure Act*].

(b) In any action or proceeding based on any alleged unlawful employment practice, no person shall be subject to any liability or punishment for or on account of (1) the commission by such person of an unlawful employment practice if he pleads and proves that the act or omission complained of was in good faith, in conformity with, and in reliance on any written interpretation or opinion of the Commission, or (2) the failure of such person to publish and file any information required by any provision of this subchapter if he pleads and proves that he failed to publish and file such information in good faith, in conformity with the instructions of the Commission issued under this subchapter regarding the filing of such information. Such a defense, if established, shall be a bar to the action or proceeding, notwithstanding that (A) after such act or omission, such interpretation or opinion is modified or rescinded or is determined by judicial authority to be invalid or of no legal effect, or (B) after publishing or filing the description and annual reports, such publication or filing is determined by judicial authority not to be in conformity with the requirements of this subchapter.

APPLICATION TO PERSONNEL OF COMMISSION OF SECTIONS 111 AND 1114 OF TITLE 18; PUNISHMENT FOR VIOLATION OF SECTION 1114 OF TITLE 18

SEC. 2000e-13. [*Section 714*]

The provisions of sections 111 and 1114, Title 18 [*United States Code*], shall apply to officers, agents, and employees of the Commission in the performance of their official duties. Notwithstanding the provisions of sections 111 and 1114 of Title 18 [*United States Code*], whoever in violation of the provisions of section 1114 of such title kills a person while engaged in or on account of the performance of his official functions under this Act shall be punished by imprisonment for any term of years or for life.

TRANSFER OF AUTHORITY

[*Administration of the duties of the Equal Employment Opportunity Coordinating Council was transferred to the Equal Employment Opportunity Commission effective July 1, 1978, under the President's Reorganization Plan of 1978.*]

EQUAL EMPLOYMENT OPPORTUNITY COORDINATING COUNCIL; ESTABLISHMENT; COMPOSITION; DUTIES; REPORT TO PRESIDENT AND CONGRESS

SEC. 2000e-14. [*Section 715*]

[*Original introductory text: There shall be established an Equal Employment Opportunity Coordinating Council (hereinafter referred to in this section as the Council) composed of the Secretary of Labor, the Chairman of the Equal Employment Opportunity Commission, the Attorney General, the Chairman of the United States Civil Service Commission, and the Chairman of the United States Civil Rights Commission, or their respective delegates.*]

The Equal Employment Opportunity Commission [*originally, Council*] shall have the responsibility for developing and implementing agreements, policies and practices designed to maximize effort, promote efficiency, and eliminate conflict, competition, duplication and inconsistency among the operations, functions and jurisdictions of the various departments, agencies and branches of the Federal Government responsible for the implementation and enforcement of equal employment opportunity legislation, orders, and policies. On or before October 1 [*originally, July 1*] of each year, the Equal Employment Opportunity Commission [*originally, Council*] shall transmit to the President and to the Congress a report of its activities, together with such recommendations for legislative or administrative changes as it concludes are desirable to further promote the purposes of this section.

PRESIDENTIAL CONFERENCES; ACQUAINTANCE OF LEADERSHIP WITH PROVISIONS FOR EMPLOYMENT RIGHTS AND OBLIGATIONS; PLANS FOR FAIR ADMINISTRATION; MEMBERSHIP

SEC. 2000e-15. [*Section 716*]

[*Original text: (a) This title shall become effective one year after the date of its enactment.*

(b) Notwithstanding subsection (a), sections of this title other than sections 703, 704, 706, and 707 shall become effective immediately.

(c)] The President shall, as soon as feasible after July 2, 1964 [*the date of enactment of this title*], convene one or more conferences for the purpose of enabling the leaders of groups whose members will be affected by this subchapter to become familiar with the rights afforded and obligations imposed by its provisions, and for the purpose of making plans which will result in the fair and effective administration of this subchapter when all of its provisions become effective. The President shall invite the participation in such conference or conferences of (1) the members of the President's Committee on Equal Employment Opportunity, (2) the members of the Commission on Civil Rights, (3) representatives of State and local agencies engaged in furthering equal employment opportunity, (4) representatives of private agencies engaged in furthering equal employment opportunity, and (5) representatives of employers, labor organizations, and employment agencies who will be subject to this subchapter.

TRANSFER OF AUTHORITY

[*Enforcement of Section 717 was transferred to the Equal Employment Opportunity Commission from the Civil Service Commission (Office of Personnel Management) effective January 1, 1979 under the President's Reorganization Plan No. 1 of 1978.*]

EMPLOYMENT BY FEDERAL GOVERNMENT

SEC. 2000e-16. [*Section 717*]

(a) Discriminatory practices prohibited; employees or applicants for employment subject to coverage

All personnel actions affecting employees or applicants for employment (except with regard to aliens employed outside the limits of the United States) in military departments as defined in section 102 of Title 5 [*United States Code*], in executive agencies [*originally, other than the General Accounting Office*] as defined in section 105 of Title 5 [*United States Code*] (including employees and applicants for employment who are paid from nonappropriated funds), in the United States Postal Service and the Postal Regulatory Commission, in those units of the Government of the District of Columbia having positions in the competitive service, and in those units of the judicial branch of the Federal Government having positions in the competitive service, in the Smithsonian Institution, and in the Government Printing Office, the Government Accountability Office, and the Library of Congress shall be made free from any discrimination based on race, color, religion, sex, or national origin.

(b) Equal Employment Opportunity Commission; enforcement powers; issuance of rules, regulations, etc.; annual review and approval of national and regional equal employment opportunity plans; review and evaluation of equal employment opportunity programs and publication of progress reports; consultations with interested parties; compliance with rules, regulations, etc.; contents of national and regional equal employment opportunity plans; authority of Librarian of Congress

Except as otherwise provided in this subsection, the Equal Employment Opportunity Commission [*originally, Civil Service Commission*] shall have authority to enforce the provisions of subsection (a) of this section through appropriate remedies, including reinstatement or hiring of employees with or without back pay, as will effectuate the policies of this section, and shall issue such rules, regulations, orders and instructions as it deems necessary and appropriate to carry out its responsibilities under this section. The Equal Employment Opportunity Commission [*originally, Civil Service Commission*] shall-

(1) be responsible for the annual review and approval of a national and regional equal employment opportunity plan which each department and agency and each appropriate unit referred to in subsection (a) of this section shall submit in order to maintain an affirmative program of equal employment opportunity for all such employees and applicants for employment;

(2) be responsible for the review and evaluation of the operation of all agency equal employment opportunity programs, periodically obtaining and publishing (on at least a semiannual basis) progress reports from each such department, agency, or unit; and

(3) consult with and solicit the recommendations of interested individuals, groups, and organizations relating to equal employment opportunity.

The head of each such department, agency, or unit shall comply with such rules, regulations, orders, and instructions which shall include a provision that an employee or applicant for employment shall be notified of any final action taken on any complaint of discrimination filed by him thereunder. The plan submitted by each department, agency, and unit shall include, but not be limited to-

(1) provision for the establishment of training and education programs designed to provide a maximum opportunity for employees to advance so as to perform at their highest potential; and

(2) a description of the qualifications in terms of training and experience relating to equal employment opportunity for the principal and operating officials of each such department, agency, or unit responsible for carrying out the equal employment opportunity program and of the allocation of personnel and resources proposed by such department, agency, or unit to carry out its equal employment opportunity program.

With respect to employment in the Library of Congress, authorities granted in this subsection to the Equal Employment Opportunity Commission [*originally, Civil Service Commission*] shall be exercised by the Librarian of Congress.

(c) Civil action by employee or applicant for employment for redress of grievances; time for bringing of action; head of department, agency, or unit as defendant

Within 90 days of receipt of notice of final action taken by a department, agency, or unit referred to in subsection (a) of this section, or by the Equal Employment Opportunity Commission [*originally, Civil Service Commission*] upon an appeal from a decision or order of such department, agency, or unit on a complaint of discrimination based on race, color, religion, sex or national origin, brought pursuant to subsection (a) of this section, Executive Order 11478 or any succeeding Executive orders, or after one hundred and eighty days from the filing of the initial charge with the department, agency, or unit or with the Equal Employment Opportunity Commission [*originally, Civil Service Commission*] on appeal from a decision or order of such department, agency, or unit until such time as final action may be taken by a department, agency, or unit, an employee or applicant for employment, if aggrieved by the final disposition of his complaint, or by the failure to take final action on his complaint, may file a civil action as provided in section 2000e-5 of this title [*section 706*], in which civil action the head of the department, agency, or unit, as appropriate, shall be the defendant.

(d) Section 2000e-5(f) through (k) of this title applicable to civil actions

The provisions of section 2000e-5(f) through (k) of this title [*section 706(f) through (k)*], as applicable, shall govern civil actions brought hereunder, and the same interest to compensate for delay in payment shall be available as in cases involving nonpublic parties.

(e) Government agency or official not relieved of responsibility to assure nondiscrimination in employment or equal employment opportunity

Nothing contained in this Act shall relieve any Government agency or official of its or his primary responsibility to assure nondiscrimination in employment as required by the Constitution and statutes or of its or his responsibilities under Executive Order 11478 relating to equal employment opportunity in the Federal Government.

(f) Section 2000e-5(e)(3) [*Section 706(e)(3)*] shall apply to complaints of discrimination in compensation under this section.

PROCEDURE FOR DENIAL, WITHHOLDING, TERMINATION, OR SUSPENSION OF GOVERNMENT CONTRACT SUBSEQUENT TO ACCEPTANCE BY GOVERNMENT OF AFFIRMATIVE ACTION PLAN OF EMPLOYER; TIME OF ACCEPTANCE OF PLAN

SEC. 2000e-17. [*Section 718*]

No Government contract, or portion thereof, with any employer, shall be denied, withheld, terminated, or suspended, by any agency or officer of the United States under any equal employment opportunity law or order, where such employer has an affirmative action plan which has previously been accepted by the Government for the same facility within the past twelve months without first according such employer full hearing and adjudication under the provisions of section 554 of Title 5 [*United States Code*], and the following pertinent sections: Provided, That if such employer has deviated substantially from such previously agreed to affirmative action plan, this section shall not apply: Provided further, That for the purposes of this section an affirmative action plan shall be deemed to have been accepted by the Government at the time the appropriate compliance agency has accepted such plan unless within forty-five days thereafter the Office of Federal Contract Compliance has disapproved such plan.

THE PREGNANCY DISCRIMINATION ACT OF 1978

An Act

To amend Title VII of the Civil Rights Act of 1964 to prohibit sex discrimination on the basis of pregnancy.

Be it enacted by the Senate and House of Representatives of the United States of America in Congress assembled,

That section 701 of the Civil Rights Act of 1964 is amended by adding at the end thereof the following new subsection:

"(k) The terms 'because of sex' or 'on the basis of sex' include, but are not limited to, because of or on the basis of pregnancy, childbirth, or related medical conditions; and women affected by pregnancy, childbirth, or related medical conditions shall be treated the same for all employment-related purposes, including receipt of benefits under fringe benefit programs, as other persons not so affected but similar in their ability or inability to work, and nothing in section 703(h) of this title shall be interpreted to permit otherwise. This subsection shall not require an employer to pay for health insurance benefits for abortion, except where the life of the mother would be endangered if the fetus were carried to term, or except where medical complications have arisen from an abortion: Provided, That nothing herein shall preclude an employer from providing abortion benefits or otherwise affect bargaining agreements in regard to abortion."

Sec. 2. (a) Except as provided in subsection (b), the amendment made by this Act shall be effective on the date of enactment.

(b) The provisions of the amendment made by the first section of this Act shall not apply to any fringe benefit program or fund, or insurance program which is in effect on the date of enactment of this Act until 180 days after enactment of this Act.

Sec. 3. Until the expiration of a period of one year from the date of enactment of this Act or, if there is an applicable collective- bargaining agreement in effect on the date of enactment of this Act, until the termination of that agreement, no person who, on the date of enactment of this Act is providing either by direct payment or by making contributions to a fringe benefit fund or insurance program, benefits in violation with this Act shall, in order to come into compliance with this Act, reduce the benefits or the compensation provided any employee on the date of enactment of this Act, either directly or by failing to provide sufficient contributions to a fringe benefit fund or insurance program: Provided, That where the costs of such benefits on the date of enactment of this Act are apportioned between employers and employees, the payments or contributions required to comply with this Act may be made by employers and employees in the same proportion: And provided further, That nothing in this section shall prevent the readjustment of benefits or compensation for reasons unrelated to compliance with this Act.

Approved October 31, 1978.

THE EQUAL PAY ACT OF 1963

Minimum Wage

SEC. 206. [*Section 6*]

(d) Prohibition of sex discrimination

(1) No employer having employees subject to any provisions of this section shall discriminate, within any establishment in which such employees are employed, between employees on the basis of sex by paying wages to employees in such establishment at a rate less than the rate at which he pays wages to employees of the opposite sex in such establishment for equal work on jobs the performance of which requires equal skill, effort, and responsibility, and which are performed under similar working conditions, except where such payment is made pursuant to (i) a seniority system; (ii) a merit system; (iii) a system which measures earnings by quantity or quality of production; or (iv) a differential based on any other factor other than sex: *Provided*, That an employer who is paying a wage rate differential in violation of this subsection shall not, in order to comply with the provisions of this subsection, reduce the wage rate of any employee.

(2) No labor organization, or its agents, representing employees of an employer having employees subject to any provisions of this section shall cause or attempt

to cause such an employer to discriminate against an employee in violation of paragraph (1) of this subsection.

(3) For purposes of administration and enforcement, any amounts owing to any employee which have been withheld in violation of this subsection shall be deemed to be unpaid minimum wages or unpaid overtime compensation under this chapter.

(4) As used in this subsection, the term "labor organization" means any organization of any kind, or any agency or employee representation committee or plan, in which employees participate and which exists for the purpose, in whole or in part, of dealing with employers concerning grievances, labor disputes, wages, rates of pay, hours of employment, or conditions of work.

ADDITIONAL PROVISIONS OF EQUAL PAY ACT OF 1963

An Act

To prohibit discrimination on account of sex in the payment of wages by employers engaged in commerce or in the production of goods for commerce.

Be it enacted by the Senate and House of Representatives of the United States of America in Congress assembled, That this Act may be cited as the "Equal Pay Act of 1963."

***

DECLARATION OF PURPOSE

Not Reprinted in U.S. Code [*Section 2*]

(a) The Congress hereby finds that the existence in industries engaged in commerce or in the production of goods for commerce of wage differentials based on sex-

(1) depresses wages and living standards for employees necessary for their health and efficiency;

(2) prevents the maximum utilization of the available labor

resources;

(3) tends to cause labor disputes, thereby burdening, affecting, and obstructing commerce;

(4) burdens commerce and the free flow of goods in commerce; and

(5) constitutes an unfair method of competition.

(b) It is hereby declared to be the policy of this Act, through exercise by Congress of its power to regulate commerce among the several States and with foreign nations, to correct the conditions above referred to in such industries.

[*Section 3 of the Equal Pay Act of 1963 amends section 6 of the Fair Labor Standards Act by adding a new subsection (d). The amendment is incorporated in the revised text of the Fair Labor Standards Act.*]

EFFECTIVE DATE

Not Reprinted in U.S. Code [*Section 4*]

The amendments made by this Act shall take effect upon the expiration of one year from the date of its enactment: Provided, That in the case of employees covered by a bona fide collective bargaining agreement in effect at least thirty days prior to the date of enactment of this Act entered into by a labor organization (as defined in section 6(d)(4) of the Fair Labor Standards Act of 1938, as amended) [subsection (d)(4) of this section], the amendments made by this Act shall take effect upon the termination of such collective bargaining agreement or upon the expiration of two years from the date of enactment of this Act, whichever shall first occur.

Approved June 10, 1963, 12 m.

[*In the following excerpts from the Fair Labor Standards Act of 1938, as amended, authority given to the Secretary of Labor is exercised by the Equal Employment Opportunity Commission for purposes of enforcing the Equal Pay Act of 1963.*]

ATTENDANCE OF WITNESSES

SEC. 209 [*Section 9*]

For the purpose of any hearing or investigation provided for in this chapter, the provisions of sections 49 and 50 of title 15 [*Federal Trade Commission Act of September 16, 1914, as amended (U.S.C., 1934 edition)*] (relating to the attendance of witnesses and the production of books, papers, and documents), are made applicable to the jurisdiction, powers, and duties of the Administrator, the Secretary of Labor, and the industry committees.

COLLECTION OF DATA

SEC. 211 [*Section 11*]

(a) Investigations and inspections

The Administrator or his designated representatives may investigate and gather data regarding the wages, hours, and other conditions and practices of employment in any industry subject to this chapter, and may enter and inspect such places and such records (and make such transcriptions thereof), question such employees, and investigate such facts, conditions, practices, or matters as he may deem necessary or appropriate to determine whether any person has violated any provision of this chapter, or which may aid in the enforcement of the provisions of this chapter. Except as provided in section 212 [*section 12*] of this title and in subsection (b) of this section, the Administrator shall utilize the bureaus and divisions of the Department of Labor for all the investigations and inspections necessary under this section. Except as provided in section 212 [*section 12*] *of this title*, the Administrator shall bring all actions under section 217 [*section 17*] of this title to restrain violations of this chapter.

(b) State and local agencies and employees

With the consent and cooperation of State agencies charged with the administration of State labor laws, the Administrator and the Secretary of Labor may, for the purpose of carrying out their respective functions and duties under this chapter, utilize the services of State and local agencies and their employees and, notwithstanding any other provision of law, may reimburse such State and local agencies and their employees for services rendered for such purposes.

(c) Records

Every employer subject to any provision of this chapter or of any order issued under this chapter shall make, keep, and preserve such records of the persons employed by him and of the wages, hours, and other conditions and practices of employment maintained by him, and shall preserve such records for such periods of time, and shall make such reports therefrom to the Administrator as he shall prescribe by regulation or order as necessary or appropriate for the enforcement of the provisions of this chapter or the regulations or orders thereunder. The employer of an employee who performs substitute work described in section 207(p)(3) [*section 7(p)(3)*] of this title may not be required under this subsection to keep a record of the hours of the substitute work.

(d) Homework regulations

The Administrator is authorized to make such regulations and orders regulating, restricting, or prohibiting industrial homework as are necessary or appropriate to prevent the circumvention or evasion of and to safeguard the minimum wage rate prescribed in this chapter, and all existing regulations or orders of the Administrator relating to industrial homework are continued in full force and effect.

EXEMPTIONS

SEC. 213 [*Section 13*]

(a) Minimum wage and maximum hour requirements

The provisions of sections 206 [*section 6*] (except subsection (d) in the case of paragraph (1) of this subsection) and section 207 [*section 7*] of this title shall not apply with respect to-

(1) any employee employed in a bona fide executive, administrative, or professional capacity (including any employee employed in the capacity of academic

administrative personnel or teacher in elementary or secondary schools), or in the capacity of outside salesman (as such terms are defined and delimited from time to time by regulations of the Secretary, subject to the provisions of subchapter II of chapter 5 of Title 5 [*the Administrative Procedure Act*], except that an employee of a retail or service establishment shall not be excluded from the definition of employee employed in a bona fide executive or administrative capacity because of the number of hours in his workweek which he devotes to activities not directly or closely related to the performance of executive or administrative activities, if less than 40 per centum of his hours worked in the workweek are devoted to such activities); or

(2) [*Repealed*]

[*Note: Section 13(a)(2) (relating to employees employed by a retail or service establishment) was repealed by Pub. L. 101-157, section 3(c)(1), November 17, 1989.*]

(3) any employee employed by an establishment which is an amusement or recreational establishment, organized camp, or religious or non-profit educational conference center, if (A) it does not operate for more than seven months in any calendar year, or (B) during the preceding calendar year, its average receipts for any six months of such year were not more than 33 1/3 per centum of its average receipts for the other six months of such year, except that the exemption from sections 206 and 207 [*sections 6 and 7*] of this title provided by this paragraph does not apply with respect to any employee of a private entity engaged in providing services or facilities (other than, in the case of the exemption from section 206 [*section 6*] of this title, a private entity engaged in providing services and facilities directly related to skiing) in a national park or a national forest, or on land in the National Wildlife Refuge System, under a contract with the Secretary of the Interior or the Secretary of Agriculture; or

(4) [*Repealed*]

[*Note: Section 13(a)(4) (relating to employees employed by an establishment which qualified as an exempt retail establishment) was repealed by Pub. L. 101-157, Section 3(c)(1), November 17, 1989.*]

(5) any employee employed in the catching, taking, propagating, harvesting, cultivating, or farming of any kind of fish, shellfish, crustacea, sponges, seaweeds, or other aquatic forms of animal and vegetable life, or in the first processing, canning or packing such marine products at sea as an incident to, or in conjunction with, such fishing operations, including the going to and returning from work and loading and unloading when performed by any such employee; or

(6) any employee employed in agriculture (A) if such employee is employed by an employer who did not, during any calendar quarter during the preceding calendar year, use more than five hundred man-days of agricultural labor, (B) if such

employee is the parent, spouse, child, or other member of his employer's immediate family, (C) if such employee (i) is employed as a hand harvest laborer and is paid on a piece rate basis in an operation which has been, and is customarily and generally recognized as having been, paid on a piece rate basis in the region of employment, (ii) commutes daily from his permanent residence to the farm on which he is so employed, and (iii) has been employed in agriculture less than thirteen weeks during the preceding calendar year, (D) if such employee (other than an employee described in clause (C) of this subsection) (i) is sixteen years of age or under and is employed as a hand harvest laborer, is paid on a piece rate basis in an operation which has been, and is customarily and generally recognized as having been, paid on a piece rate basis in the region of employment, (ii) is employed on the same farm as his parent or person standing in the place of his parent, and (iii) is paid at the same piece rate as employees over age sixteen are paid on the same farm, or (E) if such employee is principally engaged in the range production of livestock; or

(7) any employee to the extent that such employee is exempted by regulations, order, or certificate of the Secretary issued under section 214 [*section 14*] of this title; or

(8) any employee employed in connection with the publication of any weekly, semiweekly, or daily newspaper with a circulation of less than four thousand the major part of which circulation is within the county where published or counties contiguous thereto; or

(9) [*Repealed*]

[*Note: Section 13(a)(9) (relating to motion picture theater employees) was repealed by section 23 of the Fair Labor Standards Amendments of 1974. The 1974 amendments created an exemption for such employees from the overtime provisions only in section 13(b)27.*]

(10) any switchboard operator employed by an independently owned public telephone company which has not more than seven hundred and fifty stations; or

(11) [*Repealed*]

[*Note: Section 13(a)(11) (relating to telegraph agency employees) was repealed by section 10 of the Fair Labor Standards Amendments of 1974. The 1974 amendments created an exemption from the overtime provisions only in section 13(b) (23), which was repealed effective May 1, 1976.*]

(12) any employee employed as a seaman on a vessel other than an American vessel; or

(13) [*Repealed*]

[*Note: Section 13(a)(13) (relating to small logging crews) was repealed by section 23 of the Fair Labor Standards Amendments of 1974. The 1974 amendments created an exemption for such employees from the overtime provisions only in section 13(b)(28).*]

(14) [*Repealed*]

[*Note: Section 13(a)(14) (relating to employees employed in growing and harvesting of shade grown tobacco) was repealed by section 9 of the Fair Labor Standards Amendments of 1974. The 1974 amendments created an exemption for certain tobacco producing employees from the overtime provisions only in section 13(b)(22). The section 13(b)(22) exemption was repealed, effective January 1, 1978, by section 5 of the Fair Labor Standards Amendments of 1977.*]

(15) any employee employed on a casual basis in domestic service employment to provide babysitting services or any employee employed in domestic service employment to provide companionship services for individuals who (because of age or infirmity) are unable to care for themselves (as such terms are defined and delimited by regulations of the Secretary); or

(16) a criminal investigator who is paid availability pay under section 5545a of Title 5 [*Law Enforcement Availability Pay Act of 1994*]; or

(17) any employee who is a computer systems analyst, computer programmer, software engineer, or other similarly skilled worker, whose primary duty is—

(A) the application of systems analysis techniques and procedures, including consulting with users, to determine hardware, software, or system functional specifications;

(B) the design, development, documentation, analysis, creation, testing, or modification of computer systems or programs, including prototypes, based on and related to user or system design specifications;

(C) the design, documentation, testing, creation, or modification of computer programs related to machine operating systems; or

(D) a combination of duties described in subparagraphs (A), (B), and (C) the performance of which requires the same level of skills, and

who, in the case of an employee who is compensated on an hourly basis, is compensated at a rate of not less than $27.63 an hour.

***

(g) Certain employment in retail or service establishments, agriculture

The exemption from section 206 [*section 6*] of this title provided by paragraph (6) of subsection (a) of this section shall not apply with respect to any employee employed by an establishment (1) which controls, is controlled by, or is under common control with, another establishment the activities of which are not related for a common business purpose to, but materially support the activities of the establishment employing such employee; and (2) whose annual gross volume of sales made or business done, when combined with the annual gross volume of sales made or business done by each establishment which controls, is controlled by, or is under common control with, the establishment employing such employee, exceeds $10,000,000 (exclusive of excise taxes at the retail level which are separately stated).

PROHIBITED ACTS

SEC. 215 [*Section 15*]

(a) After the expiration of one hundred and twenty days from June 25, 1938 [*the date of enactment of this Act*], it shall be unlawful for any person-

(1) to transport, offer for transportation, ship, deliver, or sell in commerce, or to ship, deliver, or sell with knowledge that shipment or delivery or sale thereof in commerce is intended, any goods in the production of which any employee was employed in violation of section 206 [*section 6*] or section 207 [*section 7*] of this title, or in violation of any regulation or order of the Secretary issued under section 214 [*section 14*] of this title, except that no provision of this chapter shall impose any liability upon any common carrier for the transportation in commerce in the regular course of its business of any goods not produced by such common carrier, and no provision of this chapter shall excuse any common carrier from its obligation to accept any goods for transportation; and except that any such transportation, offer, shipment, delivery, or sale of such goods by a purchaser who acquired them in good faith in reliance on written assurance from the producer that the goods were produced in compliance with the requirements of this chapter, and who acquired such goods for value without notice of any such violation, shall not be deemed unlawful;

(2) to violate any of the provisions of section 206 [*section 6*] or section 207 [*section 7*] of this title, or any of the provisions of any regulation or order of the Secretary issued under section 214 [*section 14*] of this title;

(3) to discharge or in any other manner discriminate against any employee because such employee has filed any complaint or instituted or caused to be instituted any proceeding under or related to this chapter, or has testified or is about to testify in any such proceeding, or has served or is about to serve on an industry committee;

(4) to violate any of the provisions of section 212 [*section 12*] of this title;

(5) to violate any of the provisions of section 211(c) [*section 11(c)*] of this title, or any regulation or order made or continued in effect under the provisions of section 211(d) [*section 11(d)*] of this title, or to make any statement, report, or record filed or kept pursuant to the provisions of such section or of any regulation or order thereunder, knowing such statement, report, or record to be false in a material respect.

(b) For the purposes of subsection (a)(1) of this section proof that any employee was employed in any place of employment where goods shipped or sold in commerce were produced, within ninety days prior to the removal of the goods from such place of employment, shall be prima facie evidence that such employee was engaged in the production of such goods.

PENALTIES

SEC. 216 [*Section 16*]

(a) Fines and imprisonment

Any person who willfully violates any of the provisions of section 215 [*section 15*] of this title shall upon conviction thereof be subject to a fine of not more than $10,000, or to imprisonment for not more than six months, or both. No person shall be imprisoned under this subsection except for an offense committed after the conviction of such person for a prior offense under this subsection.

(b) Damages; right of action; attorney's fees and costs; termination of right of action

Any employer who violates the provisions of section 206 [*section 6*] or section 207 [*section 7*] of this title shall be liable to the employee or employees affected in the amount of their unpaid minimum wages, or their unpaid overtime compensation, as the case may be, and in an additional equal amount as liquidated damages. Any employer who violates the provisions of section 215(a)(3) [*section 15(a)(3)*] of this title shall be liable for such legal or equitable relief as may be appropriate to effectuate the purposes of section 215(a)(3) [*section 15(a)(3)*] of this title, including without limitation employment, reinstatement, promotion, and the payment of wages lost and an additional equal amount as liquidated damages. An action to recover the liability prescribed in either of the preceding sentences may be maintained against any employer (including a public agency) in any Federal or State court of competent jurisdiction by any one or more employees for and in behalf of himself or themselves and other employees similarly situated. No employee shall be a party plaintiff to any such action unless he gives his consent in writing to become such a party and such consent is filed in the court in which such action is brought. The court in such action shall, in addition to any judgment awarded to the plaintiff or plaintiffs, allow a reasonable attorney's fee to be paid by the defendant, and costs of the action. The right provided by this subsection to bring an action by or on behalf of any employee, and the right of any employee to become a party plaintiff to any such action, shall terminate upon the filing of a complaint by the Secretary of Labor in an action under section 217 [*section 17*] of this title in which (1) restraint is sought of any further delay in the payment of unpaid minimum wages, or the amount of unpaid overtime compensation, as the case may be, owing to such employee under section 206 [*section 6*] or section 207 [*section 7*] of this title by an employer liable therefor[*sic*] under the provisions of this subsection or (2) legal or equitable relief is sought as a result of alleged violations of section 215(a)(3) [*section 15(a)(3)*] of this title.

(c) Payment of wages and compensation; waiver of claims; actions by the Secretary; limitation of actions

The Secretary is authorized to supervise the payment of the unpaid minimum wages or the unpaid overtime compensation owing to any employee or employees under section 206 [*section 6*] or section 207 [*section 7*] of this title, and the agreement of any employee to accept such payment shall upon payment in full constitute a waiver by such employee of any right he may have under subsection (b) of this section to such unpaid minimum wages or unpaid overtime compensation and an additional equal amount as liquidated damages. The Secretary may bring an action in any court of competent jurisdiction to recover the amount of the unpaid minimum wages or overtime compensation and an equal amount as liquidated damages. The right provided by subsection (b) of this section to bring an action by or on behalf of any employee to recover the liability specified in the first sentence of such subsection and of any employee to become a party plaintiff to any such action shall terminate upon the filing of a complaint by the Secretary in an action under this subsection in which a recovery is sought of unpaid minimum wages or unpaid overtime compensation under sections 206 and 207 [*sections 6 and 7*] of this title or liquidated or other damages provided by this subsection owing to such employee by an employer liable under the provisions of subsection (b) of this section, unless such action is dismissed without prejudice on motion of the Secretary. Any sums thus recovered by the Secretary of Labor on behalf of an employee pursuant to this subsection shall be held in a special deposit account and shall be paid, on order of the Secretary of Labor, directly to the employee or employees affected. Any such sums not paid to an employee because of inability to do so within a period of three years shall be covered into the Treasury of the United States as miscellaneous receipts. In determining when an action is commenced by the Secretary of Labor under this subsection for the purposes of the statutes of limitations provided in section 255(a) of this title [*section 6(a) of the Portal-to-Portal Act of 1947*], it shall be considered to be commenced in the case of any individual claimant on the date when the complaint is filed if he is specifically named as a party plaintiff in the complaint, or if his name did not so appear, on the subsequent date on which his name is added as a party plaintiff in such action.

(d) Savings provisions

In any action or proceeding commenced prior to, on, or after August 8, 1956 [*the date of enactment of this subsection*], no employer shall be subject to any liability or punishment under this chapter or the Portal-to-Portal Act of 1947 [29 U.S.C. 251 et seq.] on account of his failure to comply with any provision or provisions of this chapter or such Act (1) with respect to work heretofore or hereafter performed in a workplace to which the exemption in section 213(f) [*section 13(f)*] of this title is applicable, (2) with respect to work performed in Guam, the Canal Zone or Wake Island before the effective date of this amendment of subsection (d), or (3) with respect to work performed in a possession named in section 206(a)(3) [*section 6(a)*

(3)] of this title at any time prior to the establishment by the Secretary, as provided therein, of a minimum wage rate applicable to such work.

(e)(1)(A) Any person who violates the provisions of sections 212 or 213(c) [*sections 12 or 13(c)*] of this title, relating to child labor, or any regulation issued pursuant to such sections, shall be subject to a civil penalty of not to exceed—

(i) $11,000 for each employee who was the subject of such a violation; or

(ii) $50,000 with regard to each such violation that causes the death or serious injury of any employee under the age of 18 years, which penalty may be doubled where the violation is a repeated or willful violation.

(B) For purposes of subparagraph (A), the term “serious injury” means—

(i) permanent loss or substantial impairment of one of the senses (sight, hearing, taste, smell, tactile sensation);

(ii) permanent loss or substantial impairment of the function of a bodily member, organ, or mental faculty, including the loss of all or part of an arm, leg, foot, hand or other body part; or

(iii) permanent paralysis or substantial impairment that causes loss of movement or mobility of an arm, leg, foot, hand or other body part.

(2) Any person who repeatedly or willfully violates section 206 or 207 [*section 6 or 7*], relating to wages, shall be subject to a civil penalty not to exceed $1,100 for each such violation.

(3) In determining the amount of any penalty under this subsection, the appropriateness of such penalty to the size of the business of the person charged and the gravity of the violation shall be considered. The amount of any penalty under this subsection, when finally determined, may be-

(A) deducted from any sums owing by the United States to the person charged;

(B) recovered in a civil action brought by the Secretary in any court of competent jurisdiction, in which litigation the Secretary shall be represented by the Solicitor of Labor; or

(C) ordered by the court, in an action brought for a violation of section 215(a)(4) [*section 15(a)(4)*] of this title or a repeated or willful violation of section 215(a)(2) [*section 15(a)(2)*] of this title, to be paid to the Secretary.

(4) Any administrative determination by the Secretary of the amount of any penalty under this subsection shall be final, unless within 15 days after receipt of notice thereof by certified mail the person charged with the violation takes exception to the determination that the violations for which the penalty is imposed occurred, in which event final determination of the penalty shall be made in an administrative

proceeding after opportunity for hearing in accordance with section 554 of Title 5 [*Administrative Procedure Act*], and regulations to be promulgated by the Secretary.

(5) Except for civil penalties collected for violations of section 212 [*section 12*] of this title, sums collected as penalties pursuant to this section shall be applied toward reimbursement of the costs of determining the violations and assessing and collecting such penalties, in accordance with the provision of section 9a of Title 29 [*An Act to authorize the Department of Labor to make special statistical studies upon payment of the cost thereof and for other purposes*]. Civil penalties collected for violations of section 212 [*section 12*] of this title shall be deposited in the general fund of the Treasury.

INJUNCTION PROCEEDINGS

SEC. 217 [*Section 17*]

The districts courts, together with the United States District Court for the District of the Canal Zone, the District Court of the Virgin Islands, and the District Court of Guam shall have jurisdiction, for cause shown, to restrain violations of section 215 [*section 15*] of this title, including in the case of violations of section 215(a)(2) of this title the restraint of any withholding of payment of minimum wages or overtime compensation found by the court to be due to employees under this chapter (except sums which employees are barred from recovering, at the time of the commencement of the action to restrain the violations, by virtue of the provisions of section 255 of this title [*section 6 of the Portal-to-Portal Act of 1947*].

RELATION TO OTHER LAWS

SEC. 218 [*Section 18*]

(a) No provision of this chapter or of any order thereunder shall

excuse noncompliance with any Federal or State law or municipal ordinance establishing a minimum wage higher than the minimum wage established under this chapter or a maximum work week lower than the maximum workweek established under this chapter, and no provision of this chapter relating to the employment of child labor shall justify noncompliance with any Federal or State law or municipal ordinance establishing a higher standard than the standard established under this chapter. No provision of this chapter shall justify any employer in reducing a wage paid by him which is in excess of the applicable minimum wage under this chapter, or justify any employer in increasing hours of employment maintained by him which are shorter than the maximum hours applicable under this chapter.

SEPARABILITY OF PROVISIONS

SEC. 219 [*Section 19*]

If any provision of this chapter or the application of such provision to any person or circumstance is held invalid, the remainder of this chapter and the application of such provision to other persons or circumstances shall not be affected thereby.

Approved June 25, 1938.

[*In the following excerpts from the Portal-to-Portal Act of 1947, the authority given to the Secretary of Labor is exercised by the Equal Employment Opportunity Commission for purposes of enforcing the Equal Pay Act of 1963.*]

PART IV - MISCELLANEOUS

STATUTE OF LIMITATIONS

SEC. 255 [*Section 6*]

Any action commenced on or after May 14, 1947 [*the date of the enactment of this Act*], to enforce any cause of action for unpaid minimum wages, unpaid overtime compensation, or liquidated damages, under the Fair Labor Standards Act of 1938, as amended, [29 U.S.C. 201 et seq.], the Walsh-Healey Act [41 U.S.C. 35 et seq.], or the Bacon-Davis Act [40 U.S.C. 276a et seq.]-

(a) if the cause of action accrues on or after May 14, 1947 [*the date of the enactment of this Act*]-may be commenced within two years after the cause of action accrued, and every such action shall be forever barred unless commenced within two years after the cause of action accrued, except that a cause of action arising out of a willful violation may be commenced within three years after the cause of action accrued;

DETERMINATION OF COMMENCEMENT OF FUTURE ACTIONS

SEC. 256 [*Section 7*]

In determining when an action is commenced for the purposes of section 255 [*section 6*] of this title, an action commenced on or after May 14, 1947 [*the date of the enactment of this Act*] under the Fair Labor Standards Act of 1938, as amended, [29 U.S.C. 201 et seq.], the Walsh-Healey Act [41 U.S.C. 35 et seq.], or the Bacon-Davis Act [40 U.S.C. 276a et seq.], shall be considered to be commenced on the date when the complaint is filed; except that in the case of a collective or class action instituted under the Fair Labor Standards Act of 1938, as amended, or the Bacon-Davis Act, it shall be considered to be commenced in the case of any individual claimant—

(a) on the date when the complaint is filed, if he is specifically named as a party plaintiff in the complaint and his written consent to become a party plaintiff is filed on such date in the court in which the action is brought; or

(b) if such written consent was not so filed or if his name did not so appear—on the subsequent date on which such written consent is filed in the court in which the action was commenced.

RELIANCE IN FUTURE ON ADMINISTRATIVE RULINGS, ETC.

SEC. 259 [*Section 10*]

(a) In any action or proceeding based on any act or omission on or after May 14, 1947 [*the date of the enactment of this Act*], no employer shall be subject to any liability or punishment for or on account of the failure of the employer to pay minimum wages or overtime compensation under the Fair Labor Standards Act of 1938, as amended, [29 U.S.C. 201 et seq.], the Walsh-Healey Act [41 U.S.C. 35 et seq.], or the Bacon-Davis Act [40 U.S.C. 276a et seq.], if he pleads and proves that the act or omission complained of was in good faith in conformity with and in reliance on any written administrative regulation, order, ruling, approval, or interpretation, of the agency of the United States specified in subsection (b) of this section, or any administrative practice or enforcement policy of such agency with respect to the class of employers to which he belonged. Such a defense, if established, shall be a bar to the action or proceeding, notwithstanding that after such act or omission, such administrative regulation, order, ruling, approval, interpretation, practice, or enforcement policy is modified or rescinded or is determined by judicial authority to be invalid or of no legal effect.

(b) The agency referred to in subsection (a) shall be-

(1) in the case of the Fair Labor Standards Act of 1938, as amended [29 U.S.C. 201 et seq.]- the Administrator of the Wage and Hour Division of the Department of Labor;

LIQUIDATED DAMAGES

SEC. 260 [*Section 11*]

In any action commenced prior to or on or after May 14, 1947 [*the date of the enactment of this Act*] to recover unpaid minimum wages, unpaid overtime compensation, or liquidated damages, under the Fair Labor Standards Act of 1938, as amended [29 U.S.C. 201 et seq.], if the employer shows to the satisfaction of the court that the act or omission giving rise to such action was in good faith and that he had reasonable grounds for believing that his act or omission was not a violation of the Fair Labor Standards Act of 1938, as amended [*29 U.S.C. 201 et seq.*],the court may, in its sound discretion, award no liquidated damages or award any amount thereof not to exceed the amount specified in section 216 [*section 16*] of this title.

DEFINITIONS

SEC. 262 [*Section 13*]

(a) When the terms "employer," "employee," and "wage" are used in this chapter in relation to the Fair Labor Standards Act of 1938, as amended [29 U.S.C. 201 et seq.], they shall have the same meaning as when used in such Act of 1938.

SEPARABILITY

Not Reprinted in U.S. Code [*Section 14*]

If any provision of this Act or the application of such provision to any person or circumstance is held invalid, the remainder of this Act and the application of such provision to other persons or circumstances shall not be affected thereby.

Approved May 14, 1947.

THE AGE DISCRIMINATION IN EMPLOYMENT ACT OF 1967

An Act

To prohibit age discrimination in employment.

Be it enacted by the Senate and House of Representatives of the United States of America in Congress assembled, that this Act may be cited as the "Age Discrimination in Employment Act of 1967."

* * *

CONGRESSIONAL STATEMENT OF FINDINGS AND PURPOSE

SEC. 621. [*Section 2*]

(a) The Congress hereby finds and declares that-

(1) in the face of rising productivity and affluence, older workers find themselves disadvantaged in their efforts to retain employment, and especially to regain employment when displaced from jobs;

(2) the setting of arbitrary age limits regardless of potential for job performance has become a common practice, and certain otherwise desirable practices may work to the disadvantage of older persons;

(3) the incidence of unemployment, especially long-term unemployment with resultant deterioration of skill, morale, and employer acceptability is, relative to the younger ages, high among older workers; their numbers are great and growing; and their employment problems grave;

(4) the existence in industries affecting commerce, of arbitrary discrimination in employment because of age, burdens commerce and the free flow of goods in commerce.

(b) It is therefore the purpose of this chapter to promote employment of older persons based on their ability rather than age; to prohibit arbitrary age discrimination in employment; to help employers and workers find ways of meeting problems arising from the impact of age on employment.

EDUCATION AND RESEARCH PROGRAM; RECOMMENDATION TO CONGRESS

SEC. 622. [*Section 3*]

(a) The EEOC [*originally, the Secretary of Labor*] shall undertake studies and provide information to labor unions, management, and the general public concerning the needs and abilities of older workers, and their potentials for continued employment and contribution to the economy. In order to achieve the purposes of this chapter, the EEOC [*originally, the Secretary of Labor*] shall carry on a continuing program of education and information, under which he may, among other measures-

(1) undertake research, and promote research, with a view to reducing barriers to the employment of older persons, and the promotion of measures for utilizing their skills;

(2) publish and otherwise make available to employers, professional societies, the various media of communication, and other interested persons the findings of studies and other materials for the promotion of employment;

(3) foster through the public employment service system and through cooperative effort the development of facilities of public and private agencies for expanding the opportunities and potentials of older persons;

(4) sponsor and assist State and community informational and educational programs.

(b) Not later than six months after the effective date of this chapter, the Secretary shall recommend to the Congress any measures he may deem desirable to change the lower or upper age limits set forth in section 631 of this title [*section 12*].

PROHIBITION OF AGE DISCRIMINATION

SEC. 623. [*Section 4*]

(a) Employer practices

It shall be unlawful for an employer-

(1) to fail or refuse to hire or to discharge any individual or otherwise discriminate against any individual with respect to his compensation, terms, conditions, or privileges of employment, because of such individual's age;

(2) to limit, segregate, or classify his employees in any way which would deprive or tend to deprive any individual of employment opportunities or otherwise adversely affect his status as an employee, because of such individual's age; or

(3) to reduce the wage rate of any employee in order to comply with this chapter.

(b) It shall be unlawful for an employment agency to fail or refuse to refer for employment, or otherwise to discriminate against, any individual because of such individual's age, or to classify or refer for employment any individual on the basis of such individual's age.

(c) Labor organization practices

It shall be unlawful for a labor organization-

(1) to exclude or to expel from its membership, or otherwise to discriminate against, any individual because of his age;

(2) to limit, segregate, or classify its membership, or to classify or fail or refuse to refer for employment any individual, in any way which would deprive or tend to deprive any individual of employment opportunities, or would limit such employment opportunities or otherwise adversely affect his status as an employee or as an applicant for employment, because of such individual's age;

(3) to cause or attempt to cause an employer to discriminate against an individual in violation of this section.

(d) Opposition to unlawful practices; participation in investigations, proceedings, or litigation

It shall be unlawful for an employer to discriminate against any of his employees or applicants for employment, for an employment agency to discriminate against any individual, or for a labor organization to discriminate against any member thereof or applicant for membership, because such individual, member or applicant for membership has opposed any practice made unlawful by this section, or because such individual, member or applicant for membership has made a charge, testified, assisted, or participated in any manner in an investigation, proceeding, or litigation under this chapter.

(e) Printing or publication of notice or advertisement indicating preference, limitation, etc.

It shall be unlawful for an employer, labor organization, or employment agency to print or publish, or cause to be printed or published, any notice or advertisement relating to employment by such an employer or membership in or any classification or referral for employment by such a labor organization, or relating to any classification or referral for employment by such an employment agency, indicating any preference, limitation, specification, or discrimination, based on age.

(f) Lawful practices; age an occupational qualification; other reasonable factors; laws of foreign workplace; seniority system; employee benefit plans; discharge or discipline for good cause

It shall not be unlawful for an employer, employment agency, or labor organization-

(1) to take any action otherwise prohibited under subsections (a), (b), (c), or (e) of this section where age is a bona fide occupational qualification reasonably necessary to the normal operation of the particular business, or where the differentiation is based on reasonable factors other than age, or where such practices involve an employee in a workplace in a foreign country, and compliance with such subsections would cause such employer, or a corporation controlled by such employer, to violate the laws of the country in which such workplace is located;

(2) to take any action otherwise prohibited under subsection (a), (b), (c), or (e) of this section—

(A) to observe the terms of a bona fide seniority system that is not intended to evade the purposes of this chapter, except that no such seniority system shall require or permit the involuntary retirement of any individual specified by section 631(a) of this title because of the age of such individual; or

(B) to observe the terms of a bona fide employee benefit plan-

(i) where, for each benefit or benefit package, the actual amount of payment made or cost incurred on behalf of an older worker is no less than that made or incurred on behalf of a younger worker, as permissible under section 1625.10, title 29, Code of Federal Regulations (as in effect on June 22, 1989); or

(ii) that is a voluntary early retirement incentive plan consistent with the relevant purpose or purposes of this chapter.

Notwithstanding clause (i) or (ii) of subparagraph (B), no such employee benefit plan or voluntary early retirement incentive plan shall excuse the failure to hire any individual, and no such employee benefit plan shall require or permit the involuntary retirement of any individual specified by section 631(a) of this title, because of the age of such individual. An employer, employment agency, or labor organization acting under subparagraph (A), or under clause (i) or (ii) of subparagraph (B), shall have the burden of proving that such actions are lawful in any civil enforcement proceeding brought under this chapter; or

(3) to discharge or otherwise discipline an individual for good cause.

(g) [*Repealed*]

(h) Practices of foreign corporations controlled by American employers; foreign employers not controlled by American employers; factors determining control

(1) If an employer controls a corporation whose place of incorporation is in a foreign country, any practice by such corporation prohibited under this section shall be presumed to be such practice by such employer.

(2) The prohibitions of this section shall not apply where the employer is a foreign person not controlled by an American employer.

(3) For the purpose of this subsection the determination of whether an employer controls a corporation shall be based upon the-

(A) interrelation of operations,

(B) common management,

(C) centralized control of labor relations, and

(D) common ownership or financial control, of the employer and the corporation.

(i) Employee pension benefit plans; cessation or reduction of benefit accrual or of allocation to employee account; distribution of benefits after attainment of normal retirement age; compliance; highly compensated employees

(1) Except as otherwise provided in this subsection, it shall be unlawful for an employer, an employment agency, a labor organization, or any combination thereof to establish or maintain an employee pension benefit plan which requires or permits—

(A) in the case of a defined benefit plan, the cessation of an employee's benefit accrual, or the reduction of the rate of an employee's benefit accrual, because of age, or

(B) in the case of a defined contribution plan, the cessation of allocations to an employee's account, or the reduction of the rate at which amounts are allocated to an employee's account, because of age.

(2) Nothing in this section shall be construed to prohibit an employer, employment agency, or labor organization from observing any provision of an employee pension benefit plan to the extent that such provision imposes (without regard to age) a limitation on the amount of benefits that the plan provides or a limitation on the number of years of service or years of participation which are taken into account for purposes of determining benefit accrual under the plan.

(3) In the case of any employee who, as of the end of any plan year under a defined benefit plan, has attained normal retirement age under such plan—

(A) if distribution of benefits under such plan with respect to such employee has commenced as of the end of such plan year, then any requirement of this subsection for continued accrual of benefits under such plan with respect to such employee during such plan year shall be treated as satisfied to the extent of the actuarial equivalent of in-service distribution of benefits, and

(B) if distribution of benefits under such plan with respect to such employee has not commenced as of the end of such year in accordance with section 1056(a)(3) of this title [*section 206(a)(3) of the Employee Retirement Income Security Act of 1974*] and section 401(a)(14)(C) of Title 26 [*the Internal Revenue Code of 1986*], and the payment of benefits under such plan with respect to such employee is not suspended during such plan year pursuant to section 1053(a)(3)(B) of this title or section 411(a)(3)(B) of Title 26 [*the Internal Revenue Code of 1986*], then any requirement of this subsection for continued accrual of benefits under such plan with respect to such employee during such plan year shall be treated as satisfied to the extent of any adjustment in the benefit payable under the plan during such plan year attributable to the delay in the distribution of benefits after the attainment of normal retirement age.

The provisions of this paragraph shall apply in accordance with regulations of the Secretary of the Treasury. Such regulations shall provide for the application of the preceding provisions of this paragraph to all employee pension benefit plans subject to this subsection and may provide for the application of such provisions, in the case of any such employee, with respect to any period of time within a plan year.

(4) Compliance with the requirements of this subsection with respect to an employee pension benefit plan shall constitute compliance with the requirements of this section relating to benefit accrual under such plan.

(5) Paragraph (1) shall not apply with respect to any employee who is a highly compensated employee (within the meaning of section 414(q) of Title 26 [*the Internal Revenue Code of 1986*]) to the extent provided in regulations prescribed by the Secretary of the Treasury for purposes of precluding discrimination in favor of highly compensated employees within the meaning of subchapter D of chapter 1 of Title 26 [*the Internal Revenue Code of 1986*].

(6) A plan shall not be treated as failing to meet the requirements of paragraph (1) solely because the subsidized portion of any early retirement benefit is disregarded in determining benefit accruals or it is a plan permitted by subsection (m) of this section.

(7) Any regulations prescribed by the Secretary of the Treasury pursuant to clause (v) of section 411(b)(1)(H) of Title 26 [*the Internal Revenue Code of 1986*] and

subparagraphs (C) and (D), of section 411(b)(2) of Title 26 [*the Internal Revenue Code of 1986*] shall apply with respect to the requirements of this subsection in the same manner and to the same extent as such regulations apply with respect to the requirements of such sections 411(b)(1)(H) and 411(b)(2).

(8) A plan shall not be treated as failing to meet the requirements of this section solely because such plan provides a normal retirement age described in section 1002(24)(B) [*section 2(24)(B) of the Employee Retirement Income Security Act of 1974*] of this title and section 411(a)(8)(B) of Title 26 [*the Internal Revenue Code of 1986*].

(9) For purposes of this subsection-

(A) The terms "employee pension benefit plan", "defined benefit plan", "defined contribution plan", and "normal retirement age" have the meanings provided such terms in section 1002 of this title [*section 3 of the Employee Retirement Income Security Act of 1974*].

(B) The term "compensation" has the meaning provided by section 414(s) of Title 26 [*the Internal Revenue Code of 1986*].

(10) Special rules relating to age

(A) Comparison to similarly situated younger individual

(i) In general—A plan shall not be treated as failing to meet the requirements of paragraph (1) if a participant's accrued benefit, as determined as of any date under the terms of the plan, would be equal to or greater than that of any similarly situated, younger individual who is or could be a participant.

(ii) Similarly situated—For purposes of this subparagraph, a participant is similarly situated to any other individual if such participant is identical to such other individual in every respect (including period of service, compensation, position, date of hire, work history, and any other respect) except for age.

(iii) Disregard of subsidized early retirement benefits—In determining the accrued benefit as of any date for purposes of this clause, the subsidized portion of any early retirement benefit or retirement-type subsidy shall be disregarded.

(iv) Accrued benefit—For purposes of this subparagraph, the accrued benefit may, under the terms of the plan, be expressed as an annuity payable at normal retirement age, the balance of a hypothetical account, or the current value of the accumulated percentage of the employee's final average compensation.

(B) Applicable defined benefit plans

(i) Interest credits

(I) In general—An applicable defined benefit plan shall be treated as failing to meet the requirements of paragraph (1) unless the terms of the plan provide that any interest credit (or an equivalent amount) for any plan year shall be at a rate which is not greater than a market rate of return. A plan shall not be treated as failing to meet the requirements of this subclause merely because the plan provides for a reasonable minimum guaranteed rate of return or for a rate of return that is equal to the greater of a fixed or variable rate of return

(II) Preservation of capital—An interest credit (or an equivalent amount) of less than zero shall in no event result in the account balance or similar amount being less than the aggregate amount of contributions credited to the account.

(III) Market rate of return—The Secretary of the Treasury may provide by regulation for rules governing the calculation of a market rate of return for purposes of subclause (I) and for permissible methods of crediting interest to the account (including fixed or variable interest rates) resulting in effective rates of return meeting the requirements of subclause (I). In the case of a governmental plan (as defined in the first sentence of section 414(d) of Title 26 [*the Internal Revenue Code of 1986*], a rate of return or a method of crediting interest established pursuant to any provision of Federal, State, or local law (including any administrative rule or policy adopted in accordance with any such law) shall be treated as a market rate of return for purposes of subclause (I) and a permissible method of crediting interest for purposes of meeting the requirements of subclause (I), except that this sentence shall only apply to a rate of return or method of crediting interest if such rate or method does not violate any other requirement of this chapter.

(ii) Special rule for plan conversions—If, after June 29, 2005, an applicable plan amendment is adopted, the plan shall be treated as failing to meet the requirements of paragraph (1)(H) unless the requirements of clause (iii) are met with respect to each individual who was a participant in the plan immediately before the adoption of the amendment.

(iii) Rate of benefit accrual—Subject to clause (iv), the requirements of this clause are met with respect to any participant if the accrued benefit of the participant under the terms of the plan as in effect after the amendment is not less than the sum of—

(I) the participant's accrued benefit for years of service before the effective date of the amendment, determined under the terms of the plan as in effect before the amendment, plus

(II) the participant's accrued benefit for years of service after the effective date of the amendment, determined under the terms of the plan as in effect after the amendment.

(iv) Special rules for early retirement subsidies—For purposes of clause (iii)(I), the plan shall credit the accumulation account or similar amount with the amount of any early retirement benefit or retirement-type subsidy for the plan year in which the participant retires if, as of such time, the participant has met the age, years of service, and other requirements under the plan for entitlement to such benefit or subsidy.

(v) Applicable plan amendment—For purposes of this subparagraph—

(I) In general—The term "applicable plan amendment" means an amendment to a defined benefit plan which has the effect of converting the plan to an applicable defined benefit plan.

(II) Special rule for coordinated benefits—If the benefits of 2 or more defined benefit plans established or maintained by an employer are coordinated in such a manner as to have the effect of the adoption of an amendment described in subclause (I), the sponsor of the defined benefit plan or plans providing for such coordination shall be treated as having adopted such a plan amendment as of the date such coordination begins.

(III) Multiple amendments—The Secretary of the Treasury shall issue regulations to prevent the avoidance of the purposes of this subparagraph through the use of 2 or more plan amendments rather than a single amendment.

(IV) Applicable defined benefit plan—For purposes of this subparagraph, the term "applicable defined benefit plan" has the meaning given such term by section 1053(f)(3) of this title [*section 203(f)(3) of the Employee Retirement Income Security Act of 1974*].

(vi) Termination requirements—An applicable defined benefit plan shall not be treated as meeting the requirements of clause (i) unless the plan provides that, upon the termination of the plan—

(I) if the interest credit rate (or an equivalent amount) under the plan is a variable rate, the rate of interest used to determine accrued benefits under the plan shall be equal to the average of the rates of interest used under the plan during the 5-year period ending on the termination date, and

(II) the interest rate and mortality table used to determine the amount of any benefit under the plan payable in the form of an annuity payable at normal retirement age shall be the rate and table specified under the plan for such purpose as of the termination date, except that if such interest rate is a variable rate, the interest rate shall be determined under the rules of subclause (I).

(C) Certain offsets permitted—A plan shall not be treated as failing to meet the requirements of paragraph (1) solely because the plan provides offsets against benefits under the plan to the extent such offsets are allowable in applying the requirements of section 401(a) of Title 26 [*the Internal Revenue Code of 1986*].

(D) Permitted disparities in plan contributions or benefits—A plan shall not be treated as failing to meet the requirements of paragraph (1) solely because the plan provides a disparity in contributions or benefits with respect to which the requirements of section 401(l) of Title 26 [*the Internal Revenue Code of 1986*] are met.

(E) Indexing permitted—

(i) In general—A plan shall not be treated as failing to meet the requirements of paragraph (1) solely because the plan provides for indexing of accrued benefits under the plan.

(ii) Protection against loss—Except in the case of any benefit provided in the form of a variable annuity, clause (i) shall not apply with respect to any indexing which results in an accrued benefit less than the accrued benefit determined without regard to such indexing.

(iii) Indexing—For purposes of this subparagraph, the term "indexing" means, in connection with an accrued benefit, the periodic adjustment of the accrued benefit by means of the application of a recognized investment index or methodology.

(F) Early retirement benefit or retirement-type subsidy—For purposes of this paragraph, the terms "early retirement benefit" and "retirement-type subsidy" have the meaning given such terms in section 1053(g)(2)(A) of this title [*section 203(g)(2)(A) of the Employee Retirement Income Security Act of 1974*].

(G) Benefit accrued to date—For purposes of this paragraph, any reference to the accrued benefit shall be a reference to such benefit accrued to date.

(j) Employment as firefighter or law enforcement officer

It shall not be unlawful for an employer which is a State, a political subdivision of a State, an agency or instrumentality of a State or a political subdivision of a State, or an interstate agency to fail or refuse to hire or to discharge any individual because of such individual's age if such action is taken-

(1) with respect to the employment of an individual as a firefighter or as a law enforcement officer, the employer has complied with section 3(d)(2) of the Age Discrimination in Employment Amendments of 1996 if the individual was discharged after the date described in such section, and the individual has attained-

(A) the age of hiring or retirement, respectively, in effect under applicable State or local law on March 3, 1983; or

(B) (i) if the individual was not hired, the age of hiring in effect on the date of such failure or refusal to hire under applicable State or local law enacted after September 30, 1996; or

(ii) if applicable State or local law was enacted after September 30, 1996, and the individual was discharged, the higher of-

(I) the age of retirement in effect on the date of such discharge under such law; and

(II) age 55; and

(2) pursuant to a bona fide hiring or retirement plan that is not a subterfuge to evade the purposes of this chapter.

(k) Seniority system or employee benefit plan; compliance

A seniority system or employee benefit plan shall comply with this chapter regardless of the date of adoption of such system or plan.

(l) Lawful practices; minimum age as condition of eligibility for retirement benefits; deductions from severance pay; reduction of long-term disability benefits

Notwithstanding clause (i) or (ii) of subsection (f)(2)(B) of this section-

(A) It shall not be a violation of subsection (a), (b), (c), or (e) of this section solely because-

(i) an employee pension benefit plan (as defined in section 1002(2) of this title [*section 2(2) of the Employee Retirement Income Security Act of 1974*]) provides for the attainment of a minimum age as a condition of eligibility for normal or early retirement benefits; or

(ii) a defined benefit plan (as defined in section 1002(35) of this title [*section 2(35) of the Employee Retirement Income Security Act*]) provides for-

(I) payments that constitute the subsidized portion of an early retirement benefit; or

(II) social security supplements for plan participants that commence before the age and terminate at the age (specified by the plan) when participants are eligible to receive reduced or unreduced old-age insurance benefits under title II of the Social Security Act (42 U.S.C. 401 et seq.), and that do not exceed such old-age insurance benefits.

(B) A voluntary early retirement incentive plan that—

(i) is maintained by—

(I) a local educational agency (as defined in section 7801 of Title 20 *the Elementary and Secondary Education Act of 1965*), or

(II) an education association which principally represents employees of 1 or more agencies described in subclause (I) and which is described in section 501(c) (5) or (6) of Title 26 [*the Internal Revenue Code of 1986*] and exempt from taxation under section 501(a) of Title 26 [*the Internal Revenue Code of 1986*], and

(ii) makes payments or supplements described in subclauses (I) and (II) of subparagraph (A)(ii) in coordination with a defined benefit plan (as so defined) maintained by an eligible employer described in section 457(e)(1) (A) of Title 26 [*the Internal Revenue Code of 1986*] or by an education association described in clause (i)(II),

shall be treated solely for purposes of subparagraph (A)(ii) as if it were a part of the defined benefit plan with respect to such payments or supplements. Payments or supplements under such a voluntary early retirement incentive plan shall not constitute severance pay for purposes of paragraph (2).

(A) It shall not be a violation of subsection (a), (b), (c), or (e) of this section solely because following a contingent event unrelated to age—

(i) the value of any retiree health benefits received by an individual eligible for an immediate pension;

(ii) the value of any additional pension benefits that are made available solely as a result of the contingent event unrelated to age and following which the individual is eligible for not less than an immediate and unreduced pension; or

(iii) the values described in both clauses (i) and (ii) are deducted from severance pay made available as a result of the contingent event unrelated to age.

(B) For an individual who receives immediate pension benefits that are actuarially reduced under subparagraph (A)(i), the amount of the deduction available pursuant to subparagraph (A)(i) shall be reduced by the same percentage as the reduction in the pension benefits.

(C) For purposes of this paragraph, severance pay shall include that portion of supplemental unemployment compensation benefits (as described in section 501(c)(17) of Title 26 [*the Internal Revenue Code of 1986*]) that-

(i) constitutes additional benefits of up to 52 weeks;

(ii) has the primary purpose and effect of continuing benefits until an individual becomes eligible for an immediate and unreduced pension; and

(iii) is discontinued once the individual becomes eligible for an immediate and unreduced pension.

(D) For purposes of this paragraph and solely in order to make the deduction authorized under this paragraph, the term "retiree health benefits" means benefits provided pursuant to a group health plan covering retirees, for which (determined as of the contingent event unrelated to age)—

(i) the package of benefits provided by the employer for the retirees who are below age 65 is at least comparable to benefits provided under title XVIII of the Social Security Act (42 U.S.C. 1395 et seq.);

(ii) the package of benefits provided by the employer for the retirees who are age 65 and above is at least comparable to that offered under a plan that provides a benefit package with one-fourth the value of benefits provided under title XVIII of such Act; or

(iii) the package of benefits provided by the employer is as described in clauses (i) and (ii).

(E)(i) If the obligation of the employer to provide retiree health benefits is of limited duration, the value for each individual shall be calculated at a rate of $3,000 per year for benefit years before age 65, and $750 per year for benefit years beginning at age 65 and above.

(ii) If the obligation of the employer to provide retiree health benefits is of unlimited duration, the value for each individual shall be calculated at a rate of $48,000 for individuals below age 65, and $24,000 for individuals age 65 and above.

(iii) The values described in clauses (i) and (ii) shall be calculated based on the age of the individual as of the date of the contingent event unrelated to age. The values are effective on October 16, 1990, and shall be adjusted on an annual basis, with respect to a contingent event that occurs subsequent to the first year after October 16, 1990, based on the medical component of the Consumer Price Index for all-urban consumers published by the Department of Labor.

(iv) If an individual is required to pay a premium for retiree health benefits, the value calculated pursuant to this subparagraph shall be reduced by whatever percentage of the overall premium the individual is required to pay.

(F) If an employer that has implemented a deduction pursuant to subparagraph (A) fails to fulfill the obligation described in subparagraph (E), any aggrieved individual may bring an action for specific performance of the obligation described in subparagraph (E). The relief shall be in addition to any other remedies provided under Federal or State law.

(3) It shall not be a violation of subsection (a), (b), (c), or (e) of this section solely because an employer provides a bona fide employee benefit plan or plans under which long-term disability benefits received by an individual are reduced by any pension benefits (other than those attributable to employee contributions)—

(A) paid to the individual that the individual voluntarily elects to receive; or

(B) for which an individual who has attained the later of age 62 or normal retirement age is eligible.

(m) Voluntary retirement incentive plans

Notwithstanding subsection (f)(2)(b) of this section, it shall not be a violation of subsection (a), (b), (c), or (e) of this section solely because a plan of an institution of higher education (as defined in section 1001 of Title 20 [*the Higher Education Act of 1965*]) offers employees who are serving under a contract of unlimited tenure (or similar arrangement providing for unlimited tenure) supplemental benefits upon voluntary retirement that are reduced or eliminated on the basis of age, if—

(1) such institution does not implement with respect to such employees any age-based reduction or cessation of benefits that are not such supplemental benefits, except as permitted by other provisions of this chapter;

(2) such supplemental benefits are in addition to any retirement or severance benefits which have been offered generally to employees serving under a contract of unlimited tenure (or similar arrangement providing for unlimited tenure), independent of any early retirement or exit-incentive plan, within the preceding 365 days; and

(3) any employee who attains the minimum age and satisfies all non-age-based conditions for receiving a benefit under the plan has an opportunity lasting not less than 180 days to elect to retire and to receive the maximum benefit that could then be elected by a younger but otherwise similarly situated employee, and the plan does not require retirement to occur sooner than 180 days after such election.

STUDY BY SECRETARY OF LABOR; REPORTS TO PRESIDENT AND CONGRESS; SCOPE OF STUDY; IMPLEMENTATION OF STUDY; TRANSMITTAL DATE OF REPORTS

SEC. 624. [*Section 5*]

(a) The EEOC [*originally, the Secretary of Labor*] is directed to undertake an appropriate study of institutional and other arrangements giving rise to involuntary retirement, and report his findings and any appropriate legislative recommendations to the President and to the Congress. Such study shall include—

(A) an examination of the effect of the amendment made by section 3(a) of the Age Discrimination in Employment Act Amendments of 1978 in raising the upper age limitation established by section 631(a) of this title [*section 1(a)*] to 70 years of age;

(B) a determination of the feasibility of eliminating such limitation;

(C) a determination of the feasibility of raising such limitation above 70 years of age; and

(D) an examination of the effect of the exemption contained in section 631(c) of this title [*section 1(c)*], relating to certain executive employees, and the exemption contained in section 631(d) of this title [*section 1(d)*], relating to tenured teaching personnel.

(2) The EEOC [*originally, the Secretary of Labor*] may undertake the study required by paragraph (1) of this subsection directly or by contract or other arrangement.

(b) The report required by subsection (a) of this section shall be transmitted to the President and to the Congress as an interim report not later than January 1, 1981, and in final form not later than January 1, 1982.

Transfer of Functions [*All functions relating to age discrimination administration and enforcement vested by Section 6 in the Secretary of Labor or the Civil Service Commission were transferred to the Equal Employment Opportunity Commission effective January 1, 1979 under the President's Reorganization Plan No. 1.*]

ADMINISTRATION

SEC. 625. [*Section 6*]

The EEOC [*originally, the Secretary of Labor*] shall have the power-

(a) Delegation of functions; appointment of personnel; technical assistance

to make delegations, to appoint such agents and employees, and to pay for technical assistance on a fee for service basis, as he deems necessary to assist him in the performance of his functions under this chapter;

(b) Cooperation with other agencies, employers, labor organizations, and employment agencies

to cooperate with regional, State, local, and other agencies, and to cooperate with and furnish technical assistance to employers, labor organizations, and employment agencies to aid in effectuating the purposes of this chapter.

RECORDKEEPING, INVESTIGATION, AND ENFORCEMENT

SEC. 626. [*Section 7*]

(a) Attendance of witnesses; investigations, inspections, records, and homework regulations

The Equal Employment Opportunity Commission shall have the power to make investigations and require the keeping of records necessary or appropriate for the administration of this chapter in accordance with the powers and procedures provided in sections 209 and 211 of this title [*sections 9 and 11 of the Fair Labor Standards Act of 1938, as amended*].

(b) Enforcement; prohibition of age discrimination under fair labor standards; unpaid minimum wages and unpaid overtime compensation; liquidated damages; judicial relief; conciliation, conference, and persuasion

The provisions of this chapter shall be enforced in accordance with the powers, remedies, and procedures provided in sections 211(b), 216 (except for subsection (a) thereof), and 217 of this title [*sections 11(b), 16 (except for subsection (a) thereof), and 17 of the Fair Labor Standards Act of 1938, as amended*], and subsection (c) of this section. Any act prohibited under section 623 of this title [*section 4*] shall be deemed to be a prohibited act under section 215 of this title [*section 15 of the Fair Labor Standards Act of 1938, as amended*]. Amounts owing to a person as a result of a violation of this chapter shall be deemed to be unpaid minimum wages or unpaid overtime compensation for purposes of sections 216 and 217 of this title [*sections 16 and 17 of the Fair Labor Standards Act of 1938, as amended*]: *Provided,* That liquidated damages shall be payable only in cases of willful violations of this chapter. In any action brought to enforce this chapter the court shall have jurisdiction to grant such legal or equitable relief as may be appropriate to effectuate the purposes of this chapter, including without limitation judgments compelling employment, reinstatement or promotion, or enforcing the liability for amounts deemed to be unpaid minimum wages or unpaid overtime compensation under this section. Before instituting any action under this section, the Equal Employment Opportunity Commission shall attempt to eliminate the discriminatory practice or practices alleged, and to effect voluntary compliance with the requirements of this chapter through informal methods of conciliation, conference, and persuasion.

(c) Civil actions; persons aggrieved; jurisdiction; judicial relief; termination of individual action upon commencement of action by Commission; jury trial

(1) Any person aggrieved may bring a civil action in any court of competent jurisdiction for such legal or equitable relief as will effectuate the purposes of this chapter: *Provided,* That the right of any person to bring such action shall terminate upon the commencement of an action by the Equal Employment Opportunity Commission to enforce the right of such employee under this chapter.

(2) In an action brought under paragraph (1), a person shall be entitled to a trial by jury of any issue of fact in any such action for recovery of amounts owing as a result of a violation of this chapter, regardless of whether equitable relief is sought by any party in such action.

(d)(1) Filing of charge with Commission; timeliness; conciliation, conference, and persuasion

No civil action may be commenced by an individual under this section until 60 days after a charge alleging unlawful discrimination has been filed with the Equal Employment Opportunity Commission. Such a charge shall be filed-

(A) within 180 days after the alleged unlawful practice occurred; or

(B) in a case to which section 633(b) of this title applies, within 300 days after the alleged unlawful practice occurred, or within 30 days after receipt by the individual of notice of termination of proceedings under State law, whichever is earlier.

(2) Upon receiving such a charge, the Commission shall promptly notify all persons named in such charge as prospective defendants in the action and shall promptly seek to eliminate any alleged unlawful practice by informal methods of conciliation, conference, and persuasion.

(3) For purposes of this section, an unlawful practice occurs, with respect to discrimination in compensation in violation of this Act, when a discriminatory compensation decision or other practice is adopted, when a person becomes subject to a discriminatory compensation decision or other practice, or when a person is affected by application of a discriminatory compensation decision or other practice, including each time wages, benefits, or other compensation is paid, resulting in whole or in part from such a decision or other practice.

(e) Reliance on administrative rulings; notice of dismissal or termination; civil action after receipt of notice

Section 259 of this title [*section 10 of the Portal to Portal Act of 1947*] shall apply to actions under this chapter. If a charge filed with the Commission under this chapter is dismissed or the proceedings of the Commission are otherwise terminated by the Commission, the Commission shall notify the person aggrieved. A civil action may be brought under this section by a person defined in section 630(a) of this title [*section 11(a)*] against the respondent named in the charge within 90 days after the date of the receipt of such notice.—

(f) Waiver

(1) An individual may not waive any right or claim under this chapter unless the waiver is knowing and voluntary. Except as provided in paragraph (2), a waiver may not be considered knowing and voluntary unless at a minimum—

(A) the waiver is part of an agreement between the individual and the employer that is written in a manner calculated to be understood by such individual, or by the average individual eligible to participate;

(B) the waiver specifically refers to rights or claims arising under this chapter;

(C) the individual does not waive rights or claims that may arise after the date the waiver is executed;

(D) the individual waives rights or claims only in exchange for consideration in addition to anything of value to which the individual already is entitled;

(E) the individual is advised in writing to consult with an attorney prior to executing the agreement;

(F)(i) the individual is given a period of at least 21 days within which to consider the agreement; or

(ii) if a waiver is requested in connection with an exit incentive or other employment termination program offered to a group or class of employees, the individual is given a period of at least 45 days within which to consider the agreement;

(G) the agreement provides that for a period of at least 7 days following the execution of such agreement, the individual may revoke the agreement, and the agreement shall not become effective or enforceable until the revocation period has expired;

(H) if a waiver is requested in connection with an exit incentive or other employment termination program offered to a group or class of employees, the employer (at the commencement of the period specified in subparagraph (F)) informs the individual in writing in a manner calculated to be understood by the average individual eligible to participate, as to—

(i) any class, unit, or group of individuals covered by such program, any eligibility factors for such program, and any time limits applicable to such program; and

(ii) the job titles and ages of all individuals eligible or selected for the program, and the ages of all individuals in the same job classification or organizational unit who are not eligible or selected for the program.

(2) A waiver in settlement of a charge filed with the Equal Employment Opportunity Commission, or an action filed in court by the individual or the individual's representative, alleging age discrimination of a kind prohibited under section 623 or 633a of this title [*section 4 or 15*] may not be considered knowing and voluntary unless at a minimum—

(A) subparagraphs (A) through (E) of paragraph (1) have been met; and

(B) the individual is given a reasonable period of time within which to consider the settlement agreement.

(3) In any dispute that may arise over whether any of the requirements, conditions, and circumstances set forth in subparagraph (A), (B), (C), (D), (E), (F), (G), or (H) of paragraph (1), or subparagraph (A) or (B) of paragraph (2), have been met, the party asserting the validity of a waiver shall have the burden of proving in a court of competent jurisdiction that a waiver was knowing and voluntary pursuant to paragraph (1) or (2).

(4) No waiver agreement may affect the Commission's rights and responsibilities to enforce this chapter. No waiver may be used to justify interfering with the protected right of an employee to file a charge or participate in an investigation or proceeding conducted by the Commission.

NOTICES TO BE POSTED

SEC. 627. [*Section 8*]

Every employer, employment agency, and labor organization shall post and keep posted in conspicuous places upon its premises a notice to be prepared or approved by the Equal Employment Opportunity Commission setting forth information as the Commission deems appropriate to effectuate the purposes of this chapter.

RULES AND REGULATIONS

SEC. 628. [*Section 9*]

In accordance with the provisions of subchapter II of chapter 5 of title 5 [*Administrative Procedures Act, 5 U.S.C. Section 551 et seq.*], the Equal Employment Opportunity Commission may issue such rules and regulations as it may consider necessary or appropriate for carrying out this chapter, and may establish such reasonable exemptions to and from any or all provisions of this chapter as it may find necessary and proper in the public interest.

CRIMINAL PENALTIES

SEC. 629. [*Section 10*]

Whoever shall forcibly resist, oppose, impede, intimidate or interfere with a duly authorized representative of the Equal Employment Opportunity Commission while it is engaged in the performance of duties under this chapter shall be punished by a fine of not more than $500 or by imprisonment for not more than one year, or by both: *Provided, however,* That no person shall be imprisoned under this section except when there has been a prior conviction hereunder.

DEFINITIONS

SEC. 630. [*Section 11*]

For the purposes of this chapter-

(a) The term "person" means one or more individuals, partnerships, associations, labor organizations, corporations, business trusts, legal representatives, or any organized groups of persons.

(b) The term "employer" means a person engaged in an industry affecting commerce who has twenty or more employees for each working day in each of twenty or more calendar weeks in the current or preceding calendar year: *Provided,* That prior to June 30, 1968, employers having fewer than fifty employees shall not be considered employers. The term also means (1) any agent of such a person, and (2) a State or political subdivision of a State and any agency or instrumentality of a State or a political subdivision of a State, and any interstate agency, but such term does not include the United States, or a corporation wholly owned by the Government of the United States.

(c) The term "employment agency" means any person regularly undertaking with or without compensation to procure employees for an employer and includes an agent of such a person; but shall not include an agency of the United States.

(d) The term "labor organization" means a labor organization engaged in an industry affecting commerce, and any agent of such an organization, and includes any organization of any kind, any agency, or employee representation committee, group, association, or plan so engaged in which employees participate and which exists for the purpose, in whole or in part, of dealing with employers concerning grievances, labor disputes, wages, rates of pay, hours, or other terms or conditions of employment, and any conference, general committee, joint or system board, or joint council so engaged which is subordinate to a national or international labor organization.

(e) A labor organization shall be deemed to be engaged in an industry affecting commerce if (1) it maintains or operates a hiring hall or hiring office which procures employees for an employer or procures for employees opportunities to work for an employer, or (2) the number of its members (or, where it is a labor organization composed of other labor organizations or their representatives, if the aggregate number of the members of such other labor organization) is fifty or more prior to July 1, 1968, or twenty-five or more on or after July 1, 1968, and such labor organization—

(1) is the certified representative of employees under the provisions of the National Labor Relations Act, as amended [*29 U.S.C. 151 et seq.*], or the Railway Labor Act, as amended [*45 U.S.C. 151 et seq.*]; or

(2) although not certified, is a national or international labor organization or a local labor organization recognized or acting as the representative of employees of an employer or employers engaged in an industry affecting commerce; or

(3) has chartered a local labor organization or subsidiary body which is representing or actively seeking to represent employees of employers within the meaning of paragraph (1) or (2); or

(4) has been chartered by a labor organization representing or actively seeking to represent employees within the meaning of paragraph (1) or (2) as the local or subordinate body through which such employees may enjoy membership or become affiliated with such labor organization; or

(5) is a conference, general committee, joint or system board, or joint council subordinate to a national or international labor organization, which includes a labor organization engaged in an industry affecting commerce within the meaning of any of the preceding paragraphs of this subsection.

(f) The term "employee" means an individual employed by any employer except that the term "employee" shall not include any person elected to public office in any State or political subdivision of any State by the qualified voters thereof, or any person chosen by such officer to be on such officer's personal staff, or an appointee on the policymaking level or an immediate adviser with respect to the exercise of the constitutional or legal powers of the office. The exemption set forth in the preceding sentence shall not include employees subject to the civil service laws of a State government, governmental agency, or political subdivision. The term "employee" includes any individual who is a citizen of the United States employed by an employer in a workplace in a foreign country.

[*The exclusion from the term "employee" of any person chosen by an elected official "to be on such official's personal staff, or an appointee on the policymaking level or an immediate advisor with respect to the exercise of the constitutional or legal powers of the office," remains in section 11(f). However, the Civil Rights Act of 1991 now provides special procedures for such persons who feel they are victims of age and other types of discrimination prohibited by EEOC enforced statutes. See section 321 of the Civil Rights Act of 1991.*]

(g) The term "commerce" means trade, traffic, commerce, transportation, transmission, or communication among the several States; or between a State and any place outside thereof; or within the District of Columbia, or a possession of the United States; or between points in the same State but through a point outside thereof.

(h) The term "industry affecting commerce" means any activity, business, or industry in commerce or in which a labor dispute would hinder or obstruct commerce or the free flow of commerce and includes any activity or industry "affecting commerce" within the meaning of the Labor-Management Reporting and Disclosure Act of 1959 [*29 U.S.C. 401 et seq.*].

(i) The term "State" includes a State of the United States, the District of Columbia, Puerto Rico, the Virgin Islands, American Samoa, Guam, Wake Island, the Canal Zone, and Outer Continental Shelf lands defined in the Outer Continental Shelf Lands Act [*43 U.S.C. 1331 et seq.*].

(j) The term "firefighter" means an employee, the duties of whose position are primarily to perform work directly connected with the control and extinguishment of fires or the maintenance and use of firefighting apparatus and equipment, including an employee engaged in this activity who is transferred to a supervisory or administrative position.

(k) The term "law enforcement officer" means an employee, the duties of whose position are primarily the investigation, apprehension, or detention of individuals suspected or convicted of offenses against the criminal laws of a State, including an employee engaged in this activity who is transferred to a supervisory or administrative position. For the purpose of this subsection, "detention" includes the duties of employees assigned to guard individuals incarcerated in any penal institution.

(l) The term "compensation, terms, conditions, or privileges of employment" encompasses all employee benefits, including such benefits provided pursuant to a bona fide employee benefit plan.

AGE LIMITS

SEC. 631. [*Section 12*]

(a) Individuals of at least 40 years of age

The prohibitions in this chapter shall be limited to individuals who are at least 40 years of age.

(b) Employees or applicants for employment in Federal Government

In the case of any personnel action affecting employees or applicants for employment which is subject to the provisions of section 633a of this title [*section 15*], the prohibitions established in section 633a of this title [*section 15*] shall be limited to individuals who are at least 40 years of age.

(c) Bona fide executives or high policymakers

(1) Nothing in this chapter shall be construed to prohibit compulsory retirement of any employee who has attained 65 years of age and who, for the 2-year period immediately before retirement, is employed in a bona fide executive or a high policymaking position, if such employee is entitled to an immediate nonforfeitable annual retirement benefit from a pension, profit-sharing, savings, or deferred compensation plan, or any combination of such plans, of the employer of such employee, which equals, in the aggregate, at least $44,000.

(2) In applying the retirement benefit test of paragraph (1) of this subsection, if any such retirement benefit is in a form other than a straight life annuity (with no ancillary benefits), or if employees contribute to any such plan or make rollover contributions, such benefit shall be adjusted in accordance with regulations prescribed by the Equal Employment Opportunity Commission, after consultation with the Secretary of the Treasury, so that the benefit is the equivalent of a straight life annuity (with no ancillary benefits) under a plan to which employees do not contribute and under which no rollover contributions are made.

ANNUAL REPORT

SEC. 632. [*Section 13*]

[*Repealed*]

FEDERAL-STATE RELATIONSHIP

SEC. 633. [*Section 14*]

(a) Federal action superseding State action

Nothing in this chapter shall affect the jurisdiction of any agency of any State performing like functions with regard to discriminatory employment practices on account of age except that upon commencement of action under this chapter such action shall supersede any State action.

(b) Limitation of Federal action upon commencement of State proceedings

In the case of an alleged unlawful practice occurring in a State which has a law prohibiting discrimination in employment because of age and establishing or authorizing a State authority to grant or seek relief from such discriminatory practice, no suit may be brought under section 626 of this title [*section 7*] before the expiration of sixty days after proceedings have been commenced under the State law, unless such proceedings have been earlier terminated: *Provided*, That such sixty-day period shall be extended to one hundred and twenty days during the first year after the effective date of such State law. If any requirement for the commencement of such proceedings is imposed by a State authority other than a requirement of the filing of a written and signed statement of the facts upon which the proceeding is based, the proceeding shall be deemed to have been commenced for the purposes of this subsection at the time such statement is sent by registered mail to the appropriate State authority.

NONDISCRIMINATION ON ACCOUNT OF AGE IN FEDERAL GOVERNMENT EMPLOYMENT

SEC. 633a. [*Section 15*]

(a) Federal agencies affected

All personnel actions affecting employees or applicants for employment who are at least 40 years of age (except personnel actions with regard to aliens employed outside the limits of the United States) in military departments as defined in section 102 of Title 5 [*5 U.S.C. Section 102*], in executive agencies as defined in section 105 of Title 5 [*5 U.S.C. Section 105*] (including employees and applicants for employment who are paid from nonappropriated funds), in the United States Postal Service and the Postal Regulatory Commission, in those units in the government of the District of Columbia having positions in the competitive service, and in those units of the judicial branch of the Federal Government having positions in the competitive service, in the Smithsonian Institution, and in the Government Printing Office, the Government Accountability Office, and the Library of Congress shall be made free from any discrimination based on age.

(b) Enforcement by Equal Employment Opportunity Commission and by Librarian of Congress in the Library of Congress; remedies; rules, regulations, orders, and instructions of Commission: compliance by Federal agencies; powers and duties of Commission; notification of final action on complaint of discrimination; exemptions: bona fide occupational qualification

Except as otherwise provided in this subsection, the Equal Employment Opportunity Commission is authorized to enforce the provisions of subsection (a) of this section through appropriate remedies, including reinstatement or hiring of employees with or without backpay, as will effectuate the policies of this section. The Equal Employment Opportunity Commission shall issue such rules, regulations, orders, and instructions as it deems necessary and appropriate to carry out its responsibilities under this section. The Equal Employment Opportunity Commission shall-

(1) be responsible for the review and evaluation of the operation of all agency programs designed to carry out the policy of this section, periodically obtaining

and publishing (on at least a semiannual basis) progress reports from each department, agency, or unit referred to in subsection (a) of this section;

(2) consult with and solicit the recommendations of interested individuals, groups, and organizations relating to nondiscrimination in employment on account of age; and

(3) provide for the acceptance and processing of complaints of discrimination in Federal employment on account of age.

The head of each such department, agency, or unit shall comply with such rules, regulations, orders, and instructions of the Equal Employment Opportunity Commission which shall include a provision that an employee or applicant for employment shall be notified of any final action taken on any complaint of discrimination filed by him thereunder. Reasonable exemptions to the provisions of this section may be established by the Commission but only when the Commission has established a maximum age requirement on the basis of a determination that age is a bona fide occupational qualification necessary to the performance of the duties of the position. With respect to employment in the Library of Congress, authorities granted in this subsection to the Equal Employment Opportunity Commission shall be exercised by the Librarian of Congress.

(c) Civil actions; jurisdiction; relief

Any person aggrieved may bring a civil action in any Federal district court of competent jurisdiction for such legal or equitable relief as will effectuate the purposes of this chapter.

(d) Notice to Commission; time of notice; Commission notification of prospective defendants; Commission elimination of unlawful practices

When the individual has not filed a complaint concerning age discrimination with the Commission, no civil action may be commenced by any individual under this section until the individual has given the Commission not less than thirty days' notice of an intent to file such action. Such notice shall be filed within one hundred and eighty days after the alleged unlawful practice occurred. Upon receiving a notice of intent to sue, the Commission shall promptly notify all persons named therein as prospective defendants in the action and take any appropriate action to assure the elimination of any unlawful practice.

(e) Duty of Government agency or official

Nothing contained in this section shall relieve any Government agency or official of the responsibility to assure nondiscrimination on account of age in employment as required under any provision of Federal law.

(f) Applicability of statutory provisions to personnel action of Federal departments, etc.

Any personnel action of any department, agency, or other entity referred to in subsection (a) of this section shall not be subject to, or affected by, any provision of this chapter, other than the provisions of sections 7(d)(3) and 631(b) of this title [*section 12(b)*] and the provisions of this section.

(g) Study and report to President and Congress by Equal Employment Opportunity Commission; scope

(1) The Equal Employment Opportunity Commission shall undertake a study relating to the effects of the amendments made to this section by the Age Discrimination in Employment Act Amendments of 1978, and the effects of section 631(b) of this title [*section 12(b)*].

(2) The Equal Employment Opportunity Commission shall transmit a report to the President and to the Congress containing the findings of the Commission resulting from the study of the Commission under paragraph (1) of this subsection. Such report shall be transmitted no later than January 1, 1980.

EFFECTIVE DATE [*Section 16 of the ADEA (not reproduced in the U.S. Code)*]

This Act shall become effective one hundred and eighty days after enactment, except (a) that the Secretary of Labor may extend the delay in effective date of any provision of this Act up to an additional ninety days thereafter if he finds that such time is necessary in permitting adjustments to the provisions hereof, and (b) that on or after the date of enactment the EEOC [*originally, the Secretary of Labor*] *is authorized to issue such rules and regulations as may be necessary to carry out its provisions.*

TITLES I AND V OF THE AMERICANS WITH DISABILITIES ACT OF 1990 (ADA)

An Act to establish a clear and comprehensive prohibition of discrimination on the basis of disability.

Be it enacted by the Senate and House of Representatives of the United States of America assembled, that this Act may be cited as the "Americans with Disabilities Act of 1990."

* * *

FINDINGS AND PURPOSES

SEC. 12101. [*Section 2*]

(a) Findings. - The Congress finds that-

(1) **physical or mental disabilities in no way diminish a person's right to fully participate in all aspects of society, yet many people with physical or mental**

disabilities have been precluded from doing so because of discrimination; others who have a record of a disability or are regarded as having a disability also have been subjected to discrimination;

(2) historically, society has tended to isolate and segregate individuals with disabilities, and, despite some improvements, such forms of discrimination against individuals with disabilities continue to be a serious and pervasive social problem;

(3) discrimination against individuals with disabilities persists in such critical areas as employment, housing, public accommodations, education, transportation, communication, recreation, institutionalization, health services, voting, and access to public services;

(4) unlike individuals who have experienced discrimination on the basis of race, color, sex, national origin, religion, or age, individuals who have experienced discrimination on the basis of disability have often had no legal recourse to redress such discrimination;

(5) individuals with disabilities continually encounter various forms of discrimination, including outright intentional exclusion, the discriminatory effects of architectural, transportation, and communication barriers, overprotective rules and policies, failure to make modifications to existing facilities and practices, exclusionary qualification standards and criteria, segregation, and relegation to lesser services, programs, activities, benefits, jobs, or other opportunities;

(6) census data, national polls, and other studies have documented that people with disabilities, as a group, occupy an inferior status in our society, and are severely disadvantaged socially, vocationally, economically, and educationally;

(7) the Nation's proper goals regarding individuals with disabilities are to assure equality of opportunity, full participation, independent living, and economic self-sufficiency for such individuals; and

(**8**) the continuing existence of unfair and unnecessary discrimination and prejudice denies people with disabilities the opportunity to compete on an equal basis and to pursue those opportunities for which our free society is justifiably famous, and costs the United States billions of dollars in unnecessary expenses resulting from dependency and nonproductivity.

(b) Purpose. - It is the purpose of this chapter-

(1) to provide a clear and comprehensive national mandate for the elimination of discrimination against individuals with disabilities;

(2) to provide clear, strong, consistent, enforceable standards addressing discrimination against individuals with disabilities;

(3) to ensure that the Federal Government plays a central role in enforcing the standards established in this chapter on behalf of individuals with disabilities; and

(4) to invoke the sweep of congressional authority, including the power to enforce the fourteenth amendment and to regulate commerce, in order to address the major areas of discrimination faced day to day by people with disabilities.

DEFINITION OF DISABILITY

SEC. 12102. [*Section 3*]

As used in this chapter:

(1) Disability. - The term "disability" means, with respect to an individual-

(A) a physical or mental impairment that substantially limits one or more major life activities of such individual;

(B) a record of such an impairment; or

(C) being regarded as having such an impairment (as described in paragraph (3)).

(2) Major life activities

A) In general

For purposes of paragraph (1), major life activities include, but are not limited to, caring for oneself, performing manual tasks, seeing, hearing, eating, sleeping, walking, standing, lifting, bending, speaking, breathing, learning, reading, concentrating, thinking, communicating, and working.

(B) Major bodily functions

For purposes of paragraph (1), a major life activity also includes the operation of a major bodily function, including but not limited to, functions of the immune system, normal cell growth, digestive, bowel, bladder, neurological, brain, respiratory, circulatory, endocrine, and reproductive functions.

(3) Regarded as having such an impairment

For purposes of paragraph (1)(C):

(A) An individual meets the requirement of "being regarded as having such an impairment" if the individual establishes that he or she has been subjected to an action prohibited under this chapter because of an actual or perceived physical or mental impairment whether or not the impairment limits or is perceived to limit a major life activity.

(B) Paragraph (1)(C) shall not apply to impairments that are transitory and minor. A transitory impairment is an impairment with an actual or expected duration of 6 months or less.

(4) Rules of construction regarding the definition of disability

The definition of "disability" in paragraph (1) shall be construed in accordance with the following:

(A) The definition of disability in this chapter shall be construed in favor of broad coverage of individuals under this chapter, to the maximum extent permitted by the terms of this chapter.

(B) The term "substantially limits" shall be interpreted consistently with the findings and purposes of the ADA Amendments Act of 2008.

(C) An impairment that substantially limits one major life activity need not limit other major life activities in order to be considered a disability.

(D) An impairment that is episodic or in remission is a disability if it would substantially limit a major life activity when active.

(E)(i) The determination of whether an impairment substantially limits a major life activity shall be made without regard to the ameliorative effects of mitigating measures such as-

(I) medication, medical supplies, equipment, or appliances, low-vision devices (which do not include ordinary eyeglasses or contact lenses), prosthetics including limbs and devices, hearing aids and cochlear implants or other implantable hearing devices, mobility devices, or oxygen therapy equipment and supplies;

(II) use of assistive technology;

(III) reasonable accommodations or auxiliary aids or services; or

(IV) learned behavioral or adaptive neurological modifications.

(ii) The ameliorative effects of the mitigating measures of ordinary eyeglasses or contact lenses shall be considered in determining whether an impairment substantially limits a major life activity.

(iii) As used in this subparagraph-

(I) the term "ordinary eyeglasses or contact lenses" means lenses that are intended to fully correct visual acuity or eliminate refractive error; and

(II) the term "low-vision devices" means devices that magnify, enhance, or otherwise augment a visual image.

ADDITIONAL DEFINITIONS

SEC. 12103. *[Section 4]*

As used in this chapter:

(1) Auxiliary aids and services. - The term "auxiliary aids and services" includes-

(A) qualified interpreters or other effective methods of making aurally delivered materials available to individuals with hearing impairments;

(B) qualified readers, taped texts, or other effective methods of making visually delivered materials available to individuals with visual impairments;

(C) acquisition or modification of equipment or devices; and

(D) other similar services and actions.

(2) State. - The term "State" means each of the several States, the District of Columbia, the Commonwealth of Puerto Rico, Guam, American Samoa, the Virgin Islands of the United States, the Trust Territory of the Pacific Islands, and the Commonwealth of the Northern Mariana Islands.

SUBCHAPTER I [*TITLE I*] - EMPLOYMENT

DEFINITIONS

SEC. 12111. [*Section 101*]

As used in this subchapter:

(1) Commission. - The term "Commission" means the Equal Employment Opportunity Commission established by section 2000e-4 of this title [*section 705 of the Civil Rights Act of 1964*].

(2) Covered entity. - The term "covered entity" means an employer, employment agency, labor organization, or joint labor management committee.

(3) Direct threat. - The term "direct threat" means a significant risk to the health or safety of others that cannot be eliminated by reasonable accommodation.

(4) Employee. - The term "employee" means an individual employed by an employer. With respect to employment in a foreign country, such term includes an individual who is a citizen of the United States.

(5) Employer. -

(A) In general. - The term "employer" means a person engaged in an industry affecting commerce who has 15 or more employees for each working day in each of 20 or more calendar weeks in the current or preceding calendar year, and any agent of such person, except that, for two years following the effective date of this subchapter, an employer means a person engaged in an industry affecting commerce who has 25 or more employees for each working day in each of 20 or more calendar weeks in the current or preceding year, and any agent of such person.

(B) Exceptions. - The term “employer” does not include-

(i) the United States, a corporation wholly owned by the government of the United States, or an Indian tribe; or

(ii) a bona fide private membership club (other than a labor organization) that is exempt from taxation under section 501(c) of Title 26 [*the Internal Revenue Code of 1986*].

(6) Illegal use of drugs. -

(A) In general. - The term “illegal use of drugs” means the use of drugs, the possession or distribution of which is unlawful under the Controlled Substances Act [21 U.S.C. 801 et seq.]. Such term does not include the use of a drug taken under supervision by a licensed health care professional, or other uses authorized by the Controlled Substances Act or other provisions of Federal law.

(B) Drugs. - The term “drug” means a controlled substance, as defined in schedules I through V of section 202 of the Controlled Substances Act [21 U.S.C. 812].

(7) Person, etc. - The terms “person,” “labor organization,” “employment agency,” “commerce,” and “industry affecting commerce,” shall have the same meaning given such terms in section 2000e of this title [*section 701 of the Civil Rights Act of 1964*].

(8) **Qualified individual. - The term “qualified individual” means an individual** who, with or without reasonable accommodation, can perform the essential functions of the employment position that such individual holds or desires. For the purposes of this subchapter, consideration shall be given to the employer’s judgment as to what functions of a job are essential, and if an employer has prepared a written description before advertising or interviewing applicants for the job, this description shall be considered evidence of the essential functions of the job.

(9) Reasonable accommodation. - The term “reasonable accommodation” may include-

(A) making existing facilities used by employees readily accessible to and usable by individuals with disabilities; and

(B) job restructuring, part-time or modified work schedules, reassignment to a vacant position, acquisition or modification of equipment or devices, appropriate adjustment or modifications of examinations, training materials or policies, the provision of qualified readers or interpreters, and other similar accommodations for individuals with disabilities.

(10) Undue hardship. -

(A) In general. - The term “undue hardship” means an action requiring significant difficulty or expense, when considered in light of the factors set forth in subparagraph (B).

(B) Factors to be considered. - In determining whether an accommodation would impose an undue hardship on a covered entity, factors to be considered include-

(i) the nature and cost of the accommodation needed under this chapter;

(ii) the overall financial resources of the facility or facilities involved in the provision of the reasonable accommodation; the number of persons employed at such facility; the effect on expenses and resources, or the impact otherwise of such accommodation upon the operation of the facility;

(iii) the overall financial resources of the covered entity; the overall size of the business of a covered entity with respect to the number of its employees; the number, type, and location of its facilities; and

(iv) the type of operation or operations of the covered entity, including the composition, structure, and functions of the workforce of such entity; the geographic separateness, administrative, or fiscal relationship of the facility or facilities in question to the covered entity.

DISCRIMINATION

SEC. 12112. [*Section 102*]

(a) General rule. - No covered entity shall discriminate against a qualified individual **on the basis of disability** in regard to job application procedures, the hiring, advancement, or discharge of employees, employee compensation, job training, and other terms, conditions, and privileges of employment.

(b) Construction. - As used in subsection (a) of this section, the term "discriminate **against a qualified individual on the basis of disability"** includes-

(1) limiting, segregating, or classifying a job applicant or employee in a way that adversely affects the opportunities or status of such applicant or employee because of the disability of such applicant or employee;

(2) participating in a contractual or other arrangement or relationship that has the effect of subjecting a covered entity's qualified applicant or employee with a disability to the discrimination prohibited by this subchapter (such relationship includes a relationship with an employment or referral agency, labor union, an organization providing fringe benefits to an employee of the covered entity, or an organization providing training and apprenticeship programs);

(3) utilizing standards, criteria, or methods of administration-

(A) that have the effect of discrimination on the basis of disability; or

(B) that perpetuate the discrimination of others who are subject to common administrative control;

(4) excluding or otherwise denying equal jobs or benefits to a qualified individual because of the known disability of an individual with whom the qualified individual is known to have a relationship or association;

(5) (A) not making reasonable accommodations to the known physical or mental limitations of an otherwise qualified individual with a disability who is an applicant or employee, unless such covered entity can demonstrate that the accommodation would impose an undue hardship on the operation of the business of such covered entity; or

(B) denying employment opportunities to a job applicant or employee who is an otherwise qualified individual with a disability, if such denial is based on the need of such covered entity to make reasonable accommodation to the physical or mental impairments of the employee or applicant;

(6) using qualification standards, employment tests or other selection criteria that screen out or tend to screen out an individual with a disability or a class of individuals with disabilities unless the standard, test or other selection criteria, as used by the covered entity, is shown to be job related for the position in question and is consistent with business necessity; and

(7) failing to select and administer tests concerning employment in the most effective manner to ensure that, when such test is administered to a job applicant or employee who has a disability that impairs sensory, manual, or speaking skills, such test results accurately reflect the skills, aptitude, or whatever other factor of such applicant or employee that such test purports to measure, rather than reflecting the impaired sensory, manual, or speaking skills of such employee or applicant (except where such skills are the factors that the test purports to measure).

(c) Covered entities in foreign countries. -

(1) In general. - It shall not be unlawful under this section for a covered entity to take any action that constitutes discrimination under this section with respect to an employee in a workplace in a foreign country if compliance with this section would cause such covered entity to violate the law of the foreign country in which such workplace is located.

(2) Control of corporation

(A) Presumption. - If an employer controls a corporation whose place of incorporation is a foreign country, any practice that constitutes discrimination under this section and is engaged in by such corporation shall be presumed to be engaged in by such employer.

(B) Exception. - This section shall not apply with respect to the foreign operations of an employer that is a foreign person not controlled by an American employer.

(C) Determination. - For purposes of this paragraph, the determination of whether an employer controls a corporation shall be based on-

(i) the interrelation of operations;

(ii) the common management;

(iii) the centralized control of labor relations; and

(iv) the common ownership or financial control, of the employer and the corporation.

(d) Medical examinations and inquiries. -

(1) In general. - The prohibition against discrimination as referred to in subsection (a) of this section shall include medical examinations and inquiries.

(2) Pre-employment. -

(A) Prohibited examination or inquiry. - Except as provided in paragraph (3), a covered entity shall not conduct a medical examination or make inquiries of a job applicant as to whether such applicant is an individual with a disability or as to the nature or severity of such disability.

(B) Acceptable inquiry. - A covered entity may make pre-employment inquiries into the ability of an applicant to perform job-related functions.

(3) Employment entrance examination. - A covered entity may require a medical examination after an offer of employment has been made to a job applicant and prior to the commencement of the employment duties of such applicant, and may condition an offer of employment on the results of such examination, if-

(A) all entering employees are subjected to such an examination regardless of disability;

(B) information obtained regarding the medical condition or history of the applicant is collected and maintained on separate forms and in separate medical files and is treated as a confidential medical record, except that-

(i) supervisors and managers may be informed regarding necessary restrictions on the work or duties of the employee and necessary accommodations;

(ii) first aid and safety personnel may be informed, when appropriate, if the disability might require emergency treatment; and

(iii) government officials investigating compliance with this chapter shall be provided relevant information on request; and

(C) the results of such examination are used only in accordance with this subchapter.

(4) Examination and inquiry. -

(A) Prohibited examinations and inquiries. - A covered entity shall not require a medical examination and shall not make inquiries of an employee as to whether such employee is an individual with a disability or as to the nature or severity of the disability, unless such examination or inquiry is shown to be job related and consistent with business necessity.

(B) Acceptable examinations and inquiries. - A covered entity may conduct voluntary medical examinations, including voluntary medical histories, which are part of an employee health program available to employees at that work site. A covered entity may make inquiries into the ability of an employee to perform job-related functions.

(C) Requirement. - Information obtained under subparagraph (B) regarding the medical condition or history of any employee are subject to the requirements of subparagraphs (B) and (C) of paragraph (3).

DEFENSES

SEC. 12113. [*Section 103*]

(a) In general. - It may be a defense to a charge of discrimination under this chapter that an alleged application of qualification standards, tests, or selection criteria that screen out or tend to screen out or otherwise deny a job or benefit to an individual with a disability has been shown to be job-related and consistent with business necessity, and such performance cannot be accomplished by reasonable accommodation, as required under this subchapter.

(b) Qualification standards. - The term "qualification standards" may include a requirement that an individual shall not pose a direct threat to the health or safety of other individuals in the workplace.

(c) Qualification standards and tests related to uncorrected vision. - Notwithstanding section 12102(4)(E)(ii) of this title, a covered entity shall not use qualification standards, employment tests, or other selection criteria based on an individual's uncorrected vision unless the standard, test, or other selection criteria, as used by the covered entity, is shown to be job-related for the position in question and consistent with business necessity.

(d) Religious entities. -

(1) In general. - This subchapter shall not prohibit a religious corporation, association, educational institution, or society from giving preference in employment to individuals of a particular religion to perform work connected with the carrying on by such corporation, association, educational institution, or society of its activities.

(2) Religious tenets requirement. - Under this subchapter, a religious organization may require that all applicants and employees conform to the religious tenets of such organization.

(e) List of infectious and communicable diseases. -

(1) In general. - The Secretary of Health and Human Services, not later than 6 months after July 26, 1990 [*the date of enactment of this Act*], shall-

(A) review all infectious and communicable diseases which may be transmitted through handling the food supply;

(B) publish a list of infectious and communicable diseases which are transmitted through handling the food supply;

(C) publish the methods by which such diseases are transmitted; and

(D) widely disseminate such information regarding the list of diseases and their modes of transmissibility to the general public. Such list shall be updated annually.

(2) Applications. - In any case in which an individual has an infectious or communicable disease that is transmitted to others through the handling of food, that is included on the list developed by the Secretary of Health and Human Services under paragraph (1), and which cannot be eliminated by reasonable accommodation, a covered entity may refuse to assign or continue to assign such individual to a job involving food handling.

(3) Construction. - Nothing in this chapter shall be construed to preempt, modify, or amend any State, county, or local law, ordinance, or regulation applicable to food handling which is designed to protect the public health from individuals who pose a significant risk to the health or safety of others, which cannot be eliminated by reasonable accommodation, pursuant to the list of infectious or communicable diseases and the modes of transmissibility published by the Secretary of Health and Human Services.

ILLEGAL USE OF DRUGS AND ALCOHOL

SEC. 12114. [*Section 104*]

(a) Qualified individual with a disability. - For purposes of this subchapter, **a qualified individual with a disability** shall not include any employee or applicant who is currently engaging in the illegal use of drugs, when the covered entity acts on the basis of such use.

(b) Rules of construction. - Nothing in subsection (a) of this section shall be construed to exclude as a qualified individual with a disability an individual who-

(1) has successfully completed a supervised drug rehabilitation program and is no longer engaging in the illegal use of drugs, or has otherwise been rehabilitated successfully and is no longer engaging in such use;

(2) is participating in a supervised rehabilitation program and is no longer engaging in such use; or

(3) is erroneously regarded as engaging in such use, but is not engaging in such use; except that it shall not be a violation of this chapter for a covered entity to adopt or administer reasonable policies or procedures, including but not limited to drug testing, designed to ensure that an individual described in paragraph (1) or (2) is no longer engaging in the illegal use of drugs.

(c) Authority of covered entity. -

A covered entity-

(1) may prohibit the illegal use of drugs and the use of alcohol at the workplace by all employees;

(2) may require that employees shall not be under the influence of alcohol or be engaging in the illegal use of drugs at the workplace;

(3) may require that employees behave in conformance with the requirements established under the Drug Free Workplace Act of 1988 (41 U.S.C. 701 et seq.);

(4) may hold an employee who engages in the illegal use of drugs or who is an alcoholic to the same qualification standards for employment or job performance and behavior that such entity holds other employees, even if any unsatisfactory performance or behavior is related to the drug use or alcoholism of such employee; and (5) may, with respect to Federal regulations regarding alcohol and the illegal use of drugs, require that-

(A) employees comply with the standards established in such regulations of the Department of Defense, if the employees of the covered entity are employed in an industry subject to such regulations, including complying with regulations (if any) that apply to employment in sensitive positions in such an industry, in the case of employees of the covered entity who are employed in such positions (as defined in the regulations of the Department of Defense);

(B) employees comply with the standards established in such regulations of the Nuclear Regulatory Commission, if the employees of the covered entity are employed in an industry subject to such regulations, including complying with regulations (if any) that apply to employment in sensitive positions in such an industry, in the case of employees of the covered entity who are employed in such positions (as defined in the regulations of the Nuclear Regulatory Commission); and

(C) employees comply with the standards established in such regulations of the Department of Transportation, if the employees of the covered entity are employed in a transportation industry subject to such regulations, including complying with such regulations (if any) that apply to employment in sensitive positions in such an industry, in the case of employees of the covered entity who are employed in such positions (as defined in the regulations of the Department of Transportation).

(d) Drug testing. -

(1) In general. - For purposes of this subchapter, a test to determine the illegal use of drugs shall not be considered a medical examination.

(2) Construction. - Nothing in this subchapter shall be construed to encourage, prohibit, or authorize the conducting of drug testing for the illegal use of drugs by job applicants or employees or making employment decisions based on such test results.

(e) Transportation employees. - Nothing in this subchapter shall be construed to encourage, prohibit, restrict, or authorize the otherwise lawful exercise by entities subject to the jurisdiction of the Department of Transportation of authority to-

(1) test employees of such entities in, and applicants for, positions involving safety sensitive duties for the illegal use of drugs and for on duty impairment by alcohol; and

(2) remove such persons who test positive for illegal use of drugs and on duty impairment by alcohol pursuant to paragraph (1) from safety sensitive duties in implementing subsection (c) of this section.

POSTING NOTICES

SEC. 12115. [*Section 105*]

Every employer, employment agency, labor organization, or joint labor management committee covered under this subchapter shall post notices in an accessible format to applicants, employees, and members describing the applicable provisions of this chapter, in the manner prescribed by section 2000e-10 of this title [*section 711 of the Civil Rights Act of 1964*].

REGULATIONS

SEC. 12116. [*Section 106*]

Not later than 1 year after July 26, 1990 [*the date of enactment of this Act*], the Commission shall issue regulations in an accessible format to carry out this subchapter in accordance with subchapter II of chapter 5 of title 5 [*United States Code*].

ENFORCEMENT

SEC. 12117. [*Section 107*]

(a) Powers, remedies, and procedures. - The powers, remedies, and procedures set forth in sections 2000e-4, 2000e-5, 2000e-6, 2000e-8, and 2000e-9 of this title [*sections 705, 706, 707, 709 and 710 of the Civil Rights Act of 1964*] shall be the powers, remedies, and procedures this subchapter provides to the Commission, to the Attorney General, or to any person alleging discrimination on the basis of disability in violation of any provision of this chapter, or regulations promulgated under section 12116 of this title [*section 106*], concerning employment.

(b) Coordination. - The agencies with enforcement authority for actions which allege employment discrimination under this subchapter and under the Rehabilitation Act of 1973 [*29 U.S.C. 701 et seq.*] shall develop procedures to ensure that administrative complaints filed under this subchapter and under the Rehabilitation Act of 1973 are dealt with in a manner that avoids duplication of effort and prevents imposition of inconsistent or conflicting standards for the same requirements under this subchapter and the Rehabilitation Act of 1973. The Commission, the Attorney General, and the Office of Federal Contract Compliance Programs shall establish such coordinating mechanisms (similar to provisions contained in the joint regulations promulgated by the Commission and the Attorney General at part 42 of title 28 and part 1691 of title 29, Code of Federal Regulations, and the Memorandum of Understanding between the Commission and the Office of Federal Contract Compliance Programs dated January 16, 1981 (46 Fed. Reg. 7435, January 23, 1981)) in regulations implementing this subchapter and Rehabilitation Act of 1973 not later than 18 months after July 26, 1990 [*the date of enactment of this Act*].

[***42 U.S.C. Section 2000e-5 note***]

> **(a) AMERICANS WITH DISABILITIES ACT OF 1990. - The amendments made by section 3 [*Lilly Ledbetter Fair Pay Act of 2009, PL 111-2, 123 Stat. 5*] shall apply to claims of discrimination in compensation brought under title I and section 503 of the Americans with Disabilities Act of 1990 (42 U.S.C. 12111 et seq., 12203), pursuant to section 107(a) of such Act (42 U.S.C. 12117(a)), which adopts the powers, remedies, and procedures set forth in section 706 of the Civil Rights Act of 1964 (42 U.S.C. 2000e-5).**

SUBCHAPTER IV [*TITLE V*] - MISCELLANEOUS PROVISIONS

CONSTRUCTION

SEC. 12201. [*Section 501*]

(a) In general. - Except as otherwise provided in this chapter, nothing in this chapter shall be construed to apply a lesser standard than the standards applied under Title V of the Rehabilitation Act of 1973 (29 U.S.C. 790 et seq.) or the regulations issued by Federal agencies pursuant to such title.

(b) Relationship to other laws. - Nothing in this chapter shall be construed to invalidate or limit the remedies, rights, and procedures of any Federal law or law of any State or political subdivision of any State or jurisdiction that provides greater or equal protection for the rights of individuals with disabilities than are afforded by this chapter. Nothing in this chapter shall be construed to preclude the prohibition of, or the imposition of restrictions on, smoking in places of employment covered by subchapter I of this chapter [*title I*], in transportation covered by subchapter II or III of this chapter [*title II or III*], or in places of public accommodation covered by subchapter III of this chapter [*title III*].

(c) Insurance. - Subchapters I through III of this chapter [*titles I through III*] and title IV of this Act shall not be construed to prohibit or restrict-

(1) an insurer, hospital or medical service company, health maintenance organization, or any agent, or entity that administers benefit plans, or similar organizations from underwriting risks, classifying risks, or administering such risks that are based on or not inconsistent with State law; or

(2) a person or organization covered by this chapter from establishing, sponsoring, observing or administering the terms of a bona fide benefit plan that are based on underwriting risks, classifying risks, or administering such risks that are based on or not inconsistent with State law; or

(3) a person or organization covered by this chapter from establishing, sponsoring, observing or administering the terms of a bona fide benefit plan that is not subject to State laws that regulate insurance.

Paragraphs (1), (2), and (3) shall not be used as a subterfuge to evade the purposes of subchapter I and III of this chapter [*titles I and III*].

(d) Accommodations and services. - Nothing in this chapter shall be construed to require an individual with a disability to accept an accommodation, aid, service, opportunity, or benefit which such individual chooses not to accept.

(e) Benefits under State worker's compensation laws

Nothing in this chapter alters the standards for determining eligibility for benefits under State worker's compensation laws or under State and Federal disability benefit programs.

(f) Fundamental alteration

Nothing in this chapter alters the provision of section 12182(b)(2)(A)(ii) of this title, specifying that reasonable modifications in policies, practices, or procedures shall be required, unless an entity can demonstrate that making such modifications in policies, practices, or procedures, including academic requirements in postsecondary education, would fundamentally alter the nature of the goods, services, facilities, privileges, advantages, or accommodations involved.

(g) Claims of no disability

Nothing in this chapter shall provide the basis for a claim by an individual without a disability that the individual was subject to discrimination because of the individual's lack of disability.

(h) Reasonable accommodations and modifications

A covered entity under subchapter I of this chapter, a public entity under subchapter II of this chapter, and any person who owns, leases (or leases to), or operates a place of public accommodation under subchapter III of this chapter, need not provide a reasonable accommodation or a reasonable modification to policies, practices, or procedures to an individual who meets the definition of disability in section 12102(1) of this title solely under subparagraph (C) of such section.

STATE IMMUNITY

SEC. 12202. [*Section 502*]

A State shall not be immune under the eleventh amendment to the Constitution of the United States from an action in Federal or State court of competent jurisdiction for a violation of this chapter. In any action against a State for a violation of the requirements of this chapter, remedies (including remedies both at law and in equity) are available for such a violation to the same extent as such remedies are available for such a violation in an action against any public or private entity other than a State.

PROHIBITION AGAINST RETALIATION AND COERCION

SEC. 12203. [*Section 503*]

(a) Retaliation. - No person shall discriminate against any individual because such individual has opposed any act or practice made unlawful by this chapter or because such individual made a charge, testified, assisted, or participated in any manner in an investigation, proceeding, or hearing under this chapter.

(b) Interference, coercion, or intimidation. - It shall be unlawful to coerce, intimidate, threaten, or interfere with any individual in the exercise or enjoyment of, or on account of his or her having exercised or enjoyed, or on account of his or her having aided or encouraged any other individual in the exercise or enjoyment of, any right granted or protected by this chapter.

(c) Remedies and procedures. - The remedies and procedures available under sections 12117, 12133, and 12188 of this title [*sections 107, 203 and 308*] shall be available to aggrieved persons for violations of subsections (a) and (b) of this section, with respect to subchapter I, subchapter II and subchapter III, respectively, of this chapter [*title I, title II and title III, respectively*].

[*42 U.S.C. Section 2000e-5 note*]

(a) AMERICANS WITH DISABILITIES ACT OF 1990. - The amendments made by section 3 [*Lilly Ledbetter Fair Pay Act of 2009, PL 111-2, 123 Stat. 5*] shall apply to claims of discrimination in compensation brought under title I and section 503 of the Americans with Disabilities Act of 1990 (42 U.S.C. 12111 et seq., 12203), pursuant to section 107(a) of such Act (42 U.S.C. 12117(a)), which adopts the powers, remedies, and procedures set forth in section 706 of the Civil Rights Act of 1964 (42 U.S.C. 2000e-5).

REGULATIONS BY THE ARCHITECTURAL AND TRANSPORTATION BARRIERS COMPLIANCE BOARD

SEC. 12204. [*Section 504*]

(a) Issuance of guidelines. - Not later than 9 months after July 26, 1990 [*the date of enactment of this Act*], the Architectural and Transportation Barriers Compliance Board shall issue minimum guidelines that shall supplement the existing Minimum Guidelines and Requirements for Accessible Design for purposes of subchapters II and III of this chapter [*titles II and III*]. (b) Contents of guidelines. - The supplemental guidelines issued under subsection (a) of this section shall establish additional requirements, consistent with this chapter, to ensure that buildings, facilities, rail passenger cars, and vehicles are accessible, in terms of architecture and design, transportation, and communication, to individuals with disabilities.

(c) Qualified historic properties. -

(1) In general. - The supplemental guidelines issued under subsection (a) of this section shall include procedures and requirements for alterations that will threaten or destroy the historic significance of qualified historic buildings and facilities as defined in 4.1.7(1)(a) of the Uniform Federal Accessibility Standards.

(2) Sites eligible for listing in National Register. - With respect to alterations of buildings or facilities that are eligible for listing in the National Register of Historic Places under the National Historic Preservation Act (16 U.S.C. 470 et seq.), the guidelines described in paragraph (1) shall, at a minimum, maintain the procedures and requirements established in 4.1.7(1) and (2) of the Uniform Federal Accessibility Standards.

(3) Other sites. - With respect to alterations of buildings or facilities designated as historic under State or local law, the guidelines described in paragraph (1) shall establish procedures equivalent to those established by 4.1.7(1) (b) and (c) of the Uniform Federal Accessibility Standards, and shall require, at a minimum, compliance with the requirements established in 4.1.7(2) of such standards.

ATTORNEY'S FEES

SEC. 12205. [*Section 505*]

In any action or administrative proceeding commenced pursuant to this chapter, the court or agency, in its discretion, may allow the prevailing party, other than the United States, a reasonable attorney's fee, including litigation expenses, and costs, and the United States shall be liable for the foregoing the same as a private individual.

Rule of construction regarding regulatory authority

SEC. 12205a. [Section 506]

The authority to issue regulations granted to the Equal Employment Opportunity Commission, the Attorney General, and the Secretary of Transportation under this chapter includes the authority to issue regulations implementing the definitions of disability in section 12102 of this title (including rules of construction) and the definitions in section 12103 of this title, consistent with the ADA Amendments Act of 2008.

TECHNICAL ASSISTANCE

SEC. 12206. [*Section 507*]

(a) Plan for assistance. -

(1) In general. - Not later than 180 days after July 26, 1990 [*the date of enactment of this Act*], the Attorney General, in consultation with the Chair of the Equal Employment Opportunity Commission, the Secretary of Transportation, the Chair of the Architectural and Transportation Barriers Compliance Board, and the Chairman of the Federal Communications Commission, shall develop a plan to assist entities covered under this chapter, and other Federal agencies, in understanding the responsibility of such entities and agencies under this chapter.

(2) Publication of plan. - The Attorney General shall publish the plan referred to in paragraph (1) for public comment in accordance with subchapter II of chapter 5 of title 5 [*United States Code*] (commonly known as the Administrative Procedure Act).

(b) Agency and public assistance. - The Attorney General may obtain the assistance of other Federal agencies in carrying out subsection (a) of this section, including the National Council on Disability, the President's Committee on Employment of People with Disabilities, the Small Business Administration, and the Department of Commerce.

(c) Implementation. -

(1) Rendering assistance. - Each Federal agency that has responsibility under paragraph (2) for implementing this chapter may render technical assistance to

individuals and institutions that have rights or duties under the respective subchapter or subchapters of this chapter for which such agency has responsibility.

(2) Implementation of subchapters. -

(A) Subchapter I [*Title I*]. - The Equal Employment Opportunity Commission and the Attorney General shall implement the plan for assistance developed under subsection (a) of this section, for subchapter I of this chapter [*title I*]. (B) Subchapter II [*Title II*]. -

(i) Part A [*Subtitle A*]. - The Attorney General shall implement such plan for assistance for part A of subchapter II of this chapter [*subtitle A of title II*].

(ii) Part B [*Subtitle B*]. - The Secretary of Transportation shall implement such plan for assistance for part B of subchapter II of this chapter [*subtitle B of title II*].

(C) Subchapter III [*Title III*]. - The Attorney General, in coordination with the Secretary of Transportation and the Chair of the Architectural Transportation Barriers Compliance Board, shall implement such plan for assistance for subchapter III of this chapter, except for section 12184 of this title [*section 304*], the plan for assistance for which shall be implemented by the Secretary of Transportation.

(D) Title IV. - The Chairman of the Federal Communications Commission, in coordination with the Attorney General, shall implement such plan for assistance for title IV.

(3) Technical assistance manuals. - Each Federal agency that has responsibility under paragraph (2) for implementing this chapter shall, as part of its implementation responsibilities, ensure the availability and provision of appropriate technical assistance manuals to individuals or entities with rights or duties under this chapter no later than six months after applicable final regulations are published under subchapters I, II, and III of this chapter [*titles I, II, and III*] and title IV.

(d) Grants and contracts. -

(1) In general. - Each Federal agency that has responsibility under subsection (c)(2) of this section for implementing this chapter may make grants or award contracts to effectuate the purposes of this section, subject to the availability of appropriations. Such grants and contracts may be awarded to individuals, institutions not organized for profit and no part of the net earnings of which inures to the benefit of any private shareholder or individual (including educational institutions), and associations representing individuals who have rights or duties under this chapter. Contracts may be awarded to entities organized for profit, but such entities may not be the recipients or grants described in this paragraph.

(2) Dissemination of information. - Such grants and contracts, among other uses, may be designed to ensure wide dissemination of information about the rights and duties established by this chapter and to provide information and technical assistance about techniques for effective compliance with this chapter.

(e) Failure to receive assistance. - An employer, public accommodation, or other entity covered under this chapter shall not be excused from compliance with the requirements of this chapter because of any failure to receive technical assistance under this section, including any failure in the development or dissemination of any technical assistance manual authorized by this section.

FEDERAL WILDERNESS AREAS

SEC. 12207. [*Section* ***508***]

(a) Study. - The National Council on Disability shall conduct a study and report on the effect that wilderness designations and wilderness land management practices have on the ability of individuals with disabilities to use and enjoy the National Wilderness Preservation System as established under the Wilderness Act (16 U.S.C. 1131 et seq.).

(b) Submission of report. - Not later than 1 year after July 26, 1990 [*the date of enactment of this Act*], the National Council on Disability shall submit the report required under subsection (a) of this section to Congress.

(c) Specific wilderness access. -

(1) In general. - Congress reaffirms that nothing in the Wilderness Act [16 U.S.C. 1131 et seq.] is to be construed as prohibiting the use of a wheelchair in a wilderness area by an individual whose disability requires use of a wheelchair, and consistent with the Wilderness Act no agency is required to provide any form of special treatment or accommodation, or to construct any facilities or modify any conditions of lands within a wilderness area in order to facilitate such use.

(2) Definition. - For purposes of paragraph (1), the term "wheelchair" means a device designed solely for use by a mobility-impaired person for locomotion, that is suitable for use in an indoor pedestrian area.

TRANSVESTITES

SEC. 12208. [*Section* ***509***]

For the purposes of this chapter, the term "disabled" or "disability" shall not apply to an individual solely because that individual is a transvestite.

COVERAGE OF CONGRESS AND THE AGENCIES OF THE LEGISLATIVE BRANCH

SEC. 12209. [*Section **510***]

(a) Coverage of the Senate. -

(1) Commitment to Rule XLII. - The Senate reaffirms its commitment to Rule XLII of the Standing Rules of the Senate which provides as follows:

"No member, officer, or employee of the Senate shall, with respect to employment by the Senate or any office thereof-

(a) fail or refuse to hire an individual;

(b) discharge an individual; or

(c) otherwise discriminate against an individual with respect to promotion, compensation, or terms, conditions, or privileges of employment on the basis of such individual's race, color, religion, sex, national origin, age, or state of physical handicap."

(2) Matters other than employment. -

(A) In general. - The rights and protections under this chapter shall, subject to subparagraph (B), apply with respect to the conduct of the Senate regarding matters other than employment.

(B) Remedies. - The Architect of the Capitol shall establish remedies and procedures to be utilized with respect to the rights and protections provided pursuant to subparagraph (A). Such remedies and procedures shall apply exclusively, after approval in accordance with subparagraph (C).

(C) Proposed remedies and procedures. - For purposes of subparagraph (B), the Architect of the Capitol shall submit proposed remedies and procedures to the Senate Committee on Rules and Administration. The remedies and procedures shall be effective upon the approval of the Committee on Rules and Administration.

(3) Exercise of rulemaking power. - Notwithstanding any other provision of law, enforcement and adjudication of the rights and protections referred to in paragraph (2)(A) shall be within the exclusive jurisdiction of the United States Senate. The provisions of paragraph (1), (2) are enacted by the Senate as an exercise of the rulemaking power of the Senate, with full recognition of the right of the Senate to change its rules, in the same manner, and to the same extent, as in the case of any other rule of the Senate.

(b) Coverage of the House of Representatives. -

(1) In general. - Notwithstanding any other provision of this chapter or of law, the purposes of this chapter shall, subject to paragraphs (2) and (3), apply in their entirety to the House of Representatives.

(2) Employment in the House. -

(A) Application. - The rights and protections under this chapter shall, subject to subparagraph (B), apply with respect to any employee in an employment position in the House of Representatives and any employing authority of the House of Representatives.

(B) Administration. -

(i) In general. - In the administration of this paragraph, the remedies and procedures made applicable pursuant to the resolution described in clause (ii) shall apply exclusively.

(ii) Resolution. - The resolution referred to in clause (i) is House Resolution 15 of the One Hundred First Congress, as agreed to January 3, 1989, or any other provision that continues in effect the provisions of, or is a successor to, the Fair Employment Practices Resolution (House Resolution 558 of the One Hundredth Congress, as agreed to October 4, 1988).

(C) Exercise of rulemaking power. - The provisions of subparagraph (B) are enacted by the House of Representatives as an exercise of the rulemaking power of the House of Representatives, with full recognition of the right of the House to change its rules, in the same manner, and to the same extent as in the case of any other rule of the House.

(3) Matters other than employment. -

(A) In general. - The rights and protections under this chapter shall, subject to subparagraph (B), apply with respect to the conduct of the House of Representatives regarding matters other than employment.

(B) Remedies. - The Architect of the Capitol shall establish remedies and procedures to be utilized with respect to the rights and protections provided pursuant to subparagraph (A). Such remedies and procedures shall apply exclusively, after approval in accordance with subparagraph (C).

(C) Approval. - For purposes of subparagraph (B), the Architect of the Capitol shall submit proposed remedies and procedures to the Speaker of the House of Representatives. The remedies and procedures shall be effective upon the approval of the Speaker, after consultation with the House Office Building Commission.

(c) Instrumentalities of Congress. -

(1) In general. - The rights and protections under this chapter shall, subject to paragraph (2), apply with respect to the conduct of each instrumentality of the Congress.

(2) Establishment of remedies and procedures by instrumentalities. - The chief official of each instrumentality of the Congress shall establish remedies and procedures to be utilized with respect to the rights and protections provided pursuant to paragraph (1). Such remedies and procedures shall apply exclusively, except for the employees who are defined as Senate employees, in section 201(c)(1) of the Civil Rights Act of 1991.

(3) Report to Congress. - The chief official of each instrumentality of the Congress shall, after establishing remedies and procedures for purposes of paragraph (2), submit to the Congress a report describing the remedies and procedures.

(4) Definition of instrumentalities. - For purposes of this section, instrumentalities of the Congress include the following: the Architect of the Capitol, the Congressional Budget Office, the General Accounting Office, the Government Printing Office, the Library of Congress, the Office of Technology Assessment, and the United States Botanic Garden.

(5) Construction. - Nothing in this section shall alter the enforcement procedures for individuals with disabilities provided in the General Accounting Office Personnel Act of 1980 [*31 U.S.C. 731 et seq.*] and regulations promulgated pursuant to that Act.

ILLEGAL USE OF DRUGS

SEC. 12210. [*Section **511***]

(a) In general. - For purposes of this chapter, the term "individual with a disability" does not include an individual who is currently engaging in the illegal use of drugs, when the covered entity acts on the basis of such use.

(b) Rules of construction. - Nothing in subsection (a) of this section shall be construed to exclude as an individual with a disability an individual who-

(1) has successfully completed a supervised drug rehabilitation program and is no longer engaging in the illegal use of drugs, or has otherwise been rehabilitated successfully and is no longer engaging in such use;

(2) is participating in a supervised rehabilitation program and is no longer engaging in such use; or

(3) is erroneously regarded as engaging in such use, but is not engaging in such use; except that it shall not be a violation of this chapter for a covered entity to

adopt or administer reasonable policies or procedures, including but not limited to drug testing, designed to ensure that an individual described in paragraph (1) or (2) is no longer engaging in the illegal use of drugs; however, nothing in this section shall be construed to encourage, prohibit, restrict, or authorize the conducting of testing for the illegal use of drugs.

(c) Health and other services. - Notwithstanding subsection (a) of this section and section 12211(b)(3) of this title [*section **512**(b)(3)*], an individual shall not be denied health services, or services provided in connection with drug rehabilitation, on the basis of the current illegal use of drugs if the individual is otherwise entitled to such services.

(d) Definition of illegal use of drugs. -

(1) In general. - The term "illegal use of drugs" means the use of drugs, the possession or distribution of which is unlawful under the Controlled Substances Act (21 U.S.C. 812). Such term does not include the use of a drug taken under supervision by a licensed health care professional, or other uses authorized by the Controlled Substances Act or other provisions of Federal law.

(2) Drugs

The term "drug" means a controlled substance, as defined in schedules I through V of section 202 of the Controlled Substances Act [*21 U.S.C. 812*].

DEFINITIONS

SEC. 12211. [*Section **512***]

(a) Homosexuality and bisexuality. - For purposes of the definition of "disability" in section 12102(2) of this title [*section 3(2)*], homosexuality and bisexuality are not impairments and as such are not disabilities under this chapter.

(b) Certain conditions. - Under this chapter, the term "disability" shall not include-

(1) transvestism, transsexualism, pedophilia, exhibitionism, voyeurism, gender identity disorders not resulting from physical impairments, or other sexual behavior disorders;

(2) compulsive gambling, kleptomania, or pyromania; or

(3) psychoactive substance use disorders resulting from current illegal use of drugs.

ALTERNATIVE MEANS OF DISPUTE RESOLUTION

SEC. 12212. [*Section **514***]

Where appropriate and to the extent authorized by law, the use of alternative means of dispute resolution, including settlement negotiations, conciliation, facilitation, mediation, factfinding, minitrials, and arbitration, is encouraged to resolve disputes arising under this chapter.

SEVERABILITY

SEC. 12213. [*Section **515***]

Should any provision in this chapter be found to be unconstitutional by a court of law, such provision shall be severed from the remainder of the chapter, and such action shall not affect the enforceability of the remaining provisions of the chapter.

THE CIVIL RIGHTS ACT OF 1991

TITLE I - FEDERAL CIVIL RIGHTS REMEDIES

DAMAGES IN CASES OF INTENTIONAL DISCRIMINATION

SEC. 102

The Revised Statutes are amended by inserting after section 1977 (42 U.S.C. 1981) the following new section:

"SEC. 1977A. DAMAGES IN CASES OF INTENTIONAL DISCRIMINATION IN EMPLOYMENT. [*42 U.S.C. 1981a*]

(a) RIGHT OF RECOVERY. -

(1) CIVIL RIGHTS. - In an action brought by a complaining party under section 706 or 717 of the Civil Rights Act of 1964 (42 U.S.C. 2000e-5) against a respondent who engaged in unlawful intentional discrimination (not an employment practice that is unlawful because of its disparate impact) prohibited under section 703, 704, or 717 of the Act (42 U.S.C. 2000e-2 or 2000e-3), and provided that the complaining party cannot recover under section 1977 of the Revised Statutes (42 U.S.C. 1981), the complaining party may recover compensatory and punitive damages as allowed in subsection (b), in addition to any relief authorized by section 706(g) of the Civil Rights Act of 1964, from the respondent.

(2) DISABILITY. - In an action brought by a complaining party under the powers, remedies, and procedures set forth in section 706 or 717 of the Civil Rights Act of 1964 (as provided in section 107(a) of the Americans with Disabilities Act of 1990 (42 U.S.C. 12117(a)), and section 505(a)(1) of the Rehabilitation Act of 1973 (29 U.S.C. 794a(a)(1)), respectively) against a respondent who engaged in unlawful intentional discrimination (not an employment practice that is unlawful because of its disparate impact) under section 501 of the Rehabilitation Act of 1973 (29 U.S.C. 791) and the regulations implementing section 501, or who violated the requirements of section 501 of the Act or the regulations implementing section 501 concerning the provision of a reasonable accommodation, or section 102 of the Americans with Disabilities Act of 1990 (42 U.S.C. 12112), or committed a

violation of section 102(b)(5) of the Act, against an individual, the complaining party may recover compensatory and punitive damages as allowed in subsection (b), in addition to any relief authorized by section 706(g) of the Civil Rights Act of 1964, from the respondent.

(3) REASONABLE ACCOMMODATION AND GOOD FAITH EFFORT. - In cases where a discriminatory practice involves the provision of a reasonable accommodation pursuant to section 102(b)(5) of the Americans with Disabilities Act of 1990 or regulations implementing section 501 of the Rehabilitation Act of 1973, damages may not be awarded under this section where the covered entity demonstrates good faith efforts, in consultation with the person with the disability who has informed the covered entity that accommodation is needed, to identify and make a reasonable accommodation that would provide such individual with an equally effective opportunity and would not cause an undue hardship on the operation of the business.

(b) COMPENSATORY AND PUNITIVE DAMAGES. -

(1) DETERMINATION OF PUNITIVE DAMAGES. - A complaining party may recover punitive damages under this section against a respondent (other than a government, government agency or political subdivision) if the complaining party demonstrates that the respondent engaged in a discriminatory practice or discriminatory practices with malice or with reckless indifference to the federally protected rights of an aggrieved individual.

(2) EXCLUSIONS FROM COMPENSATORY DAMAGES. - Compensatory damages awarded under this section shall not include backpay, interest on backpay, or any other type of relief authorized under section 706(g) of the Civil Rights Act of 1964.

(3) LIMITATIONS. - The sum of the amount of compensatory damages awarded under this section for future pecuniary losses, emotional pain, suffering, inconvenience, mental anguish, loss of enjoyment of life, and other nonpecuniary losses, and the amount of punitive damages awarded under this section, shall not exceed, for each complaining party -

(A) in the case of a respondent who has more than 14 and fewer than 101 employees in each of 20 or more calendar weeks in the current or preceding calendar year, $50,000;

(B) in the case of a respondent who has more than 100 and fewer than 201 employees in each of 20 or more calendar weeks in the current or preceding calendar year, $100,000; and

(C) in the case of a respondent who has more than 200 and fewer than 501 employees in each of 20 or more calendar weeks in the current or preceding calendar year, $200,000; and

(D) in the case of a respondent who has more than 500 employees in each of 20 or more calendar weeks in the current or preceding calendar year, $300,000.

(4) CONSTRUCTION. - Nothing in this section shall be construed to limit the scope of, or the relief available under, section 1977 of the Revised Statutes (42 U.S.C. 1981).

(c) JURY TRIAL. - If a complaining party seeks compensatory or punitive damages under this section -

(1) any party may demand a trial by jury; and

(2) the court shall not inform the jury of the limitations described in subsection (b)(3).

(d) DEFINITIONS. - As used in this section:

(1) COMPLAINING PARTY. - The term 'complaining party' means -

(A) in the case of a person seeking to bring an action under subsection (a)(1), the Equal Employment Opportunity Commission, the Attorney General, or a person who may bring an action or proceeding under title VII of the Civil Rights Act of 1964 (42 U.S.C. 2000e et seq.); or

(B) in the case of a person seeking to bring an action under subsection (a)(2), the Equal Employment Opportunity Commission, the Attorney General, a person who may bring an action or proceeding under section 505(a)(1) of the Rehabilitation Act of 1973 (29 U.S.C. 794a(a)(1)), or a person who may bring an action or proceeding under title I of the Americans with Disabilities Act of 1990 (42 U.S.C. 12101 et seq.).

(2) DISCRIMINATORY PRACTICE. - The term 'discriminatory Practice' means the discrimination described in paragraph (1), or the discrimination or the violation described in paragraph (2), of subsection (a)."

ATTORNEY'S FEES

[*This section amends section 722 of the Revised Statutes (42 U.S.C. 1988) by adding a reference to section 102 of the Civil Rights Act of 1991 to the list of civil rights actions in which reasonable attorney's fees may be awarded to the prevailing party, other than the United States.*]

SEC. 103

The last sentence of section 722 of the Revised Statutes (42 U.S.C. 1988) is amended by inserting ", 1977A" after "1977".

THE REHABILITATION ACT OF 1973

SECTIONS 501 AND 505

DEFINITIONS

SEC. **705** [*Section 7*]

For the purposes of this chapter:

* * *

(10) Drug and illegal use of drugs

(A) Drug

The term "drug" means a controlled substance, as defined in schedules I through V of section 202 of the Controlled Substances Act (21 U.S.C. 812).

(B) The term "illegal use of drugs" means the use of drugs, the possession or distribution of which is unlawful under the Controlled Substances Act [*21 U.S.C. 801 et seq.*]. Such term does not include the use of a drug taken under supervision by a licensed health care professional, or other uses authorized by the Controlled Substances Act [*21 U.S.C. 801 et seq.*] or other provisions of Federal law.

* * *

(20) Individual with a disability

(B) Certain programs; limitations on major life activities

Subject to subparagraphs (C), (D), (E), and (F), the term "individual with a disability" means, for purposes of sections 701, 711, and 712 of this title and subchapters II, IV, V, and VII of this chapter [29 U.S.C. Sections 760 et seq., 780 et seq., 790 et seq., and 796 et seq.], **any person who has a disability as defined in section 12102 of Title 42**.

(C) Rights and advocacy provisions

(i) In general; exclusion of individuals engaging in drug use

For purposes of subchapter V of this chapter [29 U.S.C. Section 790 et seq.], the term "individual with a disability" does not include an individual who is currently engaging in the illegal use of drugs, when a covered entity acts on the basis of such use.

(ii) Exception for individuals no longer engaging in drug use

Nothing in clause (i) shall be construed to exclude as an individual with a disability an individual who—

(I) has successfully completed a supervised drug rehabilitation program and is no longer engaging in the illegal use of drugs, or has otherwise been rehabilitated successfully and is no longer engaging in such use;

(II) is participating in a supervised rehabilitation program and is no longer engaging in such use; or

(III) is erroneously regarded as engaging in such use, but is not engaging in such use;

except that it shall not be a violation of this chapter [29 U.S.C. Section 701 et seq.] for a covered entity to adopt or administer reasonable policies or procedures, including but not limited to drug testing, designed to ensure that an individual described in subclause (I) or (II) is no longer engaging in the illegal use of drugs.

* * *

(E) Rights provisions; exclusion of individuals on basis of homosexuality or bisexuality For the purposes of sections 791, 793, and 794 of this title—

(i) for purposes of the application of subparagraph (B) to such sections, the term "impairment" does not include homosexuality or bisexuality; and

(ii) therefore the term "individual with a disability" does not include an individual on the basis of homosexuality or bisexuality.

(F) Rights provisions; exclusion of individuals on basis of certain disorders

For the purposes of sections 791, 793, and 794 of this title, the term "individual with a disability" does not include an individual on the basis of—

(i) transvestism, transsexualism, pedophilia, exhibitionism, voyeurism, gender identity disorders not resulting from physical impairments, or other sexual behavior disorders;

(ii) compulsive gambling, kleptomania, or pyromania; or

(iii) psychoactive substance use disorders resulting from current illegal use of drugs.

* * *

EMPLOYMENT OF INDIVIDUALS WITH DISABILITIES

SEC. 791. [*Section 501*]

(a) Interagency Committee on Employees who are Individuals with Disabilities; establishment; membership; co-chairmen; availability of other Committee resources; purpose and functions

There is established within the Federal Government an Interagency Committee on Employees who are Individuals with Disabilities (hereinafter in this section referred to as the "Committee"), comprised of such members as the President may select, including the following (or their designees whose positions are Executive Level IV or higher): the Chairman of the Equal Employment Opportunity Commission (hereafter in this section referred to as the "Commission"), the Director of the Office of Personnel Management, the Secretary of Veterans Affairs, the Secretary of Labor, the Secretary of Education, and the Secretary of Health and Human Services. Either the Director of the Office of Personnel Management and the Chairman of the Commission shall serve as co-chairpersons of the Committee or the Director or Chairman shall serve as the sole chairperson of the Committee, as the Director and Chairman jointly determine, from time to time, to be appropriate. The resources of the President's Committees on Employment of People With Disabilities and on Mental Retardation shall be made fully available to the Committee. It shall be the purpose and function of the Committee (1) to provide a focus for Federal and other employment of individuals with disabilities, and to review, on a periodic basis, in cooperation with the Commission, the adequacy of hiring, placement, and advancement practices with respect to individuals with disabilities, by each department, agency, and instrumentality in the executive branch of Government and the Smithsonian Institution, and to insure that the special needs of such individuals are being met; and (2) to consult with the Commission to assist the Commission to carry out its responsibilities under subsections (b), (c), and (d) of this section. On the basis of such review and consultation, the Committee shall periodically make to the Commission such recommendations for legislative and administrative changes as it deems necessary or desirable. The Commission shall timely transmit to the appropriate committees of Congress any such recommendations.

(b) Federal agencies; affirmative action program plans

Each department, agency, and instrumentality (including the United States Postal Service and the Postal Regulatory Commission) in the executive branch and the Smithsonian Institution shall, within one hundred and eighty days after September 26, 1973, submit to the Commission and to the Committee an affirmative action program plan for the hiring, placement, and advancement of individuals with disabilities in such department, agency, instrumentality, or Institution. Such plan shall include a description of the extent to which and methods whereby the special needs of employees who are individuals with disabilities are being met. Such plan shall be updated annually, and shall be reviewed annually and approved by the Commission, if the Commission determines, after consultation with the Committee, that such plan provides sufficient assurances, procedures and commitments to provide adequate hiring, placement, and advancement opportunities for individuals with disabilities.

(c) State agencies; rehabilitated individuals, employment

The Commission, after consultation with the Committee, shall develop and recommend to the Secretary for referral to the appropriate State agencies, policies and procedures which will facilitate the hiring, placement, and advancement in employment of individuals who have received rehabilitation services under State vocational rehabilitation programs, veterans' programs, or any other program for individuals with disabilities, including the promotion of job opportunities for such individuals. The Secretary shall encourage such State agencies to adopt and implement such policies and procedures.

(d) Report to Congressional committees

The Commission, after consultation with the Committee, shall, on June 30, 1974, and at the end of each subsequent fiscal year, make a complete report to the appropriate committees of the Congress with respect to the practices of and achievements in hiring, placement, and advancement of individuals with disabilities by each department, agency, and instrumentality and the Smithsonian Institution and the effectiveness of the affirmative action programs required by subsection (b) of this section, together with recommendations as to legislation which have been submitted to the Commission under subsection (a) of this section, or other appropriate action to insure the adequacy of such practices. Such report shall also include an evaluation by the Committee of the effectiveness of the activities of the Commission under subsections (b) and (c) of this section.

(e) Federal work experience without pay; non-Federal status

An individual who, as a part of an individualized plan for employment under a State plan approved under this chapter, participates in a program of unpaid work experience in a Federal agency, shall not, by reason thereof, be considered to be a Federal employee or to be subject to the provisions of law relating to Federal employment, including those relating to hours of work, rates of compensation, leave, unemployment compensation, and Federal employee benefits.

(f) Federal agency cooperation; special consideration for positions on President's Committee on Employment of People With Disabilities

(1) The Secretary of Labor and the Secretary of Education are authorized and directed to cooperate with the President's Committee on Employment of People With Disabilities in carrying out its functions.

(2) In selecting personnel to fill all positions on the President's Committee on Employment of People With Disabilities, special consideration shall be given to qualified individuals with disabilities.

(g)Standards used in determining violation of section

The standards used to determine whether this section has been violated in a complaint alleging nonaffirmative action employment discrimination under this section shall be the standards applied under title I of the Americans with Disabilities Act of 1990 (42 U.S.C. 12111 et seq.) and the provisions of sections 501 through 504, and 510, of the Americans with Disabilities Act of 1990 (42 U.S.C. 12201-12204 and 12210), as such sections relate to employment.

[***42 U.S.C. Section 2000e-5 note***]

(b) REHABILITATION ACT OF 1973.— The amendments made by section 3 [*Lilly Ledbetter Fair Pay Act of 2009, PL 111-2, 123 Stat. 5*] shall apply to claims of discrimination in compensation brought under sections 501 and 504 of the Rehabilitation Act of 1973 (29 U.S.C. 791, 794), pursuant to—

(1) sections 501(g) and 504(d) of such Act (29 U.S.C. 791(g), 794(d)), respectively, which adopt the standards applied under title I of the Americans with Disabilities Act of 1990 [*42 U.S.C. 12101 et seq.*] for determining whether a violation has occurred in a complaint alleging employment discrimination; and

(2) paragraphs (1) and (2) of section 505(a) of such Act (29 U.S.C. 794a(a)) (as amended by subsection (c)).

REMEDIES AND ATTORNEYS' FEES

SEC. 794a. [*Section 505*]

(a)(1) The remedies, procedures, and rights set forth in section 717 of the Civil Rights Act of 1964 (42 U.S.C. 2000e-16), including the application of sections 706(f) through 706(k) (42 U.S.C. 2000e-5(f) through (k)) **(and the application of section 706(e)(3) (42 U.S.C. 2000e-5(e)(3)) to claims of discrimination in compensation)**, shall be available, with respect to any complaint under section 791 of this title, to any employee or applicant for employment aggrieved by the final disposition of such complaint, or by the failure to take final action on such complaint. In fashioning an equitable or affirmative action remedy under such section, a court may take into account the reasonableness of the cost of any necessary work place accommodation, and the availability of alternatives therefore or other appropriate relief in order to achieve an equitable and appropriate remedy.

(2) The remedies, procedures, and rights set forth in title VI of the Civil Rights Act of 1964 **(42 U.S.C. 2000d et seq.) (and in subsection (e)(3) of section 706 of such Act (42 U.S.C. 2000e-5), applied to claims of discrimination in compensation)** shall be available to any person aggrieved by any act or failure to act by any recipient of Federal assistance or Federal provider of such assistance under section 794 of this title.

(b) In any action or proceeding to enforce or charge a violation of a provision of this subchapter, the court, in its discretion, may allow the prevailing party, other than the United States, a reasonable attorney's fee as part of the costs.

THE GENETIC INFORMATION NONDISCRIMINATION ACT OF 2008

An Act

To prohibit discrimination on the basis of genetic information with respect to health insurance and employment.

Be it enacted by the Senate and House of Representatives of the United States of America in Congress assembled,

SECTION 1. SHORT TITLE; TABLE OF CONTENTS

(a) Short Title.—This Act may be cited as the "Genetic Information Nondiscrimination Act of 2008".

(b) Table of Contents.—The table of contents of this Act is as follows:

TITLE I—GENETIC NONDISCRIMINATION IN HEALTH INSURANCE

Please note: Title I is not included here. The Departments of Labor, Health and Human Services and the Treasury have responsibility for issuing regulations for Title I of GINA, which addresses the use of genetic information in health insurance.

TITLE II—PROHIBITING EMPLOYMENT DISCRIMINATION ON THE BASIS OF GENETIC INFORMATION

TITLE III—MISCELLANEOUS PROVISIONS

Sec. 301. Severability.

Sec. 302. Child labor protections.

SEC. 2. FINDINGS.

Congress makes the following findings:

(1) Deciphering the sequence of the human genome and other advances in genetics open major new opportunities for medical progress. New knowledge about the genetic basis of illness will allow for earlier detection of illnesses, often before symptoms have begun. Genetic testing can allow individuals to take steps to reduce the likelihood that they will contract a particular disorder. New knowledge about genetics may allow for the development of better therapies that are more effective against disease or have fewer side effects than current treatments. These advances give rise to the potential misuse of genetic information to discriminate in health insurance and employment.

(2) The early science of genetics became the basis of State laws that provided for the sterilization of persons having presumed genetic "defects" such as mental retardation, mental disease, epilepsy, blindness, and hearing loss, among other conditions. The first sterilization law was enacted in the State of Indiana in 1907. By 1981, a majority of States adopted sterilization laws to "correct" apparent genetic traits or tendencies. Many of these State laws have since been repealed, and many have been modified to include essential constitutional requirements of due process and equal protection. However, the current explosion in the science of genetics, and the history of sterilization laws by the States based on early genetic science, compels Congressional action in this area.

(3) Although genes are facially neutral markers, many genetic conditions and disorders are associated with particular racial and ethnic groups and gender. Because some genetic traits are most prevalent in particular groups, members of a particular group may be stigmatized or discriminated against as a result of that genetic information. This form of discrimination was evident in the 1970s, which saw the advent of programs to screen and identify carriers of sickle cell anemia, a disease which afflicts African-Americans. Once again, State legislatures began to enact discriminatory laws in the area, and in the early 1970s began mandating genetic screening of all African Americans for sickle cell anemia, leading to discrimination and unnecessary fear. To alleviate some of this stigma, Congress in 1972 passed the National Sickle Cell Anemia Control Act, which withholds Federal funding from States unless sickle cell testing is voluntary.

(4) Congress has been informed of examples of genetic discrimination in the workplace. These include the use of pre-employment genetic screening at Lawrence Berkeley Laboratory, which led to a court decision in favor of the employees in that

case *Norman-Bloodsaw v. Lawrence Berkeley Laboratory* (135 F.3d 1260, 1269 (9th Cir. 1998)). Congress clearly has a compelling public interest in relieving the fear of discrimination and in prohibiting its actual practice in employment and health insurance.

(5) Federal law addressing genetic discrimination in health insurance and employment is incomplete in both the scope and depth of its protections. Moreover, while many States have enacted some type of genetic non-discrimination law, these laws vary widely with respect to their approach, application, and level of protection. Congress has collected substantial evidence that the American public and the medical community find the existing patchwork of State and Federal laws to be confusing and inadequate to protect them from discrimination. Therefore Federal legislation establishing a national and uniform basic standard is necessary to fully protect the public from discrimination and allay their concerns about the potential for discrimination, thereby allowing individuals to take advantage of genetic testing, technologies, research, and new therapies.

TITLE II—PROHIBITING EMPLOYMENT DISCRIMINATION ON THE BASIS OF GENETIC INFORMATION

SEC. 201. DEFINITIONS

In this title:

(1) Commission.—The term "Commission" means the Equal Employment Opportunity Commission as created by section 705 of the Civil Rights Act of 1964 (42 U.S.C. 2000e-4).

(2) Employee; employer; employment agency; labor organization; member.—

(A) In general.—The term "employee" means—

(i) an employee (including an applicant), as defined in section 701(f) of the Civil Rights Act of 1964 (42 U.S.C. 2000e(f));

(ii) a State employee (including an applicant) described in section 304(a) of the Government Employee Rights Act of 1991 (42 U.S.C. 2000e- 16c(a));

(iii) a covered employee (including an applicant), as defined in section 101 of the Congressional Accountability Act of 1995 (2 U.S.C. 1301);

(iv) a covered employee (including an applicant), as defined in section 411(c) of title 3, United States Code; or

(v) an employee or applicant to which section 717(a) of the Civil Rights Act of 1964 (42 U.S.C. 2000e-16(a)) applies.

(B) Employer.—The term "employer" means—

(i) an employer (as defined in section 701(b) of the Civil Rights Act of 1964 (42 U.S.C. 2000e(b)));

(ii) an entity employing a State employee described in section 304(a) of the Government Employee Rights Act of 1991;

(iii) an employing office, as defined in section 101 of the Congressional Accountability Act of 1995;

(iv) an employing office, as defined in section 411(c) of title 3, United States Code; or

(v) an entity to which section 717(a) of the Civil Rights Act of 1964 applies.

(C) Employment agency; labor organization.—The terms "employment agency" and "labor organization" have the meanings given the terms in section 701 of the Civil Rights Act of 1964 (42 U.S.C. 2000e).

(D) Member.—The term "member", with respect to a labor organization, includes an applicant for membership in a labor organization.

(3) Family member.—The term "family member" means, with respect to an individual—

(A) a dependent (as such term is used for purposes of section 701(f)(2) of the Employee Retirement Income Security Act of 1974) of such individual, and

(B) any other individual who is a first-degree, second-degree, third-degree, or fourth-degree relative of such individual or of an individual described in subparagraph (A).

(4) Genetic information.—

(A) In general.—The term "genetic information" means, with respect to any individual, information about—

(i) such individual's genetic tests,

(ii) the genetic tests of family members of such individual, and

(iii) the manifestation of a disease or disorder in family members of such individual.

(B) Inclusion of genetic services and participation in genetic research.—Such term includes, with respect to any individual, any request for, or receipt of, genetic services, or participation in clinical research which includes genetic services, by such individual or any family member of such individual.

(C) Exclusions.—The term "genetic information" shall not include information about the sex or age of any individual.

(5) Genetic monitoring.—The term "genetic monitoring" means the periodic examination of employees to evaluate acquired modifications to their genetic material,

such as chromosomal damage or evidence of increased occurrence of mutations, that may have developed in the course of employment due to exposure to toxic substances in the workplace, in order to identify, evaluate, and respond to the effects of or control adverse environmental exposures in the workplace.

(6) Genetic services.—The term "genetic services" means—

(A) a genetic test;

(B) genetic counseling (including obtaining, interpreting, or assessing genetic information); or

(C) genetic education.

(7) Genetic test.—

(A) In general.—The term "genetic test" means an analysis of human DNA, RNA, chromosomes, proteins, or metabolites, that detects genotypes, mutations, or chromosomal changes.

(B) Exceptions.—The term "genetic test" does not mean an analysis of proteins or metabolites that does not detect genotypes, mutations, or chromosomal changes.

SEC. 202. EMPLOYER PRACTICES

(a) Discrimination Based on Genetic Information.—It shall be an unlawful employment practice for an employer—

(1) to fail or refuse to hire, or to discharge, any employee, or otherwise to discriminate against any employee with respect to the compensation, terms, conditions, or privileges of employment of the employee, because of genetic information with respect to the employee; or

(2) to limit, segregate, or classify the employees of the employer in any way that would deprive or tend to deprive any employee of employment opportunities or otherwise adversely affect the status of the employee as an employee, because of genetic information with respect to the employee.

(b) Acquisition of Genetic Information.—It shall be an unlawful employment practice for an employer to request, require, or purchase genetic information with respect to an employee or a family member of the employee except—

(1) where an employer inadvertently requests or requires family medical history of the employee or family member of the employee;

(2) where—

(A) health or genetic services are offered by the employer, including such services offered as part of a wellness program;

(B) the employee provides prior, knowing, voluntary, and written authorization;

(C) only the employee (or family member if the family member is receiving genetic services) and the licensed health care professional or board certified genetic counselor involved in providing such services receive individually identifiable information concerning the results of such services; and

(D) any individually identifiable genetic information provided under subparagraph (C) in connection with the services provided under subparagraph (A) is only available for purposes of such services and shall not be disclosed to the employer except in aggregate terms that do not disclose the identity of specific employees;

(3) where an employer requests or requires family medical history from the employee to comply with the certification provisions of section 103 of the Family and Medical Leave Act of 1993 (29 U.S.C. 2613) or such requirements under State family and medical leave laws;

(4) where an employer purchases documents that are commercially and publicly available (including newspapers, magazines, periodicals, and books, but not including medical databases or court records) that include family medical history;

(5) where the information involved is to be used for genetic monitoring of the biological effects of toxic substances in the workplace, but only if—

(A) the employer provides written notice of the genetic monitoring to the employee;

(B)(i) the employee provides prior, knowing, voluntary, and written authorization; or

(ii) the genetic monitoring is required by Federal or State law;

(C) the employee is informed of individual monitoring results;

(D) the monitoring is in compliance with—

(i) any Federal genetic monitoring regulations, including any such regulations that may be promulgated by the Secretary of Labor pursuant to the Occupational Safety and Health Act of 1970 (29 U.S.C. 651 et seq.), the Federal Mine Safety and Health Act of 1977 (30 U.S.C. 801 et seq.), or the Atomic Energy Act of 1954 (42 U.S.C. 2011 et seq.); or

(ii) State genetic monitoring regulations, in the case of a State that is implementing genetic monitoring regulations under the authority of the Occupational Safety and Health Act of 1970 (29 U.S.C. 651 et seq.); and

(E) the employer, excluding any licensed health care professional or board certified genetic counselor that is involved in the genetic monitoring program, receives the results of the monitoring only in aggregate terms that do not disclose the identity of specific employees; or

(6) where the employer conducts DNA analysis for law enforcement purposes as a forensic laboratory or for purposes of human remains identification, and requests or requires genetic information of such employer's employees, but only to the extent that such genetic information is used for analysis of DNA identification markers for quality control to detect sample contamination.

(c) Preservation of Protections.—In the case of information to which any of paragraphs (1) through (6) of subsection (b) applies, such information may not be used in violation of paragraph (1) or (2) of subsection (a) or treated or disclosed in a manner that violates section 206.

SEC. 203. EMPLOYMENT AGENCY PRACTICES

(a) Discrimination Based on Genetic Information.—It shall be an unlawful employment practice for an employment agency—

(1) to fail or refuse to refer for employment, or otherwise to discriminate against, any individual because of genetic information with respect to the individual;

(2) to limit, segregate, or classify individuals or fail or refuse to refer for employment any individual in any way that would deprive or tend to deprive any individual of employment opportunities, or otherwise adversely affect the status of the individual as an employee, because of genetic information with respect to the individual; or

(3) to cause or attempt to cause an employer to discriminate against an individual in violation of this title.

(b) Acquisition of Genetic Information.—It shall be an unlawful employment practice for an employment agency to request, require, or purchase genetic information with respect to an individual or a family member of the individual except—

(1) where an employment agency inadvertently requests or requires family medical history of the individual or family member of the individual;

(2) where—

(A) health or genetic services are offered by the employment agency, including such services offered as part of a wellness program;

(B) the individual provides prior, knowing, voluntary, and written authorization;

(C) only the individual (or family member if the family member is receiving genetic services) and the licensed health care professional or board certified genetic counselor involved in providing such services receive individually identifiable information concerning the results of such services; and

(D) any individually identifiable genetic information provided under subparagraph (C) in connection with the services provided under subparagraph (A) is only available for purposes of such services and shall not be disclosed to the employment agency except in aggregate terms that do not disclose the identity of specific individuals;

(3) where an employment agency requests or requires family medical history from the individual to comply with the certification provisions of section 103 of the Family and Medical Leave Act of 1993 (29 U.S.C. 2613) or such requirements under State family and medical leave laws;

(4) where an employment agency purchases documents that are commercially and publicly available (including newspapers, magazines, periodicals, and books, but not including medical databases or court records) that include family medical history; or

(5) where the information involved is to be used for genetic monitoring of the biological effects of toxic substances in the workplace, but only if—

(A) the employment agency provides written notice of the genetic monitoring to the individual;

(B)(i) the individual provides prior, knowing, voluntary, and written authorization; or

(ii) the genetic monitoring is required by Federal or State law;

(C) the individual is informed of individual monitoring results;

(D) the monitoring is in compliance with—

(i) any Federal genetic monitoring regulations, including any such regulations that may be promulgated by the Secretary of Labor pursuant to the Occupational Safety and Health Act of 1970 (29 U.S.C. 651 et seq.), the Federal Mine Safety and Health Act of 1977 (30 U.S.C. 801 et seq.), or the Atomic Energy Act of 1954 (42 U.S.C. 2011 et seq.); or

(ii) State genetic monitoring regulations, in the case of a State that is implementing genetic monitoring regulations under the authority of the Occupational Safety and Health Act of 1970 (29 U.S.C. 651 et seq.); and

(E) the employment agency, excluding any licensed health care professional or board certified genetic counselor that is involved in the genetic monitoring program, receives the results of the monitoring only in aggregate terms that do not disclose the identity of specific individuals.

(c) Preservation of Protections.—In the case of information to which any of paragraphs (1) through (5) of subsection (b) applies, such information may not be used in violation of paragraph (1), (2), or (3) of subsection (a) or treated or disclosed in a manner that violates section 206.

SEC. 204. LABOR ORGANIZATION PRACTICES

(a) Discrimination Based on Genetic Information.—It shall be an unlawful employment practice for a labor organization—

(1) to exclude or to expel from the membership of the organization, or otherwise to discriminate against, any member because of genetic information with respect to the member;

(2) to limit, segregate, or classify the members of the organization, or fail or refuse to refer for employment any member, in any way that would deprive or tend to deprive any member of employment opportunities, or otherwise adversely affect the status of the member as an employee, because of genetic information with respect to the member; or

(3) to cause or attempt to cause an employer to discriminate against a member in violation of this title.

(b) Acquisition of Genetic Information.—It shall be an unlawful employment practice for a labor organization to request, require, or purchase genetic information with respect to a member or a family member of the member except—

(1) where a labor organization inadvertently requests or requires family medical history of the member or family member of the member;

(2) where—

(A) health or genetic services are offered by the labor organization, including such services offered as part of a wellness program;

(B) the member provides prior, knowing, voluntary, and written authorization;

(C) only the member (or family member if the family member is receiving genetic services) and the licensed health care professional or board certified genetic counselor involved in providing such services receive individually identifiable information concerning the results of such services; and

(D) any individually identifiable genetic information provided under subparagraph (C) in connection with the services provided under subparagraph (A) is only available for purposes of such services and shall not be disclosed to the labor organization except in aggregate terms that do not disclose the identity of specific members;

(3) where a labor organization requests or requires family medical history from the members to comply with the certification provisions of section 103 of the Family and Medical Leave Act of 1993 (29 U.S.C. 2613) or such requirements under State family and medical leave laws;

(4) where a labor organization purchases documents that are commercially and publicly available (including newspapers, magazines, periodicals, and books, but not including medical databases or court records) that include family medical history; or

(5) where the information involved is to be used for genetic monitoring of the biological effects of toxic substances in the workplace, but only if—

(A) the labor organization provides written notice of the genetic monitoring to the member;

(B)(i) the member provides prior, knowing, voluntary, and written authorization; or

(ii) the genetic monitoring is required by Federal or State law;

(C) the member is informed of individual monitoring results;

(D) the monitoring is in compliance with—

(i) any Federal genetic monitoring regulations, including any such regulations that may be promulgated by the Secretary of Labor pursuant to the Occupational Safety and Health Act of 1970 (29 U.S.C. 651 et seq.), the Federal Mine Safety and Health Act of 1977 (30 U.S.C. 801 et seq.), or the Atomic Energy Act of 1954 (42 U.S.C. 2011 et seq.); or

(ii) State genetic monitoring regulations, in the case of a State that is implementing genetic monitoring regulations under the authority of the Occupational Safety and Health Act of 1970 (29 U.S.C. 651 et seq.); and

(E) the labor organization, excluding any licensed health care professional or board certified genetic counselor that is involved in the genetic monitoring program, receives the results of the monitoring only in aggregate terms that do not disclose the identity of specific members.

(c) Preservation of Protections.—In the case of information to which any of paragraphs (1) through (5) of subsection (b) applies, such information may not be used in violation of paragraph (1), (2), or (3) of subsection (a) or treated or disclosed in a manner that violates section 206.

SEC. 205. TRAINING PROGRAMS

(a) Discrimination Based on Genetic Information.—It shall be an unlawful employment practice for any employer, labor organization, or joint labor-management committee controlling apprenticeship or other training or retraining, including on-the-job training programs—

(1) to discriminate against any individual because of genetic information with respect to the individual in admission to, or employment in, any program established to provide apprenticeship or other training or retraining;

(2) to limit, segregate, or classify the applicants for or participants in such apprenticeship or other training or retraining, or fail or refuse to refer for employment any individual, in any way that would deprive or tend to deprive any individual of employment opportunities, or otherwise adversely affect the status of the individual as an employee, because of genetic information with respect to the individual; or

(3) to cause or attempt to cause an employer to discriminate against an applicant for or a participant in such apprenticeship or other training or retraining in violation of this title.

(b) Acquisition of Genetic Information.—It shall be an unlawful employment practice for an employer, labor organization, or joint labor- management committee described in subsection (a) to request, require, or purchase genetic information with respect to an individual or a family member of the individual except—

(1) where the employer, labor organization, or joint labor- management committee inadvertently requests or requires family medical history of the individual or family member of the individual;

(2) where—

(A) health or genetic services are offered by the employer, labor organization, or joint labor-management committee, including such services offered as part of a wellness program;

(B) the individual provides prior, knowing, voluntary, and written authorization;

(C) only the individual (or family member if the family member is receiving genetic services) and the licensed health care professional or board certified genetic counselor involved in providing such services receive individually identifiable information concerning the results of such services; and

(D) any individually identifiable genetic information provided under subparagraph (C) in connection with the services provided under subparagraph (A) is only available for purposes of such services and shall not be disclosed to the employer, labor organization, or joint labor-management committee except in aggregate terms that do not disclose the identity of specific individuals;

(3) where the employer, labor organization, or joint labor- management committee requests or requires family medical history from the individual to comply with the certification provisions of section 103 of the Family and Medical Leave Act of 1993 (29 U.S.C. 2613) or such requirements under State family and medical leave laws;

(4) where the employer, labor organization, or joint labor- management committee purchases documents that are commercially and publicly available (including newspapers, magazines, periodicals, and books, but not including medical databases or court records) that include family medical history;

(5) where the information involved is to be used for genetic monitoring of the biological effects of toxic substances in the workplace, but only if—

(A) the employer, labor organization, or joint labor-management committee provides written notice of the genetic monitoring to the individual;

(B)(i) the individual provides prior, knowing, voluntary, and written authorization; or

(ii) the genetic monitoring is required by Federal or State law;

(C) the individual is informed of individual monitoring results;

(D) the monitoring is in compliance with—

(i) any Federal genetic monitoring regulations, including any such regulations that may be promulgated by the Secretary of Labor pursuant to the Occupational Safety and Health Act of 1970 (29 U.S.C. 651 et seq.), the Federal Mine Safety and Health Act of 1977 (30 U.S.C. 801 et seq.), or the Atomic Energy Act of 1954 (42 U.S.C. 2011 et seq.); or

(ii) State genetic monitoring regulations, in the case of a State that is implementing genetic monitoring regulations under the authority of the Occupational Safety and Health Act of 1970 (29 U.S.C. 651 et seq.); and

(E) the employer, labor organization, or joint labor-management committee, excluding any licensed health care professional or board certified genetic counselor that is involved in the genetic monitoring program, receives the results of the monitoring only in aggregate terms that do not disclose the identity of specific individuals; or

(6) where the employer conducts DNA analysis for law enforcement purposes as a forensic laboratory or for purposes of human remains identification, and requests or requires genetic information of such employer's apprentices or trainees, but only to the extent that such genetic information is used for analysis of DNA identification markers for quality control to detect sample contamination.

(c) Preservation of Protections.—In the case of information to which any of paragraphs (1) through (6) of subsection (b) applies, such information may not be used in violation of paragraph (1), (2), or (3) of subsection (a) or treated or disclosed in a manner that violates section 206.

SEC. 206. CONFIDENTIALITY OF GENETIC INFORMATION

(a) Treatment of Information as Part of Confidential Medical Record.—If an employer, employment agency, labor organization, or joint labor-management committee possesses genetic information about an employee or member, such information shall be maintained on separate forms and in separate medical files and be treated as a confidential medical record of the employee or member. An employer, employment agency, labor organization, or joint labor-management committee shall be considered to be in compliance with the maintenance of information requirements of this subsection with respect to genetic information subject to this subsection that is maintained with and treated as a confidential medical record under section 102(d)(3)(B) of the Americans with Disabilities Act (42 U.S.C. 12112(d)(3)(B)).

(b) Limitation on Disclosure.—An employer, employment agency, labor organization, or joint labor-management committee shall not disclose genetic information concerning an employee or member except—

(1) to the employee or member of a labor organization (or family member if the family member is receiving the genetic services) at the written request of the employee or member of such organization;

(2) to an occupational or other health researcher if the research is conducted in compliance with the regulations and protections provided for under part 46 of title 45, Code of Federal Regulations;

(3) in response to an order of a court, except that—

(A) the employer, employment agency, labor organization, or joint labor-management committee may disclose only the genetic information expressly authorized by such order; and

(B) if the court order was secured without the knowledge of the employee or member to whom the information refers, the employer, employment agency, labor organization, or joint labor-management committee shall inform the employee or member of the court order and any genetic information that was disclosed pursuant to such order;

(4) to government officials who are investigating compliance with this title if the information is relevant to the investigation;

(5) to the extent that such disclosure is made in connection with the employee's compliance with the certification provisions of section 103 of the Family and Medical Leave Act of 1993 (29 U.S.C. 2613) or such requirements under State family and medical leave laws; or

(6) to a Federal, State, or local public health agency only with regard to information that is described in section 201(4)(A)(iii) and that concerns a contagious disease that presents an imminent hazard of death or life-threatening illness, and that the employee whose family member or family members is or are the subject of a disclosure under this paragraph is notified of such disclosure.

(c) Relationship to HIPAA Regulations.—With respect to the regulations promulgated by the Secretary of Health and Human Services under part C of title XI of the Social Security Act (42 U.S.C. 1320d et seq.) and section 264 of the Health Insurance Portability and Accountability Act of 1996 (42 U.S.C. 1320d-2 note), this title does not prohibit a covered entity under such regulations from any use or disclosure of health information that is authorized for the covered entity under such regulations. The previous sentence does not affect the authority of such Secretary to modify such regulations.

SEC. 207. REMEDIES AND ENFORCEMENT

(a) Employees Covered by Title VII of the Civil Rights Act of 1964.—

(1) In general.—The powers, procedures, and remedies provided in sections 705, 706, 707, 709, 710, and 711 of the Civil Rights Act of 1964 (42 U.S.C. 2000e-4 et seq.) to the Commission, the Attorney General, or any person, alleging a violation of title VII of that Act (42 U.S.C. 2000e et seq.) shall be the powers, procedures, and remedies this title provides to the Commission, the Attorney General, or any person, respectively, alleging an unlawful employment practice in violation of this title against an employee described in section 201(2)(A)(i), except as provided in paragraphs (2) and (3).

(2) Costs and fees.—The powers, remedies, and procedures provided in subsections (b) and (c) of section 722 of the Revised Statutes of the United States (42 U.S.C. 1988), shall be powers, remedies, and procedures this title provides to the Commission, the Attorney General, or any person, alleging such a practice. (3) Damages.—The powers, remedies, and procedures provided in section 1977A of the Revised Statutes of the United States (42 U.S.C. 1981a), including the limitations contained in subsection (b)(3) of such section 1977A, shall be powers, remedies, and procedures this title provides to the Commission, the Attorney General, or any person, alleging such a practice (not an employment practice specifically excluded from coverage under section 1977A(a)(1) of the Revised Statutes of the United States).

(b) Employees Covered by Government Employee Rights Act of 1991.—

(1) In general.—The powers, remedies, and procedures provided in sections 302 and 304 of the Government Employee Rights Act of 1991 (42 U.S.C. 2000e-16b, 2000e-16c) to the Commission, or any person, alleging a violation of section 302(a)(1) of that Act (42 U.S.C. 2000e-16b(a)(1)) shall be the powers, remedies, and procedures this title provides to the Commission, or any person, respectively, alleging an unlawful employment practice in violation of this title against an employee described in section 201(2)(A)(ii), except as provided in paragraphs (2) and (3).

(2) Costs and fees.—The powers, remedies, and procedures provided in subsections (b) and (c) of section 722 of the Revised Statutes of the United States (42 U.S.C. 1988), shall be powers, remedies, and procedures this title provides to the Commission, or any person, alleging such a practice.

(3) Damages.—The powers, remedies, and procedures provided in section 1977A of the Revised Statutes of the United States (42 U.S.C. 1981a), including the limitations contained in subsection (b)(3) of such section 1977A, shall be powers, remedies,

and procedures this title provides to the Commission, or any person, alleging such a practice (not an employment practice specifically excluded from coverage under section 1977A(a)(1) of the Revised Statutes of the United States).

(c) Employees Covered by Congressional Accountability Act of 1995.—

(1) In general.—The powers, remedies, and procedures provided in the Congressional Accountability Act of 1995 (2 U.S.C. 1301 et seq.) to the Board (as defined in section 101 of that Act (2 U.S.C. 1301)), or any person, alleging a violation of section 201(a)(1) of that Act (42 U.S.C. 1311(a)(1)) shall be the powers, remedies, and procedures this title provides to that Board, or any person, alleging an unlawful employment practice in violation of this title against an employee described in section 201(2)(A)(iii), except as provided in paragraphs (2) and (3).

(2) Costs and fees.—The powers, remedies, and procedures provided in subsections (b) and (c) of section 722 of the Revised Statutes of the United States (42 U.S.C. 1988), shall be powers, remedies, and procedures this title provides to that Board, or any person, alleging such a practice.

(3) Damages.—The powers, remedies, and procedures provided in section 1977A of the Revised Statutes of the United States (42 U.S.C. 1981a), including the limitations contained in subsection (b)(3) of such section 1977A, shall be powers, remedies, and procedures this title provides to that Board, or any person, alleging such a practice (not an employment practice specifically excluded from coverage under section 1977A(a)(1) of the Revised Statutes of the United States). (4) Other applicable provisions.—With respect to a claim alleging a practice described in paragraph (1), title III of the Congressional Accountability Act of 1995 (2 U.S.C. 1381 et seq.) shall apply in the same manner as such title applies with respect to a claim alleging a violation of section 201(a)(1) of such Act (2 U.S.C. 1311(a)(1)).

(d) Employees Covered by Chapter 5 of Title 3, United States Code.—

(1) In general.—The powers, remedies, and procedures provided in chapter 5 of title 3, United States Code, to the President, the Commission, the Merit Systems Protection Board, or any person, alleging a violation of section 411(a)(1) of that title, shall be the powers, remedies, and procedures this title provides to the President, the Commission, such Board, or any person, respectively, alleging an unlawful employment practice in violation of this title against an employee described in section 201(2)(A)(iv), except as provided in paragraphs (2) and (3).

(2) Costs and fees.—The powers, remedies, and procedures provided in subsections (b) and (c) of section 722 of the Revised Statutes of the United States (42 U.S.C. 1988), shall be powers, remedies, and procedures this title provides to the President, the Commission, such Board, or any person, alleging such a practice.

(3) Damages.—The powers, remedies, and procedures provided in section 1977A of the Revised Statutes of the United States (42 U.S.C. 1981a), including the limitations contained in subsection (b)(3) of such section 1977A, shall be powers, remedies, and

procedures this title provides to the President, the Commission, such Board, or any person, alleging such a practice (not an employment practice specifically excluded from coverage under section 1977A(a)(1) of the Revised Statutes of the United States).

(e) Employees Covered by Section 717 of the Civil Rights Act of 1964.—

(1) In general.—The powers, remedies, and procedures provided in section 717 of the Civil Rights Act of 1964 (42 U.S.C. 2000e-16) to the Commission, the Attorney General, the Librarian of Congress, or any person, alleging a violation of that section shall be the powers, remedies, and procedures this title provides to the Commission, the Attorney General, the Librarian of Congress, or any person, respectively, alleging an unlawful employment practice in violation of this title against an employee or applicant described in section 201(2)(A)(v), except as provided in paragraphs (2) and (3).

(2) Costs and fees.—The powers, remedies, and procedures provided in subsections (b) and (c) of section 722 of the Revised Statutes of the United States (42 U.S.C. 1988), shall be powers, remedies, and procedures this title provides to the Commission, the Attorney General, the Librarian of Congress, or any person, alleging such a practice.

(3) Damages.—The powers, remedies, and procedures provided in section 1977A of the Revised Statutes of the United States (42 U.S.C. 1981a), including the limitations contained in subsection (b)(3) of such section 1977A, shall be powers, remedies, and procedures this title provides to the Commission, the Attorney General, the Librarian of Congress, or any person, alleging such a practice (not an employment practice specifically excluded from coverage under section 1977A(a)(1) of the Revised Statutes of the United States).

(f) Prohibition Against Retaliation.—No person shall discriminate against any individual because such individual has opposed any act or practice made unlawful by this title or because such individual made a charge, testified, assisted, or participated in any manner in an investigation, proceeding, or hearing under this title. The remedies and procedures otherwise provided for under this section shall be available to aggrieved individuals with respect to violations of this subsection.

(g) Definition.—In this section, the term "Commission" means the Equal Employment Opportunity Commission.

SEC. 208. DISPARATE IMPACT

(a) General Rule.—Notwithstanding any other provision of this Act, "disparate impact", as that term is used in section 703(k) of the Civil Rights Act of 1964 (42 U.S.C. 2000e-2(k)), on the basis of genetic information does not establish a cause of action under this Act.

(b) Commission.—On the date that is 6 years after the date of enactment of this Act, there shall be established a commission, to be known as the Genetic Nondiscrimination Study Commission (referred to in this section as the "Commission") to review the developing science of genetics and to make recommendations to Congress regarding whether to provide a disparate impact cause of action under this Act.

(c) Membership.—

(1) In general.—The Commission shall be composed of 8 members, of which—

(A) 1 member shall be appointed by the Majority Leader of the Senate;

(B) 1 member shall be appointed by the Minority Leader of the Senate;

(C) 1 member shall be appointed by the Chairman of the Committee on Health, Education, Labor, and Pensions of the Senate;

(D) 1 member shall be appointed by the ranking minority member of the Committee on Health, Education, Labor, and Pensions of the Senate;

(E) 1 member shall be appointed by the Speaker of the House of Representatives;

(F) 1 member shall be appointed by the Minority Leader of the House of Representatives;

(G) 1 member shall be appointed by the Chairman of the Committee on Education and Labor of the House of Representatives; and

(H) 1 member shall be appointed by the ranking minority member of the Committee on Education and Labor of the House of Representatives.

(2) Compensation and expenses.—The members of the Commission shall not receive compensation for the performance of services for the Commission, but shall be allowed travel expenses, including per diem in lieu of subsistence, at rates authorized for employees of agencies under subchapter I of chapter 57 of title 5, United States Code, while away from their homes or regular places of business in the performance of services for the Commission.

(d) Administrative Provisions.—

(1) Location.—The Commission shall be located in a facility maintained by the Equal Employment Opportunity Commission.

(2) Detail of government employees.—Any Federal Government employee may be detailed to the Commission without reimbursement, and such detail shall be without interruption or loss of civil service status or privilege.

(3) Information from federal agencies.—The Commission may secure directly from any Federal department or agency such information as the Commission considers necessary to carry out the provisions of this section. Upon request of the Commission, the head of such department or agency shall furnish such information to the Commission.

(4) Hearings.—The Commission may hold such hearings, sit and act at such times and places, take such testimony, and receive such evidence as the Commission considers advisable to carry out the objectives of this section, except that, to the extent possible, the Commission shall use existing data and research.

(5) Postal services.—The Commission may use the United States mails in the same manner and under the same conditions as other departments and agencies of the Federal Government.

(e) Report.—Not later than 1 year after all of the members are appointed to the Commission under subsection (c)(1), the Commission shall submit to Congress a report that summarizes the findings of the Commission and makes such recommendations for legislation as are consistent with this Act.

(f) Authorization of Appropriations.—There are authorized to be appropriated to the Equal Employment Opportunity Commission such sums as may be necessary to carry out this section. SEC. 209. CONSTRUCTION.

(a) In General.—Nothing in this title shall be construed to—

(1) limit the rights or protections of an individual under any other Federal or State statute that provides equal or greater protection to an individual than the rights or protections provided for under this title, including the protections of an individual under the Americans with Disabilities Act of 1990 (42 U.S.C. 12101 et seq.) (including coverage afforded to individuals under section 102 of such Act (42 U.S.C. 12112)), or under the Rehabilitation Act of 1973 (29 U.S.C. 701 et seq.);

(2)(A) limit the rights or protections of an individual to bring an action under this title against an employer, employment agency, labor organization, or joint labor-management committee for a violation of this title; or

(B) provide for enforcement of, or penalties for violation of, any requirement or prohibition applicable to any employer, employment agency, labor organization, or joint labor-management committee subject to enforcement for a violation under—

(i) the amendments made by title I of this Act;

(ii)(I) subsection (a) of section 701 of the Employee Retirement Income Security Act of 1974 as such section applies with respect to genetic information pursuant to subsection (b)(1)(B) of such section;

(II) section 702(a)(1)(F) of such Act; or

(III) section 702(b)(1) of such Act as such section applies with respect to genetic information as a health status-related factor;

(iii)(I) subsection (a) of section 2701 of the Public Health Service Act as such section applies with respect to genetic information pursuant to subsection (b)(1)(B) of such section;

(II) section 2702(a)(1)(F) of such Act; or

(III) section 2702(b)(1) of such Act as such section applies with respect to genetic information as a health status-related factor; or

(iv)(I) subsection (a) of section 9801 of the Internal Revenue Code of 1986 as such section applies with respect to genetic information pursuant to subsection (b)(1)(B) of such section;

(II) section 9802(a)(1)(F) of such Act; or

(III) section 9802(b)(1) of such Act as such section applies with respect to genetic information as a health status-related factor;

(3) apply to the Armed Forces Repository of Specimen Samples for the Identification of Remains;

(4) limit or expand the protections, rights, or obligations of employees or employers under applicable workers' compensation laws;

(5) limit the authority of a Federal department or agency to conduct or sponsor occupational or other health research that is conducted in compliance with the regulations contained in part 46 of title 45, Code of Federal Regulations (or any corresponding or similar regulation or rule);

(6) limit the statutory or regulatory authority of the Occupational Safety and Health Administration or the Mine Safety and Health Administration to promulgate or enforce workplace safety and health laws and regulations; or

(7) require any specific benefit for an employee or member or a family member of an employee or member under any group health plan or health insurance issuer offering group health insurance coverage in connection with a group health plan.

(b) Genetic Information of a Fetus or Embryo.—Any reference in this title to genetic information concerning an individual or family member of an individual shall—

(1) with respect to such an individual or family member of an individual who is a pregnant woman, include genetic information of any fetus carried by such pregnant woman; and

(2) with respect to an individual or family member utilizing an assisted reproductive technology, include genetic information of any embryo legally held by the individual or family member.

(c) Relation to Authorities Under Title I.—With respect to a group health plan, or a health insurance issuer offering group health insurance coverage in connection with a group health plan, this title does not prohibit any activity of such plan or issuer that is authorized for the plan or issuer under any provision of law referred to in clauses (i) through (iv) of subsection (a)(2)(B).

SEC. 210. MEDICAL INFORMATION THAT IS NOT GENETIC INFORMATION

An employer, employment agency, labor organization, or joint labor- management committee shall not be considered to be in violation of this title based on the use, acquisition, or disclosure of medical information that is not genetic information about a manifested disease, disorder, or pathological condition of an employee or member, including a manifested disease, disorder, or pathological condition that has or may have a genetic basis.

SEC. 211. REGULATIONS

Not later than 1 year after the date of enactment of this title, the Commission shall issue final regulations to carry out this title.

SEC. 212. AUTHORIZATION OF APPROPRIATIONS

There are authorized to be appropriated such sums as may be necessary to carry out this title (except for section 208).

SEC. 213. EFFECTIVE DATE

This title takes effect on the date that is 18 months after the date of enactment of this Act.

TITLE III—MISCELLANEOUS PROVISIONS

SEC. 301. SEVERABILITY

If any provision of this Act, an amendment made by this Act, or the application of such provision or amendment to any person or circumstance is held to be unconstitutional, the remainder of this Act, the amendments made by this Act, and the application of such provisions to any person or circumstance shall not be affected thereby.

Approved May 21, 2008.

Appendix 3: Family and Medical Leave Act

SUBPART A—COVERAGE UNDER THE FAMILY AND MEDICAL LEAVE ACT

SECTION 825.100 THE FAMILY AND MEDICAL LEAVE ACT

(a) The Family and Medical Leave Act of 1993, as amended, (FMLA or Act) allows eligible employees of a covered employer to take job-protected, unpaid leave, or to substitute appropriate paid leave if the employee has earned or accrued it, for up to a total of 12 workweeks in any 12 months (*see* Section 825.200(b)) because of the birth of a child and to care for the newborn child, because of the placement of a child with the employee for adoption or foster care, because the employee is needed to care for a family member (child, spouse, or parent) with a serious health condition, because the employee's own serious health condition makes the employee unable to perform the functions of his or her job, or because of any qualifying exigency arising out of the fact that the employee's spouse, son, daughter, or parent is a military member on active duty or call to covered active duty status (or has been notified of an impending call or order to covered active duty). In addition, eligible employees of a covered employer may take job-protected, unpaid leave, or substitute appropriate paid leave if the employee has earned or accrued it, for up to a total of 26 workweeks in a single 12-month period to care for a covered servicemember with a serious injury or illness. In certain cases, FMLA leave may be taken on an intermittent basis rather than all at once, or the employee may work a part-time schedule.

(b) An employee on FMLA leave is also entitled to have health benefits maintained while on leave as if the employee had continued to work instead of taking the leave. If an employee was paying all or part of the premium payments prior to leave, the employee would continue to pay his or her share during the leave period. The employer may recover its share only if the employee does not return to work for a reason other than the serious health condition of the employee or the employee's covered family member, the serious injury or illness of a covered servicemember, or another reason beyond the employee's control.

(c) An employee generally has a right to return to the same position or an equivalent position with equivalent pay, benefits, and working conditions at the conclusion of the leave. The taking of FMLA leave cannot result in the loss of any benefit that accrued prior to the start of the leave.

(d) The employer generally has a right to advance notice from the employee. In addition, the employer may require an employee to submit certification to substantiate that the leave is due to the serious health condition of the employee or the employee's covered family member, due to the serious injury or illness of a covered servicemember, or because of a qualifying exigency. Failure to comply with these requirements may result in a delay in the start of FMLA leave. Pursuant to a uniformly applied policy, the employer may also require that an employee present a certification of fitness to return to work when the absence was caused by the employee's serious health condition (*see* Sections 825.312 and 825.313). The employer may delay restoring the employee to employment without such certificate relating to the health condition which caused the employee's absence.

Section 825.101 Purpose of the Act

(a) FMLA is intended to allow employees to balance their work and family life by taking reasonable unpaid leave for medical reasons, for the birth or adoption of a child, for the care of a child, spouse, or parent who has a serious health condition, for the care of a covered servicemember with a serious injury or illness, or because of a qualifying exigency arising out of the fact that the employee's spouse, son, daughter, or parent is a military member on covered active duty or call to covered active duty status. The Act is intended to balance the demands of the workplace with the needs of families, to promote the stability and economic security of families, and to promote national interests in preserving family integrity. It was intended that the Act accomplish these purposes in a manner that accommodates the legitimate interests of employers, and in a manner consistent with the Equal Protection Clause of the Fourteenth Amendment in minimizing the potential for employment discrimination on the basis of sex, while promoting equal employment opportunity for men and women.

(b) The FMLA was predicated on two fundamental concerns—the needs of the American workforce, and the development of high-performance organizations. Increasingly, America's children and elderly are dependent upon family members who must spend long hours at work. When a family emergency arises, requiring workers to attend to seriously-ill children or parents, or to newly-born or adopted infants, or even to their own serious illness, workers need reassurance that they will not be asked to choose between continuing their employment, and meeting their personal and family obligations or tending to vital needs at home.

(c) The FMLA is both intended and expected to benefit employers as well as their employees. A direct correlation exists between stability in the family and productivity in the workplace. FMLA will encourage the development of high-performance organizations. When workers can count on durable links to their workplace they are able to make their own full commitments to their jobs. The record of hearings on family and medical leave indicate the powerful productive advantages of stable workplace relationships, and the comparatively small costs of guaranteeing that those relationships will not be dissolved while workers attend to pressing family health obligations or their own serious illness.

Section 825.102 Definitions

For purposes of this part:

Act or *FMLA* means the Family and Medical Leave Act of 1993, Public Law 103-3 (February 5, 1993), 107 Stat. 6 (29 U.S.C. 2601 *et seq., as amended*).

ADA means the Americans with Disabilities Act (42 U.S.C. 12101 *et seq.,* as amended).

Administrator means the Administrator of the Wage and Hour Division, U.S. Department of Labor, and includes any official of the Wage and Hour Division authorized to perform any of the functions of the Administrator under this part.

Airline flight crew employee means an airline flight crewmember or flight attendant as those terms are defined in regulations of the Federal Aviation Administration. *See also* Section 825.800(a).

Applicable monthly guarantee means:

(1) For an airline flight crew employee who is not on reserve status (line holder), the minimum number of hours for which an employer has agreed to *schedule* such employee for any given month; and

(2) For an airline flight crew employee who is on reserve status, the number of hours for which an employer has agreed to *pay* the employee for any given month. *See also* Section 825.801(b)(1).

COBRA means the continuation coverage requirements of Title X of the Consolidated Omnibus Budget Reconciliation Act of 1986, as amended (Pub. L. 99-272, title X, section 10002; 100 Stat 227; 29 U.S.C. 1161-1168).

Commerce and *industry or activity affecting commerce* mean any activity, business, or industry in commerce or in which a labor dispute would hinder or obstruct commerce or the free flow of commerce, and include "commerce" and any "industry affecting commerce" as defined in sections 501(1) and 501(3) of the Labor Management Relations Act of 1947, 29 U.S.C. 142(1) and (3).

Contingency operation means a military operation that:

(1) Is designated by the Secretary of Defense as an operation in which members of the Armed Forces are or may become involved in military actions, operations, or hostilities against an enemy of the United States or against an opposing military force; or

(2) Results in the call or order to, or retention on, active duty of members of the uniformed services under section 688, 12301(a), 12302, 12304, 12305, or 12406 of Title 10 of the United States Code, chapter 15 of Title 10 of the United States Code, or any

other provision of law during a war or during a national emergency declared by the President or Congress. *See also* Section 825.126(a)(2).

Continuing treatment by a health care provider means any one of the following:

(1) *Incapacity and treatment.* A period of incapacity of more than three consecutive, full calendar days, and any subsequent treatment or period of incapacity relating to the same condition, that also involves:

(i) Treatment two or more times, within 30 days of the first day of incapacity, unless extenuating circumstances exist, by a health care provider, by a nurse under direct supervision of a health care provider, or by a provider of health care services (e.g., physical therapist) under orders of, or on referral by, a health care provider; or

(ii) Treatment by a health care provider on at least one occasion, which results in a regimen of continuing treatment under the supervision of the health care provider.

(iii) The requirement in paragraphs (i) and (ii) of this definition for treatment by a health care provider means an in-person visit to a health care provider. The first in-person treatment visit must take place within seven days of the first day of incapacity.

(iv) Whether additional treatment visits or a regimen of continuing treatment is necessary within the 30-day period shall be determined by the health care provider.

(v) The term "extenuating circumstances" in paragraph (i) means circumstances beyond the employee's control that prevent the follow-up visit from occurring as planned by the health care provider. Whether a given set of circumstances are extenuating depends on the facts. *See also* Section 825.115(a)(5).

(2) *Pregnancy or prenatal care.* Any period of incapacity due to pregnancy, or for prenatal care. *See also* Section 825.120.

(3) *Chronic conditions.* Any period of incapacity or treatment for such incapacity due to a chronic serious health condition. A chronic serious health condition is one which:

(i) Requires periodic visits (defined as at least twice a year) for treatment by a health care provider, or by a nurse under direct supervision of a health care provider;

(ii) Continues over an extended period of time (including recurring episodes of a single underlying condition); and

(iii) May cause episodic rather than a continuing period of incapacity (e.g., asthma, diabetes, epilepsy, etc.).

(4) *Permanent or long-term conditions.* A period of incapacity which is permanent or long-term due to a condition for which treatment may not be effective. The employee or family member must be under the continuing supervision of, but

need not be receiving active treatment by, a health care provider. Examples include Alzheimer's, a severe stroke, or the terminal stages of a disease.

(5) *Conditions requiring multiple treatments.* Any period of absence to receive multiple treatments (including any period of recovery therefrom) by a health care provider or by a provider of health care services under orders of, or on referral by, a health care provider, for:

(i) Restorative surgery after an accident or other injury; or

(ii) A condition that would likely result in a period of incapacity of more than three consecutive full calendar days in the absence of medical intervention or treatment, such as cancer (chemotherapy, radiation, etc.), severe arthritis (physical therapy), kidney disease (dialysis).

(6) Absences attributable to incapacity under paragraphs (2) or (3) of this definition qualify for FMLA leave even though the employee or the covered family member does not receive treatment from a health care provider during the absence, and even if the absence does not last more than three consecutive full calendar days. For example, an employee with asthma may be unable to report for work due to the onset of an asthma attack or because the employee's health care provider has advised the employee to stay home when the pollen count exceeds a certain level. An employee who is pregnant may be unable to report to work because of severe morning sickness.

Covered active duty or call to covered active duty status means:

(1) In the case of a member of the Regular Armed Forces, duty during the deployment of the member with the Armed Forces to a foreign country; and,

(2) In the case of a member of the Reserve components of the Armed Forces, duty during the deployment of the member with the Armed Forces to a foreign country under a Federal call or order to active duty in support of a contingency operation pursuant to: Section 688 of Title 10 of the United States Code, which authorizes ordering to active duty retired members of the Regular Armed Forces and members of the retired Reserve who retired after completing at least 20 years of active service; Section 12301(a) of Title 10 of the United States Code, which authorizes ordering all reserve component members to active duty in the case of war or national emergency; Section 12302 of Title 10 of the United States Code, which authorizes ordering any unit or unassigned member of the Ready Reserve to active duty; Section 12304 of Title 10 of the United States Code, which authorizes ordering any unit or unassigned member of the Selected Reserve and certain members of the Individual Ready Reserve to active duty; Section 12305 of Title 10 of the United States Code, which authorizes the suspension of promotion, retirement or separation rules for certain Reserve components; Section 12406 of Title 10 of the United States Code, which authorizes calling the National Guard into Federal service in certain circumstances; chapter 15 of Title 10 of the United States Code, which authorizes calling the National Guard and state military into Federal service in the case of insurrections

and national emergencies; or any other provision of law during a war or during a national emergency declared by the President or Congress so long as it is in support of a contingency operation. *See* 10 U.S.C. 101(a)(13)(B). *See also* Section 825.126(a).

Covered servicemember means:

(1) A current member of the Armed Forces, including a member of the National Guard or Reserves, who is undergoing medical treatment, recuperation, or therapy, is otherwise in outpatient status, or is otherwise on the temporary disability retired list, for a serious injury or illness, or

(2) A covered veteran who is undergoing medical treatment, recuperation, or therapy for a serious injury or illness.

Covered veteran means an individual who was a member of the Armed Forces (including a member of the National Guard or Reserves), and was discharged or released under conditions other than dishonorable at any time during the five-year period prior to the first date the eligible employee takes FMLA leave to care for the covered veteran. *See* Section 825.127(b)(2).

Eligible employee means:

(1) An employee who has been employed for a total of at least 12 months by the employer on the date on which any FMLA leave is to commence, except that an employer need not consider any period of previous employment that occurred more than seven years before the date of the most recent hiring of the employee, *unless:*

(i) The break in service is occasioned by the fulfillment of the employee's Uniformed Services Employment and Reemployment Rights Act (USERRA), 38 U.S.C. 4301, *et seq.,* covered service obligation (the period of absence from work due to or necessitated by USERRA-covered service must be also counted in determining whether the employee has been employed for at least 12 months by the employer, but this section does not provide any greater entitlement to the employee than would be available under the USERRA; or

(ii) A written agreement, including a collective bargaining agreement, exists concerning the employer's intention to rehire the employee after the break in service (e.g., for purposes of the employee furthering his or her education or for childrearing purposes); and

(2) Who, on the date on which any FMLA leave is to commence, has met the hours of service requirement by having been employed for at least 1,250 hours of service with such employer during the previous 12-month period, or for an airline flight crew employee, in the previous 12 months, having worked or been paid for not less than 60 percent of the applicable total monthly guarantee and having worked or been paid for not less than 504 hours, not counting personal commute time, or vacation, medical or sick leave (*see* Section 825.801(b)), *except that:*

(i) An employee returning from fulfilling his or her USERRA-covered service obligation shall be credited with the hours of service that would have been performed but for the period of absence from work due to or necessitated by USERRA-covered service in determining whether the employee met the hours of service requirement (accordingly, a person reemployed following absence from work due to or necessitated by USERRA-covered service has the hours that would have been worked for the employer (or, for an airline flight crew employee, would have been worked for or paid by the employer) added to any hours actually worked (or, for an airline flight crew employee, actually worked or paid) during the previous 12-month period to meet the hours of service requirement); and

(ii) To determine the hours that would have been worked (or, for an airline flight crew employee, would have been worked or paid) during the period of absence from work due to or necessitated by USERRA-covered service, the employee's pre-service work schedule can generally be used for calculations; and

(3) Who is employed in any State of the United States, the District of Columbia or any Territories or possession of the United States.

(4) Excludes any Federal officer or employee covered under subchapter V of chapter 63 of title 5, United States Code.

(5) Excludes any employee of the United States House of Representatives or the United States Senate covered by the Congressional Accountability Act of 1995, 2 U.S.C. 1301.

(6) Excludes any employee who is employed at a worksite at which the employer employs fewer than 50 employees if the total number of employees employed by that employer within 75 miles of that worksite is also fewer than 50.

(7) Excludes any employee employed in any country other than the United States or any Territory or possession of the United States.

Employ means to suffer or permit to work.

Employee has the meaning given the same term as defined in section 3(e) of the Fair Labor Standards Act, 29 U.S.C. 203(e), as follows:

(1) The term *employee* means any individual employed by an employer;

(2) In the case of an individual employed by a public agency, *employee* means—

(i) Any individual employed by the Government of the United States—

(A) As a civilian in the military departments (as defined in section 102 of Title 5, United States Code),

(B) In any executive agency (as defined in section 105 of Title 5, United States Code), excluding any Federal officer or employee covered under subchapter V of chapter 63 of Title 5, United States Code,

(C) In any unit of the legislative or judicial branch of the Government which has positions in the competitive service, excluding any employee of the United States House of Representatives or the United States Senate who is covered by the Congressional Accountability Act of 1995,

(D) In a nonappropriated fund instrumentality under the jurisdiction of the Armed Forces, or

(ii) Any individual employed by the United States Postal Service or the Postal Regulatory Commission; and

(iii) Any individual employed by a State, political subdivision of a State, or an interstate governmental agency, other than such an individual—

(A) Who is not subject to the civil service laws of the State, political subdivision, or agency which employs the employee; and

(B) Who—

(1) Holds a public elective office of that State, political subdivision, or agency,

(2) Is selected by the holder of such an office to be a member of his personal staff,

(3) Is appointed by such an officeholder to serve on a policymaking level,

(4) Is an immediate adviser to such an officeholder with respect to the constitutional or legal powers of the office of such officeholder, or

(5) Is an employee in the legislative branch or legislative body of that State, political subdivision, or agency and is not employed by the legislative library of such State, political subdivision, or agency.

Employee employed in an instructional capacity. See the definition of *Teacher* in this section.

Employer means any person engaged in commerce or in an industry or activity affecting commerce who employs 50 or more employees for each working day during each of 20 or more calendar workweeks in the current or preceding calendar year, and includes—

(1) Any person who acts, directly or indirectly, in the interest of an employer to any of the employees of such employer;

(2) Any successor in interest of an employer; and

(3) Any public agency.

Employment benefits means all benefits provided or made available to employees by an employer, including group life insurance, health insurance, disability insurance, sick leave, annual leave, educational benefits, and pensions, regardless of whether such benefits are provided by a practice or written policy of an employer or through an employee benefit plan as defined in section 3(3) of the Employee Retirement Income Security Act of 1974, 29 U.S.C. 1002(3). The term does not include non-employment related obligations paid by employees through voluntary deductions such as supplemental insurance coverage. *See also* Section 825.209(a).

FLSA means the Fair Labor Standards Act (29 U.S.C. 201 *et seq.*).

Group health plan means any plan of, or contributed to by, an employer (including a self-insured plan) to provide health care (directly or otherwise) to the employer's employees, former employees, or the families of such employees or former employees. For purposes of FMLA the term group health plan shall not include an insurance program providing health coverage under which employees purchase individual policies from insurers provided that:

(1) No contributions are made by the employer;

(2) Participation in the program is completely voluntary for employees;

(3) The sole functions of the employer with respect to the program are, without endorsing the program, to permit the insurer to publicize the program to employees, to collect premiums through payroll deductions and to remit them to the insurer;

(4) The employer receives no consideration in the form of cash or otherwise in connection with the program, other than reasonable compensation, excluding any profit, for administrative services actually rendered in connection with payroll deduction; and,

(5) The premium charged with respect to such coverage does not increase in the event the employment relationship terminates.

Health care provider means:

(1) The Act defines health care provider as:

(i) A doctor of medicine or osteopathy who is authorized to practice medicine or surgery (as appropriate) by the State in which the doctor practices; or

(ii) Any other person determined by the Secretary to be capable of providing health care services.

(2) Others "capable of providing health care services" include only:

(i) Podiatrists, dentists, clinical psychologists, optometrists, and chiropractors (limited to treatment consisting of manual manipulation of the spine to correct a subluxation as demonstrated by X-ray to exist) authorized to practice in the State and performing within the scope of their practice as defined under State law;

(ii) Nurse practitioners, nurse-midwives, clinical social workers and physician assistants who are authorized to practice under State law and who are performing within the scope of their practice as defined under State law;

(iii) Christian Science Practitioners listed with the First Church of Christ, Scientist in Boston, Massachusetts. Where an employee or family member is receiving treatment from a Christian Science practitioner, an employee may not object to any requirement from an employer that the employee or family member submit to examination (though not treatment) to obtain a second or third certification from a health care provider other than a Christian Science practitioner except as otherwise provided under applicable State or local law or collective bargaining agreement.

(iv) Any health care provider from whom an employer or the employer's group health plan's benefits manager will accept certification of the existence of a serious health condition to substantiate a claim for benefits; and

(v) A health care provider listed above who practices in a country other than the United States, who is authorized to practice in accordance with the law of that country, and who is performing within the scope of his or her practice as defined under such law.

(3) The phrase "authorized to practice in the State" as used in this section means that the provider must be authorized to diagnose and treat physical or mental health conditions.

Incapable of self-care means that the individual requires active assistance or supervision to provide daily self-care in several of the "activities of daily living" (ADLs) or "instrumental activities of daily living" (IADLs). Activities of daily living include adaptive activities such as caring appropriately for one's grooming and hygiene, bathing, dressing and eating. Instrumental activities of daily living include cooking, cleaning, shopping, taking public transportation, paying bills, maintaining a residence, using telephones and directories, using a post office, etc.

Instructional employee: See the definition of *Teacher* in this section.

Intermittent leave means leave taken in separate periods of time due to a single illness or injury, rather than for one continuous period of time, and may include leave of periods from an hour or more to several weeks. Examples of intermittent leave would include leave taken on an occasional basis for medical appointments, or leave taken several days at a time spread over a period of six months, such as for chemotherapy.

Invitational travel authorization (ITA) or *Invitational travel order* (ITO) are orders issued by the Armed Forces to a family member to join an injured or ill servicemember at his or her bedside. *See also* Section 825.310(e).

Key employee means a salaried FMLA-eligible employee who is among the highest paid 10 percent of all the employees employed by the employer within 75 miles of the employee's worksite. *See also* Section 825.217.

Mental disability: See the definition of *Physical or mental disability* in this section.

Military caregiver leave means leave taken to care for a covered servicemember with a serious injury or illness under the Family and Medical Leave Act of 1993. *See also* Section 825.127.

Next of kin of a covered servicemember means the nearest blood relative other than the covered servicemember's spouse, parent, son, or daughter, in the following order of priority: blood relatives who have been granted legal custody of the covered servicemember by court decree or statutory provisions, brothers and sisters, grandparents, aunts and uncles, and first cousins, unless the covered servicemember has specifically designated in writing another blood relative as his or her nearest blood relative for purposes of military caregiver leave under the FMLA. When no such designation is made, and there are multiple family members with the same level of relationship to the covered servicemember, all such family members shall be considered the covered servicemember's next of kin and may take FMLA leave to provide care to the covered servicemember, either consecutively or simultaneously. When such designation has been made, the designated individual shall be deemed to be the covered servicemember's only next of kin. *See also* Section 825.127(d)(3).

Outpatient status means, with respect to a covered servicemember who is a current member of the Armed Forces, the status of a member of the Armed Forces assigned to either a military medical treatment facility as an outpatient; or a unit established for the purpose of providing command and control of members of the Armed Forces receiving medical care as outpatients. *See also* Section 825.127(b)(1).

Parent means a biological, adoptive, step or foster father or mother, or any other individual who stood in loco parentis to the employee when the employee was a son or daughter as defined below. This term does not include parents "in law."

Parent of a covered servicemember means a covered servicemember's biological, adoptive, step or foster father or mother, or any other individual who stood in loco parentis to the covered servicemember. This term does not include parents "in law." *See also* Section 825.127(d)(2).

Person means an individual, partnership, association, corporation, business trust, legal representative, or any organized group of persons, and includes a public agency for purposes of this part.

Physical or mental disability means a physical or mental impairment that substantially limits one or more of the major life activities of an individual. Regulations at 29 CFR part 1630, issued by the Equal Employment Opportunity Commission under the Americans with Disabilities Act (ADA), 42 U.S.C. 12101 *et seq.,* as amended, define these terms.

Public agency means the government of the United States; the government of a State or political subdivision thereof; any agency of the United States (including the United States Postal Service and Postal Regulatory Commission), a State, or a political subdivision of a State, or any interstate governmental agency. Under section 101(5)(B) of the Act, a public agency is considered to be a "person" engaged in commerce or in an industry or activity affecting commerce within the meaning of the Act.

Reduced leave schedule means a leave schedule that reduces the usual number of hours per workweek, or hours per workday, of an employee.

Reserve components of the Armed Forces, for purposes of qualifying exigency leave, include the Army National Guard of the United States, Army Reserve, Navy Reserve, Marine Corps Reserve, Air National Guard of the United States, Air Force Reserve, and Coast Guard Reserve, and retired members of the Regular Armed Forces or Reserves who are called up in support of a contingency operation. *See also* Section 825.126(a)(2)(i).

Secretary means the Secretary of Labor or authorized representative.

Serious health condition means an illness, injury, impairment or physical or mental condition that involves inpatient care as defined in Section 825.114 or continuing treatment by a health care provider as defined in Section 825.115. Conditions for which cosmetic treatments are administered (such as most treatments for acne or plastic surgery) are not serious health conditions unless inpatient hospital care is required or unless complications develop. Restorative dental or plastic surgery after an injury or removal of cancerous growths are serious health conditions provided all the other conditions of this regulation are met. Mental illness or allergies may be serious health conditions, but only if all the conditions of Section 825.113 are met.

Serious injury or illness means: (1) In the case of a current member of the Armed Forces, including a member of the National Guard or Reserves, an injury or illness that was incurred by the covered servicemember in the line of duty on active duty in the Armed Forces or that existed before the beginning of the member's active duty and was aggravated by service in the line of duty on active duty in the Armed Forces and that may render the servicemember medically unfit to perform the duties of the member's office, grade, rank, or rating; and

(2) In the case of a covered veteran, an injury or illness that was incurred by the member in the line of duty on active duty in the Armed Forces (or existed before the beginning of the member's active duty and was aggravated by service in the line of duty on active duty in the Armed Forces) and manifested itself before or after the member became a veteran, and is:

(i) A continuation of a serious injury or illness that was incurred or aggravated when the covered veteran was a member of the Armed Forces and rendered the servicemember unable to perform the duties of the servicemember's office, grade, rank, or rating; or

(ii) A physical or mental condition for which the covered veteran has received a U.S. Department of Veterans Affairs Service-Related Disability Rating (VASRD) of 50 percent or greater, and such VASRD rating is based, in whole or in part, on the condition precipitating the need for military caregiver leave; or

(iii) A physical or mental condition that substantially impairs the covered veteran's ability to secure or follow a substantially gainful occupation by reason of a disability or disabilities related to military service, or would do so absent treatment; or

(iv) An injury, including a psychological injury, on the basis of which the covered veteran has been enrolled in the Department of Veterans Affairs Program of Comprehensive Assistance for Family Caregivers. *See also* Section 825.127(c).

Son or daughter means a biological, adopted, or foster child, a stepchild, a legal ward, or a child of a person standing in loco parentis, who is either under age 18, or age 18 or older and "incapable of self-care because of a mental or physical disability" at the time that FMLA leave is to commence.

Son or daughter of a covered servicemember means a covered servicemember's biological, adopted, or foster child, stepchild, legal ward, or a child for whom the covered servicemember stood in loco parentis, and who is of any age. *See also* Section 825.127(d)(1).

Son or daughter on covered active duty or call to covered active duty status means the employee's biological, adopted, or foster child, stepchild, legal ward, or a child for whom the employee stood in loco parentis, who is on covered active duty or call to covered active duty status, and who is of any age. *See also* Section 825.126(a)(5).

Spouse, as defined in the statute, means a husband or wife. For purposes of this definition, husband or wife refers to the other person with whom an individual entered into marriage as defined or recognized under state law for purposes of marriage in the State in which the marriage was entered into or, in the case of a marriage entered into outside of any State, if the marriage is valid in the place where entered into and could have been entered into in at least one State. This definition includes an individual in a same-sex or common law marriage that either:

(1) Was entered into in a State that recognizes such marriages; or

(2) If entered into outside of any State, is valid in the place where entered into and could have been entered into in at least one State.

State means any State of the United States or the District of Columbia or any Territory or possession of the United States.

Teacher (or employee employed in an instructional capacity, or instructional employee) means an employee employed principally in an instructional capacity by an educational agency or school whose principal function is to teach and instruct students in a class, a small group, or an individual setting, and includes athletic coaches, driving instructors, and special education assistants such as signers for the hearing impaired. The term does not include teacher assistants or aides who do not have as their principal function actual teaching or instructing, nor auxiliary personnel such as counselors, psychologists, curriculum specialists, cafeteria workers, maintenance workers, bus drivers, or other primarily noninstructional employees.

TRICARE is the health care program serving active duty servicemembers, National Guard and Reserve members, retirees, their families, survivors, and certain former spouses worldwide.

[78 FR 8902, Feb. 6, 2013, as amended at 80 FR 10000, Feb. 25, 2015]

Section 825.103 [Reserved]

Section 825.104 Covered employer

(a) An employer covered by FMLA is any person engaged in commerce or in any industry or activity affecting commerce, who employs 50 or more employees for each working day during each of 20 or more calendar workweeks in the current or preceding calendar year. Employers covered by FMLA also include any person acting, directly or indirectly, in the interest of a covered employer to any of the employees of the employer, any successor in interest of a covered employer, and any public agency. Public agencies are covered employers without regard to the number of employees employed. Public as well as private elementary and secondary schools are also covered employers without regard to the number of employees employed. *See* Section 825.600.

(b) The terms commerce and industry affecting commerce are defined in accordance with section 501(1) and (3) of the Labor Management Relations Act of 1947 (LMRA) (29 U.S.C. 142(1) and (3)), as set forth in the definitions at Section 825.102 of this part. For purposes of the FMLA, employers who meet the 50-employee coverage test are deemed to be engaged in commerce or in an industry or activity affecting commerce.

(c) Normally the legal entity which employs the employee is the employer under FMLA. Applying this principle, a corporation is a single employer rather than its separate establishments or divisions.

(1) Where one corporation has an ownership interest in another corporation, it is a separate employer unless it meets the joint employment test discussed in Section 825.106, or the integrated employer test contained in paragraph (c)(2) of this section.

(2) Separate entities will be deemed to be parts of a single employer for purposes of FMLA if they meet the integrated employer test. Where this test is met, the employees of all entities making up the integrated employer will be counted in determining employer coverage and employee eligibility. A determination of whether or not separate entities are an integrated employer is not determined by the application of any single criterion, but rather the entire relationship is to be reviewed in its totality. Factors considered in determining whether two or more entities are an integrated employer include:

(i) Common management;

(ii) Interrelation between operations;

(iii) Centralized control of labor relations; and

(iv) Degree of common ownership/financial control.

(d) An employer includes any person who acts directly or indirectly in the interest of an employer to any of the employer's employees. The definition of employer in section 3(d) of the Fair Labor Standards Act (FLSA), 29 U.S.C. 203(d), similarly includes any person acting directly or indirectly in the interest of an employer in relation to an employee. As under the FLSA, individuals such as corporate officers "acting in the interest of an employer" are individually liable for any violations of the requirements of FMLA.

[78 FR 8902, Feb. 6, 2013, as amended at 82 FR 2230, Jan. 9, 2017]

Section 825.105 Counting employees for determining coverage

(a) The definition of employ for purposes of FMLA is taken from the Fair Labor Standards Act, Section 3(g), 29 U.S.C. 203(g). The courts have made it clear that the employment relationship under the FLSA is broader than the traditional common law concept of master and servant. The difference between the employment relationship under the FLSA and that under the common law arises from the fact that the term "employ" as defined in the Act includes "to suffer or permit to work." The courts have indicated that, while "to permit" requires a more positive action than "to suffer," both terms imply much less positive action than required by the common law. Mere knowledge by an employer of work done for the employer by another is sufficient to create the employment relationship under the Act. The courts have said that there is no definition that solves all problems as to the limitations of the employer-employee relationship under the Act; and that determination of the relation cannot be based on isolated factors or upon a single characteristic or technical concepts, but depends "upon the circumstances of the whole activity" including the underlying "economic reality." In general an employee, as distinguished from an independent contractor who is engaged in a business of his/her own, is one who "follows the usual path of an employee" and is dependent on the business which he/she serves.

(b) Any employee whose name appears on the employer's payroll will be considered employed each working day of the calendar week, and must be counted whether or not any compensation is received for the week. However, the FMLA applies only to employees who are employed within any State of the United States, the District of Columbia or any Territory or possession of the United States. Employees who are employed outside these areas are not counted for purposes of determining employer coverage or employee eligibility.

(c) Employees on paid or unpaid leave, including FMLA leave, leaves of absence, disciplinary suspension, etc., are counted as long as the employer has a reasonable expectation that the employee will later return to active employment. If there is no employer/employee relationship (as when an employee is laid off, whether temporarily or permanently) such individual is not counted. Part-time employees, like full-time employees, are considered to be employed each working day of the calendar week, as long as they are maintained on the payroll.

(d) An employee who does not begin to work for an employer until after the first working day of a calendar week, or who terminates employment before the last working day of a calendar week, is not considered employed on each working day of that calendar week.

(e) A private employer is covered if it maintained 50 or more employees on the payroll during 20 or more calendar workweeks (not necessarily consecutive workweeks) in either the current or the preceding calendar year.

(f) Oncc a private employer meets the 50 employees/20 workweeks threshold, the employer remains covered until it reaches a future point where it no longer has employed 50 employees for 20 (nonconsecutive) workweeks in the current and preceding calendar year. For example, if an employer who met the 50 employees/20 workweeks test in the calendar year as of September 1, 2008, subsequently dropped below 50 employees before the end of 2008 and continued to employ fewer than 50 employees in all workweeks throughout calendar year 2009, the employer would continue to be covered throughout calendar year 2009 because it met the coverage criteria for 20 workweeks of the preceding (*i.e.*, 2008) calendar year.

Section 825.106 Joint employer coverage

(a) Where two or more businesses exercise some control over the work or working conditions of the employee, the businesses may be joint employers under FMLA. Joint employers may be separate and distinct entities with separate owners, managers, and facilities. Where the employee performs work which simultaneously benefits two or more employers, or works for two or more employers at different times during the workweek, a joint employment relationship generally will be considered to exist in situations such as:

(1) Where there is an arrangement between employers to share an employee's services or to interchange employees;

(2) Where one employer acts directly or indirectly in the interest of the other employer in relation to the employee; or,

(3) Where the employers are not completely disassociated with respect to the employee's employment and may be deemed to share control of the employee, directly or indirectly, because one employer controls, is controlled by, or is under common control with the other employer.

(b)(1) A determination of whether or not a joint employment relationship exists is not determined by the application of any single criterion, but rather the entire relationship is to be viewed in its totality. For example, joint employment will ordinarily be found to exist when a temporary placement agency supplies employees to a second employer.

(2) A type of company that is often called a Professional Employer Organization (PEO) contracts with client employers to perform administrative functions such as payroll, benefits, regulatory paperwork, and updating employment policies. The determination of whether a PEO is a joint employer also turns on the economic realities of the situation and must be based upon all the facts and circumstances. A PEO does not enter into a joint employment relationship with the employees of its client companies when it merely performs such administrative functions. On the other hand, if in a particular fact situation, a PEO has the right to hire, fire, assign, or direct and control the client's employees, or benefits from the work that the employees perform, such rights may lead to a determination that the PEO would be a joint employer with the client employer, depending upon all the facts and circumstances.

(c) In joint employment relationships, only the primary employer is responsible for giving required notices to its employees, providing FMLA leave, and maintenance of health benefits. Factors considered in determining which is the primary employer include authority/responsibility to hire and fire, assign/place the employee, make payroll, and provide employment benefits. For employees of temporary placement agencies, for example, the placement agency most commonly would be the primary employer. Where a PEO is a joint employer, the client employer most commonly would be the primary employer.

(d) Employees jointly employed by two employers must be counted by both employers, whether or not maintained on one of the employer's payroll, in determining employer coverage and employee eligibility. For example, an employer who jointly employs 15 workers from a temporary placement agency and 40 permanent workers is covered by FMLA. (A special rule applies to employees jointly employed who physically work at a facility of the secondary employer for a period of at least one year. *See* Section 825.111(a)(3).) An employee on leave who is working for a secondary employer is considered employed by the secondary employer, and must be counted for coverage and eligibility purposes, as long as the employer has a reasonable expectation that that employee will return to employment with that employer. In those cases in which a PEO is determined to be a joint employer of a client employer's employees, the client employer would only be required to count employees of the PEO (or employees of other clients of the PEO) if the client employer jointly employed those employees.

(e) Job restoration is the primary responsibility of the primary employer. The secondary employer is responsible for accepting the employee returning from FMLA leave in place of the replacement employee if the secondary employer continues to utilize an employee from the temporary placement agency, and the agency chooses to place the employee with the secondary employer. A secondary employer is also responsible for compliance with the prohibited acts provisions with respect to its jointly employed employees, whether or not the secondary employer is covered by FMLA. *See* Section 825.220(a). The prohibited acts include prohibitions against interfering with an employee's attempt to exercise rights under the Act, or discharging or discriminating against an employee for opposing a practice which is unlawful under FMLA. A covered secondary employer will be responsible for compliance with all the provisions of the FMLA with respect to its regular, permanent workforce.

Section 825.107 Successor in interest coverage

(a) For purposes of FMLA, in determining whether an employer is covered because it is a "successor in interest" to a covered employer, the factors used under Title VII of the Civil Rights Act and the Vietnam Era Veterans' Adjustment Act will be considered. However, unlike Title VII, whether the successor has notice of the employee's claim is not a consideration. Notice may be relevant, however, in determining successor liability for violations of the predecessor. The factors to be considered include:

(1) Substantial continuity of the same business operations;

(2) Use of the same plant;

(3) Continuity of the work force;

(4) Similarity of jobs and working conditions;

(5) Similarity of supervisory personnel;

(6) Similarity in machinery, equipment, and production methods;

(7) Similarity of products or services; and

(8) The ability of the predecessor to provide relief.

(b) A determination of whether or not a successor in interest exists is not determined by the application of any single criterion, but rather the entire circumstances are to be viewed in their totality.

(c) When an employer is a successor in interest, employees' entitlements are the same as if the employment by the predecessor and successor were continuous employment by a single employer. For example, the successor, whether or not it meets FMLA coverage criteria, must grant leave for eligible employees who had provided appropriate notice to the predecessor, or continue leave begun while employed by the predecessor, including maintenance of group health benefits during the leave and job restoration at the conclusion of the leave. A successor which meets FMLA's coverage criteria must count periods of employment and hours of service with the predecessor for purposes of determining employee eligibility for FMLA leave.

Section 825.108 Public agency coverage

(a) An employer under FMLA includes any public agency, as defined in section 3(x) of the Fair Labor Standards Act, 29 U.S.C. 203(x). Section 3(x) of the FLSA defines *public agency* as the government of the United States; the government of a

State or political subdivision of a State; or an agency of the United States, a State, or a political subdivision of a State, or any interstate governmental agency. *State* is further defined in Section 3(c) of the FLSA to include any State of the United States, the District of Columbia, or any Territory or possession of the United States.

(b) The determination of whether an entity is a public agency, as distinguished from a private employer, is determined by whether the agency has taxing authority, or whether the chief administrative officer or board, etc., is elected by the voters-at-large or their appointment is subject to approval by an elected official.

(c)(1) A State or a political subdivision of a State constitutes a single public agency and, therefore, a single employer for purposes of determining employee eligibility. For example, a State is a single employer; a county is a single employer; a city or town is a single employer. Whether two agencies of the same State or local government constitute the same public agency can only be determined on a case-by-case basis. One factor that would support a conclusion that two agencies are separate is whether they are treated separately for statistical purposes in the Census of Governments issued by the Bureau of the Census, U.S. Department of Commerce.

(2) The Census Bureau takes a census of governments at five-year intervals. Volume I, Government Organization, contains the official counts of the number of State and local governments. It includes tabulations of governments by State, type of government, size, and county location. Also produced is a universe list of governmental units, classified according to type of government. Copies of Volume I, Government Organization, and subsequent volumes are available from the Superintendent of Documents, U.S. Government Printing Office, Washington, DC 20402, U.S. Department of Commerce District Offices, or can be found in Regional and selective depository libraries, or online at *http://www.census.gov/govs/www/index.html.* For a list of all depository libraries, write to the Government Printing Office, 710 N. Capitol St. NW., Washington, DC 20402.

(d) All public agencies are covered by the FMLA regardless of the number of employees; they are not subject to the coverage threshold of 50 employees carried on the payroll each day for 20 or more weeks in a year. However, employees of public agencies must meet all of the requirements of eligibility, including the requirement that the employer (e.g., State) employ 50 employees at the worksite or within 75 miles.

Section 825.109 Federal agency coverage

(a) Most employees of the government of the United States, if they are covered by the FMLA, are covered under Title II of the FMLA (incorporated in Title V, Chapter 63, Subchapter 5 of the United States Code) which is administered by the U.S. Office of Personnel Management (OPM). OPM has separate regulations at 5 CFR Part 630, Subpart L. Employees of the Government Printing Office are covered by Title II. While employees of the Government Accountability Office and the Library of Congress are covered by Title I of the FMLA, the Comptroller General of the United States and the Librarian of Congress, respectively, have responsibility for the administration of the FMLA with respect to these employees. Other legislative branch employees, such as employees of the Senate and House of Representatives, are covered by the Congressional Accountability Act of 1995, 2 U.S.C. 1301.

(b) The Federal Executive Branch employees within the jurisdiction of these regulations include:

(1) Employees of the Postal Service;

(2) Employees of the Postal Regulatory Commission;

(3) A part-time employee who does not have an established regular tour of duty during the administrative workweek; and,

(4) An employee serving under an intermittent appointment or temporary appointment with a time limitation of one year or less.

(c) Employees of other Federal executive agencies are also covered by these regulations if they are not covered by Title II of FMLA.

(d) Employees of the judicial branch of the United States are covered by these regulations only if they are employed in a unit which has employees in the competitive service. For example, employees of the U.S. Tax Court are covered by these regulations.

(e) For employees covered by these regulations, the U.S. Government constitutes a single employer for purposes of determining employee eligibility. These employees must meet all of the requirements for eligibility, including the requirement that the Federal Government employ 50 employees at the worksite or within 75 miles.

Section 825.110 Eligible employee

(a) An eligible employee is an employee of a covered employer who:

(1) Has been employed by the employer for at least 12 months, and

(2) Has been employed for at least 1,250 hours of service during the 12-month period immediately preceding the commencement of the leave (*see* Section 825.801 for special hours of service requirements for airline flight crew employees), and

(3) Is employed at a worksite where 50 or more employees are employed by the employer within 75 miles of that worksite. *See* Section 825.105(b) regarding employees who work outside the U.S.

(b) The 12 months an employee must have been employed by the employer need not be consecutive months, *provided*

(1) Subject to the exceptions provided in paragraph (b)(2) of this section, employment periods prior to a break in service of seven years or more need not be counted

in determining whether the employee has been employed by the employer for at least 12 months.

(2) Employment periods preceding a break in service of more than seven years must be counted in determining whether the employee has been employed by the employer for at least 12 months where:

(i) The employee's break in service is occasioned by the fulfillment of his or her Uniformed Services Employment and Reemployment Rights Act (USERRA), 38 U.S.C. 4301, *et seq.,* covered service obligation. The period of absence from work due to or necessitated by USERRA-covered service must be also counted in determining whether the employee has been employed for at least 12 months by the employer. However, this section does not provide any greater entitlement to the employee than would be available under the USERRA; or

(ii) A written agreement, including a collective bargaining agreement, exists concerning the employer's intention to rehire the employee after the break in service (e.g., for purposes of the employee furthering his or her education or for childrearing purposes).

(3) If an employee is maintained on the payroll for any part of a week, including any periods of paid or unpaid leave (sick, vacation) during which other benefits or compensation are provided by the employer (e.g., workers' compensation, group health plan benefits, etc.), the week counts as a week of employment. For purposes of determining whether intermittent/occasional/casual employment qualifies as at least 12 months, 52 weeks is deemed to be equal to 12 months.

(4) Nothing in this section prevents employers from considering employment prior to a continuous break in service of more than seven years when determining whether an employee has met the 12-month employment requirement. However, if an employer chooses to recognize such prior employment, the employer must do so uniformly, with respect to all employees with similar breaks in service.

(c)(1) Except as provided in paragraph (c)(2) of this section and in Section 825.801 containing the special hours of service requirement for airline flight crew employees, whether an employee has worked the minimum 1,250 hours of service is determined according to the principles established under the Fair Labor Standards Act (FLSA) for determining compensable hours of work. *See* 29 CFR part 785. The determining factor is the number of hours an employee has worked for the employer within the meaning of the FLSA. The determination is not limited by methods of recordkeeping, or by compensation agreements that do not accurately reflect all of the hours an employee has worked for or been in service to the employer. Any accurate accounting of actual hours worked under FLSA's principles may be used.

(2) An employee returning from USERRA-covered service shall be credited with the hours of service that would have been performed but for the period of absence from work due to or necessitated by USERRA-covered service in determining the

employee's eligibility for FMLA-qualifying leave. Accordingly, a person reemployed following USERRA-covered service has the hours that would have been worked for the employer added to any hours actually worked during the previous 12-month period to meet the hours of service requirement. In order to determine the hours that would have been worked during the period of absence from work due to or necessitated by USERRA-covered service, the employee's pre-service work schedule can generally be used for calculations. *See* Section 825.801(c) for special rules applicable to airline flight crew employees.

(3) In the event an employer does not maintain an accurate record of hours worked by an employee, including for employees who are exempt from FLSA's requirement that a record be kept of their hours worked (e.g., bona fide executive, administrative, and professional employees as defined in FLSA Regulations, 29 CFR part 541), the employer has the burden of showing that the employee has not worked the requisite hours. An employer must be able to clearly demonstrate, for example, that full-time teachers (*see* Section 825.102 for definition) of an elementary or secondary school system, or institution of higher education, or other educational establishment or institution (who often work outside the classroom or at their homes) did not work 1,250 hours during the previous 12 months in order to claim that the teachers are not eligible for FMLA leave. *See* Section 825.801(d) for special rules applicable to airline flight crew employees.

(d) The determination of whether an employee meets the hours of service requirement and has been employed by the employer for a total of at least 12 months must be made as of the date the FMLA leave is to start. An employee may be on non-FMLA leave at the time he or she meets the 12-month eligibility requirement, and in that event, any portion of the leave taken for an FMLA-qualifying reason after the employee meets the eligibility requirement would be FMLA leave. *See* Section 825.300(b) for rules governing the content of the eligibility notice given to employees.

(e) Whether 50 employees are employed within 75 miles to ascertain an employee's eligibility for FMLA benefits is determined when the employee gives notice of the need for leave. Whether the leave is to be taken at one time or on an intermittent or reduced leave schedule basis, once an employee is determined eligible in response to that notice of the need for leave, the employee's eligibility is not affected by any subsequent change in the number of employees employed at or within 75 miles of the employee's worksite, for that specific notice of the need for leave. Similarly, an employer may not terminate employee leave that has already started if the employee count drops below 50. For example, if an employer employs 60 employees in August, but expects that the number of employees will drop to 40 in December, the employer must grant FMLA benefits to an otherwise eligible employee who gives notice of the need for leave in August for a period of leave to begin in December.

Section 825.111 Determining whether 50 employees are employed within 75 miles

(a) Generally, a worksite can refer to either a single location or a group of contiguous locations. Structures which form a campus or industrial park, or separate facilities in proximity with one another, may be considered a single site of employment. On the other hand, there may be several single sites of employment within a single building, such as an office building, if separate employers conduct activities within the building. For example, an office building with 50 different businesses as tenants will contain 50 sites of employment. The offices of each employer will be considered separate sites of employment for purposes of FMLA. An employee's worksite under FMLA will ordinarily be the site the employee reports to or, if none, from which the employee's work is assigned.

(1) Separate buildings or areas which are not directly connected or in immediate proximity are a single worksite if they are in reasonable geographic proximity, are used for the same purpose, and share the same staff and equipment. For example, if an employer manages a number of warehouses in a metropolitan area but regularly shifts or rotates the same employees from one building to another, the multiple warehouses would be a single worksite.

(2) For employees with no fixed worksite, e.g., construction workers, transportation workers (e.g., truck drivers, seamen, pilots), salespersons, etc., the worksite is the site to which they are assigned as their home base, from which their work is assigned, or to which they report. For example, if a construction company headquartered in New Jersey opened a construction site in Ohio, and set up a mobile trailer on the construction site as the company's on-site office, the construction site in Ohio would be the worksite for any employees hired locally who report to the mobile trailer/company office daily for work assignments, etc. If that construction company also sent personnel such as job superintendents, foremen, engineers, an office manager, etc., from New Jersey to the job site in Ohio, those workers sent from New Jersey continue to have the headquarters in New Jersey as their worksite. The workers who have New Jersey as their worksite would not be counted in determining eligibility of employees whose home base is the Ohio worksite, but would be counted in determining eligibility of employees whose home base is New Jersey. For transportation employees, their worksite is the terminal to which they are assigned, report for work, depart, and return after completion of a work assignment. For example, an airline pilot may work for an airline with headquarters in New York, but the pilot regularly reports for duty and originates or begins flights from the company's facilities located in an airport in Chicago and returns to Chicago at the completion of one or more flights to go off duty. The pilot's worksite is the facility in Chicago. An employee's personal residence is not a worksite in the case of employees, such as salespersons, who travel a sales territory and who generally leave to work and return from work to their personal residence, or employees who work at home, as under the concept of flexiplace or telecommuting. Rather, their worksite is the office to which they report and from which assignments are made.

(3) For purposes of determining that employee's eligibility, when an employee is jointly employed by two or more employers (*see* Section 825.106), the employee's worksite is the primary employer's office from which the employee is assigned or reports, unless the employee has physically worked for at least one year at a facility of a secondary employer, in which case the employee's worksite is that location. The

employee is also counted by the secondary employer to determine eligibility for the secondary employer's full-time or permanent employees.

(b) The 75-mile distance is measured by surface miles, using surface transportation over public streets, roads, highways and waterways, by the shortest route from the facility where the employee needing leave is employed. Absent available surface transportation between worksites, the distance is measured by using the most frequently utilized mode of transportation (e.g., airline miles).

(c) The determination of how many employees are employed within 75 miles of the worksite of an employee is based on the number of employees maintained on the payroll. Employees of educational institutions who are employed permanently or who are under contract are maintained on the payroll during any portion of the year when school is not in session. *See* Section 825.105(c).

Section 825.112 Qualifying reasons for leave, general rule

(a) *Circumstances qualifying for leave.* Employers covered by FMLA are required to grant leave to eligible employees:

(1) For birth of a son or daughter, and to care for the newborn child (*see* Section 825.120);

(2) For placement with the employee of a son or daughter for adoption or foster care (*see* Section 825.121);

(3) To care for the employee's spouse, son, daughter, or parent with a serious health condition (*see* Sections 825.113 and 825.122);

(4) Because of a serious health condition that makes the employee unable to perform the functions of the employee's job (*see* Sections 825.113 and 825.123);

(5) Because of any qualifying exigency arising out of the fact that the employee's spouse, son, daughter, or parent is a military member on covered active duty (or has been notified of an impending call or order to covered active duty status) (*see* Sections 825.122 and 825.126); and

(6) To care for a covered servicemember with a serious injury or illness if the employee is the spouse, son, daughter, parent, or next of kin of the covered servicemember. *See* Sections 825.122 and 825.127.

(b) *Equal application.* The right to take leave under FMLA applies equally to male and female employees. A father, as well as a mother, can take family leave for the birth, placement for adoption, or foster care of a child.

(c) *Active employee.* In situations where the employer/employee relationship has been interrupted, such as an employee who has been on layoff, the employee must be recalled or otherwise be re-employed before being eligible for FMLA leave. Under such circumstances, an eligible employee is immediately entitled to further FMLA leave for a qualifying reason.

Section 825.113 Serious health condition

(a) For purposes of FMLA, *serious health condition* entitling an employee to FMLA leave means an illness, injury, impairment or physical or mental condition that involves inpatient care as defined in Section 825.114 or continuing treatment by a health care provider as defined in Section 825.115.

(b) The term *incapacity* means inability to work, attend school or perform other regular daily activities due to the serious health condition, treatment therefore, or recovery therefrom.

(c) The term treatment includes (but is not limited to) examinations to determine if a serious health condition exists and evaluations of the condition. Treatment does not include routine physical examinations, eye examinations, or dental examinations. A regimen of continuing treatment includes, for example, a course of prescription medication (e.g., an antibiotic) or therapy requiring special equipment to resolve or alleviate the health condition (e.g., oxygen). A regimen of continuing treatment that includes the taking of over-the-counter medications such as aspirin, antihistamines, or salves; or bed-rest, drinking fluids, exercise, and other similar activities that can be initiated without a visit to a health care provider, is not, by itself, sufficient to constitute a regimen of continuing treatment for purposes of FMLA leave.

(d) Conditions for which cosmetic treatments are administered (such as most treatments for acne or plastic surgery) are not serious health conditions unless inpatient hospital care is required or unless complications develop. Ordinarily, unless complications arise, the common cold, the flu, ear aches, upset stomach, minor ulcers, headaches other than migraine, routine dental or orthodontia problems, periodontal disease, etc., are examples of conditions that do not meet the definition of a serious health condition and do not qualify for FMLA leave. Restorative dental or plastic surgery after an injury or removal of cancerous growths are serious health conditions provided all the other conditions of this regulation are met. Mental illness or allergies may be serious health conditions, but only if all the conditions of this section are met.

Section 825.114 Inpatient care

Inpatient care means an overnight stay in a hospital, hospice, or residential medical care facility, including any period of incapacity as defined in Section 825.113(b), or any subsequent treatment in connection with such inpatient care.

Section 825.115 Continuing treatment

A serious health condition involving continuing treatment by a health care provider includes any one or more of the following:

(a) *Incapacity and treatment.* A period of incapacity of more than three consecutive, full calendar days, and any subsequent treatment or period of incapacity relating to the same condition, that also involves:

(1) Treatment two or more times, within 30 days of the first day of incapacity, unless extenuating circumstances exist, by a health care provider, by a nurse under direct supervision of a health care provider, or by a provider of health care services (e.g., physical therapist) under orders of, or on referral by, a health care provider; or

(2) Treatment by a health care provider on at least one occasion, which results in a regimen of continuing treatment under the supervision of the health care provider.

(3) The requirement in paragraphs (a)(1) and (2) of this section for treatment by a health care provider means an in-person visit to a health care provider. The first (or only) in-person treatment visit must take place within seven days of the first day of incapacity.

(4) Whether additional treatment visits or a regimen of continuing treatment is necessary within the 30-day period shall be determined by the health care provider.

(5) The term *extenuating circumstances* in paragraph (a)(1) of this section means circumstances beyond the employee's control that prevent the follow-up visit from occurring as planned by the health care provider. Whether a given set of circumstances are extenuating depends on the facts. For example, extenuating circumstances exist if a health care provider determines that a second in-person visit is needed within the 30-day period, but the health care provider does not have any available appointments during that time period.

(b) *Pregnancy or prenatal care.* Any period of incapacity due to pregnancy, or for prenatal care. *See also* Section 825.120.

(c) *Chronic conditions.* Any period of incapacity or treatment for such incapacity due to a chronic serious health condition. A chronic serious health condition is one which:

(1) Requires periodic visits (defined as at least twice a year) for treatment by a health care provider, or by a nurse under direct supervision of a health care provider;

(2) Continues over an extended period of time (including recurring episodes of a single underlying condition); and

(3) May cause episodic rather than a continuing period of incapacity (e.g., asthma, diabetes, epilepsy, etc.).

(d) *Permanent or long-term conditions.* A period of incapacity which is permanent or long-term due to a condition for which treatment may not be effective. The employee or family member must be under the continuing supervision of, but need not be receiving active treatment by, a health care provider. Examples include Alzheimer's, a severe stroke, or the terminal stages of a disease.

(e) *Conditions requiring multiple treatments.* Any period of absence to receive multiple treatments (including any period of recovery therefrom) by a health care provider or by a provider of health care services under orders of, or on referral by, a health care provider, for:

(1) Restorative surgery after an accident or other injury; or

(2) A condition that would likely result in a period of incapacity of more than three consecutive, full calendar days in the absence of medical intervention or treatment, such as cancer (chemotherapy, radiation, etc.), severe arthritis (physical therapy), or kidney disease (dialysis).

(f) Absences attributable to incapacity under paragraph (b) or (c) of this section qualify for FMLA leave even though the employee or the covered family member does not receive treatment from a health care provider during the absence, and even if the absence does not last more than three consecutive, full calendar days. For example, an employee with asthma may be unable to report for work due to the onset of an asthma attack or because the employee's health care provider has advised the employee to stay home when the pollen count exceeds a certain level. An employee who is pregnant may be unable to report to work because of severe morning sickness.

Sections 825.116-825.118 [Reserved]

Section 825.119 Leave for treatment of substance abuse

(a) Substance abuse may be a serious health condition if the conditions of Sections 825.113 through 825.115 are met. However, FMLA leave may only be taken for treatment for substance abuse by a health care provider or by a provider of health care services on referral by a health care provider. On the other hand, absence because of the employee's use of the substance, rather than for treatment, does not qualify for FMLA leave.

(b) Treatment for substance abuse does not prevent an employer from taking employment action against an employee. The employer may not take action against the employee because the employee has exercised his or her right to take FMLA leave for treatment. However, if the employer has an established policy, applied in a non-discriminatory manner that has been communicated to all employees, that provides under certain circumstances an employee may be terminated for substance abuse, pursuant to that policy the employee may be terminated whether or not the employee is presently taking FMLA leave. An employee may also take FMLA leave to care for a covered family member who is receiving treatment for substance abuse. The employer may not take action against an employee who is providing care for a covered family member receiving treatment for substance abuse.

Section 825.120 Leave for pregnancy or birth

(a) *General rules.* Eligible employees are entitled to FMLA leave for pregnancy or birth of a child as follows:

(1) Both parents are entitled to FMLA leave for the birth of their child.

(2) Both parents are entitled to FMLA leave to be with the healthy newborn child (*i.e.,* bonding time) during the 12-month period beginning on the date of birth. An employee's entitlement to FMLA leave for a birth expires at the end of the 12-month period beginning on the date of the birth. If state law allows, or the employer permits, bonding leave to be taken beyond this period, such leave will not qualify as FMLA leave. *See* Section 825.701 regarding non-FMLA leave which may be available under applicable State laws. Under this section, both parents are entitled to FMLA leave even if the newborn does not have a serious health condition.

(3) Spouses who are eligible for FMLA leave and are employed by the same covered employer may be limited to a combined total of 12 weeks of leave during any 12-month period if the leave is taken for birth of the employee's son or daughter or to care for the child after birth, for placement of a son or daughter with the employee for adoption or foster care or to care for the child after placement, or to care for the employee's parent with a serious health condition. This limitation on the total weeks of leave applies to leave taken for the reasons specified as long as the spouses are employed by the same employer. It would apply, for example, even though the spouses are employed at two different worksites of an employer located more than 75 miles from each other, or by two different operating divisions of the same company. On the other hand, if one spouse is ineligible for FMLA leave, the other spouse would be entitled to a full 12 weeks of FMLA leave. Where spouses both use a portion of the total 12-week FMLA leave entitlement for either the birth of a child, for placement for adoption or foster care, or to care for a parent, the spouses would each be entitled to the difference between the amount he or she has taken individually and 12 weeks for FMLA leave for other purposes. For example, if each spouse took six weeks of leave to care for a healthy, newborn child, each could use an additional six weeks due to his or her own serious health condition or to care for a child with a serious health condition. Note, too, that many state pregnancy disability laws specify a period of disability either before or after the birth of a child; such periods would also be considered FMLA leave for a serious health condition of the birth mother, and would not be subject to the combined limit.

(4) The expectant mother is entitled to FMLA leave for incapacity due to pregnancy, for prenatal care, or for her own serious health condition following the birth of the child. Circumstances may require that FMLA leave begin before the actual date of birth of a child. An expectant mother may take FMLA leave before the birth of the child for prenatal care or if her condition makes her unable to work. The expectant mother is entitled to leave for incapacity due to pregnancy even though she does not receive treatment from a health care provider during the absence, and even if the absence does not last for more than three consecutive calendar days. For example, a pregnant employee may be unable to report to work because of severe morning sickness.

(5) A spouse is entitled to FMLA leave if needed to care for a pregnant spouse who is incapacitated or if needed to care for her during her prenatal care, or if needed to

care for her following the birth of a child if she has a serious health condition. *See* Section 825.124.

(6) Both parents are entitled to FMLA leave if needed to care for a child with a serious health condition if the requirements of Sections 825.113 through 825.115 and 825.122(d) are met. Thus, spouses may each take 12 weeks of FMLA leave if needed to care for their newborn child with a serious health condition, even if both are employed by the same employer, provided they have not exhausted their entitlements during the applicable 12-month FMLA leave period.

(b) *Intermittent and reduced schedule leave.* An eligible employee may use intermittent or reduced schedule leave after the birth to be with a healthy newborn child only if the employer agrees. For example, an employer and employee may agree to a part-time work schedule after the birth. If the employer agrees to permit intermittent or reduced schedule leave for the birth of a child, the employer may require the employee to transfer temporarily, during the period the intermittent or reduced leave schedule is required, to an available alternative position for which the employee is qualified and which better accommodates recurring periods of leave than does the employee's regular position. Transfer to an alternative position may require compliance with any applicable collective bargaining agreement, Federal law (such as the Americans with Disabilities Act), and State law. Transfer to an alternative position may include altering an existing job to better accommodate the employee's need for intermittent or reduced leave. The employer's agreement is not required for intermittent leave required by the serious health condition of the expectant mother or newborn child. *See* Sections 825.202—825.205 for general rules governing the use of intermittent and reduced schedule leave. *See* Section 825.121 for rules governing leave for adoption or foster care. *See* Section 825.601 for special rules applicable to instructional employees of schools. *See* Section 825.802 for special rules applicable to airline flight crew employees.

[78 FR 8902, Feb. 6, 2013, as amended at 80 FR 10000, Feb. 25, 2015]

Section 825.121 Leave for adoption or foster care

(a) *General rules.* Eligible employees are entitled to FMLA leave for placement with the employee of a son or daughter for adoption or foster care as follows:

(1) Employees may take FMLA leave before the actual placement or adoption of a child if an absence from work is required for the placement for adoption or foster care to proceed. For example, the employee may be required to attend counseling sessions, appear in court, consult with his or her attorney or the doctor(s) representing the birth parent, submit to a physical examination, or travel to another country to complete an adoption. The source of an adopted child (e.g., whether from a licensed placement agency or otherwise) is not a factor in determining eligibility for leave for this purpose.

(2) An employee's entitlement to leave for adoption or foster care expires at the end of the 12-month period beginning on the date of the placement. If state law allows,

or the employer permits, leave for adoption or foster care to be taken beyond this period, such leave will not qualify as FMLA leave. *See* Section 825.701 regarding non-FMLA leave which may be available under applicable State laws. Under this section, the employee is entitled to FMLA leave even if the adopted or foster child does not have a serious health condition.

(3) Spouses who are eligible for FMLA leave and are employed by the same covered employer may be limited to a combined total of 12 weeks of leave during any 12-month period if the leave is taken for the placement of the employee's son or daughter or to care for the child after placement, for the birth of the employee's son or daughter or to care for the child after birth, or to care for the employee's parent with a serious health condition. This limitation on the total weeks of leave applies to leave taken for the reasons specified as long as the spouses are employed by the same employer. It would apply, for example, even though the spouses are employed at two different worksites of an employer located more than 75 miles from each other, or by two different operating divisions of the same company. On the other hand, if one spouse is ineligible for FMLA leave, the other spouse would be entitled to a full 12 weeks of FMLA leave. Where spouses both use a portion of the total 12-week FMLA leave entitlement for either the birth of a child, for placement for adoption or foster care, or to care for a parent, the spouses would each be entitled to the difference between the amount he or she has taken individually and 12 weeks for FMLA leave for other purposes. For example, if each spouse took six weeks of leave to care for a healthy, newly placed child, each could use an additional six weeks due to his or her own serious health condition or to care for a child with a serious health condition.

(4) An eligible employee is entitled to FMLA leave in order to care for an adopted or foster child with a serious health condition if the requirements of Sections 825.113 through 825.115 and 825.122(d) are met. Thus, spouses may each take 12 weeks of FMLA leave if needed to care for an adopted or foster child with a serious health condition, even if both are employed by the same employer, provided they have not exhausted their entitlements during the applicable 12-month FMLA leave period.

(b) *Use of intermittent and reduced schedule leave.* An eligible employee may use intermittent or reduced schedule leave after the placement of a healthy child for adoption or foster care only if the employer agrees. Thus, for example, the employer and employee may agree to a part-time work schedule after the placement for bonding purposes. If the employer agrees to permit intermittent or reduced schedule leave for the placement for adoption or foster care, the employer may require the employee to transfer temporarily, during the period the intermittent or reduced leave schedule is required, to an available alternative position for which the employee is qualified and which better accommodates recurring periods of leave than does the employee's regular position. Transfer to an alternative position may require compliance with any applicable collective bargaining agreement, federal law (such as the Americans with Disabilities Act), and State law. Transfer to an alternative position may include altering an existing job to better accommodate the employee's need for intermittent or reduced leave. The employer's agreement is not required for intermittent leave required by the serious health condition of the adopted or foster child. *See*

Sections 825.202-825.205 for general rules governing the use of intermittent and reduced schedule leave. *See* Section 825.120 for general rules governing leave for pregnancy and birth of a child. *See* Section 825.601 for special rules applicable to instructional employees of schools. *See* Section 825.802 for special rules applicable to airline flight crew employees.

[78 FR 8902, Feb. 6, 2013, as amended at 80 FR 10000, Feb. 25, 2015]

Section 825.122 Definitions of covered servicemember, spouse, parent, son or daughter, next of kin of a covered servicemember, adoption, foster care, son or daughter on covered active duty or call to covered active duty status, son or daughter of a covered servicemember, and parent of a covered servicemember

(a) Covered servicemember means: (1) A current member of the Armed Forces, including a member of the National Guard or Reserves, who is undergoing medical treatment, recuperation or therapy, is otherwise in outpatient status, or is otherwise on the temporary disability retired list, for a serious injury or illness; or

(2) A covered veteran who is undergoing medical treatment, recuperation, or therapy for a serious injury or illness. *Covered veteran* means an individual who was a member of the Armed Forces (including a member of the National Guard or Reserves), and was discharged or released under conditions other than dishonorable at any time during the five-year period prior to the first date the eligible employee takes FMLA leave to care for the covered veteran. *See* Section 825.127(b)(2).

(b) *Spouse, as defined in the statute,* means a husband or wife. For purposes of this definition, husband or wife refers to the other person with whom an individual entered into marriage as defined or recognized under state law for purposes of marriage in the State in which the marriage was entered into or, in the case of a marriage entered into outside of any State, if the marriage is valid in the place where entered into and could have been entered into in at least one State. This definition includes an individual in a same-sex or common law marriage that either:

(1) Was entered into in a State that recognizes such marriages; or

(2) If entered into outside of any State, is valid in the place where entered into and could have been entered into in at least one State.

(c) *Parent.* Parent means a biological, adoptive, step or foster father or mother, or any other individual who stood in loco parentis to the employee when the employee was a son or daughter as defined in paragraph (d) of this section. This term does not include parents "in law."

(d) *Son or daughter.* For purposes of FMLA leave taken for birth or adoption, or to care for a family member with a serious health condition, son or daughter means a biological, adopted, or foster child, a stepchild, a legal ward, or a child of a person standing in loco parentis, who is either under age 18, or age 18 or older and "incapable of self-care because of a mental or physical disability" at the time that FMLA leave is to commence.

(1) *Incapable of self-care* means that the individual requires active assistance or supervision to provide daily self-care in three or more of the activities of daily living (ADLs) or instrumental activities of daily living (IADLs). Activities of daily living include adaptive activities such as caring appropriately for one's grooming and hygiene, bathing, dressing and eating. Instrumental activities of daily living include cooking, cleaning, shopping, taking public transportation, paying bills, maintaining a residence, using telephones and directories, using a post office, etc.

(2) *Physical or mental disability* means a physical or mental impairment that substantially limits one or more of the major life activities of an individual. Regulations at 29 CFR 1630.2(h), (i), and (j), issued by the Equal Employment Opportunity Commission under the Americans with Disabilities Act (ADA), 42 U.S.C. 12101 *et seq.,* define these terms.

(3) Persons who are "in loco parentis" include those with day-to-day responsibilities to care for and financially support a child, or, in the case of an employee, who had such responsibility for the employee when the employee was a child. A biological or legal relationship is not necessary.

(e) *Next of kin of a covered servicemember* means the nearest blood relative other than the covered servicemember's spouse, parent, son, or daughter, in the following order of priority: blood relatives who have been granted legal custody of the covered servicemember by court decree or statutory provisions, brothers and sisters, grandparents, aunts and uncles, and first cousins, unless the covered servicemember has specifically designated in writing another blood relative as his or her nearest blood relative for purposes of military caregiver leave under the FMLA. When no such designation is made, and there are multiple family members with the same level of relationship to the covered servicemember, all such family members shall be considered the covered servicemember's next of kin and may take FMLA leave to provide care to the covered servicemember, either consecutively or simultaneously. When such designation has been made, the designated individual shall be deemed to be the covered servicemember's only next of kin. *See* Section 825.127(d)(3).

(f) *Adoption* means legally and permanently assuming the responsibility of raising a child as one's own. The source of an adopted child (e.g., whether from a licensed placement agency or otherwise) is not a factor in determining eligibility for FMLA leave. *See* Section 825.121 for rules governing leave for adoption.

(g) *Foster care* means 24-hour care for children in substitution for, and away from, their parents or guardian. Such placement is made by or with the agreement of the State as a result of a voluntary agreement between the parent or guardian that the child be removed from the home, or pursuant to a judicial determination of the necessity for foster care, and involves agreement between the State and foster family that the foster family will take care of the child. Although foster care may be with relatives of the child, State action is involved in the removal of the child from parental custody. *See* Section 825.121 for rules governing leave for foster care.

(h) *Son or daughter on covered active duty or call to covered active duty status* means the employee's biological, adopted, or foster child, stepchild, legal ward, or a child for whom the employee stood in loco parentis, who is on covered active duty or call to covered active duty status, and who is of any age. *See* Section 825.126(a)(5).

(i) *Son or daughter of a covered servicemember* means the covered servicemember's biological, adopted, or foster child, stepchild, legal ward, or a child for whom the covered servicemember stood in loco parentis, and who is of any age. *See* Section 825.127(d)(1).

(j) *Parent of a covered servicemember* means a covered servicemember's biological, adoptive, step or foster father or mother, or any other individual who stood in loco parentis to the covered servicemember. This term does not include parents "in law." *See* Section 825.127(d)(2).

(k) *Documenting relationships.* For purposes of confirmation of family relationship, the employer may require the employee giving notice of the need for leave to provide reasonable documentation or statement of family relationship. This documentation may take the form of a simple statement from the employee, or a child's birth certificate, a court document, etc. The employer is entitled to examine documentation such as a birth certificate, etc., but the employee is entitled to the return of the official document submitted for this purpose.

[78 FR 8902, Feb. 6, 2013, as amended at 80 FR 10001, Feb. 25, 2015]

Section 825.123 Unable to perform the functions of the position

(a) *Definition.* An employee is unable to perform the functions of the position where the health care provider finds that the employee is unable to work at all or is unable to perform any one of the essential functions of the employee's position within the meaning of the Americans with Disabilities Act (ADA), as amended, 42 U.S.C. 12101 *et seq.*, and the regulations at 29 CFR 1630.2(n). An employee who must be absent from work to receive medical treatment for a serious health condition is considered to be unable to perform the essential functions of the position during the absence for treatment.

(b) *Statement of functions.* An employer has the option, in requiring certification from a health care provider, to provide a statement of the essential functions of the employee's position for the health care provider to review. A sufficient medical certification must specify what functions of the employee's position the employee is unable to perform so that the employer can then determine whether the employee is unable to perform one or more essential functions of the employee's position. For purposes of FMLA, the essential functions of the employee's position are to be determined with reference to the position the employee held at the time notice is given or leave commenced, whichever is earlier. *See* Section 825.306.

Section 825.124 Needed to care for a family member or covered servicemember

(a) The medical certification provision that an employee is needed to care for a family member or covered servicemember encompasses both physical and psychological care. It includes situations where, for example, because of a serious health condition, the family member is unable to care for his or her own basic medical, hygienic, or nutritional needs or safety, or is unable to transport himself or herself to the doctor. The term also includes providing psychological comfort and reassurance which would be beneficial to a child, spouse or parent with a serious health condition who is receiving inpatient or home care.

(b) The term also includes situations where the employee may be needed to substitute for others who normally care for the family member or covered servicemember, or to make arrangements for changes in care, such as transfer to a nursing home. The employee need not be the only individual or family member available to care for the family member or covered servicemember.

(c) An employee's intermittent leave or a reduced leave schedule necessary to care for a family member or covered servicemember includes not only a situation where the condition of the family member or covered servicemember itself is intermittent, but also where the employee is only needed intermittently—such as where other care is normally available, or care responsibilities are shared with another member of the family or a third party. *See* Sections 825.202-825.205 for rules governing the use of intermittent or reduced schedule leave.

Section 825.125 Definition of health care provider

(a) The Act defines *health care provider* as:

(1) A doctor of medicine or osteopathy who is authorized to practice medicine or surgery (as appropriate) by the State in which the doctor practices; or

(2) Any other person determined by the Secretary to be capable of providing health care services.

(b) Others capable of providing health care services include only:

(1) Podiatrists, dentists, clinical psychologists, optometrists, and chiropractors (limited to treatment consisting of manual manipulation of the spine to correct a subluxation as demonstrated by X-ray to exist) authorized to practice in the State and performing within the scope of their practice as defined under State law;

(2) Nurse practitioners, nurse-midwives, clinical social workers and physician assistants who are authorized to practice under State law and who are performing within the scope of their practice as defined under State law;

(3) Christian Science Practitioners listed with the First Church of Christ, Scientist in Boston, Massachusetts. Where an employee or family member is receiving treatment from a Christian Science practitioner, an employee may not object to any requirement from an employer that the employee or family member submit to examination (though not treatment) to obtain a second or third certification from a health care provider other than a Christian Science practitioner except as otherwise provided under applicable State or local law or collective bargaining agreement;

(4) Any health care provider from whom an employer or the employer's group health plan's benefits manager will accept certification of the existence of a serious health condition to substantiate a claim for benefits; and

(5) A health care provider listed above who practices in a country other than the United States, who is authorized to practice in accordance with the law of that country, and who is performing within the scope of his or her practice as defined under such law.

(c) The phrase authorized to practice in the State as used in this section means that the provider must be authorized to diagnose and treat physical or mental health conditions.

Section 825.126 Leave because of a qualifying exigency

(a) Eligible employees may take FMLA leave for a qualifying exigency while the employee's spouse, son, daughter, or parent (the military member or member) is on covered active duty or call to covered active duty status (or has been notified of an impending call or order to covered active duty).

(1) *Covered active duty or call to covered active duty status* in the case of a member of the Regular Armed Forces means duty during the deployment of the member with the Armed Forces to a foreign country. The active duty orders of a member of the Regular components of the Armed Forces will generally specify if the member is deployed to a foreign country.

(2) *Covered active duty or call to covered active duty status* in the case of a member of the Reserve components of the Armed Forces means duty during the deployment of the member with the Armed Forces to a foreign country under a Federal call or order to active duty in support of a contingency operation pursuant to: Section 688 of Title 10 of the United States Code, which authorizes ordering to active duty retired members of the Regular Armed Forces and members of the retired Reserve who retired after completing at least 20 years of active service; Section 12301(a) of Title 10 of the United States Code, which authorizes ordering all reserve component members to active duty in the case of war or national emergency; Section 12302 of Title 10 of the United States Code, which authorizes ordering any unit or unassigned member of the Ready Reserve to active duty; Section 12304 of Title 10 of the United States Code, which authorizes ordering any unit or unassigned member of the Selected Reserve and certain members of the Individual Ready Reserve to active

duty; Section 12305 of Title 10 of the United States Code, which authorizes the suspension of promotion, retirement or separation rules for certain Reserve components; Section 12406 of Title 10 of the United States Code, which authorizes calling the National Guard into Federal service in certain circumstances; chapter 15 of Title 10 of the United States Code, which authorizes calling the National Guard and state military into Federal service in the case of insurrections and national emergencies; or any other provision of law during a war or during a national emergency declared by the President or Congress so long as it is in support of a contingency operation. *See* 10 U.S.C. 101(a)(13)(B).

(i) For purposes of covered active duty or call to covered active duty status, the Reserve components of the Armed Forces include the Army National Guard of the United States, Army Reserve, Navy Reserve, Marine Corps Reserve, Air National Guard of the United States, Air Force Reserve and Coast Guard Reserve, and retired members of the Regular Armed Forces or Reserves who are called up in support of a contingency operation pursuant to one of the provisions of law identified in paragraph (a)(2).

(ii) The active duty orders of a member of the Reserve components will generally specify if the military member is serving in support of a contingency operation by citation to the relevant section of Title 10 of the United States Code and/or by reference to the specific name of the contingency operation and will specify that the deployment is to a foreign country.

(3) *Deployment of the member with the Armed Forces to a foreign country* means deployment to areas outside of the United States, the District of Columbia, or any Territory or possession of the United States, including international waters.

(4) A call to covered active duty for purposes of leave taken because of a qualifying exigency refers to a Federal call to active duty. State calls to active duty are not covered unless under order of the President of the United States pursuant to one of the provisions of law identified in paragraph (a)(2) of this section.

(5) *Son or daughter on covered active duty or call to covered active duty status* means the employee's biological, adopted, or foster child, stepchild, legal ward, or a child for whom the employee stood in loco parentis, who is on covered active duty or call to covered active duty status, and who is of any age.

(b) An eligible employee may take FMLA leave for one or more of the following qualifying exigencies:

(1) *Short-notice deployment.* (i) To address any issue that arises from the fact that the military member is notified of an impending call or order to covered active duty seven or less calendar days prior to the date of deployment;

(ii) Leave taken for this purpose can be used for a period of seven calendar days beginning on the date the military member is notified of an impending call or order to covered active duty;

(2) *Military events and related activities.* (i) To attend any official ceremony, program, or event sponsored by the military that is related to the covered active duty or call to covered active duty status of the military member; and

(ii) To attend family support or assistance programs and informational briefings sponsored or promoted by the military, military service organizations, or the American Red Cross that are related to the covered active duty or call to covered active duty status of the military member;

(3) *Childcare and school activities.* For the purposes of leave for childcare and school activities listed in (i) through (iv) of this paragraph, a child of the military member must be the military member's biological, adopted, or foster child, stepchild, legal ward, or child for whom the military member stands in loco parentis, who is either under 18 years of age or 18 years of age or older and incapable of self-care because of a mental or physical disability at the time that FMLA leave is to commence. As with all instances of qualifying exigency leave, the military member must be the spouse, son, daughter, or parent of the employee requesting qualifying exigency leave.

(i) To arrange for alternative childcare for a child of the military member when the covered active duty or call to covered active duty status of the military member necessitates a change in the existing childcare arrangement;

(ii) To provide childcare for a child of the military member on an urgent, immediate need basis (but not on a routine, regular, or everyday basis) when the need to provide such care arises from the covered active duty or call to covered active duty status of the military member;

(iii) To enroll in or transfer to a new school or day care facility a child of the military member when enrollment or transfer is necessitated by the covered active duty or call to covered active duty status of the military member; and

(iv) To attend meetings with staff at a school or a daycare facility, such as meetings with school officials regarding disciplinary measures, parent-teacher conferences, or meetings with school counselors, for a child of the military member, when such meetings are necessary due to circumstances arising from the covered active duty or call to covered active duty status of the military member;

(4) *Financial and legal arrangements.* (i) To make or update financial or legal arrangements to address the military member's absence while on covered active duty or call to covered active duty status, such as preparing and executing financial and healthcare powers of attorney, transferring bank account signature authority, enrolling in the Defense Enrollment Eligibility Reporting System (DEERS), obtaining military identification cards, or preparing or updating a will or living trust; and

(ii) To act as the military member's representative before a federal, state, or local agency for purposes of obtaining, arranging, or appealing military service benefits while the military member is on covered active duty or call to covered active duty status, and for a period of 90 days following the termination of the military member's covered active duty status;

(5) *Counseling.* To attend counseling provided by someone other than a health care provider, for oneself, for the military member, or for the biological, adopted, or foster child, a stepchild, or a legal ward of the military member, or a child for whom the military member stands in loco parentis, who is either under age 18, or age 18 or older and incapable of self-care because of a mental or physical disability at the time that FMLA leave is to commence, provided that the need for counseling arises from the covered active duty or call to covered active duty status of the military member;

(6) *Rest and Recuperation.* (i) To spend time with the military member who is on short-term, temporary, Rest and Recuperation leave during the period of deployment;

(ii) Leave taken for this purpose can be used for a period of 15 calendar days beginning on the date the military member commences each instance of Rest and Recuperation leave;

(7) *Post-deployment activities.* (i) To attend arrival ceremonies, reintegration briefings and events, and any other official ceremony or program sponsored by the military for a period of 90 days following the termination of the military member's covered active duty status; and

(ii) To address issues that arise from the death of the military member while on covered active duty status, such as meeting and recovering the body of the military member, making funeral arrangements, and attending funeral services;

(8) *Parental care.* For purposes of leave for parental care listed in (i) through (iv) of this paragraph, the parent of the military member must be incapable of self-care and must be the military member's biological, adoptive, step, or foster father or mother, or any other individual who stood in loco parentis to the military member when the member was under 18 years of age. A parent who is incapable of self-care means that the parent requires active assistance or supervision to provide daily self-care in three or more of the activities of daily living or instrumental activities of daily living. Activities of daily living include adaptive activities such as caring appropriately for one's grooming and hygiene, bathing, dressing, and eating. Instrumental activities of daily living include cooking, cleaning, shopping, taking public transportation, paying bills, maintaining a residence, using telephones and directories, using a post office, etc. As with all instances of qualifying exigency leave, the military member must be the spouse, son, daughter, or parent of the employee requesting qualifying exigency leave.

(i) To arrange for alternative care for a parent of the military member when the parent is incapable of self-care and the covered active duty or call to covered active duty status of the military member necessitates a change in the existing care arrangement for the parent;

(ii) To provide care for a parent of the military member on an urgent, immediate need basis (but not on a routine, regular, or everyday basis) when the parent is incapable of self-care and the need to provide such care arises from the covered active duty or call to covered active duty status of the military member;

(iii) To admit to or transfer to a care facility a parent of the military member when admittance or transfer is necessitated by the covered active duty or call to covered active duty status of the military member; and

(iv) To attend meetings with staff at a care facility, such as meetings with hospice or social service providers for a parent of the military member, when such meetings are necessary due to circumstances arising from the covered active duty or call to covered active duty status of the military member but not for routine or regular meetings;

(9) *Additional activities.* To address other events which arise out of the military member's covered active duty or call to covered active duty status provided that the employer and employee agree that such leave shall qualify as an exigency, and agree to both the timing and duration of such leave.

Section 825.127 Leave to Care for a Covered Servicemember with a Serious Injury or Illness (Military Caregiver Leave)

(a) Eligible employees are entitled to FMLA leave to care for a covered servicemember with a serious illness or injury.

(b) *Covered servicemember* means:

(1) A current member of the Armed Forces, including a member of the National Guard or Reserves, who is undergoing medical treatment, recuperation, or therapy, is otherwise in outpatient status; or is otherwise on the temporary disability retired list, for a serious injury or illness. *Outpatient status* means the status of a member of the Armed Forces assigned to either a military medical treatment facility as an outpatient or a unit established for the purpose of providing command and control of members of the Armed Forces receiving medical care as outpatients.

(2) A covered veteran who is undergoing medical treatment, recuperation or therapy for a serious injury or illness. *Covered veteran* means an individual who was a member of the Armed Forces (including a member of the National Guard or Reserves), and was discharged or released under conditions other than dishonorable at any time during the five-year period prior to the first date the eligible employee takes FMLA leave to care for the covered veteran. An eligible employee must commence leave to care for a covered veteran within five years of the veteran's active duty service, but the single 12-month period described in paragraph (e)(1) of this section may extend beyond the five-year period.

(i) For an individual who was a member of the Armed Forces (including a member of the National Guard or Reserves) and who was discharged or released under conditions other than dishonorable prior to the effective date of this Final Rule, the period between October 28, 2009 and the effective date of this Final Rule shall not count towards the determination of the five-year period for covered veteran status.

(c) A *serious injury or illness* means:

(1) In the case of a current member of the Armed Forces, including a member of the National Guard or Reserves, means an injury or illness that was incurred by the covered servicemember in the line of duty on active duty in the Armed Forces or that existed before the beginning of the member's active duty and was aggravated by service in the line of duty on active duty in the Armed Forces, and that may render the member medically unfit to perform the duties of the member's office, grade, rank or rating; and,

(2) In the case of a covered veteran, means an injury or illness that was incurred by the member in the line of duty on active duty in the Armed Forces (or existed before the beginning of the member's active duty and was aggravated by service in the line of duty on active duty in the Armed Forces) and manifested itself before or after the member became a veteran, and is:

(i) a continuation of a serious injury or illness that was incurred or aggravated when the covered veteran was a member of the Armed Forces and rendered the servicemember unable to perform the duties of the servicemember's office, grade, rank, or rating; or

(ii) a physical or mental condition for which the covered veteran has received a U.S. Department of Veterans Affairs Service-Related Disability Rating (VASRD) of 50 percent or greater, and such VASRD rating is based, in whole or in part, on the condition precipitating the need for military caregiver leave; or

(iii) a physical or mental condition that substantially impairs the covered veteran's ability to secure or follow a substantially gainful occupation by reason of a disability or disabilities related to military service, or would do so absent treatment; or

(iv) an injury, including a psychological injury, on the basis of which the covered veteran has been enrolled in the Department of Veterans Affairs Program of Comprehensive Assistance for Family Caregivers.

(d) In order to care for a covered servicemember, an eligible employee must be the spouse, son, daughter, or parent, or next of kin of a covered servicemember.

(1) *Son or daughter of a covered servicemember* means the covered servicemember's biological, adopted, or foster child, stepchild, legal ward, or a child for whom the covered servicemember stood in loco parentis, and who is of any age.

(2) *Parent of a covered servicemember* means a covered servicemember's biological, adoptive, step or foster father or mother, or any other individual who stood in loco parentis to the covered servicemember. This term does not include parents "in law."

(3) *Next of kin of a covered servicemember* means the nearest blood relative, other than the covered servicemember's spouse, parent, son, or daughter, in the following order of priority: blood relatives who have been granted legal custody of the servicemember by court decree or statutory provisions, brothers and sisters, grandparents, aunts and uncles, and first cousins, unless the covered servicemember has specifically designated in writing another blood relative as his or her nearest blood relative for purposes of military caregiver leave under the FMLA. When no such designation is made, and there are multiple family members with the same level of relationship to the covered servicemember, all such family members shall be considered the covered servicemember's next of kin and may take FMLA leave to provide care to the covered servicemember, either consecutively or simultaneously. When such designation has been made, the designated individual shall be deemed to be the covered servicemember's only next of kin. For example, if a covered servicemember has three siblings and has not designated a blood relative to provide care, all three siblings would be considered the covered servicemember's next of kin. Alternatively, where a covered servicemember has a sibling(s) and designates a cousin as his or her next of kin for FMLA purposes, then only the designated cousin is eligible as the covered servicemember's next of kin. An employer is permitted to require an employee to provide confirmation of covered family relationship to the covered servicemember pursuant to Section 825.122(k).

(e) An eligible employee is entitled to 26 workweeks of leave to care for a covered servicemember with a serious injury or illness during a single 12-month period.

(1) The single 12-month period described in paragraph (e) of this section begins on the first day the eligible employee takes FMLA leave to care for a covered servicemember and ends 12 months after that date, regardless of the method used by the employer to determine the employee's 12 workweeks of leave entitlement for other FMLA-qualifying reasons. If an eligible employee does not take all of his or her 26 workweeks of leave entitlement to care for a covered servicemember during this single 12-month period, the remaining part of his or her 26 workweeks of leave entitlement to care for the covered servicemember is forfeited.

(2) The leave entitlement described in paragraph (e) of this section is to be applied on a per-covered-servicemember, per-injury basis such that an eligible employee may be entitled to take more than one period of 26 workweeks of leave if the leave is to care for different covered servicemembers or to care for the same servicemember with a subsequent serious injury or illness, except that no more than 26 workweeks of leave may be taken within any single 12-month period. An eligible employee may take more than one period of 26 workweeks of leave to care for a covered servicemember with more than one serious injury or illness only when the serious injury or illness is a subsequent serious injury or illness. When an eligible employee takes leave to care for more than one covered servicemember or for a subsequent serious injury or illness of the same covered servicemember, and the single 12-month periods corresponding to the different military caregiver leave entitlements overlap, the employee is limited to taking no more than 26 workweeks of leave in each single 12-month period.

(3) An eligible employee is entitled to a combined total of 26 workweeks of leave for any FMLA-qualifying reason during the single 12-month period described in paragraph (e) of this section, provided that the employee is entitled to no more than 12 workweeks of leave for one or more of the following: because of the birth of a son or daughter of the employee and in order to care for such son or daughter; because of the placement of a son or daughter with the employee for adoption or foster care; in order to care for the spouse, son, daughter, or parent with a serious health condition; because of the employee's own serious health condition; or because of a qualifying exigency. Thus, for example, an eligible employee may, during the single 12-month period, take 16 workweeks of FMLA leave to care for a covered servicemember and 10 workweeks of FMLA leave to care for a newborn child. However, the employee may not take more than 12 weeks of FMLA leave to care for the newborn child during the single 12-month period, even if the employee takes fewer than 14 workweeks of FMLA leave to care for a covered servicemember.

(4) In all circumstances, including for leave taken to care for a covered servicemember, the employer is responsible for designating leave, paid or unpaid, as FMLA-qualifying, and for giving notice of the designation to the employee as provided in Section 825.300. In the case of leave that qualifies as both leave to care for a covered servicemember and leave to care for a family member with a serious health condition during the single 12-month period described in paragraph (e) of this section, the employer must designate such leave as leave to care for a covered servicemember in the first instance. Leave that qualifies as both leave to care for a covered servicemember and leave taken to care for a family member with a serious health condition during the single 12-month period described in paragraph (e) of this section must not be designated and counted as both leave to care for a covered servicemember and leave to care for a family member with a serious health condition. As is the case with leave taken for other qualifying reasons, employers may retroactively designate leave as leave to care for a covered servicemember pursuant to Section 825.301(d).

(f) Spouses who are eligible for FMLA leave and are employed by the same covered employer may be limited to a combined total of 26 workweeks of leave during the single 12-month period described in paragraph (e) of this section if the leave is taken for birth of the employee's son or daughter or to care for the child after birth, for placement of a son or daughter with the employee for adoption or foster care, or to care for the child after placement, to care for the employee's parent with a serious health condition, or to care for a covered servicemember with a serious injury or illness. This limitation on the total weeks of leave applies to leave taken for the reasons specified as long as the spouses are employed by the same employer. It would apply, for example, even though the spouses are employed at two different worksites of an employer located more than 75 miles from each other, or by two different operating divisions of the same company. On the other hand, if one spouse is ineligible for FMLA leave, the other spouse would be entitled to a full 26 workweeks of FMLA leave.

[78 FR 8902, Feb. 6, 2013, as amended at 80 FR 10001, Feb. 25, 2015]

SUBPART B—EMPLOYEE LEAVE ENTITLEMENTS UNDER THE FAMILY AND MEDICAL LEAVE ACT

Section 825.200 Amount of leave

(a) Except in the case of leave to care for a covered servicemember with a serious injury or illness, an eligible employee's FMLA leave entitlement is limited to a total of 12 workweeks of leave during any 12-month period for any one, or more, of the following reasons:

(1) The birth of the employee's son or daughter, and to care for the newborn child;

(2) The placement with the employee of a son or daughter for adoption or foster care, and to care for the newly placed child;

(3) To care for the employee's spouse, son, daughter, or parent with a serious health condition;

(4) Because of a serious health condition that makes the employee unable to perform one or more of the essential functions of his or her job; and,

(5) Because of any qualifying exigency arising out of the fact that the employee's spouse, son, daughter, or parent is a military member on covered active duty status (or has been notified of an impending call or order to covered active duty).

(b) An employer is permitted to choose any one of the following methods for determining the 12-month period in which the 12 weeks of leave entitlement described in paragraph (a) of this section occurs:

(1) The calendar year;

(2) Any fixed 12-month leave year, such as a fiscal year, a year required by State law, or a year starting on an employee's anniversary date;

(3) The 12-month period measured forward from the date any employee's first FMLA leave under paragraph (a) begins; or,

(4) A "rolling" 12-month period measured backward from the date an employee uses any FMLA leave as described in paragraph (a).

(c) Under methods in paragraphs (b)(1) and (b)(2) of this section an employee would be entitled to up to 12 weeks of FMLA leave at any time in the fixed 12-month period selected. An employee could, therefore, take 12 weeks of leave at the end of the year and 12 weeks at the beginning of the following year. Under the method in paragraph (b)(3) of this section, an employee would be entitled to 12 weeks of leave during the year beginning on the first date FMLA leave is taken; the next 12-month period would begin the first time FMLA leave is taken after completion of any previous 12-month period. Under the method in paragraph (b)(4) of this section, the "rolling"

12-month period, each time an employee takes FMLA leave the remaining leave entitlement would be any balance of the 12 weeks which has not been used during the immediately preceding 12 months. For example, if an employee has taken eight weeks of leave during the past 12 months, an additional four weeks of leave could be taken. If an employee used four weeks beginning February 1, 2008, four weeks beginning June 1, 2008, and four weeks beginning December 1, 2008, the employee would not be entitled to any additional leave until February 1, 2009. However, beginning on February 1, 2009, the employee would again be eligible to take FMLA leave, recouping the right to take the leave in the same manner and amounts in which it was used in the previous year. Thus, the employee would recoup (and be entitled to use) one additional day of FMLA leave each day for four weeks, commencing February 1, 2009. The employee would also begin to recoup additional days beginning on June 1, 2009, and additional days beginning on December 1, 2009. Accordingly, employers using the rolling 12-month period may need to calculate whether the employee is entitled to take FMLA leave each time that leave is requested, and employees taking FMLA leave on such a basis may fall in and out of FMLA protection based on their FMLA usage in the prior 12 months. For example, in the example above, if the employee needs six weeks of leave for a serious health condition commencing February 1, 2009, only the first four weeks of the leave would be FMLA protected.

(d)(1) Employers will be allowed to choose any one of the alternatives in paragraph (b) of this section for the leave entitlements described in paragraph (a) of this section provided the alternative chosen is applied consistently and uniformly to all employees. An employer wishing to change to another alternative is required to give at least 60 days notice to all employees, and the transition must take place in such a way that the employees retain the full benefit of 12 weeks of leave under whichever method affords the greatest benefit to the employee. Under no circumstances may a new method be implemented in order to avoid the Act's leave requirements.

(2) An exception to this required uniformity would apply in the case of a multi-State employer who has eligible employees in a State which has a family and medical leave statute. The State may require a single method of determining the period during which use of the leave entitlement is measured. This method may conflict with the method chosen by the employer to determine any 12 months for purposes of the Federal statute. The employer may comply with the State provision for all employees employed within that State, and uniformly use another method provided by this regulation for the leave entitlements described in paragraph (a) for all other employees.

(e) If an employer fails to select one of the options in paragraph (b) of this section for measuring the 12-month period for the leave entitlements described in paragraph (a), the option that provides the most beneficial outcome for the employee will be used. The employer may subsequently select an option only by providing the 60-day notice to all employees of the option the employer intends to implement. During the running of the 60-day period any other employee who needs FMLA leave may use the option providing the most beneficial outcome to that employee. At the conclusion of the 60-day period the employer may implement the selected option.

(f) An eligible employee's FMLA leave entitlement is limited to a total of 26 workweeks of leave during a single 12-month period to care for a covered servicemember with a serious injury or illness. An employer shall determine the single 12-month period in which the 26-weeks-of-leave-entitlement described in this paragraph occurs using the 12-month period measured forward from the date an employee's first FMLA leave to care for the covered servicemember begins. *See* Section 825.127(e)(1).

(g) During the single 12-month period described in paragraph (f), an eligible employee's FMLA leave entitlement is limited to a combined total of 26 workweeks of FMLA leave for any qualifying reason. *See* Section 825.127(e)(3).

(h) For purposes of determining the amount of leave used by an employee, the fact that a holiday may occur within the week taken as FMLA leave has no effect; the week is counted as a week of FMLA leave. However, if an employee is using FMLA leave in increments of less than one week, the holiday will not count against the employee's FMLA entitlement unless the employee was otherwise scheduled and expected to work during the holiday. Similarly, if for some reason the employer's business activity has temporarily ceased and employees generally are not expected to report for work for one or more weeks (e.g., a school closing two weeks for the Christmas/New Year holiday or the summer vacation or an employer closing the plant for retooling or repairs), the days the employer's activities have ceased do not count against the employee's FMLA leave entitlement. Methods for determining an employee's 12-week leave entitlement are also described in Section 825.205. *See* Section 825.802 for special calculation of leave rules applicable to airline flight crew employees.

Section 825.201 Leave to care for a parent

(a) *General rule.* An eligible employee is entitled to FMLA leave if needed to care for the employee's parent with a serious health condition. Care for parents-in-law is not covered by the FMLA. *See* Section 825.122(c) for definition of parent.

(b) *Same employer limitation.* Spouses who are eligible for FMLA leave and are employed by the same covered employer may be limited to a combined total of 12 weeks of leave during any 12-month period if the leave is taken to care for the employee's parent with a serious health condition, for the birth of the employee's son or daughter or to care for the child after the birth, or for placement of a son or daughter with the employee for adoption or foster care or to care for the child after placement. This limitation on the total weeks of leave applies to leave taken for the reasons specified as long as the spouses are employed by the same employer. It would apply, for example, even though the spouses are employed at two different worksites of an employer located more than 75 miles from each other, or by two different operating divisions of the same company. On the other hand, if one spouse is ineligible for FMLA leave, the other spouse would be entitled to a full 12 weeks of FMLA leave. Where the spouses both use a portion of the total 12-week FMLA leave entitlement for either the birth of a child, for placement for adoption or foster care, or to care for a parent, the spouses would each be entitled to the difference between the amount he or she has taken individually and 12 weeks for FMLA leave for other purposes. For example, if each spouse took six weeks of leave to care for a parent, each could use an additional six weeks due to his or her own serious health condition or to care for a child with a serious health condition. *See also* Section 825.127(d).

[78 FR 8902, Feb. 6, 2013, as amended at 80 FR 10001, Feb. 25, 2015]

Section 825.202 Intermittent leave or reduced leave schedule

(a) *Definition.* FMLA leave may be taken intermittently or on a reduced leave schedule under certain circumstances. *Intermittent leave* is FMLA leave taken in separate blocks of time due to a single qualifying reason. A *reduced leave schedule* is a leave schedule that reduces an employee's usual number of working hours per workweek, or hours per workday. A reduced leave schedule is a change in the employee's schedule for a period of time, normally from full-time to part-time.

(b) *Medical necessity.* For intermittent leave or leave on a reduced leave schedule taken because of one's own serious health condition, to care for a spouse, parent, son, or daughter with a serious health condition, or to care for a covered servicemember with a serious injury or illness, there must be a medical need for leave and it must be that such medical need can be best accommodated through an intermittent or reduced leave schedule. The treatment regimen and other information described in the certification of a serious health condition and in the certification of a serious injury or illness, if required by the employer, addresses the medical necessity of intermittent leave or leave on a reduced leave schedule. *See* Sections 825.306, 825.310. Leave may be taken intermittently or on a reduced leave schedule when medically necessary for planned and/or unanticipated medical treatment of a serious health condition or of a covered servicemember's serious injury or illness, or for recovery from treatment or recovery from a serious health condition or a covered servicemember's serious injury or illness. It may also be taken to provide care or psychological comfort to a covered family member with a serious health condition or a covered servicemember with a serious injury or illness.

(1) Intermittent leave may be taken for a serious health condition of a spouse, parent, son, or daughter, for the employee's own serious health condition, or a serious injury or illness of a covered servicemember which requires treatment by a health care provider periodically, rather than for one continuous period of time, and may include leave of periods from an hour or more to several weeks. Examples of intermittent leave would include leave taken on an occasional basis for medical appointments, or leave taken several days at a time spread over a period of six months, such as for chemotherapy. A pregnant employee may take leave intermittently for prenatal examinations or for her own condition, such as for periods of severe morning sickness. An example of an employee taking leave on a reduced leave schedule is an employee who is recovering from a serious health condition and is not strong enough to work a full-time schedule.

(2) Intermittent or reduced schedule leave may be taken for absences where the employee or family member is incapacitated or unable to perform the essential functions of the position because of a chronic serious health condition or a serious injury or illness of a covered servicemember, even if he or she does not receive treatment by a health care provider. *See* Sections 825.113 and 825.127.

(c) *Birth or placement.* When leave is taken after the birth of a healthy child or placement of a healthy child for adoption or foster care, an employee may take leave intermittently or on a reduced leave schedule only if the employer agrees. Such a schedule reduction might occur, for example, where an employee, with the employer's agreement, works part-time after the birth of a child, or takes leave in several segments. The employer's agreement is not required, however, for leave during which the expectant mother has a serious health condition in connection with the birth of her child or if the newborn child has a serious health condition. *See* Section 825.204 for rules governing transfer to an alternative position that better accommodates intermittent leave. *See also* Section 825.120 (pregnancy) and Section 825.121 (adoption and foster care).

(d) *Qualifying exigency.* Leave due to a qualifying exigency may be taken on an intermittent or reduced leave schedule basis.

[78 FR 8902, Feb. 6, 2013, as amended at 80 FR 10001, Feb. 25, 2015]

Section 825.203 Scheduling of intermittent or reduced schedule leave

Eligible employees may take FMLA leave on an intermittent or reduced schedule basis when medically necessary due to the serious health condition of a covered family member or the employee or the serious injury or illness of a covered servicemember. *See* Section 825.202. Eligible employees may also take FMLA leave on an intermittent or reduced schedule basis when necessary because of a qualifying exigency. If an employee needs leave intermittently or on a reduced leave schedule for planned medical treatment, then the employee must make a reasonable effort to schedule the treatment so as not to disrupt unduly the employer's operations.

Section 825.204 Transfer of an employee to an alternative position during intermittent leave or reduced schedule leave

(a) *Transfer or reassignment.* If an employee needs intermittent leave or leave on a reduced leave schedule that is foreseeable based on planned medical treatment for the employee, a family member, or a covered servicemember, including during a period of recovery from one's own serious health condition, a serious health condition of a spouse, parent, son, or daughter, or a serious injury or illness of a covered servicemember, or if the employer agrees to permit intermittent or reduced schedule leave for the birth of a child or for placement of a child for adoption or foster care, the employer may require the employee to transfer temporarily, during the period that the intermittent or reduced leave schedule is required, to an available alternative position for which the employee is qualified and which better accommodates recurring periods of leave than does the employee's regular position. *See* Section 825.601 for special rules applicable to instructional employees of schools.

(b) *Compliance.* Transfer to an alternative position may require compliance with any applicable collective bargaining agreement, Federal law (such as the Americans with Disabilities Act), and State law. Transfer to an alternative position may include altering an existing job to better accommodate the employee's need for intermittent or reduced schedule leave.

(c) *Equivalent pay and benefits.* The alternative position must have equivalent pay and benefits. An alternative position for these purposes does not have to have equivalent duties. The employer may increase the pay and benefits of an existing alternative position, so as to make them equivalent to the pay and benefits of the employee's regular job. The employer may also transfer the employee to a part-time job with the same hourly rate of pay and benefits, provided the employee is not required to take more leave than is medically necessary. For example, an employee desiring to take leave in increments of four hours per day could be transferred to a half-time job, or could remain in the employee's same job on a part-time schedule, paying the same hourly rate as the employee's previous job and enjoying the same benefits. The employer may not eliminate benefits which otherwise would not be provided to part-time employees; however, an employer may proportionately reduce benefits such as vacation leave where an employer's normal practice is to base such benefits on the number of hours worked.

(d) *Employer limitations.* An employer may not transfer the employee to an alternative position in order to discourage the employee from taking leave or otherwise work a hardship on the employee. For example, a white collar employee may not be assigned to perform laborer's work; an employee working the day shift may not be reassigned to the graveyard shift; an employee working in the headquarters facility may not be reassigned to a branch a significant distance away from the employee's normal job location. Any such attempt on the part of the employer to make such a transfer will be held to be contrary to the prohibited acts of the FMLA.

(e) *Reinstatement of employee.* When an employee who is taking leave intermittently or on a reduced leave schedule and has been transferred to an alternative position no longer needs to continue on leave and is able to return to full-time work, the employee must be placed in the same or equivalent job as the job he or she left when the leave commenced. An employee may not be required to take more leave than necessary to address the circumstance that precipitated the need for leave.

Section 825.205 Increments of FMLA leave for intermittent or reduced schedule leave

(a) *Minimum increment.* (1) When an employee takes FMLA leave on an intermittent or reduced leave schedule basis, the employer must account for the leave using an increment no greater than the shortest period of time that the employer uses to account for use of other forms of leave provided that it is not greater than one hour and provided further that an employee's FMLA leave entitlement may not be reduced by more than the amount of leave actually taken. An employer may not require an employee to take more leave than is necessary to address the circumstances that precipitated the need for the leave, provided that the leave is counted using the shortest increment of leave used to account for any other type of leave. *See also* Section 825.205(a)(2) for the physical impossibility exception, Sections 825.600 and 825.601 for special rules applicable to employees of schools, and Section 825.802 for special rules applicable to airline flight crew employees. If an employer uses different increments to account for different types of leave, the employer must account for FMLA leave in the smallest increment used to account for any other type of leave. For example, if an employer accounts for the use of annual leave in increments of one hour and the use of sick leave in increments of one-half hour, then FMLA leave use must be accounted for using increments no larger than one-half hour. If an employer accounts for use of leave in varying increments at different times of the day or shift, the employer may also account for FMLA leave in varying increments, provided that the increment used for FMLA leave is no greater than the smallest increment used for any other type of leave during the period in which the FMLA leave is taken. If an employer accounts for other forms of leave use in increments greater than one hour, the employer must account for FMLA leave use in increments no greater than one hour. An employer may account for FMLA leave in shorter increments than used for other forms of leave. For example, an employer that accounts for other forms of leave in one hour increments may account for FMLA leave in a shorter increment when the employee arrives at work several minutes late, and the employer wants the employee to begin work immediately. Such accounting for FMLA leave will not alter the increment considered to be the shortest period used to account for other forms of leave or the use of FMLA leave in other circumstances. In all cases, employees may not be charged FMLA leave for periods during which they are working.

(2) Where it is physically impossible for an employee using intermittent leave or working a reduced leave schedule to commence or end work mid-way through a shift, such as where a flight attendant or a railroad conductor is scheduled to work aboard an airplane or train, or a laboratory employee is unable to enter or leave a sealed "clean room" during a certain period of time and no equivalent position is available, the entire period that the employee is forced to be absent is designated as FMLA leave and counts against the employee's FMLA entitlement. The period of the physical impossibility is limited to the period during which the employer is unable to permit the employee to work prior to a period of FMLA leave or return the employee to the same or equivalent position due to the physical impossibility after a period of FMLA leave. *See* Section 825.214.

(b) *Calculation of leave.* (1) When an employee takes leave on an intermittent or reduced leave schedule, only the amount of leave actually taken may be counted toward the employee's leave entitlement. The actual workweek is the basis of leave entitlement. Therefore, if an employee who would otherwise work 40 hours a week takes off eight hours, the employee would use one-fifth ($\frac{1}{5}$) of a week of FMLA leave. Similarly, if a full-time employee who would otherwise work eight hour days works four-hour days under a reduced leave schedule, the employee would use one-half ($\frac{1}{2}$) week of FMLA leave. Where an employee works a part-time schedule or variable hours, the amount of FMLA leave that an employee uses is determined on a pro rata or proportional basis. If an employee who would otherwise work 30 hours per week, but works only 20 hours a week under a reduced leave schedule, the employee's 10 hours of leave would constitute one-third ($\frac{1}{3}$) of a week of FMLA leave for each week the employee works the reduced leave schedule. An employer may convert these fractions to their hourly equivalent so long as the conversion equitably reflects the employee's total normally scheduled hours. An employee does not accrue FMLA-protected leave at any particular hourly rate. An eligible employee is entitled to up to a total of 12 workweeks of leave, or 26 workweeks in the case of military caregiver leave, and the total number of hours contained in those workweeks is necessarily dependent on the specific hours the employee would have worked but for the use of leave. *See also* Sections 825.601 and 825.602, special rules for schools and Section 825.802, special rules for airline flight crew employees.

(2) If an employer has made a permanent or long-term change in the employee's schedule (for reasons other than FMLA, and prior to the notice of need for FMLA leave), the hours worked under the new schedule are to be used for making this calculation.

(3) If an employee's schedule varies from week to week to such an extent that an employer is unable to determine with any certainty how many hours the employee would otherwise have worked (but for the taking of FMLA leave), a weekly average of the hours scheduled over the 12 months prior to the beginning of the leave period (including any hours for which the employee took leave of any type) would be used for calculating the employee's leave entitlement.

(c) *Overtime.* If an employee would normally be required to work overtime, but is unable to do so because of a FMLA-qualifying reason that limits the employee's ability to work overtime, the hours which the employee would have been required to work may be counted against the employee's FMLA entitlement. In such a case, the employee is using intermittent or reduced schedule leave. For example, if an employee would normally be required to work for 48 hours in a particular week, but due to a serious health condition the employee is unable to work more than 40 hours that week, the employee would utilize eight hours of FMLA-protected leave out of the 48-hour workweek, or one-sixth ($\frac{1}{6}$) of a week of FMLA leave. Voluntary overtime hours that an employee does not work due to an FMLA-qualifying reason may not be counted against the employee's FMLA leave entitlement.

Section 825.206 Interaction with the FLSA

(a) Leave taken under FMLA may be unpaid. If an employee is otherwise exempt from minimum wage and overtime requirements of the Fair Labor Standards Act (FLSA) as a salaried executive, administrative, professional, or computer employee (under regulations issued by the Secretary, 29 CFR part 541), providing unpaid

FMLA-qualifying leave to such an employee will not cause the employee to lose the FLSA exemption. *See*29 CFR 541.602(b)(7). This means that under regulations currently in effect, where an employee meets the specified duties test, is paid on a salary basis, and is paid a salary of at least the amount specified in the regulations, the employer may make deductions from the employee's salary for any hours taken as intermittent or reduced FMLA leave within a workweek, without affecting the exempt status of the employee. The fact that an employer provides FMLA leave, whether paid or unpaid, and maintains records required by this part regarding FMLA leave, will not be relevant to the determination whether an employee is exempt within the meaning of 29 CFR part 541.

(b) For an employee paid in accordance with the fluctuating workweek method of payment for overtime (*see*29 CFR 778.114), the employer, during the period in which intermittent or reduced schedule FMLA leave is scheduled to be taken, may compensate an employee on an hourly basis and pay only for the hours the employee works, including time and one-half the employee's regular rate for overtime hours. The change to payment on an hourly basis would include the entire period during which the employee is taking intermittent leave, including weeks in which no leave is taken. The hourly rate shall be determined by dividing the employee's weekly salary by the employee's normal or average schedule of hours worked during weeks in which FMLA leave is not being taken. If an employer chooses to follow this exception from the fluctuating workweek method of payment, the employer must do so uniformly, with respect to all employees paid on a fluctuating workweek basis for whom FMLA leave is taken on an intermittent or reduced leave schedule basis. If an employer does not elect to convert the employee's compensation to hourly pay, no deduction may be taken for FMLA leave absences. Once the need for intermittent or reduced scheduled leave is over, the employee may be restored to payment on a fluctuating workweek basis.

(c) This special exception to the salary basis requirements of the FLSA exemption or fluctuating workweek payment requirements applies only to employees of covered employers who are eligible for FMLA leave, and to leave which qualifies as FMLA leave. Hourly or other deductions which are not in accordance with 29 CFR part 541 or 29 CFR 778.114 may not be taken, for example, from the salary of an employee who works for an employer with fewer than 50 employees, or where the employee has not worked long enough to be eligible for FMLA leave without potentially affecting the employee's eligibility for exemption. Nor may deductions which are not permitted by 29 CFR part 541 or 29 CFR 778.114 be taken from such an employee's salary for any leave which does not qualify as FMLA leave, for example, deductions from an employee's pay for leave required under State law or under an employer's policy or practice for a reason which does not qualify as FMLA leave, e.g., leave to care for a grandparent or for a medical condition which does not qualify as a serious health condition or serious injury or illness; or for leave which is more generous than provided by FMLA. Employers may comply with State law or the employer's own policy/practice under these circumstances and maintain the employee's eligibility for exemption or for the fluctuating workweek method of pay by not taking hourly deductions from the employee's pay, in accordance with FLSA requirements, or may take such deductions, treating the employee as an hourly employee and pay overtime premium pay for hours worked over 40 in a workweek.

Section 825.207 Substitution of paid leave

(a) Generally, FMLA leave is unpaid leave. However, under the circumstances described in this section, FMLA permits an eligible employee to choose to substitute accrued paid leave for FMLA leave. If an employee does not choose to substitute accrued paid leave, the employer may require the employee to substitute accrued paid leave for unpaid FMLA leave. The term substitute means that the paid leave provided by the employer, and accrued pursuant to established policies of the employer, will run concurrently with the unpaid FMLA leave. Accordingly, the employee receives pay pursuant to the employer's applicable paid leave policy during the period of otherwise unpaid FMLA leave. An employee's ability to substitute accrued paid leave is determined by the terms and conditions of the employer's normal leave policy. When an employee chooses, or an employer requires, substitution of accrued paid leave, the employer must inform the employee that the employee must satisfy any procedural requirements of the paid leave policy only in connection with the receipt of such payment. *See* Section 825.300(c). If an employee does not comply with the additional requirements in an employer's paid leave policy, the employee is not entitled to substitute accrued paid leave, but the employee remains entitled to take unpaid FMLA leave. Employers may not discriminate against employees on FMLA leave in the administration of their paid leave policies.

(b) If neither the employee nor the employer elects to substitute paid leave for unpaid FMLA leave under the above conditions and circumstances, the employee will remain entitled to all the paid leave which is earned or accrued under the terms of the employer's plan.

(c) If an employee uses paid leave under circumstances which do not qualify as FMLA leave, the leave will not count against the employee's FMLA leave entitlement. For example, paid sick leave used for a medical condition which is not a serious health condition or serious injury or illness does not count against the employee's FMLA leave entitlement.

(d) Leave taken pursuant to a disability leave plan would be considered FMLA leave for a serious health condition and counted in the leave entitlement permitted under FMLA if it meets the criteria set forth above in Sections 825.112 through 825.115. In such cases, the employer may designate the leave as FMLA leave and count the leave against the employee's FMLA leave entitlement. Because leave pursuant to a disability benefit plan is not unpaid, the provision for substitution of the employee's accrued paid leave is inapplicable, and neither the employee nor the employer may require the substitution of paid leave. However, employers and employees may agree, where state law permits, to have paid leave supplement the disability plan benefits, such as in the case where a plan only provides replacement income for two-thirds of an employee's salary.

(e) The Act provides that a serious health condition may result from injury to the employee on or off the job. If the employer designates the leave as FMLA leave in accordance with Section 825.300(d), the leave counts against the employee's FMLA leave entitlement. Because the workers' compensation absence is not unpaid, the provision for substitution of the employee's accrued paid leave is not applicable, and neither the employee nor the employer may require the substitution of paid leave. However, employers and employees may agree, where state law permits, to have paid leave supplement workers' compensation benefits, such as in the case where workers' compensation only provides replacement income for two-thirds of an employee's salary. If the health care provider treating the employee for the workers' compensation injury certifies the employee is able to return to a light duty job but is unable to return to the same or equivalent job, the employee may decline the employer's offer of a light duty job. As a result the employee may lose workers' compensation payments, but is entitled to remain on unpaid FMLA leave until the employee's FMLA leave entitlement is exhausted. As of the date workers' compensation benefits cease, the substitution provision becomes applicable and either the employee may elect or the employer may require the use of accrued paid leave. *See also* Sections 825.210(f), 825.216(d), 825.220(d), 825.307(a) and 825.702(d)(1) and (2) regarding the relationship between workers' compensation absences and FMLA leave.

(f) Section 7(o) of the Fair Labor Standards Act (FLSA) permits public employers under prescribed circumstances to substitute compensatory time off accrued at one and one-half hours for each overtime hour worked in lieu of paying cash to an employee when the employee works overtime hours as prescribed by the Act. This section of the FLSA limits the number of hours of compensatory time an employee may accumulate depending upon whether the employee works in fire protection or law enforcement (480 hours) or elsewhere for a public agency (240 hours). In addition, under the FLSA, an employer always has the right to cash out an employee's compensatory time or to require the employee to use the time. Therefore, if an employee requests and is permitted to use accrued compensatory time to receive pay for time taken off for an FMLA reason, or if the employer requires such use pursuant to the FLSA, the time taken may be counted against the employee's FMLA leave entitlement.

Section 825.208 [Reserved]

Section 825.209 Maintenance of employee benefits

(a) During any FMLA leave, an employer must maintain the employee's coverage under any group health plan (as defined in the Internal Revenue Code of 1986 at 26 U.S.C. 5000(b)(1)) on the same conditions as coverage would have been provided if the employee had been continuously employed during the entire leave period. All employers covered by FMLA, including public agencies, are subject to the Act's requirements to maintain health coverage. The definition of group health plan is set forth in Section 825.102. For purposes of FMLA, the term group health plan shall not include an insurance program providing health coverage under which employees purchase individual policies from insurers provided that:

(1) No contributions are made by the employer;

(2) Participation in the program is completely voluntary for employees;

(3) The sole functions of the employer with respect to the program are, without endorsing the program, to permit the insurer to publicize the program to employees, to collect premiums through payroll deductions and to remit them to the insurer;

(4) The employer receives no consideration in the form of cash or otherwise in connection with the program, other than reasonable compensation, excluding any profit, for administrative services actually rendered in connection with payroll deduction; and,

(5) The premium charged with respect to such coverage does not increase in the event the employment relationship terminates.

(b) The same group health plan benefits provided to an employee prior to taking FMLA leave must be maintained during the FMLA leave. For example, if family member coverage is provided to an employee, family member coverage must be maintained during the FMLA leave. Similarly, benefit coverage during FMLA leave for medical care, surgical care, hospital care, dental care, eye care, mental health counseling, substance abuse treatment, etc., must be maintained during leave if provided in an employer's group health plan, including a supplement to a group health plan, whether or not provided through a flexible spending account or other component of a cafeteria plan.

(c) If an employer provides a new health plan or benefits or changes health benefits or plans while an employee is on FMLA leave, the employee is entitled to the new or changed plan/benefits to the same extent as if the employee were not on leave. For example, if an employer changes a group health plan so that dental care becomes covered under the plan, an employee on FMLA leave must be given the same opportunity as other employees to receive (or obtain) the dental care coverage. Any other plan changes (e.g., in coverage, premiums, deductibles, etc.) which apply to all employees of the workforce would also apply to an employee on FMLA leave.

(d) Notice of any opportunity to change plans or benefits must also be given to an employee on FMLA leave. If the group health plan permits an employee to change from single to family coverage upon the birth of a child or otherwise add new family members, such a change in benefits must be made available while an employee is on FMLA leave. If the employee requests the changed coverage it must be provided by the employer.

(e) An employee may choose not to retain group health plan coverage during FMLA leave. However, when an employee returns from leave, the employee is entitled to be reinstated on the same terms as prior to taking the leave, including family or dependent coverages, without any qualifying period, physical examination, exclusion of pre-existing conditions, etc. *See* Section 825.212(c).

(f) Except as required by the Consolidated Omnibus Budget Reconciliation Act of 1986 (COBRA) and for key employees (as discussed below), an employer's obligation to maintain health benefits during leave (and to restore the employee to the same or equivalent employment) under FMLA ceases if and when the employment relationship would have terminated if the employee had not taken FMLA leave (e.g., if the employee's position is eliminated as part of a nondiscriminatory reduction in force and the employee would not have been transferred to another position); an employee informs the employer of his or her intent not to return from leave (including before starting the leave if the employer is so informed before the leave starts); or the employee fails to return from leave or continues on leave after exhausting his or her FMLA leave entitlement in the 12-month period.

(g) If a key employee (*see* Section 825.218) does not return from leave when notified by the employer that substantial or grievous economic injury will result from his or her reinstatement, the employee's entitlement to group health plan benefits continues unless and until the employee advises the employer that the employee does not desire restoration to employment at the end of the leave period, or the FMLA leave entitlement is exhausted, or reinstatement is actually denied.

(h) An employee's entitlement to benefits other than group health benefits during a period of FMLA leave (e.g., holiday pay) is to be determined by the employer's established policy for providing such benefits when the employee is on other forms of leave (paid or unpaid, as appropriate).

[78 FR 8902, Feb. 6, 2013, as amended at 82 FR 2230, Jan. 9, 2017]

Section 825.210 Employee payment of group health benefit premiums

(a) Group health plan benefits must be maintained on the same basis as coverage would have been provided if the employee had been continuously employed during the FMLA leave period. Therefore, any share of group health plan premiums which had been paid by the employee prior to FMLA leave must continue to be paid by the employee during the FMLA leave period. If premiums are raised or lowered, the employee would be required to pay the new premium rates. Maintenance of health insurance policies which are not a part of the employer's group health plan, as described in Section 825.209(a), are the sole responsibility of the employee. The employee and the insurer should make necessary arrangements for payment of premiums during periods of unpaid FMLA leave.

(b) If the FMLA leave is substituted paid leave, the employee's share of premiums must be paid by the method normally used during any paid leave, presumably as a payroll deduction.

(c) If FMLA leave is unpaid, the employer has a number of options for obtaining payment from the employee. The employer may require that payment be made to the employer or to the insurance carrier, but no additional charge may be added to the employee's premium payment for administrative expenses. The employer may require employees to pay their share of premium payments in any of the following ways:

(1) Payment would be due at the same time as it would be made if by payroll deduction;

(2) Payment would be due on the same schedule as payments are made under COBRA;

(3) Payment would be prepaid pursuant to a cafeteria plan at the employee's option;

(4) The employer's existing rules for payment by employees on leave without pay would be followed, provided that such rules do not require prepayment (*i.e.*, prior to the commencement of the leave) of the premiums that will become due during a period of unpaid FMLA leave or payment of higher premiums than if the employee had continued to work instead of taking leave; or,

(5) Another system voluntarily agreed to between the employer and the employee, which may include prepayment of premiums (e.g., through increased payroll deductions when the need for the FMLA leave is foreseeable).

(d) The employer must provide the employee with advance written notice of the terms and conditions under which these payments must be made. *See* Section 825.300(c).

(e) An employer may not require more of an employee using unpaid FMLA leave than the employer requires of other employees on leave without pay.

(f) An employee who is receiving payments as a result of a workers' compensation injury must make arrangements with the employer for payment of group health plan benefits when simultaneously taking FMLA leave. *See* Section 825.207(e).

Section 825.211 Maintenance of benefits under multi-employer health plans

(a) A multi-employer health plan is a plan to which more than one employer is required to contribute, and which is maintained pursuant to one or more collective bargaining agreements between employee organization(s) and the employers.

(b) An employer under a multi-employer plan must continue to make contributions on behalf of an employee using FMLA leave as though the employee had been continuously employed, unless the plan contains an explicit FMLA provision for maintaining coverage such as through pooled contributions by all employers party to the plan.

(c) During the duration of an employee's FMLA leave, coverage by the group health plan, and benefits provided pursuant to the plan, must be maintained at the level of coverage and benefits which were applicable to the employee at the time FMLA leave commenced.

(d) An employee using FMLA leave cannot be required to use banked hours or pay a greater premium than the employee would have been required to pay if the employee had been continuously employed.

(e) As provided in Section 825.209(f) of this part, group health plan coverage must be maintained for an employee on FMLA leave until:

(1) The employee's FMLA leave entitlement is exhausted;

(2) The employer can show that the employee would have been laid off and the employment relationship terminated; or,

(3) The employee provides unequivocal notice of intent not to return to work.

SECTION 825.212 EMPLOYEE FAILURE TO PAY HEALTH PLAN PREMIUM PAYMENTS

(a)(1) In the absence of an established employer policy providing a longer grace period, an employer's obligations to maintain health insurance coverage cease under FMLA if an employee's premium payment is more than 30 days late. In order to drop the coverage for an employee whose premium payment is late, the employer must provide written notice to the employee that the payment has not been received. Such notice must be mailed to the employee at least 15 days before coverage is to cease, advising that coverage will be dropped on a specified date at least 15 days after the date of the letter unless the payment has been received by that date. If the employer has established policies regarding other forms of unpaid leave that provide for the employer to cease coverage retroactively to the date the unpaid premium payment was due, the employer may drop the employee from coverage retroactively in accordance with that policy, provided the 15-day notice was given. In the absence of such a policy, coverage for the employee may be terminated at the end of the 30-day grace period, where the required 15-day notice has been provided.

(2) An employer has no obligation regarding the maintenance of a health insurance policy which is not a group health plan. *See* Section 825.209(a).

(3) All other obligations of an employer under FMLA would continue; for example, the employer continues to have an obligation to reinstate an employee upon return from leave.

(b) The employer may recover the employee's share of any premium payments missed by the employee for any FMLA leave period during which the employer maintains health coverage by paying the employee's share after the premium payment is missed.

(c) If coverage lapses because an employee has not made required premium payments, upon the employee's return from FMLA leave the employer must still restore the employee to coverage/benefits equivalent to those the employee would have had if leave had not been taken and the premium payment(s) had not been missed, including family or dependent coverage. *See* Section 825.215(d)(1)-(5). In such case, an employee may not be required to meet any qualification requirements imposed by the plan, including any new preexisting condition waiting period, to wait for an open season, or to pass a medical examination to obtain reinstatement of coverage. If an employer terminates an employee's insurance in accordance with this section and fails to restore the employee's health insurance as required by this section upon the employee's return, the employer may be liable for benefits lost by reason of the violation, for other actual monetary losses sustained as a direct result of the violation, and for appropriate equitable relief tailored to the harm suffered.

Section 825.213 Employer recovery of benefit costs

(a) In addition to the circumstances discussed in Section 825.212(b), an employer may recover its share of health plan premiums during a period of unpaid FMLA leave from an employee if the employee fails to return to work after the employee's FMLA leave entitlement has been exhausted or expires, unless the reason the employee does not return is due to:

(1) The continuation, recurrence, or onset of either a serious health condition of the employee or the employee's family member, or a serious injury or illness of a covered servicemember, which would otherwise entitle the employee to leave under FMLA; or

(2) Other circumstances beyond the employee's control. Examples of other circumstances beyond the employee's control are necessarily broad. They include such situations as where a parent chooses to stay home with a newborn child who has a serious health condition; an employee's spouse is unexpectedly transferred to a job location more than 75 miles from the employee's worksite; a relative or individual other than a covered family member has a serious health condition and the employee is needed to provide care; the employee is laid off while on leave; or, the employee is a key employee who decides not to return to work upon being notified of the employer's intention to deny restoration because of substantial and grievous economic injury to the employer's operations and is not reinstated by the employer. Other circumstances beyond the employee's control would not include a situation where an employee desires to remain with a parent in a distant city even though the parent no longer requires the employee's care, or a parent chooses not to return to work to stay home with a well, newborn child.

(3) When an employee fails to return to work because of the continuation, recurrence, or onset of either a serious health condition of the employee or employee's family member, or a serious injury or illness of a covered servicemember, thereby precluding the employer from recovering its (share of) health benefit premium payments made on the employee's behalf during a period of unpaid FMLA leave, the employer may require medical certification of the employee's or the family member's serious health condition or the covered servicemember's serious injury or illness. Such certification is not required unless requested by the employer. The cost of the certification shall be borne by the employee, and the employee is not entitled to

be paid for the time or travel costs spent in acquiring the certification. The employee is required to provide medical certification in a timely manner which, for purposes of this section, is within 30 days from the date of the employer's request. For purposes of medical certification, the employee may use the optional DOL forms developed for these purposes. *See* Sections 825.306(b), 825.310(c)-(d). If the employer requests medical certification and the employee does not provide such certification in a timely manner (within 30 days), or the reason for not returning to work does not meet the test of other circumstances beyond the employee's control, the employer may recover 100 percent of the health benefit premiums it paid during the period of unpaid FMLA leave.

(b) Under some circumstances an employer may elect to maintain other benefits, e.g., life insurance, disability insurance, etc., by paying the employee's (share of) premiums during periods of unpaid FMLA leave. For example, to ensure the employer can meet its responsibilities to provide equivalent benefits to the employee upon return from unpaid FMLA leave, it may be necessary that premiums be paid continuously to avoid a lapse of coverage. If the employer elects to maintain such benefits during the leave, at the conclusion of leave, the employer is entitled to recover only the costs incurred for paying the employee's share of any premiums whether or not the employee returns to work.

(c) An employee who returns to work for at least 30 calendar days is considered to have returned to work. An employee who transfers directly from taking FMLA leave to retirement, or who retires during the first 30 days after the employee returns to work, is deemed to have returned to work.

(d) When an employee elects or an employer requires paid leave to be substituted for FMLA leave, the employer may not recover its (share of) health insurance or other non-health benefit premiums for any period of FMLA leave covered by paid leave. Because paid leave provided under a plan covering temporary disabilities (including workers' compensation) is not unpaid, recovery of health insurance premiums does not apply to such paid leave.

(e) The amount that self-insured employers may recover is limited to only the employer's share of allowable premiums as would be calculated under COBRA, excluding the two percent fee for administrative costs.

(f) When an employee fails to return to work, any health and non-health benefit premiums which this section of the regulations permits an employer to recover are a debt owed by the non-returning employee to the employer. The existence of this debt caused by the employee's failure to return to work does not alter the employer's responsibilities for health benefit coverage and, under a self-insurance plan, payment of claims incurred during the period of FMLA leave. To the extent recovery is allowed, the employer may recover the costs through deduction from any sums due to the employee (e.g., unpaid wages, vacation pay, profit sharing, etc.), provided such deductions do not otherwise violate applicable Federal or State wage payment or other laws. Alternatively, the employer may initiate legal action against the employee to recover such costs.

Section 825.214 Employee right to reinstatement

General rule. On return from FMLA leave, an employee is entitled to be returned to the same position the employee held when leave commenced, or to an equivalent position with equivalent benefits, pay, and other terms and conditions of employment. An employee is entitled to such reinstatement even if the employee has been replaced or his or her position has been restructured to accommodate the employee's absence. *See also* Section 825.106(e) for the obligations of joint employers.

Section 825.215 Equivalent position

(a) *Equivalent position.* An equivalent position is one that is virtually identical to the employee's former position in terms of pay, benefits and working conditions, including privileges, perquisites and status. It must involve the same or substantially similar duties and responsibilities, which must entail substantially equivalent skill, effort, responsibility, and authority.

(b) *Conditions to qualify.* If an employee is no longer qualified for the position because of the employee's inability to attend a necessary course, renew a license, fly a minimum number of hours, etc., as a result of the leave, the employee shall be given a reasonable opportunity to fulfill those conditions upon return to work.

(c) *Equivalent pay.* (1) An employee is entitled to any unconditional pay increases which may have occurred during the FMLA leave period, such as cost of living increases. Pay increases conditioned upon seniority, length of service, or work performed must be granted in accordance with the employer's policy or practice with respect to other employees on an equivalent leave status for a reason that does not qualify as FMLA leave. An employee is entitled to be restored to a position with the same or equivalent pay premiums, such as a shift differential. If an employee departed from a position averaging ten hours of overtime (and corresponding overtime pay) each week, an employee is ordinarily entitled to such a position on return from FMLA leave.

(2) Equivalent pay includes any bonus or payment, whether it is discretionary or non-discretionary, made to employees consistent with the provisions of paragraph (c)(1) of this section. However, if a bonus or other payment is based on the achievement of a specified goal such as hours worked, products sold or perfect attendance, and the employee has not met the goal due to FMLA leave, then the payment may be denied, unless otherwise paid to employees on an equivalent leave status for a reason that does not qualify as FMLA leave. For example, if an employee who used paid vacation leave for a non-FMLA purpose would receive the payment, then the employee who used paid vacation leave for an FMLA-protected purpose also must receive the payment.

(d) *Equivalent benefits.* Benefits include all benefits provided or made available to employees by an employer, including group life insurance, health insurance, disability insurance, sick leave, annual leave, educational benefits, and pensions, regardless of whether such benefits are provided by a practice or written policy of an employer through an employee benefit plan as defined in Section 3(3) of the Employee Retirement Income Security Act of 1974, 29 U.S.C. 1002(3).

(1) At the end of an employee's FMLA leave, benefits must be resumed in the same manner and at the same levels as provided when the leave began, and subject to any changes in benefit levels that may have taken place during the period of FMLA leave affecting the entire workforce, unless otherwise elected by the employee. Upon return from FMLA leave, an employee cannot be required to requalify for any benefits the employee enjoyed before FMLA leave began (including family or dependent coverages). For example, if an employee was covered by a life insurance policy before taking leave but is not covered or coverage lapses during the period of unpaid FMLA leave, the employee cannot be required to meet any qualifications, such as taking a physical examination, in order to requalify for life insurance upon return from leave. Accordingly, some employers may find it necessary to modify life insurance and other benefits programs in order to restore employees to equivalent benefits upon return from FMLA leave, make arrangements for continued payment of costs to maintain such benefits during unpaid FMLA leave, or pay these costs subject to recovery from the employee on return from leave. *See* Section 825.213(b).

(2) An employee may, but is not entitled to, accrue any additional benefits or seniority during unpaid FMLA leave. Benefits accrued at the time leave began, however, (e.g., paid vacation, sick or personal leave to the extent not substituted for FMLA leave) must be available to an employee upon return from leave.

(3) If, while on unpaid FMLA leave, an employee desires to continue life insurance, disability insurance, or other types of benefits for which he or she typically pays, the employer is required to follow established policies or practices for continuing such benefits for other instances of leave without pay. If the employer has no established policy, the employee and the employer are encouraged to agree upon arrangements before FMLA leave begins.

(4) With respect to pension and other retirement plans, any period of unpaid FMLA leave shall not be treated as or counted toward a break in service for purposes of vesting and eligibility to participate. Also, if the plan requires an employee to be employed on a specific date in order to be credited with a year of service for vesting, contributions or participation purposes, an employee on unpaid FMLA leave on that date shall be deemed to have been employed on that date. However, unpaid FMLA leave periods need not be treated as credited service for purposes of benefit accrual, vesting and eligibility to participate.

(5) Employees on unpaid FMLA leave are to be treated as if they continued to work for purposes of changes to benefit plans. They are entitled to changes in benefits plans, except those which may be dependent upon seniority or accrual during the

leave period, immediately upon return from leave or to the same extent they would have qualified if no leave had been taken. For example, if the benefit plan is predicated on a pre-established number of hours worked each year and the employee does not have sufficient hours as a result of taking unpaid FMLA leave, the benefit is lost. (In this regard, Section 825.209 addresses health benefits.)

(e) *Equivalent terms and conditions of employment.* An equivalent position must have substantially similar duties, conditions, responsibilities, privileges and status as the employee's original position.

(1) The employee must be reinstated to the same or a geographically proximate worksite (*i.e.,* one that does not involve a significant increase in commuting time or distance) from where the employee had previously been employed. If the employee's original worksite has been closed, the employee is entitled to the same rights as if the employee had not been on leave when the worksite closed. For example, if an employer transfers all employees from a closed worksite to a new worksite in a different city, the employee on leave is also entitled to transfer under the same conditions as if he or she had continued to be employed.

(2) The employee is ordinarily entitled to return to the same shift or the same or an equivalent work schedule.

(3) The employee must have the same or an equivalent opportunity for bonuses, profit-sharing, and other similar discretionary and non-discretionary payments.

(4) FMLA does not prohibit an employer from accommodating an employee's request to be restored to a different shift, schedule, or position which better suits the employee's personal needs on return from leave, or to offer a promotion to a better position. However, an employee cannot be induced by the employer to accept a different position against the employee's wishes.

(f) *De minimis exception.* The requirement that an employee be restored to the same or equivalent job with the same or equivalent pay, benefits, and terms and conditions of employment does not extend to de minimis, intangible, or unmeasurable aspects of the job.

Section 825.216 Limitations on an employee's right to reinstatement

(a) An employee has no greater right to reinstatement or to other benefits and conditions of employment than if the employee had been continuously employed during the FMLA leave period. An employer must be able to show that an employee would not otherwise have been employed at the time reinstatement is requested in order to deny restoration to employment. For example:

(1) If an employee is laid off during the course of taking FMLA leave and employment is terminated, the employer's responsibility to continue FMLA leave, maintain group health plan benefits and restore the employee cease at the time the employee is laid off, provided the employer has no continuing obligations under a collective bargaining

agreement or otherwise. An employer would have the burden of proving that an employee would have been laid off during the FMLA leave period and, therefore, would not be entitled to restoration. Restoration to a job slated for lay-off when the employee's original position is not would not meet the requirements of an equivalent position.

(2) If a shift has been eliminated, or overtime has been decreased, an employee would not be entitled to return to work that shift or the original overtime hours upon restoration. However, if a position on, for example, a night shift has been filled by another employee, the employee is entitled to return to the same shift on which employed before taking FMLA leave.

(3) If an employee was hired for a specific term or only to perform work on a discrete project, the employer has no obligation to restore the employee if the employment term or project is over and the employer would not otherwise have continued to employ the employee. On the other hand, if an employee was hired to perform work on a contract, and after that contract period the contract was awarded to another contractor, the successor contractor may be required to restore the employee if it is a successor employer. *See* Section 825.107.

(b) In addition to the circumstances explained above, an employer may deny job restoration to salaried eligible employees (key employees, as defined in Section 825.217(c)), if such denial is necessary to prevent substantial and grievous economic injury to the operations of the employer; or may delay restoration to an employee who fails to provide a fitness-for-duty certificate to return to work under the conditions described in Section 825.312.

(c) If the employee is unable to perform an essential function of the position because of a physical or mental condition, including the continuation of a serious health condition or an injury or illness also covered by workers' compensation, the employee has no right to restoration to another position under the FMLA. The employer's obligations may, however, be governed by the Americans with Disabilities Act (ADA), as amended. *See* Section 825.702, state leave laws, or workers' compensation laws.

(d) An employee who fraudulently obtains FMLA leave from an employer is not protected by FMLA's job restoration or maintenance of health benefits provisions.

(e) If the employer has a uniformly-applied policy governing outside or supplemental employment, such a policy may continue to apply to an employee while on FMLA leave. An employer which does not have such a policy may not deny benefits to which an employee is entitled under FMLA on this basis unless the FMLA leave was fraudulently obtained as in paragraph (d) of this section.

Section 825.217 Key employee, general rule

(a) A *key employee* is a salaried FMLA-eligible employee who is among the highest paid 10 percent of all the employees employed by the employer within 75 miles of the employee's worksite.

(b) The term *salaried* means paid on a salary basis, as defined in 29 CFR 541.602. This is the Department of Labor regulation defining employees who may qualify as exempt from the minimum wage and overtime requirements of the FLSA as executive, administrative, professional, and computer employees.

(c) A key employee must be among the highest paid 10 percent of all the employees—both salaried and non-salaried, eligible and ineligible—who are employed by the employer within 75 miles of the worksite.

(1) In determining which employees are among the highest paid 10 percent, year-to-date earnings are divided by weeks worked by the employee (including weeks in which paid leave was taken). Earnings include wages, premium pay, incentive pay, and non-discretionary and discretionary bonuses. Earnings do not include incentives whose value is determined at some future date, e.g., stock options, or benefits or perquisites.

(2) The determination of whether a salaried employee is among the highest paid 10 percent shall be made at the time the employee gives notice of the need for leave. No more than 10 percent of the employer's employees within 75 miles of the worksite may be key employees.

Section 825.218 Substantial and grievous economic injury

(a) In order to deny restoration to a key employee, an employer must determine that the restoration of the employee to employment will cause substantial and grievous economic injury to the operations of the employer, not whether the absence of the employee will cause such substantial and grievous injury.

(b) An employer may take into account its ability to replace on a temporary basis (or temporarily do without) the employee on FMLA leave. If permanent replacement is unavoidable, the cost of then reinstating the employee can be considered in evaluating whether substantial and grievous economic injury will occur from restoration; in other words, the effect on the operations of the company of reinstating the employee in an equivalent position.

(c) A precise test cannot be set for the level of hardship or injury to the employer which must be sustained. If the reinstatement of a key employee threatens the economic viability of the firm, that would constitute substantial and grievous economic injury. A lesser injury which causes substantial, long-term economic injury would also be sufficient. Minor inconveniences and costs that the employer would experience in the normal course of doing business would certainly not constitute substantial and grievous economic injury.

(d) FMLA's substantial and grievous economic injury standard is different from and more stringent than the undue hardship test under the ADA. *See also* Section 825.702.

Section 825.219 Rights of a key employee

(a) An employer who believes that reinstatement may be denied to a key employee, must give written notice to the employee at the time the employee gives notice of the need for FMLA leave (or when FMLA leave commences, if earlier) that he or she qualifies as a key employee. At the same time, the employer must also fully inform the employee of the potential consequences with respect to reinstatement and maintenance of health benefits if the employer should determine that substantial and grievous economic injury to the employer's operations will result if the employee is reinstated from FMLA leave. If such notice cannot be given immediately because of the need to determine whether the employee is a key employee, it shall be given as soon as practicable after being notified of a need for leave (or the commencement of leave, if earlier). It is expected that in most circumstances there will be no desire that an employee be denied restoration after FMLA leave and, therefore, there would be no need to provide such notice. However, an employer who fails to provide such timely notice will lose its right to deny restoration even if substantial and grievous economic injury will result from reinstatement.

(b) As soon as an employer makes a good faith determination, based on the facts available, that substantial and grievous economic injury to its operations will result if a key employee who has given notice of the need for FMLA leave or is using FMLA leave is reinstated, the employer shall notify the employee in writing of its determination, that it cannot deny FMLA leave, and that it intends to deny restoration to employment on completion of the FMLA leave. It is anticipated that an employer will ordinarily be able to give such notice prior to the employee starting leave. The employer must serve this notice either in person or by certified mail. This notice must explain the basis for the employer's finding that substantial and grievous economic injury will result, and, if leave has commenced, must provide the employee a reasonable time in which to return to work, taking into account the circumstances, such as the length of the leave and the urgency of the need for the employee to return.

(c) If an employee on leave does not return to work in response to the employer's notification of intent to deny restoration, the employee continues to be entitled to maintenance of health benefits and the employer may not recover its cost of health benefit premiums. A key employee's rights under FMLA continue unless and until the employee either gives notice that he or she no longer wishes to return to work, or the employer actually denies reinstatement at the conclusion of the leave period.

(d) After notice to an employee has been given that substantial and grievous economic injury will result if the employee is reinstated to employment, an employee is still entitled to request reinstatement at the end of the leave period even if the employee did not return to work in response to the employer's notice. The employer must then again determine whether there will be substantial and grievous economic injury from reinstatement, based on the facts at that time. If it is determined that substantial and grievous economic injury will result, the employer shall notify the employee in writing (in person or by certified mail) of the denial of restoration.

Section 825.220 Protection for employees who request leave or otherwise assert FMLA rights

(a) The FMLA prohibits interference with an employee's rights under the law, and with legal proceedings or inquiries relating to an employee's rights. More specifically, the law contains the following employee protections:

(1) An employer is prohibited from interfering with, restraining, or denying the exercise of (or attempts to exercise) any rights provided by the Act.

(2) An employer is prohibited from discharging or in any other way discriminating against any person (whether or not an employee) for opposing or complaining about any unlawful practice under the Act.

(3) All persons (whether or not employers) are prohibited from discharging or in any other way discriminating against any person (whether or not an employee) because that person has—

(i) Filed any charge, or has instituted (or caused to be instituted) any proceeding under or related to this Act;

(ii) Given, or is about to give, any information in connection with an inquiry or proceeding relating to a right under this Act;

(iii) Testified, or is about to testify, in any inquiry or proceeding relating to a right under this Act.

(b) Any violations of the Act or of these regulations constitute interfering with, restraining, or denying the exercise of rights provided by the Act. An employer may be liable for compensation and benefits lost by reason of the violation, for other actual monetary losses sustained as a direct result of the violation, and for appropriate equitable or other relief, including employment, reinstatement, promotion, or any other relief tailored to the harm suffered. *See* Section 825.400(c). Interfering with the exercise of an employee's rights would include, for example, not only refusing to authorize FMLA leave, but discouraging an employee from using such leave. It would also include manipulation by a covered employer to avoid responsibilities under FMLA, for example:

(1) Transferring employees from one worksite to another for the purpose of reducing worksites, or to keep worksites, below the 50-employee threshold for employee eligibility under the Act;

(2) Changing the essential functions of the job in order to preclude the taking of leave;

(3) Reducing hours available to work in order to avoid employee eligibility.

(c) The Act's prohibition against interference prohibits an employer from discriminating or retaliating against an employee or prospective employee for having exercised or attempted to exercise FMLA rights. For example, if an employee on leave without pay would otherwise be entitled to full benefits (other than health benefits), the same benefits would be required to be provided to an employee on unpaid FMLA leave. By the same token, employers cannot use the taking of FMLA leave as a negative factor in employment actions, such as hiring, promotions or disciplinary actions; nor can FMLA leave be counted under no fault attendance policies. *See* Section 825.215.

(d) Employees cannot waive, nor may employers induce employees to waive, their prospective rights under FMLA. For example, employees (or their collective bargaining representatives) cannot trade off the right to take FMLA leave against some other benefit offered by the employer. This does not prevent the settlement or release of FMLA claims by employees based on past employer conduct without the approval of the Department of Labor or a court. Nor does it prevent an employee's voluntary and uncoerced acceptance (not as a condition of employment) of a light duty assignment while recovering from a serious health condition. *See* Section 825.702(d). An employee's acceptance of such light duty assignment does not constitute a waiver of the employee's prospective rights, including the right to be restored to the same position the employee held at the time the employee's FMLA leave commenced or to an equivalent position. The employee's right to restoration, however, ceases at the end of the applicable 12-month FMLA leave year.

(e) Individuals, and not merely employees, are protected from retaliation for opposing (e.g., filing a complaint about) any practice which is unlawful under the Act. They are similarly protected if they oppose any practice which they reasonably believe to be a violation of the Act or regulations.

SUBPART C—EMPLOYEE AND EMPLOYER RIGHTS AND OBLIGATIONS UNDER THE ACT

Section 825.300 Employer notice requirements

(a) *General notice.* (1) Every employer covered by the FMLA is required to post and keep posted on its premises, in conspicuous places where employees are employed, a notice explaining the Act's provisions and providing information concerning the procedures for filing complaints of violations of the Act with the Wage and Hour Division. The notice must be posted prominently where it can be readily seen by employees and applicants for employment. The poster and the text must be large enough to be easily read and contain fully legible text. Electronic posting is sufficient to meet this posting requirement as long as it otherwise meets the requirements of this section. An employer that willfully violates the posting requirement may be assessed a civil money penalty by the Wage and Hour Division not to exceed $169 for each separate offense.

(2) Covered employers must post this general notice even if no employees are eligible for FMLA leave.

(3) If an FMLA-covered employer has any eligible employees, it shall also provide this general notice to each employee by including the notice in employee handbooks or other written guidance to employees concerning employee benefits or leave rights, if such written materials exist, or by distributing a copy of the general notice to each new employee upon hiring. In either case, distribution may be accomplished electronically.

(4) To meet the requirements of paragraph (a)(3) of this section, employers may duplicate the text of the Department's prototype notice (WHD Publication 1420) or may use another format so long as the information provided includes, at a minimum, all of the information contained in that notice. Where an employer's workforce is comprised of a significant portion of workers who are not literate in English, the employer shall provide the general notice in a language in which the employees are literate. Prototypes are available from the nearest office of the Wage and Hour Division or on the Internet at www.dol.gov/whd. Employers furnishing FMLA notices to sensory-impaired individuals must also comply with all applicable requirements under Federal or State law.

(b) *Eligibility notice.* (1) When an employee requests FMLA leave, or when the employer acquires knowledge that an employee's leave may be for an FMLA-qualifying reason, the employer must notify the employee of the employee's eligibility to take FMLA leave within five business days, absent extenuating circumstances. *See* Section 825.110 for definition of an eligible employee and Section 825.801 for special hours of service eligibility requirements for airline flight crews. Employee eligibility is determined (and notice must be provided) at the commencement of the first instance of leave for each FMLA-qualifying reason in the applicable 12-month period. *See* Sections 825.127(c) and 825.200(b). All FMLA absences for the same qualifying reason are considered a single leave and employee eligibility as to that reason for leave does not change during the applicable 12-month period.

(2) The eligibility notice must state whether the employee is eligible for FMLA leave as defined in Section 825.110. If the employee is not eligible for FMLA leave, the notice must state at least one reason why the employee is not eligible, including as applicable the number of months the employee has been employed by the employer, the hours of service with the employer during the 12-month period, and whether the employee is employed at a worksite where 50 or more employees are employed by the employer within 75 miles of that worksite. Notification of eligibility may be oral or in writing; employers may use optional Form WH-381 (Notice of Eligibility and Rights and Responsibility) to provide such notification to employees. Prototypes are available from the nearest office of the Wage and Hour Division or on the Internet at www.dol.gov/whd. The employer is obligated to translate this notice in any situation in which it is obligated to do so in Section 825.300(a)(4).

(3) If, at the time an employee provides notice of a subsequent need for FMLA leave during the applicable 12-month period due to a different FMLA-qualifying reason, and the employee's eligibility status has not changed, no additional eligibility notice is required. If, however, the employee's eligibility status has changed (e.g., if the employee has not met the hours of service requirement in the 12 months preceding the commencement of leave for the subsequent qualifying reason or the size of the workforce at the worksite has dropped below 50 employees), the employer must notify the employee of the change in eligibility status within five business days, absent extenuating circumstances.

(c) *Rights and responsibilities notice.* (1) Employers shall provide written notice detailing the specific expectations and obligations of the employee and explaining any consequences of a failure to meet these obligations. The employer is obligated to translate this notice in any situation in which it is obligated to do so in Section 825.300(a)(4). This notice shall be provided to the employee each time the eligibility notice is provided pursuant to paragraph (b) of this section. If leave has already begun, the notice should be mailed to the employee's address of record. Such specific notice must include, as appropriate:

(i) That the leave may be designated and counted against the employee's annual FMLA leave entitlement if qualifying (*see* Sections 825.300(c) and 825.301) and the applicable 12-month period for FMLA entitlement (*see* Sections 825.127(c), 825.200(b), (f), and (g));

(ii) Any requirements for the employee to furnish certification of a serious health condition, serious injury or illness, or qualifying exigency arising out of covered active duty or call to covered active duty status, and the consequences of failing to do so (*see* Sections 825.305, 825.309, 825.310, 825.313);

(iii) The employee's right to substitute paid leave, whether the employer will require the substitution of paid leave, the conditions related to any substitution, and the employee's entitlement to take unpaid FMLA leave if the employee does not meet the conditions for paid leave (*see* Section 825.207);

(iv) Any requirement for the employee to make any premium payments to maintain health benefits and the arrangements for making such payments (*see* Section 825.210), and the possible consequences of failure to make such payments on a timely basis (*i.e.,* the circumstances under which coverage may lapse);

(v) The employee's status as a key employee and the potential consequence that restoration may be denied following FMLA leave, explaining the conditions required for such denial (*see* Section 825.218);

(vi) The employee's rights to maintenance of benefits during the FMLA leave and restoration to the same or an equivalent job upon return from FMLA leave (*see* Sections 825.214 and 825.604); and

(vii) The employee's potential liability for payment of health insurance premiums paid by the employer during the employee's unpaid FMLA leave if the employee fails to return to work after taking FMLA leave (*see* Section 825.213).

(2) The notice of rights and responsibilities may include other information—e.g., whether the employer will require periodic reports of the employee's status and intent to return to work—but is not required to do so.

(3) The notice of rights and responsibilities may be accompanied by any required certification form.

(4) If the specific information provided by the notice of rights and responsibilities changes, the employer shall, within five business days of receipt of the employee's first notice of need for leave subsequent to any change, provide written notice referencing the prior notice and setting forth any of the information in the notice of rights and responsibilities that has changed. For example, if the initial leave period was paid leave and the subsequent leave period would be unpaid leave, the employer may need to give notice of the arrangements for making premium payments.

(5) Employers are also expected to responsively answer questions from employees concerning their rights and responsibilities under the FMLA.

(6) A prototype notice of rights and responsibilities may be obtained from local offices of the Wage and Hour Division or from the Internet at *www.dol.gov/whd.* Employers may adapt the prototype notice as appropriate to meet these notice requirements. The notice of rights and responsibilities may be distributed electronically so long as it otherwise meets the requirements of this section.

(d) *Designation notice.* (1) The employer is responsible in all circumstances for designating leave as FMLA-qualifying, and for giving notice of the designation to the employee as provided in this section. When the employer has enough information to determine whether the leave is being taken for a FMLA-qualifying reason (e.g., after receiving a certification), the employer must notify the employee whether the leave will be designated and will be counted as FMLA leave within five business days absent extenuating circumstances. Only one notice of designation is required for each FMLA-qualifying reason per applicable 12-month period, regardless of whether the leave taken due to the qualifying reason will be a continuous block of leave or intermittent or reduced schedule leave. If the employer determines that the leave will not be designated as FMLA-qualifying (e.g., if the leave is not for a reason covered by FMLA or the FMLA leave entitlement has been exhausted), the employer must notify the employee of that determination. If the employer requires paid leave to be substituted for unpaid FMLA leave, or that paid leave taken under an existing leave plan be counted as FMLA leave, the employer must inform the employee of this designation at the time of designating the FMLA leave.

(2) If the employer has sufficient information to designate the leave as FMLA leave immediately after receiving notice of the employee's need for leave, the employer may provide the employee with the designation notice at that time.

(3) If the employer will require the employee to present a fitness-for-duty certification to be restored to employment, the employer must provide notice of such requirement with the designation notice. If the employer will require that the fitness-for-duty certification address the employee's ability to perform the essential functions of the employee's position, the employer must so indicate in the designation notice, and must include a list of the essential functions of the employee's position. *See* Section 825.312. If the employer handbook or other written documents (if any) describing the employer's leave policies clearly provide that a fitness-for-duty certification will be required in specific circumstances (e.g., by stating that fitness-for-duty certification will be required in all cases of back injuries for employees in a certain occupation), the employer is not required to provide written notice of the requirement with the designation notice, but must provide oral notice no later than with the designation notice.

(4) The designation notice must be in writing. A prototype designation notice may be obtained from local offices of the Wage and Hour Division or from the Internet at *www.dol.gov/whd.* If the leave is not designated as FMLA leave because it does not meet the requirements of the Act, the notice to the employee that the leave is not designated as FMLA leave may be in the form of a simple written statement.

(5) If the information provided by the employer to the employee in the designation notice changes (e.g., the employee exhausts the FMLA leave entitlement), the employer shall provide, within five business days of receipt of the employee's first notice of need for leave subsequent to any change, written notice of the change.

(6) The employer must notify the employee of the amount of leave counted against the employee's FMLA leave entitlement. If the amount of leave needed is known at the time the employer designates the leave as FMLA-qualifying, the employer must notify the employee of the number of hours, days, or weeks that will be counted against the employee's FMLA leave entitlement in the designation notice. If it is not possible to provide the hours, days, or weeks that will be counted against the employee's FMLA leave entitlement (such as in the case of unforeseeable intermittent leave), then the employer must provide notice of the amount of leave counted against the employee's FMLA leave entitlement upon the request by the employee, but no more often than once in a 30-day period and only if leave was taken in that period. The notice of the amount of leave counted against the employee's FMLA entitlement may be oral or in writing. If such notice is oral, it shall be confirmed in writing, no later than the following payday (unless the payday is less than one week after the oral notice, in which case the notice must be no later than the subsequent payday). Such written notice may be in any form, including a notation on the employee's pay stub.

(e) *Consequences of failing to provide notice.* Failure to follow the notice requirements set forth in this section may constitute an interference with, restraint, or denial of the exercise of an employee's FMLA rights. An employer may be liable for compensation and benefits lost by reason of the violation, for other actual monetary losses sustained as a direct result of the violation, and for appropriate equitable or other relief, including employment, reinstatement, promotion, or any other relief tailored to the harm suffered *See* Section 825.400(c).

[78 FR 8902, Feb. 6, 2013, as amended at 81 FR 43452, July 1, 2016; 82 FR 5382, Jan. 18, 2017; 82 FR 14, Jan. 2, 2018]

Section 825.301 Designation of FMLA leave

(a) *Employer responsibilities.* The employer's decision to designate leave as FMLA-qualifying must be based only on information received from the employee or the employee's spokesperson (e.g., if the employee is incapacitated, the employee's spouse, adult child, parent, doctor, etc., may provide notice to the employer of the need to take FMLA leave). In any circumstance where the employer does not have sufficient information about the reason for an employee's use of leave, the employer should inquire further of the employee or the spokesperson to ascertain whether leave is potentially FMLA-qualifying. Once the employer has acquired knowledge that the leave is being taken for a FMLA-qualifying reason, the employer must notify the employee as provided in Section 825.300(d).

(b) *Employee responsibilities.* An employee giving notice of the need for FMLA leave does not need to expressly assert rights under the Act or even mention the FMLA to meet his or her obligation to provide notice, though the employee would need to state a qualifying reason for the needed leave and otherwise satisfy the notice requirements set forth in Section 825.302 or Section 825.303 depending on whether the need for leave is foreseeable or unforeseeable. An employee giving notice of the need for FMLA leave must explain the reasons for the needed leave so as to allow the employer to determine whether the leave qualifies under the Act. If the employee fails to explain the reasons, leave may be denied. In many cases, in explaining the reasons for a request to use leave, especially when the need for the leave was unexpected or unforeseen, an employee will provide sufficient information for the employer to designate the leave as FMLA leave. An employee using accrued paid leave may in some cases not spontaneously explain the reasons or their plans for using their accrued leave. However, if an employee requesting to use paid leave for a FMLA-qualifying reason does not explain the reason for the leave and the employer denies the employee's request, the employee will need to provide sufficient information to establish a FMLA-qualifying reason for the needed leave so that the employer is aware that the leave may not be denied and may designate that the paid leave be appropriately counted against (substituted for) the employee's FMLA leave entitlement. Similarly, an employee using accrued paid vacation leave who seeks an extension of unpaid leave for a FMLA-qualifying reason will need to state the reason. If this is due to an event which occurred during the period of paid leave, the employer may count the leave used after the FMLA-qualifying reason against the employee's FMLA leave entitlement.

(c) *Disputes.* If there is a dispute between an employer and an employee as to whether leave qualifies as FMLA leave, it should be resolved through discussions between the employee and the employer. Such discussions and the decision must be documented.

(d) *Retroactive designation.* If an employer does not designate leave as required by Section 825.300, the employer may retroactively designate leave as FMLA leave with appropriate notice to the employee as required by Section 825.300 provided that the employer's failure to timely designate leave does not cause harm or injury to the employee. In all cases where leave would qualify for FMLA protections, an employer and an employee can mutually agree that leave be retroactively designated as FMLA leave.

(e) *Remedies.* If an employer's failure to timely designate leave in accordance with Section 825.300 causes the employee to suffer harm, it may constitute an interference with, restraint of, or denial of the exercise of an employee's FMLA rights. An employer may be liable for compensation and benefits lost by reason of the violation, for other actual monetary losses sustained as a direct result of the violation, and for appropriate equitable or other relief, including employment, reinstatement, promotion, or any other relief tailored to the harm suffered. *See* Section 825.400(c). For example, if an employer that was put on notice that an employee needed FMLA leave failed to designate the leave properly, but the employee's own serious health condition prevented him or her from returning to work during that time period regardless of the designation, an employee may not be able to show that the employee suffered harm as a result of the employer's actions. However, if an employee took leave to provide care for a son or daughter with a serious health condition believing it would not count toward his or her FMLA entitlement, and the employee planned to later use that FMLA leave to provide care for a spouse who would need assistance when recovering from surgery planned for a later date, the employee may be able to show that harm has occurred as a result of the employer's failure to designate properly. The employee might establish this by showing that he or she would have arranged for an alternative caregiver for the seriously ill son or daughter if the leave had been designated timely.

Section 825.302 Employee notice requirements for foreseeable FMLA leave

(a) *Timing of notice.* An employee must provide the employer at least 30 days advance notice before FMLA leave is to begin if the need for the leave is foreseeable based on an expected birth, placement for adoption or foster care, planned medical treatment for a serious health condition of the employee or of a family member, or the planned medical treatment for a serious injury or illness of a covered servicemember. If 30 days notice is not practicable, such as because of a lack of knowledge of approximately when leave will be required to begin, a change in circumstances, or a medical emergency, notice must be given as soon as practicable. For example, an employee's health condition may require leave to commence earlier than anticipated before the birth of a child. Similarly, little opportunity for notice may be given before placement for adoption. For foreseeable leave due to a qualifying exigency notice must be provided as soon as practicable, regardless of how far in advance such leave is foreseeable. Whether FMLA leave is to be continuous or is to be taken intermittently or on a reduced schedule basis, notice need only be given one time, but the employee shall advise the employer as soon as practicable if dates of scheduled leave change or are extended, or were initially unknown. In those cases where the employee is required to provide at least 30 days notice of foreseeable leave and does not do so, the employee shall explain the reasons why such notice was not practicable upon a request from the employer for such information.

(b) *As soon as practicable* means as soon as both possible and practical, taking into account all of the facts and circumstances in the individual case. When an employee becomes aware of a need for FMLA leave less than 30 days in advance, it should be practicable for the employee to provide notice of the need for leave either the same day or the next business day. In all cases, however, the determination of when an employee could practicably provide notice must take into account the individual facts and circumstances.

(c) *Content of notice.* An employee shall provide at least verbal notice sufficient to make the employer aware that the employee needs FMLA-qualifying leave, and the anticipated timing and duration of the leave. Depending on the situation, such information may include that a condition renders the employee unable to perform the functions of the job; that the employee is pregnant or has been hospitalized overnight; whether the employee or the employee's family member is under the continuing care of a health care provider; if the leave is due to a qualifying exigency, that a military member is on covered active duty or call to covered active duty status (or has been notified of an impending call or order to covered active duty), and that the requested leave is for one of the reasons listed in Section 825.126(b); if the leave is for a family member, that the condition renders the family member unable to perform daily activities, or that the family member is a covered servicemember with a serious injury or illness; and the anticipated duration of the absence, if known. When an employee seeks leave for the first time for a FMLA-qualifying reason, the employee need not expressly assert rights under the FMLA or even mention the FMLA. When an employee seeks leave due to a FMLA-qualifying reason, for which the employer has previously provided FMLA-protected leave, the employee must specifically reference the qualifying reason for leave or the need for FMLA leave. In all cases, the employer should inquire further of the employee if it is necessary to have more information about whether FMLA leave is being sought by the employee, and obtain the necessary details of the leave to be taken. In the case of medical conditions, the employer may find it necessary to inquire further to determine if the leave is because of a serious health condition and may request medical certification to support the need for such leave. *See* Section 825.305. An employer may also request certification to support the need for leave for a qualifying exigency or for military caregiver leave. *See* Sections 825.309, 825.310). When an employee has been previously certified for leave due to more than one FMLA-qualifying reason, the employer may need to inquire further to determine for which qualifying reason the leave is needed. An employee has an obligation to respond to an employer's questions designed to determine whether an absence is potentially FMLA-qualifying. Failure to respond to reasonable employer inquiries regarding the leave request may result in denial of FMLA protection if the employer is unable to determine whether the leave is FMLA-qualifying.

(d) *Complying with employer policy.* An employer may require an employee to comply with the employer's usual and customary notice and procedural requirements for requesting leave, absent unusual circumstances. For example, an employer may require that written notice set forth the reasons for the requested leave, the anticipated duration of the leave, and the anticipated start of the leave. An employee also may be required by an employer's policy to contact a specific individual. Unusual circumstances would include situations such as when an employee is unable to comply with the employer's policy that requests for leave should be made by contacting a specific number because on the day the employee needs to provide notice of his or her need for FMLA leave there is no one to answer the call-in number and the voice mail box is full. Where an employee does not comply with the employer's usual notice and procedural requirements, and no unusual circumstances justify the failure to comply, FMLA-protected leave may be delayed or denied. However, FMLA-protected leave may not be delayed or denied where the employer's policy requires notice to be given sooner than set forth in paragraph (a) of this section and the employee provides timely notice as set forth in paragraph (a) of this section.

(e) *Scheduling planned medical treatment.* When planning medical treatment, the employee must consult with the employer and make a reasonable effort to schedule the treatment so as not to disrupt unduly the employer's operations, subject to the approval of the health care provider. Employees are ordinarily expected to consult with their employers prior to the scheduling of treatment in order to work out a treatment schedule which best suits the needs of both the employer and the employee. For example, if an employee who provides notice of the need to take FMLA leave on an intermittent basis for planned medical treatment neglects to consult with the employer to make a reasonable effort to arrange the schedule of treatments so as not to unduly disrupt the employer's operations, the employer may initiate discussions with the employee and require the employee to attempt to make such arrangements, subject to the approval of the health care provider. *See* Sections 825.203 and 825.205.

(f) Intermittent leave or leave on a reduced leave schedule must be medically necessary due to a serious health condition or a serious injury or illness. An employee shall advise the employer, upon request, of the reasons why the intermittent/reduced leave schedule is necessary and of the schedule for treatment, if applicable. The employee and employer shall attempt to work out a schedule for such leave that meets the employee's needs without unduly disrupting the employer's operations, subject to the approval of the health care provider.

(g) An employer may waive employees' FMLA notice requirements. *See* Section 825.304.

Section 825.303 Employee notice requirements for unforeseeable FMLA leave

(a) *Timing of notice.* When the approximate timing of the need for leave is not foreseeable, an employee must provide notice to the employer as soon as practicable under the facts and circumstances of the particular case. It generally should be practicable for the employee to provide notice of leave that is unforeseeable within the time prescribed by the employer's usual and customary notice requirements applicable to such leave. *See* Section 825.303(c). Notice may be given by the employee's spokesperson (e.g., spouse, adult family member, or other responsible party) if the employee is unable to do so personally. For example, if an employee's child has a severe asthma attack and the employee takes the child to the emergency room, the employee would not be required to leave his or her child in order to report the absence while the child is receiving emergency treatment. However, if the child's asthma attack required only the use of an inhaler at home followed by a period of rest, the employee would be expected to call the employer promptly after ensuring the child has used the inhaler.

(b) *Content of notice.* An employee shall provide sufficient information for an employer to reasonably determine whether the FMLA may apply to the leave request. Depending on the situation, such information may include that a condition renders the employee unable to perform the functions of the job; that the employee is pregnant or has been hospitalized overnight; whether the employee or the employee's family member is under the continuing care of a health care provider; if the leave is due to a qualifying exigency, that a military member is on covered active duty or call to covered active duty status (or has been notified of an impending call or order to covered active duty), that the requested leave is for one of the reasons listed in Section 825.126(b), and the anticipated duration of the absence; or if the leave is for a family member that the condition renders the family member unable to perform daily activities or that the family member is a covered servicemember with a serious injury or illness; and the anticipated duration of the absence, if known. When an employee seeks leave for the first time for a FMLA-qualifying reason, the employee need not expressly assert rights under the FMLA or even mention the FMLA. When an employee seeks leave due to a qualifying reason, for which the employer has previously provided the employee FMLA-protected leave, the employee must specifically reference either the qualifying reason for leave or the need for FMLA leave. Calling in "sick" without providing more information will not be considered sufficient notice to trigger an employer's obligations under the Act. The employer will be expected to obtain any additional required information through informal means. An employee has an obligation to respond to an employer's questions designed to determine whether an absence is potentially FMLA-qualifying. Failure to respond to reasonable employer inquiries regarding the leave request may result in denial of FMLA protection if the employer is unable to determine whether the leave is FMLA-qualifying.

(c) *Complying with employer policy.* When the need for leave is not foreseeable, an employee must comply with the employer's usual and customary notice and procedural requirements for requesting leave, absent unusual circumstances. For example, an employer may require employees to call a designated number or a specific individual to request leave. However, if an employee requires emergency medical treatment, he or she would not be required to follow the call-in procedure until his or her condition is stabilized and he or she has access to, and is able to use, a phone. Similarly, in the case of an emergency requiring leave because of a FMLA-qualifying reason, written advance notice pursuant to an employer's internal rules and procedures may not be required when FMLA leave is involved. If an employee does not comply with the employer's usual notice and procedural requirements, and no unusual circumstances justify the failure to comply, FMLA-protected leave may be delayed or denied.

Section 825.304 Employee failure to provide notice

(a) *Proper notice required.* In all cases, in order for the onset of an employee's FMLA leave to be delayed due to lack of required notice, it must be clear that the employee had actual notice of the FMLA notice requirements. This condition would be satisfied by the employer's proper posting of the required notice at the worksite where the employee is employed and the employer's provision of the required notice in either an employee handbook or employee distribution, as required by Section 825.300.

(b) *Foreseeable leave—30 days.* When the need for FMLA leave is foreseeable at least 30 days in advance and an employee fails to give timely advance notice with no reasonable excuse, the employer may delay FMLA coverage until 30 days after the date the employee provides notice. The need for leave and the approximate date leave would be taken must have been clearly foreseeable to the employee 30 days in advance of the leave. For example, knowledge that an employee would receive a telephone call about the availability of a child for adoption at some unknown point in the future would not be sufficient to establish the leave was clearly foreseeable 30 days in advance.

(c) *Foreseeable leave—less than 30 days.* When the need for FMLA leave is foreseeable fewer than 30 days in advance and an employee fails to give notice as soon as practicable under the particular facts and circumstances, the extent to which an employer may delay FMLA coverage for leave depends on the facts of the particular case. For example, if an employee reasonably should have given the employer two weeks notice but instead only provided one week notice, then the employer may delay FMLA-protected leave for one week (thus, if the employer elects to delay FMLA coverage and the employee nonetheless takes leave one week after providing the notice (*i.e.,* a week before the two week notice period has been met) the leave will not be FMLA-protected).

(d) *Unforeseeable leave.* When the need for FMLA leave is unforeseeable and an employee fails to give notice in accordance with Section 825.303, the extent to which an employer may delay FMLA coverage for leave depends on the facts of the particular case. For example, if it would have been practicable for an employee to have given the employer notice of the need for leave very soon after the need arises consistent with the employer's policy, but instead the employee provided notice two days after the leave began, then the employer may delay FMLA coverage of the leave by two days.

(e) *Waiver of notice.* An employer may waive employees' FMLA notice obligations or the employer's own internal rules on leave notice requirements. If an employer does not waive the employee's obligations under its internal leave rules, the employer may take appropriate action under its internal rules and procedures for failure to follow its usual and customary notification rules, absent unusual circumstances, as long as the actions are taken in a manner that does not discriminate against employees taking FMLA leave and the rules are not inconsistent with Section 825.303(a).

Section 825.305 Certification, general rule

(a) *General.* An employer may require that an employee's leave to care for the employee's covered family member with a serious health condition, or due to the employee's own serious health condition that makes the employee unable to perform one or more of the essential functions of the employee's position, be supported by a certification issued by the health care provider of the employee or the employee's family member. An employer may also require that an employee's leave because of a qualifying exigency or to care for a covered servicemember with a serious injury or illness be supported by a certification, as described in Sections 825.309 and 825.310, respectively. An employer must give notice of a requirement for certification each time a certification is required; such notice must be written notice whenever required by Section 825.300(c). An employer's oral request to an employee to furnish any subsequent certification is sufficient.

(b) *Timing.* In most cases, the employer should request that an employee furnish certification at the time the employee gives notice of the need for leave or within five business days thereafter, or, in the case of unforeseen leave, within five business days after the leave commences. The employer may request certification at some later date if the employer later has reason to question the appropriateness of the leave or its duration. The employee must provide the requested certification to the employer within 15 calendar days after the employer's request, unless it is not practicable under the particular circumstances to do so despite the employee's diligent, good faith efforts or the employer provides more than 15 calendar days to return the requested certification.

(c) *Complete and sufficient certification.* The employee must provide a complete and sufficient certification to the employer if required by the employer in accordance with Sections 825.306, 825.309, and 825.310. The employer shall advise an employee whenever the employer finds a certification incomplete or insufficient, and shall state in writing what additional information is necessary to make the certification complete and sufficient. A certification is considered incomplete if the employer receives a certification, but one or more of the applicable entries have not been completed. A certification is considered insufficient if the employer receives a complete certification, but the information provided is vague, ambiguous, or non-responsive. The employer must provide the employee with seven calendar days (unless not practicable under the particular circumstances despite the employee's diligent good faith efforts) to cure any such deficiency. If the deficiencies specified by the employer are not cured in the resubmitted certification, the employer may deny the taking of

FMLA leave, in accordance with Section 825.313. A certification that is not returned to the employer is not considered incomplete or insufficient, but constitutes a failure to provide certification.

(d) *Consequences.* At the time the employer requests certification, the employer must also advise an employee of the anticipated consequences of an employee's failure to provide adequate certification. If the employee fails to provide the employer with a complete and sufficient certification, despite the opportunity to cure the certification as provided in paragraph (c) of this section, or fails to provide any certification, the employer may deny the taking of FMLA leave, in accordance with Section 825.313. It is the employee's responsibility either to furnish a complete and sufficient certification or to furnish the health care provider providing the certification with any necessary authorization from the employee or the employee's family member in order for the health care provider to release a complete and sufficient certification to the employer to support the employee's FMLA request. This provision will apply in any case where an employer requests a certification permitted by these regulations, whether it is the initial certification, a recertification, a second or third opinion, or a fitness for duty certificate, including any clarifications necessary to determine if such certifications are authentic and sufficient. *See* Sections 825.306, 825.307, 825.308, and 825.312.

(e) *Annual medical certification.* Where the employee's need for leave due to the employee's own serious health condition, or the serious health condition of the employee's covered family member, lasts beyond a single leave year (as defined in Section 825.200), the employer may require the employee to provide a new medical certification in each subsequent leave year. Such new medical certifications are subject to the provisions for authentication and clarification set forth in Section 825.307, including second and third opinions.

Section 825.306 Content of Medical Certification for Leave Taken Because of an Employee's Own Serious Health Condition or the Serious Health Condition of a Family Member

(a) *Required information.* When leave is taken because of an employee's own serious health condition, or the serious health condition of a family member, an employer may require an employee to obtain a medical certification from a health care provider that sets forth the following information:

(1) The name, address, telephone number, and fax number of the health care provider and type of medical practice/specialization;

(2) The approximate date on which the serious health condition commenced, and its probable duration;

(3) A statement or description of appropriate medical facts regarding the patient's health condition for which FMLA leave is requested. The medical facts must be sufficient to support the need for leave. Such medical facts may include information

on symptoms, diagnosis, hospitalization, doctor visits, whether medication has been prescribed, any referrals for evaluation or treatment (physical therapy, for example), or any other regimen of continuing treatment;

(4) If the employee is the patient, information sufficient to establish that the employee cannot perform the essential functions of the employee's job as well as the nature of any other work restrictions, and the likely duration of such inability (*see* Section 825.123(b) and (c));

(5) If the patient is a covered family member with a serious health condition, information sufficient to establish that the family member is in need of care, as described in Section 825.124, and an estimate of the frequency and duration of the leave required to care for the family member;

(6) If an employee requests leave on an intermittent or reduced schedule basis for planned medical treatment of the employee's or a covered family member's serious health condition, information sufficient to establish the medical necessity for such intermittent or reduced schedule leave and an estimate of the dates and duration of such treatments and any periods of recovery;

(7) If an employee requests leave on an intermittent or reduced schedule basis for the employee's serious health condition, including pregnancy, that may result in unforeseeable episodes of incapacity, information sufficient to establish the medical necessity for such intermittent or reduced schedule leave and an estimate of the frequency and duration of the episodes of incapacity; and

(8) If an employee requests leave on an intermittent or reduced schedule basis to care for a covered family member with a serious health condition, a statement that such leave is medically necessary to care for the family member, as described in Sections 825.124 and 825.203(b), which can include assisting in the family member's recovery, and an estimate of the frequency and duration of the required leave.

(b) DOL has developed two optional forms (Form WH-380E and Form WH-380F, as revised) for use in obtaining medical certification, including second and third opinions, from health care providers that meets FMLA's certification requirements. Optional form WH-380E is for use when the employee's need for leave is due to the employee's own serious health condition. Optional form WH-380F is for use when the employee needs leave to care for a family member with a serious health condition. These optional forms reflect certification requirements so as to permit the health care provider to furnish appropriate medical information. Form WH-380-E and WH-380-F, as revised, or another form containing the same basic information, may be used by the employer; however, no information may be required beyond that specified in Sections 825.306, 825.307, and 825.308. In all instances the information on the form must relate only to the serious health condition for which the current need for leave exists. Prototype forms WH-380-E and WH-380-F may be obtained from local offices of the Wage and Hour Division or from the Internet at *www.dol.gov/whd.*

(c) If an employee is on FMLA leave running concurrently with a workers' compensation absence, and the provisions of the workers' compensation statute permit the employer or the employer's representative to request additional information from the employee's workers' compensation health care provider, the FMLA does not prevent the employer from following the workers' compensation provisions and information received under those provisions may be considered in determining the employee's entitlement to FMLA-protected leave. Similarly, an employer may request additional information in accordance with a paid leave policy or disability plan that requires greater information to qualify for payments or benefits, provided that the employer informs the employee that the additional information only needs to be provided in connection with receipt of such payments or benefits. Any information received pursuant to such policy or plan may be considered in determining the employee's entitlement to FMLA-protected leave. If the employee fails to provide the information required for receipt of such payments or benefits, such failure will not affect the employee's entitlement to take unpaid FMLA leave. *See* Section 825.207(a).

(d) If an employee's serious health condition may also be a disability within the meaning of the Americans with Disabilities Act (ADA), as amended, the FMLA does not prevent the employer from following the procedures for requesting medical information under the ADA. Any information received pursuant to these procedures may be considered in determining the employee's entitlement to FMLA-protected leave.

(e) While an employee may choose to comply with the certification requirement by providing the employer with an authorization, release, or waiver allowing the employer to communicate directly with the health care provider of the employee or his or her covered family member, the employee may not be required to provide such an authorization, release, or waiver. In all instances in which certification is requested, it is the employee's responsibility to provide the employer with complete and sufficient certification and failure to do so may result in the denial of FMLA leave. *See* Section 825.305(d).

Section 825.307 Authentication and Clarification of Medical Certification for Leave Taken Because of an Employee's Own Serious Health Condition or the Serious Health Condition of a Family Member; Second and Third Opinions

(a) *Clarification and authentication.* If an employee submits a complete and sufficient certification signed by the health care provider, the employer may not request additional information from the health care provider. However, the employer may contact the health care provider for purposes of clarification and authentication of the medical certification (whether initial certification or recertification) after the employer has given the employee an opportunity to cure any deficiencies as set forth in Section 825.305(c). To make such contact, the employer must use a health care provider, a human resources professional, a leave administrator, or a management official. Under no circumstances, however, may the employee's direct supervisor contact the employee's health care provider. For purposes of these regulations, *authentication* means providing the health care provider with a copy of the certification and requesting verification that the information contained on the certification form was completed and/or authorized by the health care provider who signed the document; no additional medical information may be requested. *Clari fication* means contacting the health care provider to understand the handwriting on the medical certification or to understand the meaning of a response. Employers may not ask health care providers for additional information beyond that required by the certification form. The requirements of the Health Insurance Portability and Accountability Act (HIPAA) Privacy Rule (*see* 45 CFR parts 160 and 164), which governs the privacy of individually-identifiable health information created or held by HIPAA-covered entities, must be satisfied when individually-identifiable health information of an employee is shared with an employer by a HIPAA-covered health care provider. If an employee chooses not to provide the employer with authorization allowing the employer to clarify the certification with the health care provider, and does not otherwise clarify the certification, the employer may deny the taking of FMLA leave if the certification is unclear. *See* Section 825.305(d). It is the employee's responsibility to provide the employer with a complete and sufficient certification and to clarify the certification if necessary.

(b) *Second opinion.* (1) An employer who has reason to doubt the validity of a medical certification may require the employee to obtain a second opinion at the employer's expense. Pending receipt of the second (or third) medical opinion, the employee is provisionally entitled to the benefits of the Act, including maintenance of group health benefits. If the certifications do not ultimately establish the employee's entitlement to FMLA leave, the leave shall not be designated as FMLA leave and may be treated as paid or unpaid leave under the employer's established leave policies. In addition, the consequences set forth in Section 825.305(d) will apply if the employee or the employee's family member fails to authorize his or her health care provider to release all relevant medical information pertaining to the serious health condition at issue if requested by the health care provider designated to provide a second opinion in order to render a sufficient and complete second opinion.

(2) The employer is permitted to designate the health care provider to furnish the second opinion, but the selected health care provider may not be employed on a regular basis by the employer. The employer may not regularly contract with or otherwise regularly utilize the services of the health care provider furnishing the second opinion unless the employer is located in an area where access to health care is extremely limited (e.g., a rural area where no more than one or two doctors practice in the relevant specialty in the vicinity).

(c) *Third opinion.* If the opinions of the employee's and the employer's designated health care providers differ, the employer may require the employee to obtain certification from a third health care provider, again at the employer's expense. This third opinion shall be final and binding. The third health care provider must be designated or approved jointly by the employer and the employee. The employer and the employee must each act in good faith to attempt to reach agreement on whom to select for the third opinion provider. If the employer does not attempt in good faith to reach agreement, the employer will be bound by the first certification. If the employee does not attempt in good faith to reach agreement, the employee will be bound by the second certification. For example, an employee who refuses to agree to see a doctor in the specialty in question may be failing to act in good faith. On the other hand, an employer that refuses to agree to any doctor on a list of specialists in the appropriate field provided by the employee and whom the employee has not previously consulted may be failing to act in good faith. In addition, the consequences set forth in Section 825.305(d) will apply if the employee or the employee's family member fails to authorize his or her health care provider to release all relevant medical information pertaining to the serious health condition at issue if requested by the health care provider designated to provide a third opinion in order to render a sufficient and complete third opinion.

(d) *Copies of opinions.* The employer is required to provide the employee with a copy of the second and third medical opinions, where applicable, upon request by the employee. Requested copies are to be provided within five business days unless extenuating circumstances prevent such action.

(e) *Travel expenses.* If the employer requires the employee to obtain either a second or third opinion the employer must reimburse an employee or family member for any reasonable "out of pocket" travel expenses incurred to obtain the second and third medical opinions. The employer may not require the employee or family member to travel outside normal commuting distance for purposes of obtaining the second or third medical opinions except in very unusual circumstances.

(f) *Medical certification abroad.* In circumstances in which the employee or a family member is visiting in another country, or a family member resides in another country, and a serious health condition develops, the employer shall accept a medical certification as well as second and third opinions from a health care provider who practices in that country. Where a certification by a foreign health care provider is in a language other than English, the employee must provide the employer with a written translation of the certification upon request.

Section 825.308 Recertifications for Leave Taken Because of an Employee's Own Serious Health Condition or the Serious Health Condition of a Family Member

(a) *30-day rule.* An employer may request recertification no more often than every 30 days and only in connection with an absence by the employee, unless paragraphs (b) or (c) of this section apply.

(b) *More than 30 days.* If the medical certification indicates that the minimum duration of the condition is more than 30 days, an employer must wait until that minimum duration expires before requesting a recertification, unless paragraph (c) of this section applies. For example, if the medical certification states that an employee will be unable to work, whether continuously or on an intermittent basis, for 40 days, the employer must wait 40 days before requesting a recertification. In all cases, an employer may request a recertification of a medical condition every six months in connection with an absence by the employee. Accordingly, even if the medical certification indicates that the employee will need intermittent or reduced schedule leave for a period in excess of six months (e.g., for a lifetime condition), the employer would be permitted to request recertification every six months in connection with an absence.

(c) *Less than 30 days.* An employer may request recertification in less than 30 days if:

(1) The employee requests an extension of leave;

(2) Circumstances described by the previous certification have changed significantly (e.g., the duration or frequency of the absence, the nature or severity of the illness, complications). For example, if a medical certification stated that an employee would need leave for one to two days when the employee suffered a migraine headache and the employee's absences for his or her last two migraines lasted four days each, then the increased duration of absence might constitute a significant change in circumstances allowing the employer to request a recertification in less than 30 days. Likewise, if an employee had a pattern of using unscheduled FMLA leave for migraines in conjunction with his or her scheduled days off, then the timing of the absences also might constitute a significant change in circumstances sufficient for an employer to request a recertification more frequently than every 30 days; or

(3) The employer receives information that casts doubt upon the employee's stated reason for the absence or the continuing validity of the certification. For example, if an employee is on FMLA leave for four weeks due to the employee's knee surgery, including recuperation, and the employee plays in company softball league games during the employee's third week of FMLA leave, such information might be sufficient to cast doubt upon the continuing validity of the certification allowing the employer to request a recertification in less than 30 days.

(d) *Timing.* The employee must provide the requested recertification to the employer within the time frame requested by the employer (which must allow at least 15 calendar days after the employer's request), unless it is not practicable under the particular circumstances to do so despite the employee's diligent, good faith efforts.

(e) *Content.* The employer may ask for the same information when obtaining recertification as that permitted for the original certification as set forth in Section 825.306. The employee has the same obligations to participate and cooperate (including providing a complete and sufficient certification or adequate authorization to the health care provider) in the recertification process as in the initial certification process. *See* Section 825.305(d). As part of the information allowed to be obtained on recertification for leave taken because of a serious health condition, the employer may provide the health care provider with a record of the employee's absence pattern and ask the health care provider if the serious health condition and need for leave is consistent with such a pattern.

(f) Any recertification requested by the employer shall be at the employee's expense unless the employer provides otherwise. No second or third opinion on recertification may be required.

Section 825.309 Certification for Leave Taken Because of a Qualifying Exigency

(a) *Active Duty Orders.* The first time an employee requests leave because of a qualifying exigency arising out of the covered active duty or call to covered active duty status (or notification of an impending call or order to covered active duty)of a military member (*see* Section 825.126(a)), an employer may require the employee to provide a copy of the military member's active duty orders or other documentation issued by the military which indicates that the military member is on covered active duty or call to covered active duty status, and the dates of the military member's covered active duty service. This information need only be provided to the employer once. A copy of new active duty orders or other documentation issued by the military may be required by the employer if the need for leave because of a qualifying exigency arises out of a different covered active duty or call to covered active duty status (or notification of an impending call or order to covered active duty) of the same or a different military member;

(b) *Required information.* An employer may require that leave for any qualifying exigency specified in Section 825.126 be supported by a certification from the employee that sets forth the following information:

(1) A statement or description, signed by the employee, of appropriate facts regarding the qualifying exigency for which FMLA leave is requested. The facts must be sufficient to support the need for leave. Such facts should include information on the type of qualifying exigency for which leave is requested and any available written documentation which supports the request for leave; such documentation, for example, may include a copy of a meeting announcement for informational briefings sponsored by the military, a document confirming an appointment with a counselor or school official, or a copy of a bill for services for the handling of legal or financial affairs;

(2) The approximate date on which the qualifying exigency commenced or will commence;

(3) If an employee requests leave because of a qualifying exigency for a single, continuous period of time, the beginning and end dates for such absence;

(4) If an employee requests leave because of a qualifying exigency on an intermittent or reduced schedule basis, an estimate of the frequency and duration of the qualifying exigency;

(5) If the qualifying exigency involves meeting with a third party, appropriate contact information for the individual or entity with whom the employee is meeting (such as the name, title, organization, address, telephone number, fax number, and email address) and a brief description of the purpose of the meeting; and

(6) If the qualifying exigency involves *Rest and Recuperation* leave, a copy of the military member's Rest and Recuperation orders, or other documentation issued by the military which indicates that the military member has been granted Rest and Recuperation leave, and the dates of the military member's Rest and Recuperation leave.

(c) DOL has developed an optional form (Form WH-384) for employees' use in obtaining a certification that meets FMLA's certification requirements. Form WH-384 may be obtained from local offices of the Wage and Hour Division or from the Internet at *www.dol.gov/whd.* This optional form reflects certification requirements so as to permit the employee to furnish appropriate information to support his or her request for leave because of a qualifying exigency. Form WH-384, or another form containing the same basic information, may be used by the employer; however, no information may be required beyond that specified in this section.

(d) *Verification.* If an employee submits a complete and sufficient certification to support his or her request for leave because of a qualifying exigency, the employer may not request additional information from the employee. However, if the qualifying exigency involves meeting with a third party, the employer may contact the individual or entity with whom the employee is meeting for purposes of verifying a meeting or appointment schedule and the nature of the meeting between the employee and the specified individual or entity. The employee's permission is not required in order to verify meetings or appointments with third parties, but no additional information may be requested by the employer. An employer also may contact an appropriate unit of the Department of Defense to request verification that a military member is on covered active duty or call to covered active duty status (or has been notified of an impending call or order to covered active duty); no additional information may be requested and the employee's permission is not required.

Section 825.310 Certification for Leave Taken to Care for a Covered Servicemember (Military Caregiver Leave)

(a) *Required information from health care provider.* When leave is taken to care for a covered servicemember with a serious injury or illness, an employer may require an employee to obtain a certification completed by an authorized health care provider of the covered servicemember. For purposes of leave taken to care for a covered servicemember, any one of the following health care providers may complete such a certification:

(1) A United States Department of Defense ("DOD") health care provider;

(2) A United States Department of Veterans Affairs ("VA") health care provider;

(3) A DOD TRICARE network authorized private health care provider;

(4) A DOD non-network TRICARE authorized private health care provider; or

(5) Any health care provider as defined in Section 825.125.

(b) If the authorized health care provider is unable to make certain military-related determinations outlined below, the authorized health care provider may rely on determinations from an authorized DOD representative (such as a DOD Recovery Care Coordinator) or an authorized VA representative. An employer may request that the health care provider provide the following information:

(1) The name, address, and appropriate contact information (telephone number, fax number, and/or email address) of the health care provider, the type of medical practice, the medical specialty, and whether the health care provider is one of the following:

(i) A DOD health care provider;

(ii) A VA health care provider;

(iii) A DOD TRICARE network authorized private health care provider;

(iv) A DOD non-network TRICARE authorized private health care provider; or

(v) A health care provider as defined in Section 825.125.

(2) Whether the covered servicemember's injury or illness was incurred in the line of duty on active duty or, if not, whether the covered servicemember's injury or illness existed before the beginning of the servicemember's active duty and was aggravated by service in the line of duty on active duty;

(3) The approximate date on which the serious injury or illness commenced, or was aggravated, and its probable duration;

(4) A statement or description of appropriate medical facts regarding the covered servicemember's health condition for which FMLA leave is requested. The medical facts must be sufficient to support the need for leave.

(i) In the case of a current member of the Armed Forces, such medical facts must include information on whether the injury or illness may render the covered servicemember medically unfit to perform the duties of the servicemember's office, grade, rank, or rating and whether the member is receiving medical treatment, recuperation, or therapy.

(ii) In the case of a covered veteran, such medical facts must include:

(A) Information on whether the veteran is receiving medical treatment, recuperation, or therapy for an injury or illness that is the continuation of an injury or illness that was incurred or aggravated when the covered veteran was a member of the Armed

Forces and rendered the servicemember medically unfit to perform the duties of the servicemember's office, grade, rank, or rating; or

(B) Information on whether the veteran is receiving medical treatment, recuperation, or therapy for an injury or illness that is a physical or mental condition for which the covered veteran has received a U.S. Department of Veterans Affairs Service-Related Disability Rating (VASRD) of 50 percent or greater, and that such VASRD rating is based, in whole or in part, on the condition precipitating the need for military caregiver leave; or

(C) Information on whether the veteran is receiving medical treatment, recuperation, or therapy for an injury or illness that is a physical or mental condition that substantially impairs the covered veteran's ability to secure or follow a substantially gainful occupation by reason of a disability or disabilities related to military service, or would do so absent treatment; or

(D) Documentation of enrollment in the Department of Veterans Affairs Program of Comprehensive Assistance for Family Caregivers.

(5) Information sufficient to establish that the covered servicemember is in need of care, as described in Section 825.124, and whether the covered servicemember will need care for a single continuous period of time, including any time for treatment and recovery, and an estimate as to the beginning and ending dates for this period of time;

(6) If an employee requests leave on an intermittent or reduced schedule basis for planned medical treatment appointments for the covered servicemember, whether there is a medical necessity for the covered servicemember to have such periodic care and an estimate of the treatment schedule of such appointments;

(7) If an employee requests leave on an intermittent or reduced schedule basis to care for a covered servicemember other than for planned medical treatment (e.g., episodic flare-ups of a medical condition), whether there is a medical necessity for the covered servicemember to have such periodic care, which can include assisting in the covered servicemember's recovery, and an estimate of the frequency and duration of the periodic care.

(c) *Required information from employee and/or covered servicemember.* In addition to the information that may be requested under Section 825.310(b), an employer may also request that such certification set forth the following information provided by an employee and/or covered servicemember:

(1) The name and address of the employer of the employee requesting leave to care for a covered servicemember, the name of the employee requesting such leave, and the name of the covered servicemember for whom the employee is requesting leave to care;

(2) The relationship of the employee to the covered servicemember for whom the employee is requesting leave to care;

(3) Whether the covered servicemember is a current member of the Armed Forces, the National Guard or Reserves, and the covered servicemember's military branch, rank, and current unit assignment;

(4) Whether the covered servicemember is assigned to a military medical facility as an outpatient or to a unit established for the purpose of providing command and control of members of the Armed Forces receiving medical care as outpatients (such as a medical hold or warrior transition unit), and the name of the medical treatment facility or unit;

(5) Whether the covered servicemember is on the temporary disability retired list;

(6) Whether the covered servicemember is a veteran, the date of separation from military service, and whether the separation was other than dishonorable. The employer may require the employee to provide documentation issued by the military which indicates that the covered servicemember is a veteran, the date of separation, and that the separation is other than dishonorable. Where an employer requires such documentation, an employee may provide a copy of the veteran's Certificate of Release or Discharge from Active Duty issued by the U.S. Department of Defense (DD Form 214) or other proof of veteran status. *See* Section 825.127(c)(2).

(7) A description of the care to be provided to the covered servicemember and an estimate of the leave needed to provide the care.

(d) DOL has developed optional forms (WH-385, WH-385-V) for employees' use in obtaining certification that meets FMLA's certification requirements, which may be obtained from local offices of the Wage and Hour Division or on the Internet at www.dol.gov/whd. These optional forms reflect certification requirements so as to permit the employee to furnish appropriate information to support his or her request for leave to care for a covered servicemember with a serious injury or illness. WH-385, WH-385-V, or another form containing the same basic information, may be used by the employer; however, no information may be required beyond that specified in this section. In all instances the information on the certification must relate only to the serious injury or illness for which the current need for leave exists. An employer may seek authentication and/or clarification of the certification under Section 825.307. Second and third opinions under Section 825.307 are not permitted for leave to care for a covered servicemember when the certification has been completed by one of the types of health care providers identified in Section 825.310(a)(1)-(4). However, second and third opinions under Section 825.307 are permitted when the certification has been completed by a health care provider as defined in Section 825.125 that is not one of the types identified in Section 825.310(a)(1)-(4). Additionally, recertifications under Section 825.308 are not permitted for leave to care for a covered servicemember. An employer may require an employee to provide confirmation of covered family relationship to the seriously injured or ill servicemember pursuant to Section 825.122(k) of the FMLA.

(e) An employer requiring an employee to submit a certification for leave to care for a covered servicemember must accept as sufficient certification, in lieu of the

Department's optional certification forms (WH-385) or an employer's own certification form, invitational travel orders (ITOs) or invitational travel authorizations (ITAs) issued to any family member to join an injured or ill servicemember at his or her bedside. An ITO or ITA is sufficient certification for the duration of time specified in the ITO or ITA. During that time period, an eligible employee may take leave to care for the covered servicemember in a continuous block of time or on an intermittent basis. An eligible employee who provides an ITO or ITA to support his or her request for leave may not be required to provide any additional or separate certification that leave taken on an intermittent basis during the period of time specified in the ITO or ITA is medically necessary. An ITO or ITA is sufficient certification for an employee entitled to take FMLA leave to care for a covered servicemember regardless of whether the employee is named in the order or authorization.

(1) If an employee will need leave to care for a covered servicemember beyond the expiration date specified in an ITO or ITA, an employer may request that the employee have one of the authorized health care providers listed under Section 825.310(a) complete the DOL optional certification form (WH-385) or an employer's own form, as requisite certification for the remainder of the employee's necessary leave period.

(2) An employer may seek authentication and clarification of the ITO or ITA under Section 825.307. An employer may not utilize the second or third opinion process outlined in Section 825.307 or the recertification process under Section 825.308 during the period of time in which leave is supported by an ITO or ITA.

(3) An employer may require an employee to provide confirmation of covered family relationship to the seriously injured or ill servicemember pursuant to Section 825.122(k) when an employee supports his or her request for FMLA leave with a copy of an ITO or ITA.

(f) An employer requiring an employee to submit a certification for leave to care for a covered servicemember must accept as sufficient certification of the servicemember's serious injury or illness documentation indicating the servicemember's enrollment in the Department of Veterans Affairs Program of Comprehensive Assistance for Family Caregivers. Such documentation is sufficient certification of the servicemember's serious injury or illness to support the employee's request for military caregiver leave regardless of whether the employee is the named caregiver in the enrollment documentation.

(1) An employer may seek authentication and clarification of the documentation indicating the servicemember's enrollment in the Department of Veterans Affairs Program of Comprehensive Assistance for Family Caregivers under Section 825.307. An employer may not utilize the second or third opinion process outlined in Section 825.307 or the recertification process under Section 825.308 when the servicemember's serious injury or illness is shown by documentation of enrollment in this program.

(2) An employer may require an employee to provide confirmation of covered family relationship to the seriously injured or ill servicemember pursuant to

Section 825.122(k) when an employee supports his or her request for FMLA leave with a copy of such enrollment documentation. An employer may also require an employee to provide documentation, such as a veteran's Form DD-214, showing that the discharge was other than dishonorable and the date of the veteran's discharge.

(g) Where medical certification is requested by an employer, an employee may not be held liable for administrative delays in the issuance of military documents, despite the employee's diligent, good-faith efforts to obtain such documents. *See* Section 825.305(b). In all instances in which certification is requested, it is the employee's responsibility to provide the employer with complete and sufficient certification and failure to do so may result in the denial of FMLA leave. *See* Section 825.305(d).

Section 825.311 Intent to return to work

(a) An employer may require an employee on FMLA leave to report periodically on the employee's status and intent to return to work. The employer's policy regarding such reports may not be discriminatory and must take into account all of the relevant facts and circumstances related to the individual employee's leave situation.

(b) If an employee gives unequivocal notice of intent not to return to work, the employer's obligations under FMLA to maintain health benefits (subject to COBRA requirements) and to restore the employee cease. However, these obligations continue if an employee indicates he or she may be unable to return to work but expresses a continuing desire to do so.

(c) It may be necessary for an employee to take more leave than originally anticipated. Conversely, an employee may discover after beginning leave that the circumstances have changed and the amount of leave originally anticipated is no longer necessary. An employee may not be required to take more FMLA leave than necessary to resolve the circumstance that precipitated the need for leave. In both of these situations, the employer may require that the employee provide the employer reasonable notice (*i.e.,* within two business days) of the changed circumstances where foreseeable. The employer may also obtain information on such changed circumstances through requested status reports.

Section 825.312 Fitness-for-duty certification

(a) As a condition of restoring an employee whose FMLA leave was occasioned by the employee's own serious health condition that made the employee unable to perform the employee's job, an employer may have a uniformly-applied policy or practice that requires all similarly-situated employees (*i.e.,* same occupation, same serious health condition) who take leave for such conditions to obtain and present certification from the employee's health care provider that the employee is able to resume work. The employee has the same obligations to participate and cooperate (including providing a complete and sufficient certification or providing sufficient authorization to the health care provider to provide the information directly to the employer) in the fitness-for-duty certification process as in the initial certification process. *See* Section 825.305(d).

(b) An employer may seek a fitness-for-duty certification only with regard to the particular health condition that caused the employee's need for FMLA leave. The certification from the employee's health care provider must certify that the employee is able to resume work. Additionally, an employer may require that the certification specifically address the employee's ability to perform the essential functions of the employee's job. In order to require such a certification, an employer must provide an employee with a list of the essential functions of the employee's job no later than with the designation notice required by Section 825.300(d), and must indicate in the designation notice that the certification must address the employee's ability to perform those essential functions. If the employer satisfies these requirements, the employee's health care provider must certify that the employee can perform the identified essential functions of his or her job. Following the procedures set forth in Section 825.307(a), the employer may contact the employee's health care provider for purposes of clarifying and authenticating the fitness-for-duty certification. Clarification may be requested only for the serious health condition for which FMLA leave was taken. The employer may not delay the employee's return to work while contact with the health care provider is being made. No second or third opinions on a fitness-for-duty certification may be required.

(c) The cost of the certification shall be borne by the employee, and the employee is not entitled to be paid for the time or travel costs spent in acquiring the certification.

(d) The designation notice required in Section 825.300(d) shall advise the employee if the employer will require a fitness-for-duty certification to return to work and whether that fitness-for-duty certification must address the employee's ability to perform the essential functions of the employee's job.

(e) An employer may delay restoration to employment until an employee submits a required fitness-for-duty certification unless the employer has failed to provide the notice required in paragraph (d) of this section. If an employer provides the notice required, an employee who does not provide a fitness-for-duty certification or request additional FMLA leave is no longer entitled to reinstatement under the FMLA. *See* Section 825.313(d).

(f) An employer is not entitled to a certification of fitness to return to duty for each absence taken on an intermittent or reduced leave schedule. However, an employer is entitled to a certification of fitness to return to duty for such absences up to once every 30 days if reasonable safety concerns exist regarding the employee's ability to perform his or her duties, based on the serious health condition for which the employee took such leave. If an employer chooses to require a fitness-for-duty certification under such circumstances, the employer shall inform the employee at the same time it issues the designation notice that for each subsequent instance of intermittent or reduced schedule leave, the employee will be required to submit a fitness-for-duty certification unless one has already been submitted within the past

30 days. Alternatively, an employer can set a different interval for requiring a fitness-for-duty certification as long as it does not exceed once every 30 days and as long as the employer advises the employee of the requirement in advance of the employee taking the intermittent or reduced schedule leave. The employer may not terminate the employment of the employee while awaiting such a certification of fitness to return to duty for an intermittent or reduced schedule leave absence. *Reasonable safety concerns* means a reasonable belief of significant risk of harm to the individual employee or others. In determining whether reasonable safety concerns exist, an employer should consider the nature and severity of the potential harm and the likelihood that potential harm will occur.

(g) If State or local law or the terms of a collective bargaining agreement govern an employee's return to work, those provisions shall be applied.

(h) Requirements under the Americans with Disabilities Act (ADA), as amended, apply. After an employee returns from FMLA leave, the ADA requires any medical examination at an employer's expense by the employer's health care provider be job-related and consistent with business necessity. For example, an attorney could not be required to submit to a medical examination or inquiry just because her leg had been amputated. The essential functions of an attorney's job do not require use of both legs; therefore such an inquiry would not be job related. An employer may require a warehouse laborer, whose back impairment affects the ability to lift, to be examined by an orthopedist, but may not require this employee to submit to an HIV test where the test is not related to either the essential functions of his or her job or to his/her impairment. If an employee's serious health condition may also be a disability within the meaning of the ADA, the FMLA does not prevent the employer from following the procedures for requesting medical information under the ADA.

SECTION 825.313 FAILURE TO PROVIDE CERTIFICATION

(a) *Foreseeable leave.* In the case of foreseeable leave, if an employee fails to provide certification in a timely manner as required by Section 825.305, then an employer may deny FMLA coverage until the required certification is provided. For example, if an employee has 15 days to provide a certification and does not provide the certification for 45 days without sufficient reason for the delay, the employer can deny FMLA protections for the 30-day period following the expiration of the 15-day time period, if the employee takes leave during such period.

(b) *Unforeseeable leave.* In the case of unforeseeable leave, an employer may deny FMLA coverage for the requested leave if the employee fails to provide a certification within 15 calendar days from receipt of the request for certification unless not practicable due to extenuating circumstances. For example, in the case of a medical emergency, it may not be practicable for an employee to provide the required certification within 15 calendar days. Absent such extenuating circumstances, if the employee fails to timely return the certification, the employer can deny FMLA protections for the leave following the expiration of the 15-day time period until a sufficient certification is provided. If the employee never produces the certification, the leave is not FMLA leave.

(c) *Recertification.* An employee must provide recertification within the time requested by the employer (which must allow at least 15 calendar days after the request) or as soon as practicable under the particular facts and circumstances. If an employee fails to provide a recertification within a reasonable time under the particular facts and circumstances, then the employer may deny continuation of the FMLA leave protections until the employee produces a sufficient recertification. If the employee never produces the recertification, the leave is not FMLA leave. Recertification does not apply to leave taken for a qualifying exigency or to care for a covered servicemember.

(d) *Fitness-for-duty certification.* When requested by the employer pursuant to a uniformly applied policy for similarly-situated employees, the employee must provide medical certification, at the time the employee seeks reinstatement at the end of FMLA leave taken for the employee's serious health condition, that the employee is fit for duty and able to return to work (*see* Section 825.312(a)) if the employer has provided the required notice (*see* Section 825.300(e)); the employer may delay restoration until the certification is provided. Unless the employee provides either a fitness-for-duty certification or a new medical certification for a serious health condition at the time FMLA leave is concluded, the employee may be terminated. *See also* Section 825.213(a)(3).

SUBPART D—ENFORCEMENT MECHANISMS

Section 825.400 Enforcement, general rules

(a) The employee has the choice of:

(1) Filing, or having another person file on his or her behalf, a complaint with the Secretary of Labor, or

(2) Filing a private lawsuit pursuant to section 107 of FMLA.

(b) If the employee files a private lawsuit, it must be filed within two years after the last action which the employee contends was in violation of the Act, or three years if the violation was willful.

(c) If an employer has violated one or more provisions of FMLA, and if justified by the facts of a particular case, an employee may receive one or more of the following: wages, employment benefits, or other compensation denied or lost to such employee by reason of the violation; or, where no such tangible loss has occurred, such as when FMLA leave was unlawfully denied, any actual monetary loss sustained by the employee as a direct result of the violation, such as the cost of providing care, up to a sum equal to 26 weeks of wages for the employee in a case involving leave to care for a covered servicemember or 12 weeks of wages for the employee in a case involving leave for any other FMLA qualifying reason. In addition, the employee may be entitled to interest on such sum, calculated at the prevailing rate. An amount equaling the preceding sums may also be awarded as liquidated damages unless such amount is reduced by the court because the violation was in good faith and the employer had reasonable grounds for believing the employer had not violated the Act. When appropriate, the employee may also obtain appropriate equitable relief, such as employment, reinstatement and promotion. When the employer is found in violation, the employee may recover a reasonable attorney's fee, reasonable expert witness fees, and other costs of the action from the employer in addition to any judgment awarded by the court.

Section 825.401 Filing a complaint with the Federal Government

(a) A complaint may be filed in person, by mail or by telephone, with the Wage and Hour Division, U.S. Department of Labor. A complaint may be filed at any local office of the Wage and Hour Division; the address and telephone number of local offices may be found in telephone directories or on the Department's Web site.

(b) A complaint filed with the Secretary of Labor should be filed within a reasonable time of when the employee discovers that his or her FMLA rights have been violated. In no event may a complaint be filed more than two years after the action which is alleged to be a violation of FMLA occurred, or three years in the case of a willful violation.

(c) No particular form of complaint is required, except that a complaint must be reduced to writing and should include a full statement of the acts and/or omissions, with pertinent dates, which are believed to constitute the violation.

[78 FR 8902, Feb. 6, 2013, as amended at 82 FR 2230, Jan. 9, 2017]

Section 825.402 Violations of the posting requirement

Section 825.300 describes the requirements for covered employers to post a notice for employees that explains the Act's provisions. If a representative of the Department of Labor determines that an employer has committed a willful violation of this posting requirement, and that the imposition of a civil money penalty for such violation is appropriate, the representative may issue and serve a notice of penalty on such employer in person or by certified mail. Where service by certified mail is not accepted, notice shall be deemed received on the date of attempted delivery. Where service is not accepted, the notice may be served by regular mail.

Section 825.403 Appealing the assessment of a penalty for willful violation of the posting requirement

(a) An employer may obtain a review of the assessment of penalty from the Wage and Hour Regional Administrator for the region in which the alleged violation(s) occurred. If the employer does not seek such a review or fails to do so in a timely manner, the notice of the penalty constitutes the final ruling of the Secretary of Labor.

(b) To obtain review, an employer may file a petition with the Wage and Hour Regional Administrator for the region in which the alleged violations occurred. No particular form of petition for review is required, except that the petition must be in writing, should contain the legal and factual bases for the petition, and must be mailed to the Regional Administrator within 15 days of receipt of the notice of penalty. The employer may request an oral hearing which may be conducted by telephone.

(c) The decision of the Regional Administrator constitutes the final order of the Secretary.

Section 825.404 Consequences for an employer when not paying the penalty assessment after a final order is issued

The Regional Administrator may seek to recover the unpaid penalty pursuant to the Debt Collection Act (DCA), 31 U.S.C. 3711 *et seq.,* and, in addition to seeking recovery of the unpaid final order, may seek interest and penalties as provided under the DCA. The final order may also be referred to the Solicitor of Labor for collection. The Secretary may file suit in any court of competent jurisdiction to recover the monies due as a result of the unpaid final order, interest, and penalties.

SUBPART E—RECORDKEEPING REQUIREMENTS

Section 825.500 Recordkeeping requirements

(a) FMLA provides that covered employers shall make, keep, and preserve records pertaining to their obligations under the Act in accordance with the recordkeeping requirements of section 11(c) of the Fair Labor Standards Act (FLSA) and in accordance with these regulations. FMLA also restricts the authority of the Department of Labor to require any employer or plan, fund, or program to submit books or records more than once during any 12-month period unless the Department has reasonable cause to believe a violation of FMLA exists or the Department is investigating a complaint. These regulations establish no requirement for the submission of any records unless specifically requested by a Departmental official.

(b) No particular order or form of records is required. These regulations establish no requirement that any employer revise its computerized payroll or personnel records systems to comply. However, employers must keep the records specified by these regulations for no less than three years and make them available for inspection, copying, and transcription by representatives of the Department of Labor upon request. The records may be maintained and preserved on microfilm or other basic source document of an automated data processing memory provided that adequate projection or viewing equipment is available, that the reproductions are clear and identifiable by date or pay period, and that extensions or transcriptions of the information required herein can be and are made available upon request. Records kept in computer form must be made available for transcription or copying.

(c) Covered employers who have eligible employees must maintain records that must disclose the following:

(1) Basic payroll and identifying employee data, including name, address, and occupation; rate or basis of pay and terms of compensation; daily and weekly hours worked per pay period; additions to or deductions from wages; and total compensation paid.

(2) Dates FMLA leave is taken by FMLA eligible employees (e.g., available from time records, requests for leave, etc., if so designated). Leave must be designated in records as FMLA leave; leave so designated may not include leave required under State law or an employer plan which is not also covered by FMLA.

(3) If FMLA leave is taken by eligible employees in increments of less than one full day, the hours of the leave.

(4) Copies of employee notices of leave furnished to the employer under FMLA, if in writing, and copies of all written notices given to employees as required under FMLA and these regulations *See* Section 825.300(b)-(c). Copies may be maintained in employee personnel files.

(5) Any documents (including written and electronic records) describing employee benefits or employer policies and practices regarding the taking of paid and unpaid leaves.

(6) Premium payments of employee benefits.

(7) Records of any dispute between the employer and an eligible employee regarding designation of leave as FMLA leave, including any written statement from the employer or employee of the reasons for the designation and for the disagreement.

(d) Covered employers with no eligible employees must maintain the records set forth in paragraph (c)(1) of this section.

(e) Covered employers in a joint employment situation (*see* Section 825.106) must keep all the records required by paragraph (c) of this section with respect to any primary employees, and must keep the records required by paragraph (c)(1) with respect to any secondary employees.

(f) If FMLA-eligible employees are not subject to FLSA's recordkeeping regulations for purposes of minimum wage or overtime compliance (*i.e.,* not covered by or exempt from FLSA), an employer need not keep a record of actual hours worked (as otherwise required under FLSA, 29 CFR 516.2(a)(7)), provided that:

(1) Eligibility for FMLA leave is presumed for any employee who has been employed for at least 12 months; and

(2) With respect to employees who take FMLA leave intermittently or on a reduced leave schedule, the employer and employee agree on the employee's normal schedule or average hours worked each week and reduce their agreement to a written record maintained in accordance with paragraph (b) of this section.

(g) Records and documents relating to certifications, recertifications or medical histories of employees or employees' family members, created for purposes of FMLA, shall be maintained as confidential medical records in separate files/records from the usual personnel files. If the Genetic Information Nondiscrimination Act of 2008 (GINA) is applicable, records and documents created for purposes of FMLA containing family medical history or genetic information as defined in GINA shall be maintained in accordance with the confidentiality requirements of Title II of GINA (*see* 29 CFR 1635.9), which permit such information to be disclosed consistent with the requirements of FMLA. If the ADA, as amended, is also applicable, such records shall be maintained in conformance with ADA confidentiality requirements (*see* 29 CFR 1630.14(c)(1)), except that:

(1) Supervisors and managers may be informed regarding necessary restrictions on the work or duties of an employee and necessary accommodations;

(2) First aid and safety personnel may be informed (when appropriate) if the employee's physical or medical condition might require emergency treatment; and

(3) Government officials investigating compliance with FMLA (or other pertinent law) shall be provided relevant information upon request.

(h) Special rules regarding recordkeeping apply to employers of airline flight crew employees. *See* Section 825.803.

SUBPART F—SPECIAL RULES APPLICABLE TO EMPLOYEES OF SCHOOLS

Section 825.600 Special rules for school employees, definitions

(a) Certain special rules apply to employees of local educational agencies, including public school boards and elementary and secondary schools under their jurisdiction, and private elementary and secondary schools. The special rules do not apply to other kinds of educational institutions, such as colleges and universities, trade schools, and preschools.

(b) Educational institutions are covered by FMLA (and these special rules) and the Act's 50-employee coverage test does not apply. The usual requirements for employees to be eligible do apply, however, including employment at a worksite where at least 50 employees are employed within 75 miles. For example, employees of a rural school would not be eligible for FMLA leave if the school has fewer than 50 employees and there are no other schools under the jurisdiction of the same employer (usually, a school board) within 75 miles.

(c) The special rules affect the taking of intermittent leave or leave on a reduced leave schedule, or leave near the end of an academic term (semester), by instructional employees. *Instructional employees* are those whose principal function is to teach and instruct students in a class, a small group, or an individual setting. This term includes not only teachers, but also athletic coaches, driving instructors, and special education assistants such as signers for the hearing impaired. It does not include, and the special rules do not apply to, teacher assistants or aides who do not have as their principal job actual teaching or instructing, nor does it include auxiliary personnel such as counselors, psychologists, or curriculum specialists. It also does not include cafeteria workers, maintenance workers, or bus drivers.

(d) Special rules which apply to restoration to an equivalent position apply to all employees of local educational agencies.

Section 825.601 Special rules for school employees, limitations on intermittent leave

(a) Leave taken for a period that ends with the school year and begins the next semester is leave taken consecutively rather than intermittently. The period during the summer vacation when the employee would not have been required to report for duty is not counted against the employee's FMLA leave entitlement. An instructional employee who is on FMLA leave at the end of the school year must be provided with any benefits over the summer vacation that employees would normally receive if they had been working at the end of the school year.

(1) If an eligible instructional employee needs intermittent leave or leave on a reduced leave schedule to care for a family member with a serious health condition, to care for a covered servicemember, or for the employee's own serious health condition, which is foreseeable based on planned medical treatment, and the employee would be on leave for more than 20 percent of the total number of working days over the period the leave would extend, the employer may require the employee to choose either to:

(i) Take leave for a period or periods of a particular duration, not greater than the duration of the planned treatment; or

(ii) Transfer temporarily to an available alternative position for which the employee is qualified, which has equivalent pay and benefits and which better accommodates recurring periods of leave than does the employee's regular position.

(2) These rules apply only to a leave involving more than 20 percent of the working days during the period over which the leave extends. For example, if an instructional employee who normally works five days each week needs to take two days of FMLA leave per week over a period of several weeks, the special rules would apply. Employees taking leave which constitutes 20 percent or less of the working days during the leave period would not be subject to transfer to an alternative position. *Periods*

of a particular duration means a block, or blocks, of time beginning no earlier than the first day for which leave is needed and ending no later than the last day on which leave is needed, and may include one uninterrupted period of leave.

(b) If an instructional employee does not give required notice of foreseeable FMLA leave (*see* Section 825.302) to be taken intermittently or on a reduced leave schedule, the employer may require the employee to take leave of a particular duration, or to transfer temporarily to an alternative position. Alternatively, the employer may require the employee to delay the taking of leave until the notice provision is met.

SECTION 825.602 SPECIAL RULES FOR SCHOOL EMPLOYEES, LIMITATIONS ON LEAVE NEAR THE END OF AN ACADEMIC TERM

(a) There are also different rules for instructional employees who begin leave more than five weeks before the end of a term, less than five weeks before the end of a term, and less than three weeks before the end of a term. Regular rules apply except in circumstances when:

(1) An instructional employee begins leave more than five weeks before the end of a term. The employer may require the employee to continue taking leave until the end of the term if—

(i) The leave will last at least three weeks, and

(ii) The employee would return to work during the three-week period before the end of the term.

(2) The employee begins leave during the five-week period before the end of a term because of the birth of a son or daughter; the placement of a son or daughter for adoption or foster care; to care for a spouse, son, daughter, or parent with a serious health condition; or to care for a covered servicemember. The employer may require the employee to continue taking leave until the end of the term if—

(i) The leave will last more than two weeks, and

(ii) The employee would return to work during the two-week period before the end of the term.

(3) The employee begins leave during the three-week period before the end of a term because of the birth of a son or daughter; the placement of a son or daughter for adoption or foster care; to care for a spouse, son, daughter, or parent with a serious health condition; or to care for a covered servicemember. The employer may require the employee to continue taking leave until the end of the term if the leave will last more than five working days.

(b) For purposes of these provisions, *academic term* means the school semester, which typically ends near the end of the calendar year and the end of spring each school year. In no case may a school have more than two academic terms or semesters each year for purposes of FMLA. An example of leave falling within these provisions would be where an employee plans two weeks of leave to care for a family member which will begin three weeks before the end of the term. In that situation, the employer could require the employee to stay out on leave until the end of the term.

Section 825.603 Special rules for school employees, duration of FMLA leave

(a) If an employee chooses to take leave for periods of a particular duration in the case of intermittent or reduced schedule leave, the entire period of leave taken will count as FMLA leave.

(b) In the case of an employee who is required to take leave until the end of an academic term, only the period of leave until the employee is ready and able to return to work shall be charged against the employee's FMLA leave entitlement. The employer has the option not to require the employee to stay on leave until the end of the school term. Therefore, any additional leave required by the employer to the end of the school term is not counted as FMLA leave; however, the employer shall be required to maintain the employee's group health insurance and restore the employee to the same or equivalent job including other benefits at the conclusion of the leave.

Section 825.604 Special rules for school employees, restoration to an equivalent position

The determination of how an employee is to be restored to an equivalent position upon return from FMLA leave will be made on the basis of "established school board policies and practices, private school policies and practices, and collective bargaining agreements." The "established policies" and collective bargaining agreements used as a basis for restoration must be in writing, must be made known to the employee prior to the taking of FMLA leave, and must clearly explain the employee's restoration rights upon return from leave. Any established policy which is used as the basis for restoration of an employee to an equivalent position must provide substantially the same protections as provided in the Act for reinstated employees. *See* Section 825.215. In other words, the policy or collective bargaining agreement must provide for restoration to an equivalent position with equivalent employment benefits, pay, and other terms and conditions of employment. For example, an employee may not be restored to a position requiring additional licensure or certification.

SUBPART G—EFFECT OF OTHER LAWS, EMPLOYER PRACTICES, AND COLLECTIVE BARGAINING AGREEMENTS ON EMPLOYEE RIGHTS UNDER FMLA

Section 825.700 Interaction with employer's policies

(a) An employer must observe any employment benefit program or plan that provides greater family or medical leave rights to employees than the rights established by the

FMLA. Conversely, the rights established by the Act may not be diminished by any employment benefit program or plan. For example, a provision of a CBA which provides for reinstatement to a position that is not equivalent because of seniority (e.g., provides lesser pay) is superseded by FMLA. If an employer provides greater unpaid family leave rights than are afforded by FMLA, the employer is not required to extend additional rights afforded by FMLA, such as maintenance of health benefits (other than through COBRA), to the additional leave period not covered by FMLA.

(b) Nothing in this Act prevents an employer from amending existing leave and employee benefit programs, provided they comply with FMLA. However, nothing in the Act is intended to discourage employers from adopting or retaining more generous leave policies.

Section 825.701 Interaction with State laws

(a) Nothing in FMLA supersedes any provision of State or local law that provides greater family or medical leave rights than those provided by FMLA. The Department of Labor will not, however, enforce State family or medical leave laws, and States may not enforce the FMLA. Employees are not required to designate whether the leave they are taking is FMLA leave or leave under State law, and an employer must comply with the appropriate (applicable) provisions of both. An employer covered by one law and not the other has to comply only with the law under which it is covered. Similarly, an employee eligible under only one law must receive benefits in accordance with that law. If leave qualifies for FMLA leave and leave under State law, the leave used counts against the employee's entitlement under both laws. Examples of the interaction between FMLA and State laws include:

(1) If State law provides 16 weeks of leave entitlement over two years, an employee needing leave due to his or her own serious health condition would be entitled to take 16 weeks one year under State law and 12 weeks the next year under FMLA. Health benefits maintenance under FMLA would be applicable only to the first 12 weeks of leave entitlement each year. If the employee took 12 weeks the first year, the employee would be entitled to a maximum of 12 weeks the second year under FMLA (not 16 weeks). An employee would not be entitled to 28 weeks in one year.

(2) If State law provides half-pay for employees temporarily disabled because of pregnancy for six weeks, the employee would be entitled to an additional six weeks of unpaid FMLA leave (or accrued paid leave).

(3) If State law provides six weeks of leave, which may include leave to care for a seriously-ill grandparent or a "spouse equivalent," and leave was used for that purpose, the employee is still entitled to his or her full FMLA leave entitlement, as the leave used was provided for a purpose not covered by FMLA. If FMLA leave is used first for a purpose also provided under State law, and State leave has thereby been exhausted, the employer would not be required to provide additional leave to care for the grandparent or "spouse equivalent."

(4) If State law prohibits mandatory leave beyond the actual period of pregnancy disability, an instructional employee of an educational agency subject to special FMLA rules may not be required to remain on leave until the end of the academic term, as permitted by FMLA under certain circumstances. *See* Subpart F of this part.

(b) [Reserved]

Section 825.702 Interaction with Federal and State anti-discrimination laws

(a) Nothing in FMLA modifies or affects any Federal or State law prohibiting discrimination on the basis of race, religion, color, national origin, sex, age, or disability (e.g., Title VII of the Civil Rights Act of 1964, as amended by the Pregnancy Discrimination Act). FMLA's legislative history explains that FMLA is "not intended to modify or affect the Rehabilitation Act of 1973, as amended, the regulations concerning employment which have been promulgated pursuant to that statute, or the Americans with Disabilities Act of 1990 [as amended] or the regulations issued under that act. Thus, the leave provisions of the [FMLA] are wholly distinct from the reasonable accommodation obligations of employers covered under the [ADA], employers who receive Federal financial assistance, employers who contract with the Federal government, or the Federal government itself. The purpose of the FMLA is to make leave available to eligible employees and employers within its coverage, and not to limit already existing rights and protection." S. Rep. No. 103-3, at 38 (1993). An employer must therefore provide leave under whichever statutory provision provides the greater rights to employees. When an employer violates both FMLA and a discrimination law, an employee may be able to recover under either or both statutes (double relief may not be awarded for the same loss; when remedies coincide a claimant may be allowed to utilize whichever avenue of relief is desired. *Laffey* v. *Northwest Airlines, Inc.,* 567 F.2d 429, 445 (D.C. Cir. 1976), *cert. denied,* 434 U.S. 1086 (1978).

(b) If an employee is a qualified individual with a disability within the meaning of the ADA, the employer must make reasonable accommodations, etc., barring undue hardship, in accordance with the ADA. At the same time, the employer must afford an employee his or her FMLA rights. ADA's "disability" and FMLA's "serious health condition" are different concepts, and must be analyzed separately. FMLA entitles eligible employees to 12 weeks of leave in any 12-month period due to their own serious health condition, whereas the ADA allows an indeterminate amount of leave, barring undue hardship, as a reasonable accommodation. FMLA requires employers to maintain employees' group health plan coverage during FMLA leave on the same conditions as coverage would have been provided if the employee had been continuously employed during the leave period, whereas ADA does not require maintenance of health insurance unless other employees receive health insurance during leave under the same circumstances.

(c)(1) A reasonable accommodation under the ADA might be accomplished by providing an individual with a disability with a part-time job with no health benefits, assuming the employer did not ordinarily provide health insurance for part-time employees. However, FMLA would permit an employee to work a reduced leave schedule until the equivalent of 12 workweeks of leave were used, with group health benefits maintained during this period. FMLA permits an employer to temporarily transfer an employee who is taking leave intermittently or on a reduced leave schedule for planned medical treatment to an alternative position, whereas the ADA allows an accommodation of reassignment to an equivalent, vacant position only if the employee cannot perform the essential functions of the employee's present position and an accommodation is not possible in the employee's present position, or an accommodation in the employee's present position would cause an undue hardship. The examples in the following paragraphs of this section demonstrate how the two laws would interact with respect to a qualified individual with a disability.

(2) A qualified individual with a disability who is also an eligible employee entitled to FMLA leave requests 10 weeks of medical leave as a reasonable accommodation, which the employer grants because it is not an undue hardship. The employer advises the employee that the 10 weeks of leave is also being designated as FMLA leave and will count towards the employee's FMLA leave entitlement. This designation does not prevent the parties from also treating the leave as a reasonable accommodation and reinstating the employee into the same job, as required by the ADA, rather than an equivalent position under FMLA, if that is the greater right available to the employee. At the same time, the employee would be entitled under FMLA to have the employer maintain group health plan coverage during the leave, as that requirement provides the greater right to the employee.

(3) If the same employee needed to work part-time (a reduced leave schedule) after returning to his or her same job, the employee would still be entitled under FMLA to have group health plan coverage maintained for the remainder of the two-week equivalent of FMLA leave entitlement, notwithstanding an employer policy that part-time employees do not receive health insurance. This employee would be entitled under the ADA to reasonable accommodations to enable the employee to perform the essential functions of the part-time position. In addition, because the employee is working a part-time schedule as a reasonable accommodation, the FMLA's provision for temporary assignment to a different alternative position would not apply. Once the employee has exhausted his or her remaining FMLA leave entitlement while working the reduced (part-time) schedule, if the employee is a qualified individual with a disability, and if the employee is unable to return to the same full-time position at that time, the employee might continue to work part-time as a reasonable accommodation, barring undue hardship; the employee would then be entitled to only those employment benefits ordinarily provided by the employer to part-time employees.

(4) At the end of the FMLA leave entitlement, an employer is required under FMLA to reinstate the employee in the same or an equivalent position, with equivalent pay and benefits, to that which the employee held when leave commenced. The employer's FMLA obligations would be satisfied if the employer offered the

employee an equivalent full-time position. If the employee were unable to perform the essential functions of that equivalent position even with reasonable accommodation, because of a disability, the ADA may require the employer to make a reasonable accommodation at that time by allowing the employee to work part-time or by reassigning the employee to a vacant position, barring undue hardship.

(d)(1) If FMLA entitles an employee to leave, an employer may not, in lieu of FMLA leave entitlement, require an employee to take a job with a reasonable accommodation. However, ADA may require that an employer offer an employee the opportunity to take such a position. An employer may not change the essential functions of the job in order to deny FMLA leave. *See* Section 825.220(b).

(2) An employee may be on a workers' compensation absence due to an on-the-job injury or illness which also qualifies as a serious health condition under FMLA. The workers' compensation absence and FMLA leave may run concurrently (subject to proper notice and designation by the employer). At some point the health care provider providing medical care pursuant to the workers' compensation injury may certify the employee is able to return to work in a light duty position. If the employer offers such a position, the employee is permitted but not required to accept the position. *See* Section 825.220(d). As a result, the employee may no longer qualify for payments from the workers' compensation benefit plan, but the employee is entitled to continue on unpaid FMLA leave either until the employee is able to return to the same or equivalent job the employee left or until the 12-week FMLA leave entitlement is exhausted. *See* Section 825.207(e). If the employee returning from the workers' compensation injury is a qualified individual with a disability, he or she will have rights under the ADA.

(e) If an employer requires certifications of an employee's fitness for duty to return to work, as permitted by FMLA under a uniform policy, it must comply with the ADA requirement that a fitness for duty physical be job-related and consistent with business necessity.

(f) Under Title VII of the Civil Rights Act of 1964, as amended by the Pregnancy Discrimination Act, an employer should provide the same benefits for women who are pregnant as the employer provides to other employees with short-term disabilities. Because Title VII does not require employees to be employed for a certain period of time to be protected, an employee employed for less than 12 months by the employer (and, therefore, not an eligible employee under FMLA) may not be denied maternity leave if the employer normally provides short-term disability benefits to employees with the same tenure who are experiencing other short-term disabilities.

(g) Under the Uniformed Services Employment and Reemployment Rights Act (USERRA), 38 U.S.C. 4301, *et seq.,* veterans are entitled to receive all rights and benefits of employment that they would have obtained if they had been continuously employed. Therefore, under USERRA, a returning servicemember would be eligible for FMLA leave if the months and hours that he or she would have worked (or, for airline flight crew employees, would have worked or been paid) for the civilian employer during the period of absence due to or necessitated by USERRA-covered service, combined with the months employed and the hours actually worked (or, for airline flight crew employees, actually worked or paid), meet the FMLA eligibility threshold of 12 months of employment and the hours of service requirement. *See* Sections 825.110(b)(2)(i) and (c)(2) and 825802(c).

(h) For further information on Federal antidiscrimination laws, including Title VII and the ADA, individuals are encouraged to contact the nearest office of the U.S. Equal Employment Opportunity Commission.

SUBPART H—SPECIAL RULES APPLICABLE TO AIRLINE FLIGHT CREW EMPLOYEES

SECTION 825.800 SPECIAL RULES FOR AIRLINE FLIGHT CREW EMPLOYEES, GENERAL

(a) Certain special rules apply only to airline flight crew employees as defined in Section 825.102. These special rules affect the hours of service requirement for determining the eligibility of airline flight crew employees, the calculation of leave for those employees, and the recordkeeping requirements for employers of those employees, and are issued pursuant to the Airline Flight Crew Technical Corrections Act (AFCTCA), Public Law 111-119.

(b) Except as otherwise provided in this subpart, FMLA leave for airline flight crew employees is subject to the requirements of the FMLA as set forth in Part 825, Subparts A through E, and G.

SECTION 825.801 SPECIAL RULES FOR AIRLINE FLIGHT CREW EMPLOYEES, HOURS OF SERVICE REQUIREMENT

(a) An airline flight crew employee's eligibility for FMLA leave is to be determined in accordance with Section 825.110 except that whether an airline flight crew employee meets the hours of service requirement is to be determined as provided below.

(b) Except as provided in paragraph (c) of this section, whether an airline flight crew employee meets the hours of service requirement is determined by assessing the number of hours the employee has worked or been paid over the previous 12 months. An airline flight crew employee will meet the hours of service requirement during the previous 12-month period if he or she has worked or been paid for not less than 60 percent of the employee's applicable monthly guarantee and has worked or been paid for not less than 504 hours.

(1) The *applicable monthly guarantee* for an airline flight crew employee who is not on reserve status is the minimum number of hours for which an employer has agreed to schedule such employee for any given month. The *applicable monthly guarantee* for an airline flight crew employee who is on reserve status is the number of hours for which an employer has agreed to pay the employee for any given month.

(2) The hours an airline flight crew employee has worked for purposes of the hours of service requirement is the employee's duty hours during the previous 12-month period. The hours an airline flight crew employee has been paid is the number of hours for which an employee received wages during the previous 12-month period. The 504 hours do not include personal commute time or time spent on vacation, medical, or sick leave.

(c) An airline flight crew employee returning from USERRA-covered service shall be credited with the hours of service that would have been performed but for the period of absence from work due to or necessitated by USERRA-covered service in determining the employee's eligibility for FMLA-qualifying leave. Accordingly, an airline flight crew employee re-employed following USERRA-covered service has the hours that would have been worked for or paid by the employer added to any hours actually worked or paid during the previous 12-month period to meet the hours of service requirement. In order to determine the hours that would have been worked or paid during the period of absence from work due to or necessitated by USERRA-covered service, the employee's pre-service work schedule can generally be used for calculations.

(d) In the event an employer of airline flight crew employees does not maintain an accurate record of hours worked or hours paid, the employer has the burden of showing that the employee has not worked or been paid for the requisite hours. Specifically, an employer must be able to clearly demonstrate that an airline flight crew employee has not worked or been paid for 60 percent of his or her applicable monthly guarantee or for 504 hours during the previous 12 months in order to claim that the airline flight crew employee is not eligible for FMLA leave.

Section 825.802 Special rules for airline flight crew employees, calculation of leave

(a) *Amount of leave.* (1) An eligible airline flight crew employee is entitled to 72 days of FMLA leave during any 12-month period for one, or more, of the FMLA-qualifying reasons set forth in Sections 825.112(a)(1)-(5). This entitlement is based on a uniform six-day workweek for all airline flight crew employees, regardless of time actually worked or paid, multiplied by the statutory 12-workweek entitlement for FMLA leave. For example, if an employee took six weeks of leave for an FMLA-qualifying reason, the employee would use 36 days (6 days × 6 weeks) of the employee's 72-day entitlement.

(2) An eligible airline flight crew employee is entitled to 156 days of military caregiver leave during a single 12-month period to care for a covered servicemember with a serious injury or illness under Section 825.112(a)(6). This entitlement is based on a uniform six-day workweek for all airline flight crew employees, regardless of time actually worked or paid, multiplied by the statutory 26-workweek entitlement for military caregiver leave.

(b) *Increments of FMLA leave for intermittent or reduced schedule leave.* When an airline flight crew employee takes FMLA leave on an intermittent or reduced schedule basis, the employer must account for the leave using an increment no greater than one day. For example, if an airline flight crew employee needs to take FMLA leave for a two-hour physical therapy appointment, the employer may require the employee to use a full day of FMLA leave. The entire amount of leave actually taken (in this example, one day) is designated as FMLA leave and counts against the employee's FMLA entitlement.

(c) *Application of Section 825.205.* The rules governing calculation of intermittent or reduced schedule FMLA leave set forth in Section 825.205 do not apply to airline flight crew employees except that airline flight crew employees are subject to Section 825.205(a)(2), the physical impossibility provision.

Section 825.803 Special rules for airline flight crew employees, recordkeeping requirements

(a) Employers of eligible airline flight crew employees shall make, keep, and preserve records in accordance with the requirements of Subpart E of this Part (Section 825.500).

(b) Covered employers of airline flight crew employees are required to maintain certain additional records "on file with the Secretary." To comply with this requirement, those employers shall maintain:

(1) Records and documents containing information specifying the applicable monthly guarantee with respect to each category of employee to whom such guarantee applies, including copies of any relevant collective bargaining agreements or employer policy documents; and

(2) Records of hours worked and hours paid, as those terms are defined in Section 825.801(b)(2).

Appendix 4: Occupational Safety and Health Act of 1970

Public Law 91-596
84 STAT. 1590
91st Congress, S.2193
December 29, 1970,
as amended through January 1, 2004. (1)

AN ACT

To assure safe and healthful working conditions for working men and women; by authorizing enforcement of the standards developed under the Act; by assisting and encouraging the States in their efforts to assure safe and healthful working conditions; by providing for research, information, education, and training in the field of occupational safety and health; and for other purposes.

Be it enacted by the Senate and House of Representatives of the United States of America in Congress assembled, **That this Act may be cited as the "Occupational Safety and Health Act of 1970."**

Footnote (1) See Historical notes at the end of this document for changes and amendments affecting the OSH Act since its passage in 1970 through January 1, 2004.

SEC.
2.
Congressional Findings and Purpose

(a)
29 U.S.C. 651

The Congress finds that personal injuries and illnesses arising out of work situations impose a substantial burden upon, and are a hindrance to, interstate commerce in terms of lost production, wage loss, medical expenses, and disability compensation payments.

(b)
The Congress declares it to be its purpose and policy, through the exercise of its powers to regulate commerce among the several States and with foreign nations and to provide for the general welfare, to assure so far as possible every working man and woman in the Nation safe and healthful working conditions and to preserve our human resources—

(1)
by encouraging employers and employees in their efforts to reduce the number of occupational safety and health hazards at their places of employment, and to stimulate employers and employees to institute new and to perfect existing programs for providing safe and healthful working conditions;

(2)
by providing that employers and employees have separate but dependent responsibilities and rights with respect to achieving safe and healthful working conditions;

(3)
by authorizing the Secretary of Labor to set mandatory occupational safety and health standards applicable to businesses affecting interstate commerce, and by creating an Occupational Safety and Health Review Commission for carrying out adjudicatory functions under the Act;

(4)
by building upon advances already made through employer and employee initiative for providing safe and healthful working conditions;

(5)
by providing for research in the field of occupational safety and health, including the psychological factors involved, and by developing innovative methods, techniques, and approaches for dealing with occupational safety and health problems;

(6)
by exploring ways to discover latent diseases, establishing causal connections between diseases and work in environmental conditions, and conducting other research relating to health problems, in recognition of the fact that occupational health standards present problems often different from those involved in occupational safety;

(7)
by providing medical criteria which will assure insofar as practicable that no employee will suffer diminished health, functional capacity, or life expectancy as a result of his work experience;

(8)
by providing for training programs to increase the number and competence of personnel engaged in the field of occupational safety and health; affecting the OSH Act since its passage in 1970 through January 1, 2004.

(9)
by providing for the development and promulgation of occupational safety and health standards;

(10)
by providing an effective enforcement program which shall include a prohibition against giving advance notice of any inspection and sanctions for any individual violating this prohibition;

(11)
by encouraging the States to assume the fullest responsibility for the administration and enforcement of their occupational safety and health laws by providing grants to the States to assist in identifying their needs and responsibilities in the area of occupational safety and health, to develop plans in accordance with the provisions of this Act, to improve the administration and enforcement of State occupational safety and health laws, and to conduct experimental and demonstration projects in connection therewith;

(12)
by providing for appropriate reporting procedures with respect to occupational safety and health which procedures will help achieve the objectives of this Act and accurately describe the nature of the occupational safety and health problem;

(13)
by encouraging joint labor-management efforts to reduce injuries and disease arising out of employment.

SEC.
3.
Definitions

29 U.S.C. 652
For the purposes of this Act—

(1)
The term "Secretary" means the Secretary of Labor.

(2)
The term "Commission" means the Occupational Safety and Health Review Commission established under this Act.

(3)
For Trust Territory coverage, including the Northern Mariana Islands, see Historical notes
The term "commerce" means trade, traffic, commerce, transportation, or communication among the several States, or between a State and any place outside thereof, or within the District of Columbia, or a possession of the United States (other than the Trust Territory of the Pacific Islands), or between points in the same State but through a point outside thereof.

(4)
The term "person" means one or more individuals, partnerships, associations, corporations, business trusts, legal representatives, or any organized group of persons.

(5)
Pub. L. 105-241 United States Postal Service is an employer subject to the Act. See Historical notes.

The term "employer" means a person engaged in a business affecting commerce who has employees, but does not include the United States (not including the United States Postal Service) or any State or political subdivision of a State.

(6)
The term "employee" means an employee of an employer who is employed in a business of his employer which affects commerce.

(7)
The term "State" includes a State of the United States, the District of Columbia, Puerto Rico, the Virgin Islands, American Samoa, Guam, and the Trust Territory of the Pacific Islands.

(8)
The term "occupational safety and health standard" means a standard which requires conditions, or the adoption or use of one or more practices, means, methods, operations, or processes, reasonably necessary or appropriate to provide safe or healthful employment and places of employment.

(9)
The term "national consensus standard" means any occupational safety and health standard or modification thereof which (1), has been adopted and promulgated by a nationally recognized standards-producing organization under procedures whereby it can be determined by the Secretary that persons interested and affected by the scope or provisions of the standard have reached substantial agreement on its adoption, (2) was formulated in a manner which afforded an opportunity for diverse views to be considered and (3) has been designated as such a standard by the Secretary, after consultation with other appropriate Federal agencies.

(10)
The term "established Federal standard" means any operative occupational safety and health standard established by any agency of the United States and presently in effect, or contained in any Act of Congress in force on the date of enactment of this Act.

(11)
The term "Committee" means the National Advisory Committee on Occupational Safety and Health established under this Act.

(12)
The term "Director" means the Director of the National Institute for Occupational Safety and Health.

(13)
The term "Institute" means the National Institute for Occupational Safety and Health established under this Act.

(14)
The term "Workmen's Compensation Commission" means the National Commission on State Workmen's Compensation Laws established under this Act.

SEC.
4.
Applicability of This Act

(a)
29 U.S.C. 653

For Canal Zone and Trust Territory coverage, including the Northern Mariana Islands, see Historical notes.

This Act shall apply with respect to employment performed in a workplace in a State, the District of Columbia, the Commonwealth of Puerto Rico, the Virgin Islands, American Samoa, Guam, the Trust Territory of the Pacific Islands, Wake Island, Outer Continental Shelf Lands defined in the Outer Continental Shelf Lands Act, Johnston Island, and the Canal Zone. The Secretary of the Interior shall, by regulation, provide for judicial enforcement of this Act by the courts established for areas in which there are no United States district courts having jurisdiction.

(b)
(1)
Nothing in this Act shall apply to working conditions of employees with respect to which other Federal agencies, and State agencies acting under section 274 of the Atomic Energy Act of 1954, as amended (42 U.S.C. 2021), exercise statutory authority to prescribe or enforce standards or regulations affecting occupational safety or health.

(2)
The safety and health standards promulgated under the Act of June 30, 1936, commonly known as the Walsh-Healey Act (41 U.S.C. 35 et seq.), the Service Contract Act of 1965 (41 U.S.C. 351 et seq.), Public Law 91-54, Act of August 9, 1969 (40 U.S.C. 333), Public Law 85-742, Act of August 23, 1958 (33 U.S.C. 941), and the National Foundation on Arts and Humanities Act (20 U.S.C. 951 et seq.) are superseded on the effective date of corresponding standards, promulgated under this Act, which are determined by the Secretary to be more effective. Standards issued under the laws listed in this paragraph and in effect on or after the effective date of this Act shall be deemed to be occupational safety and health standards issued under this Act, as well as under such other Acts.

(3)
The Secretary shall, within three years after the effective date of this Act, report to the Congress his recommendations for legislation to avoid unnecessary duplication and to achieve coordination between this Act and other Federal laws.

(4)
Nothing in this Act shall be construed to supersede or in any manner affect any workmen's compensation law or to enlarge or diminish or affect in any other manner the common law or statutory rights, duties, or liabilities of employers and employees under any law with respect to injuries, diseases, or death of employees arising out of, or in the course of, employment.

SEC.
5.
Duties

(a)
Each employer—

(1)
29 U.S.C. 654

shall furnish to each of his employees employment and a place of employment which are free from recognized hazards that are causing or are likely to cause death or serious physical harm to his employees;

(2)
shall comply with occupational safety and health standards promulgated under this Act.

(b)
Each employee shall comply with occupational safety and health standards and all rules, regulations, and orders issued pursuant to this Act which are applicable to his own actions and conduct.

SEC.
6.
Occupational Safety and Health Standards

(a)
29 U.S.C. 655

Without regard to chapter 5 of title 5, United States Code, or to the other subsections of this section, the Secretary shall, as soon as practicable during the period beginning with the effective date of this Act and ending two years after such date, by rule promulgate as an occupational safety or health standard any national consensus standard, and any established Federal standard, unless he determines that the promulgation of such a standard would not result in improved safety or health for specifically designated employees. In the event of conflict among any such standards, the Secretary shall promulgate the standard which assures the greatest protection of the safety or health of the affected employees.

(b)
The Secretary may by rule promulgate, modify, or revoke any occupational safety or health standard in the following manner:

(1)
Whenever the Secretary, upon the basis of information submitted to him in writing by an interested person, a representative of any organization of employers or employees, a nationally recognized standards-producing organization, the Secretary of Health and Human Services, the National Institute for Occupational Safety and Health, or a State or political subdivision, or on the basis of information developed by the Secretary or otherwise available to him, determines that a rule should be promulgated in order to serve the objectives of this Act, the Secretary may request the recommendations of an advisory committee appointed under section 7 of this Act. The Secretary shall provide such an advisory committee with any proposals of his own or of the Secretary of Health and Human Services, together with all pertinent factual information developed by the Secretary or the Secretary of Health and Human Services, or otherwise available, including the results of research, demonstrations, and experiments. An advisory committee shall submit to the Secretary its recommendations regarding the rule to be promulgated within ninety days from the date of its appointment or within such longer or shorter period as may be prescribed by the Secretary, but in no event for a period which is longer than two hundred and seventy days.

(2)
The Secretary shall publish a proposed rule promulgating, modifying, or revoking an occupational safety or health standard in the Federal Register and shall afford interested persons a period of thirty days after publication to submit written data or comments. Where an advisory committee is appointed and the Secretary determines that a rule should be issued, he shall publish the proposed rule within sixty days after the submission of the advisory committee's recommendations or the expiration of the period prescribed by the Secretary for such submission.

(3)
On or before the last day of the period provided for the submission of written data or comments under paragraph (2), any interested person may file with the Secretary written objections to the proposed rule, stating the grounds therefor and requesting a public hearing on such objections. Within thirty days after the last day for filing such objections, the Secretary shall publish in the Federal Register a notice specifying the occupational safety or health standard to which objections have been filed and a hearing requested, and specifying a time and place for such hearing.

(4)
Within sixty days after the expiration of the period provided for the submission of written data or comments under paragraph (2), or within sixty days after the completion of any hearing held under paragraph (3), the Secretary shall issue a rule promulgating, modifying, or revoking an occupational safety or health standard or make a determination that a rule should not be issued. Such a rule may contain a provision delaying its effective date for such period (not in excess of ninety days) as the Secretary determines may be necessary to insure that affected employers and employees will be informed of the existence of the standard and of its terms and that employers affected are given an opportunity to

familiarize themselves and their employees with the existence of the requirements of the standard.

(5)
The Secretary, in promulgating standards dealing with toxic materials or harmful physical agents under this subsection, shall set the standard which most adequately assures, to the extent feasible, on the basis of the best available evidence, that no employee will suffer material impairment of health or functional capacity even if such employee has regular exposure to the hazard dealt with by such standard for the period of his working life. Development of standards under this subsection shall be based upon research, demonstrations, experiments, and such other information as may be appropriate. In addition to the attainment of the highest degree of health and safety protection for the employee, other considerations shall be the latest available scientific data in the field, the feasibility of the standards, and experience gained under this and other health and safety laws. Whenever practicable, the standard promulgated shall be expressed in terms of objective criteria and of the performance desired.

(6)
(A)
Any employer may apply to the Secretary for a temporary order granting a variance from a standard or any provision thereof promulgated under this section. Such temporary order shall be granted only if the employer files an application which meets the requirements of clause (B) and establishes that—

(i)
he is unable to comply with a standard by its effective date because of unavailability of professional or technical personnel or of materials and equipment needed to come into compliance with the standard or because necessary construction or alteration of facilities cannot be completed by the effective date,

(ii)
he is taking all available steps to safeguard his employees against the hazards covered by the standard, and

(iii)
he has an effective program for coming into compliance with the standard as quickly as practicable.

Any temporary order issued under this paragraph shall prescribe the practices, means, methods, operations, and processes which the employer must adopt and use while the order is in effect and state in detail his program for coming into compliance with the standard. Such a temporary order may be granted only after notice to employees and an opportunity for a hearing: *Provided*, That the Secretary may issue one interim order to be effective until a decision is made on the basis of the hearing. No temporary order may be in effect for longer than the period needed by the employer to achieve compliance with the standard or one year, whichever is shorter, except that such an order may be renewed not more than twice (I) so long as the requirements of this paragraph are met and (II) if an application for renewal is filed at least 90 days prior to the expiration date of the order. No interim renewal of an order may remain in effect for longer than 180 days.

(B)
An application for temporary order under this paragraph (6) shall contain:

(i)
a specification of the standard or portion thereof from which the employer seeks a variance,

(ii)
a representation by the employer, supported by representations from qualified persons having firsthand knowledge of the facts represented, that he is unable to comply with the standard or portion thereof and a detailed statement of the reasons therefor,

(iii)
a statement of the steps he has taken and will take (with specific dates) to protect employees against the hazard covered by the standard,

(iv)
a statement of when he expects to be able to comply with the standard and what steps he has taken and what steps he will take (with dates specified) to come into compliance with the standard, and

(v)
a certification that he has informed his employees of the application by giving a copy thereof to their authorized representative, posting a statement giving a summary of the application and specifying where a copy may be examined at the place or places where notices to employees are normally posted, and by other appropriate means. A description of how employees have been informed shall be contained in the certification. The information to employees shall also inform them of their right to petition the Secretary for a hearing.

(C)
The Secretary is authorized to grant a variance from any standard or portion thereof whenever he determines, or the Secretary of Health and Human Services certifies, that such variance is necessary to permit an employer to participate in an experiment approved by him or the Secretary of Health and Human Services designed to demonstrate or validate new and improved techniques to safeguard the health or safety of workers.

(7)
Any standard promulgated under this subsection shall prescribe the use of labels or other appropriate forms of warning as are necessary to insure that employees are apprised of all hazards to which they are exposed, relevant symptoms and appropriate emergency treatment, and proper conditions and precautions of safe use or exposure. Where appropriate, such standard shall also prescribe suitable protective equipment

and control or technological procedures to be used in connection with such hazards and shall provide for monitoring or measuring employee exposure at such locations and intervals, and in such manner as may be necessary for the protection of employees. In addition, where appropriate, any such standard shall prescribe the type and frequency of medical examinations or other tests which shall be made available, by the employer or at his cost, to employees exposed to such hazards in order to most effectively determine whether the health of such employees is adversely affected by such exposure. In the event such medical examinations are in the nature of research, as determined by the Secretary of Health and Human Services, such examinations may be furnished at the expense of the Secretary of Health and Human Services. The results of such examinations or tests shall be furnished only to the Secretary or the Secretary of Health and Human Services, and, at the request of the employee, to his physician. The Secretary, in consultation with the Secretary of Health and Human Services, may by rule promulgated pursuant to section 553 of title 5, United States Code, make appropriate modifications in the foregoing requirements relating to the use of labels or other forms of warning, monitoring or measuring, and medical examinations, as may be warranted by experience, information, or medical or technological developments acquired subsequent to the promulgation of the relevant standard.

(8)
Whenever a rule promulgated by the Secretary differs substantially from an existing national consensus standard, the Secretary shall, at the same time, publish in the Federal Register a statement of the reasons why the rule as adopted will better effectuate the purposes of this Act than the national consensus standard.

(c)
(1)
The Secretary shall provide, without regard to the requirements of chapter 5, title 5, Unites States Code, for an emergency temporary standard to take immediate effect upon publication in the Federal Register if he determines—

(A)
that employees are exposed to grave danger from exposure to substances or agents determined to be toxic or physically harmful or from new hazards, and

(B)
that such emergency standard is necessary to protect employees from such danger.

(2)
Such standard shall be effective until superseded by a standard promulgated in accordance with the procedures prescribed in paragraph (3) of this subsection.

(3)
Upon publication of such standard in the Federal Register the Secretary shall commence a proceeding in accordance with section 6 (b) of this Act, and the standard as published shall also serve as a proposed rule for the proceeding. The Secretary shall promulgate a standard under this paragraph no later than six months after publication of the emergency standard as provided in paragraph (2) of this subsection.

(d)
Any affected employer may apply to the Secretary for a rule or order for a variance from a standard promulgated under this section. Affected employees shall be given notice of each such application and an opportunity to participate in a hearing. The Secretary shall issue such rule or order if he determines on the record, after opportunity for an inspection where appropriate and a hearing, that the proponent of the variance has demonstrated by a preponderance of the evidence that the conditions, practices, means, methods, operations, or processes used or proposed to be used by an employer will provide employment and places of employment to his employees which are as safe and healthful as those which would prevail if he complied with the standard. The rule or order so issued shall prescribe the conditions the employer must maintain, and the practices, means, methods, operations, and processes which he must adopt and utilize to the extent they differ from the standard in question. Such a rule or order may be modified or revoked upon application by an employer, employees, or by the Secretary on his own motion, in the manner prescribed for its issuance under this subsection at any time after six months from its issuance.

(e)
Whenever the Secretary promulgates any standard, makes any rule, order, or decision, grants any exemption or extension of time, or compromises, mitigates, or settles any penalty assessed under this Act, he shall include a statement of the reasons for such action, which shall be published in the Federal Register.

(f)
Any person who may be adversely affected by a standard issued under this section may at any time prior to the sixtieth day after such standard is promulgated file a petition challenging the validity of such standard with the United States court of appeals for the circuit wherein such person resides or has his principal place of business, for a judicial review of such standard. A copy of the petition shall be forthwith transmitted by the clerk of the court to the Secretary. The filing of such petition shall not, unless otherwise ordered by the court, operate as a stay of the standard. The determinations of the Secretary shall be conclusive if supported by substantial evidence in the record considered as a whole.

(g)
In determining the priority for establishing standards under this section, the Secretary shall give due regard to the urgency of the need for mandatory safety and health standards for particular industries, trades, crafts, occupations, businesses, workplaces or work environments. The Secretary shall also give due regard to the recommendations of the Secretary of Health and Human Services regarding the need for mandatory standards in determining the priority for establishing such standards.

SEC.
7.
Advisory Committees; Administration

(a)
29 U.S.C. 656

(1)
There is hereby established a National Advisory Committee on Occupational Safety and Health consisting of twelve members appointed by the Secretary, four of whom are to be designated by the Secretary of Health and Human Services, without regard to the provisions of title 5, United States Code, governing appointments in the competitive service, and composed of representatives of management, labor, occupational safety and occupational health professions, and of the public. The Secretary shall designate one of the public members as Chairman. The members shall be selected upon the basis of their experience and competence in the field of occupational safety and health.

(2)
The Committee shall advise, consult with, and make recommendations to the Secretary and the Secretary of Health and Human Services on matters relating to the administration of the Act. The Committee shall hold no fewer than two meetings during each calendar year. All meetings of the Committee shall be open to the public and a transcript shall be kept and made available for public inspection.

(3)
The members of the Committee shall be compensated in accordance with the provisions of section 3109 of title 5, United States Code.

(4)
The Secretary shall furnish to the Committee an executive secretary and such secretarial, clerical, and other services as are deemed necessary to the conduct of its business.

(b)
An advisory committee may be appointed by the Secretary to assist him in his standard-setting functions under section 6 of this Act. Each such committee shall consist of not more than fifteen members and shall include as a member one or more designees of the Secretary of Health and Human Services, and shall include among its members an equal number of persons qualified by experience and affiliation to present the viewpoint of the employers involved, and of persons similarly qualified to present the viewpoint of the workers involved, as well as one or more representatives of health and safety agencies of the States. An advisory committee may also include such other persons as the Secretary may appoint who are qualified by knowledge and experience to make a useful contribution to the work of such committee, including one or more representatives of professional organizations of technicians or professionals specializing in occupational safety or health, and one or more representatives of nationally recognized standards-producing organizations, but the number of persons so appointed to any such advisory committee shall not exceed the number appointed to such committee as representatives of Federal and State agencies. Persons appointed to advisory committees from private life shall be compensated in the same manner as consultants or experts under section 3109 of title 5, United States Code. The Secretary shall pay to any State which is the employer of a member of such a committee who is a representative of the health

or safety agency of that State, reimbursement sufficient to cover the actual cost to the State resulting from such representative's membership on such committee. Any meeting of such committee shall be open to the public and an accurate record shall be kept and made available to the public. No member of such committee (other than representatives of employers and employees) shall have an economic interest in any proposed rule.

(c)
In carrying out his responsibilities under this Act, the Secretary is authorized to—

(1)
use, with the consent of any Federal agency, the services, facilities, and personnel of such agency, with or without reimbursement, and with the consent of any State or political subdivision thereof, accept and use the services, facilities, and personnel of any agency of such State or subdivision with reimbursement; and

(2)
employ experts and consultants or organizations thereof as authorized by section 3109 of title 5, United States Code, except that contracts for such employment may be renewed annually; compensate individuals so employed at rates not in excess of the rate specified at the time of service for grade GS-18 under section 5332 of title 5, United States Code, including travel time, and allow them while away from their homes or regular places of business, travel expenses (including per diem in lieu of subsistence) as authorized by section 5703 of title 5, United States Code, for persons in the Government service employed intermittently, while so employed.

SEC.
8.
Inspections, Investigations, and Recordkeeping

(a)
29 U.S.C. 657

In order to carry out the purposes of this Act, the Secretary, upon presenting appropriate credentials to the owner, operator, or agent in charge, is authorized—

(1)
to enter without delay and at reasonable times any factory, plant, establishment, construction site, or other area, workplace or environment where work is performed by an employee of an employer; and

(2)
to inspect and investigate during regular working hours and at other reasonable times, and within reasonable limits and in a reasonable manner, any such place of employment and all pertinent conditions, structures, machines, apparatus, devices, equipment, and materials therein, and to question privately any such employer, owner, operator, agent or employee.

(b)
In making his inspections and investigations under this Act the Secretary may require the attendance and testimony of witnesses and the production of evidence under oath. Witnesses shall be paid the same fees and mileage that are paid witnesses in the courts of the United States. In case of a contumacy, failure, or refusal of any person to obey such an order, any district court of the United States or the United States courts of any territory or possession, within the jurisdiction of which such person is found, or resides or transacts business, upon the application by the Secretary, shall have jurisdiction to issue to such person an order requiring such person to appear to produce evidence if, as, and when so ordered, and to give testimony relating to the matter under investigation or in question, and any failure to obey such order of the court may be punished by said court as a contempt thereof.

(c)
(1)
Each employer shall make, keep and preserve, and make available to the Secretary or the Secretary of Health and Human Services, such records regarding his activities relating to this Act as the Secretary, in cooperation with the Secretary of Health and Human Services, may prescribe by regulation as necessary or appropriate for the enforcement of this Act or for developing information regarding the causes and prevention of occupational accidents and illnesses. In order to carry out the provisions of this paragraph such regulations may include provisions requiring employers to conduct periodic inspections. The Secretary shall also issue regulations requiring that employers, through posting of notices or other appropriate means, keep their employees informed of their protections and obligations under this Act, including the provisions of applicable standards.

(2)
The Secretary, in cooperation with the Secretary of Health and Human Services, shall prescribe regulations requiring employers to maintain accurate records of, and to make periodic reports on, work-related deaths, injuries and illnesses other than minor injuries requiring only first aid treatment and which do not involve medical treatment, loss of consciousness, restriction of work or motion, or transfer to another job.

(3)
The Secretary, in cooperation with the Secretary of Health and Human Services, shall issue regulations requiring employers to maintain accurate records of employee exposures to potentially toxic materials or harmful physical agents which are required to be monitored or measured under section 6. Such regulations shall provide employees or their representatives with an opportunity to observe such monitoring or measuring, and to have access to the records thereof. Such regulations shall also make appropriate provision for each employee or former employee to have access to such records as will indicate his own exposure to toxic materials or harmful physical agents. Each employer shall promptly notify any employee who has been or is being exposed to toxic materials or harmful physical agents in concentrations

or at levels which exceed those prescribed by an applicable occupational safety and health standard promulgated under section 6, and shall inform any employee who is being thus exposed of the corrective action being taken.

(d)
Any information obtained by the Secretary, the Secretary of Health and Human Services, or a State agency under this Act shall be obtained with a minimum burden upon employers, especially those operating small businesses. Unnecessary duplication of efforts in obtaining information shall be reduced to the maximum extent feasible.

(e)
Subject to regulations issued by the Secretary, a representative of the employer and a representative authorized by his employees shall be given an opportunity to accompany the Secretary or his authorized representative during the physical inspection of any workplace under subsection (a) for the purpose of aiding such inspection. Where there is no authorized employee representative, the Secretary or his authorized representative shall consult with a reasonable number of employees concerning matters of health and safety in the workplace.

(f)
(1)
Any employees or representative of employees who believe that a violation of a safety or health standard exists that threatens physical harm, or that an imminent danger exists, may request an inspection by giving notice to the Secretary or his authorized representative of such violation or danger. Any such notice shall be reduced to writing, shall set forth with reasonable particularity the grounds for the notice, and shall be signed by the employees or representative of employees, and a copy shall be provided the employer or his agent no later than at the time of inspection, except that, upon the request of the person giving such notice, his name and the names of individual employees referred to therein shall not appear in such copy or on any record published, released, or made available pursuant to subsection (g) of this section. If upon receipt of such notification the Secretary determines there are reasonable grounds to believe that such violation or danger exists, he shall make a special inspection in accordance with the provisions of this section as soon as practicable, to determine if such violation or danger exists. If the Secretary determines there are no reasonable grounds to believe that a violation or danger exists he shall notify the employees or representative of the employees in writing of such determination.

(2)
Prior to or during any inspection of a workplace, any employees or representative of employees employed in such workplace may notify the Secretary or any representative of the Secretary responsible for conducting the inspection, in writing, of any violation of this Act which they have reason to believe exists in such workplace. The Secretary shall, by regulation, establish procedures for informal review of any refusal by a representative of the Secretary to issue a citation with respect to any such alleged violation and shall furnish the employees or representative of employees

requesting such review a written statement of the reasons for the Secretary's final disposition of the case.

(g)
(1)
The Secretary and Secretary of Health and Human Services are authorized to compile, analyze, and publish, either in summary or detailed form, all reports or information obtained under this section.

(2)
The Secretary and the Secretary of Health and Human Services shall each prescribe such rules and regulations as he may deem necessary to carry out their responsibilities under this Act, including rules and regulations dealing with the inspection of an employer's establishment.

(h)
Pub. L. 105-198 added subsection (h).
The Secretary shall not use the results of enforcement activities, such as the number of citations issued or penalties assessed, to evaluate employees directly involved in enforcement activities under this Act or to impose quotas or goals with regard to the results of such activities.

SEC.
9.
Citations

(a)
29 U.S.C. 658

If, upon inspection or investigation, the Secretary or his authorized representative believes that an employer has violated a requirement of section 5 of this Act, of any standard, rule or order promulgated pursuant to section 6 of this Act, or of any regulations prescribed pursuant to this Act, he shall with reasonable promptness issue a citation to the employer. Each citation shall be in writing and shall describe with particularity the nature of the violation, including a reference to the provision of the Act, standard, rule, regulation, or order alleged to have been violated. In addition, the citation shall fix a reasonable time for the abatement of the violation. The Secretary may prescribe procedures for the issuance of a notice in lieu of a citation with respect to de minimis violations which have no direct or immediate relationship to safety or health.

(b)
Each citation issued under this section, or a copy or copies thereof, shall be prominently posted, as prescribed in regulations issued by the Secretary, at or near each place a violation referred to in the citation occurred.

(c)
No citation may be issued under this section after the expiration of six months following the occurrence of any violation.

SEC.
10.
Procedure for Enforcement

(a)
29 U.S.C. 659

If, after an inspection or investigation, the Secretary issues a citation under section 9(a), he shall, within a reasonable time after the termination of such inspection or investigation, notify the employer by certified mail of the penalty, if any, proposed to be assessed under section 17 and that the employer has fifteen working days within which to notify the Secretary that he wishes to contest the citation or proposed assessment of penalty. If, within fifteen working days from the receipt of the notice issued by the Secretary the employer fails to notify the Secretary that he intends to contest the citation or proposed assessment of penalty, and no notice is filed by any employee or representative of employees under subsection (c) within such time, the citation and the assessment, as proposed, shall be deemed a final order of the Commission and not subject to review by any court or agency.

(b)
If the Secretary has reason to believe that an employer has failed to correct a violation for which a citation has been issued within the period permitted for its correction (which period shall not begin to run until the entry of a final order by the Commission in the case of any review proceedings under this section initiated by the employer in good faith and not solely for delay or avoidance of penalties), the Secretary shall notify the employer by certified mail of such failure and of the penalty proposed to be assessed under section 17 by reason of such failure, and that the employer has fifteen working days within which to notify the Secretary that he wishes to contest the Secretary's notification or the proposed assessment of penalty. If, within fifteen working days from the receipt of notification issued by the Secretary, the employer fails to notify the Secretary that he intends to contest the notification or proposed assessment of penalty, the notification and assessment, as proposed, shall be deemed a final order of the Commission and not subject to review by any court or agency.

(c)
If an employer notifies the Secretary that he intends to contest a citation issued under section 9(a) or notification issued under subsection (a) or (b) of this section, or if, within fifteen working days of the issuance of a citation under section 9(a), any employee or representative of employees files a notice with the Secretary alleging that the period of time fixed in the citation for the abatement of the violation is unreasonable, the Secretary shall immediately advise the Commission of such notification, and the Commission shall afford an opportunity for a hearing (in accordance with section 554 of title 5, United States Code, but without regard to subsection (a)(3) of such section). The Commission shall thereafter issue an order, based on findings of fact, affirming, modifying, or vacating the Secretary's citation or proposed penalty, or directing other appropriate relief, and such order shall become final thirty days after its issuance. Upon a showing by an employer of a good faith effort to comply with the abatement

requirements of a citation, and that abatement has not been completed because of factors beyond his reasonable control, the Secretary, after an opportunity for a hearing as provided in this subsection, shall issue an order affirming or modifying the abatement requirements in such citation. The rules of procedure prescribed by the Commission shall provide affected employees or representatives of affected employees an opportunity to participate as parties to hearings under this subsection.

SEC.
11.
Judicial Review

(a)
29 U.S.C. 660

Any person adversely affected or aggrieved by an order of the Commission issued under subsection (c) of section 10 may obtain a review of such order in any United States court of appeals for the circuit in which the violation is alleged to have occurred or where the employer has its principal office, or in the Court of Appeals for the District of Columbia Circuit, by filing in such court within sixty days following the issuance of such order a written petition praying that the order be modified or set aside. A copy of such petition shall be forthwith transmitted by the clerk of the court to the Commission and to the other parties, and thereupon the Commission shall file in the court the record in the proceeding as provided in section 2112 of title 28, United States Code. Upon such filing, the court shall have jurisdiction of the proceeding and of the question determined therein, and shall have power to grant such temporary relief or restraining order as it deems just and proper, and to make and enter upon the pleadings, testimony, and proceedings set forth in such record a decree affirming, modifying, or setting aside in whole or in part, the order of the Commission and enforcing the same to the extent that such order is affirmed or modified. The commencement of proceedings under this subsection shall not, unless ordered by the court, operate as a stay of the order of the Commission. No objection that has not been urged before the Commission shall be considered by the court, unless the failure or neglect to urge such objection shall be excused because of extraordinary circumstances. The findings of the Commission with respect to questions of fact, if supported by substantial evidence on the record considered as a whole, shall be conclusive. If any party shall apply to the court for leave to adduce additional evidence and shall show to the satisfaction of the court that such additional evidence is material and that there were reasonable grounds for the failure to adduce such evidence in the hearing before the Commission, the court may order such additional evidence to be taken before the Commission and to be made a part of the record. The Commission may modify its findings as to the facts, or make new findings, by reason of additional evidence so taken and filed, and it shall file such modified or new findings, which findings with respect to questions of fact, if supported by substantial evidence on the record considered as a whole, shall be conclusive, and its recommendations, if any, for the modification or setting aside of its original order. Upon the filing of the record with it, the jurisdiction of the court shall be exclusive and its judgment and decree shall be final, except that the same shall be subject to review by the Supreme Court of the United States, as provided in section 1254 of title 28, United States Code.

(b)
Pub. L. 98-620

The Secretary may also obtain review or enforcement of any final order of the Commission by filing a petition for such relief in the United States court of appeals for the circuit in which the alleged violation occurred or in which the employer has its principal office, and the provisions of subsection (a) shall govern such proceedings to the extent applicable. If no petition for review, as provided in subsection (a), is filed within sixty days after service of the Commission's order, the Commission's findings of fact and order shall be conclusive in connection with any petition for enforcement which is filed by the Secretary after the expiration of such sixty-day period. In any such case, as well as in the case of a noncontested citation or notification by the Secretary which has become a final order of the Commission under subsection (a) or (b) of section 10, the clerk of the court, unless otherwise ordered by the court, shall forthwith enter a decree enforcing the order and shall transmit a copy of such decree to the Secretary and the employer named in the petition. In any contempt proceeding brought to enforce a decree of a court of appeals entered pursuant to this subsection or subsection (a), the court of appeals may assess the penalties provided in section 17, in addition to invoking any other available remedies.

(c)
(1)
No person shall discharge or in any manner discriminate against any employee because such employee has filed any complaint or instituted or caused to be instituted any proceeding under or related to this Act or has testified or is about to testify in any such proceeding or because of the exercise by such employee on behalf of himself or others of any right afforded by this Act.

(2)
Any employee who believes that he has been discharged or otherwise discriminated against by any person in violation of this subsection may, within thirty days after such violation occurs, file a complaint with the Secretary alleging such discrimination. Upon receipt of such complaint, the Secretary shall cause such investigation to be made as he deems appropriate. If upon such investigation, the Secretary determines that the provisions of this subsection have been violated, he shall bring an action in any appropriate United States district court against such person. In any such action the United States district courts shall have jurisdiction, for cause shown to restrain violations of paragraph (1) of this subsection and order all appropriate relief including rehiring or reinstatement of the employee to his former position with back pay.

(3)
Within 90 days of the receipt of a complaint filed under this subsection the Secretary shall notify the complainant of his determination under paragraph 2 of this subsection.

SEC.
12.
The Occupational Safety and Health Review Commission

(a)
29 U.S.C. 661

The Occupational Safety and Health Review Commission is hereby established. The Commission shall be composed of three members who shall be appointed by the President, by and with the advice and consent of the Senate, from among persons who by reason of training, education, or experience are qualified to carry out the functions of the Commission under this Act. The President shall designate one of the members of the Commission to serve as Chairman.

(b)
The terms of members of the Commission shall be six years except that

(1)
the members of the Commission first taking office shall serve, as designated by the President at the time of appointment, one for a term of two years, one for a term of four years, and one for a term of six years, and

(2)
a vacancy caused by the death, resignation, or removal of a member prior to the expiration of the term for which he was appointed shall be filled only for the remainder of such unexpired term.
A member of the Commission may be removed by the President for inefficiency, neglect of duty, or malfeasance in office.

(c)
See notes on omitted text.

(Text omitted.)

(d)
The principal office of the Commission shall be in the District of Columbia. Whenever the Commission deems that the convenience of the public or of the parties may be promoted, or delay or expense may be minimized, it may hold hearings or conduct other proceedings at any other place.

(e)
Pub. L. 95-251
The Chairman shall be responsible on behalf of the Commission for the administrative operations of the Commission and shall appoint such administrative law judges and other employees as he deems necessary to assist in the performance of the Commission's functions and to fix their compensation in accordance with the provisions of chapter 51 and subchapter III of chapter 53 of title 5, United States Code, relating to classification and General Schedule pay rates: *Provided*, That assignment,

removal and compensation of administrative law judges shall be in accordance with sections 3105, 3344, 5372, and 7521 of title 5, United States Code.

(f)
For the purpose of carrying out its functions under this Act, two members of the Commission shall constitute a quorum and official action can be taken only on the affirmative vote of at least two members.

(g)
Every official act of the Commission shall be entered of record, and its hearings and records shall be open to the public. The Commission is authorized to make such rules as are necessary for the orderly transaction of its proceedings. Unless the Commission has adopted a different rule, its proceedings shall be in accordance with the Federal Rules of Civil Procedure.

(h)
The Commission may order testimony to be taken by deposition in any proceedings pending before it at any state of such proceeding. Any person may be compelled to appear and depose, and to produce books, papers, or documents, in the same manner as witnesses may be compelled to appear and testify and produce like documentary evidence before the Commission. Witnesses whose depositions are taken under this subsection, and the persons taking such depositions, shall be entitled to the same fees as are paid for like services in the courts of the United States.

(i)
For the purpose of any proceeding before the Commission, the provisions of section 11 of the National Labor Relations Act (29 U.S.C. 161) are hereby made applicable to the jurisdiction and powers of the Commission.

(j)
An administrative law judge appointed by the Commission shall hear, and make a determination upon, any proceeding instituted before the Commission and any motion in connection therewith, assigned to such administrative law judge by the Chairman of the Commission, and shall make a report of any such determination which constitutes his final disposition of the proceedings. The report of the administrative law judge shall become the final order of the Commission within thirty days after such report by the administrative law judge, unless within such period any Commission member has directed that such report shall be reviewed by the Commission.

(k)
Except as otherwise provided in this Act, the administrative law judges shall be subject to the laws governing employees in the classified civil service, except that appointments shall be made without regard to section 5108 of title 5, United States Code. Each administrative law judge shall receive compensation at a rate not less than that prescribed for GS-16 under section 5332 of title 5, United States Code.

SEC.
13.
Procedures to Counteract Imminent Dangers

(a)
29 U.S.C. 662

The United States district courts shall have jurisdiction, upon petition of the Secretary, to restrain any conditions or practices in any place of employment which are such that a danger exists which could reasonably be expected to cause death or serious physical harm immediately or before the imminence of such danger can be eliminated through the enforcement procedures otherwise provided by this Act. Any order issued under this section may require such steps to be taken as may be necessary to avoid, correct, or remove such imminent danger and prohibit the employment or presence of any individual in locations or under conditions where such imminent danger exists, except individuals whose presence is necessary to avoid, correct, or remove such imminent danger or to maintain the capacity of a continuous process operation to resume normal operations without a complete cessation of operations, or where a cessation of operations is necessary, to permit such to be accomplished in a safe and orderly manner.

(b)
Upon the filing of any such petition the district court shall have jurisdiction to grant such injunctive relief or temporary restraining order pending the outcome of an enforcement proceeding pursuant to this Act. The proceeding shall be as provided by Rule 65 of the Federal Rules, Civil Procedure, except that no temporary restraining order issued without notice shall be effective for a period longer than five days.

(c)
Whenever and as soon as an inspector concludes that conditions or practices described in subsection (a) exist in any place of employment, he shall inform the affected employees and employers of the danger and that he is recommending to the Secretary that relief be sought.

(d)
If the Secretary arbitrarily or capriciously fails to seek relief under this section, any employee who may be injured by reason of such failure, or the representative of such employees, might bring an action against the Secretary in the United States district court for the district in which the imminent danger is alleged to exist or the employer has its principal office, or for the District of Columbia, for a writ of mandamus to compel the Secretary to seek such an order and for such further relief as may be appropriate.

SEC.
14.
Representation in Civil Litigation

29 U.S.C. 663
Except as provided in section 518(a) of title 28, United States Code, relating to litigation before the Supreme Court, the Solicitor of Labor may appear for and represent

the Secretary in any civil litigation brought under this Act but all such litigation shall be subject to the direction and control of the Attorney General.

SEC.

15.

Confidentiality of Trade Secrets

29 U.S.C. 664

All information reported to or otherwise obtained by the Secretary or his representative in connection with any inspection or proceeding under this Act which contains or which might reveal a trade secret referred to in section 1905 of title 18 of the United States Code shall be considered confidential for the purpose of that section, except that such information may be disclosed to other officers or employees concerned with carrying out this Act or when relevant in any proceeding under this Act. In any such proceeding the Secretary, the Commission, or the court shall issue such orders as may be appropriate to protect the confidentiality of trade secrets.

SEC.

16.

Variations, Tolerances, and Exemptions

29 U.S.C. 665

The Secretary, on the record, after notice and opportunity for a hearing may provide such reasonable limitations and may make such rules and regulations allowing reasonable variations, tolerances, and exemptions to and from any or all provisions of this Act as he may find necessary and proper to avoid serious impairment of the national defense. Such action shall not be in effect for more than six months without notification to affected employees and an opportunity being afforded for a hearing.

SEC.

17.

Penalties

(a)

29 U.S.C. 666

Pub. L. 101-508 increased the civil penalties in subsections (a)-(d) & (i). See Historical notes.

Any employer who willfully or repeatedly violates the requirements of section 5 of this Act, any standard, rule, or order promulgated pursuant to section 6 of this Act, or regulations prescribed pursuant to this Act, may be assessed a civil penalty of not more than $70,000 for each violation, but not less than $5,000 for each willful violation.

(b)

Any employer who has received a citation for a serious violation of the requirements of section 5 of this Act, of any standard, rule, or order promulgated pursuant to section 6 of this Act, or of any regulations prescribed pursuant to this Act, shall be assessed a civil penalty of up to $7,000 for each such violation.

(c)
Any employer who has received a citation for a violation of the requirements of section 5 of this Act, of any standard, rule, or order promulgated pursuant to section 6 of this Act, or of regulations prescribed pursuant to this Act, and such violation is specifically determined not to be of a serious nature, may be assessed a civil penalty of up to $7,000 for each violation.

(d)
Any employer who fails to correct a violation for which a citation has been issued under section 9(a) within the period permitted for its correction (which period shall not begin to run until the date of the final order of the Commission in the case of any review proceeding under section 10 initiated by the employer in good faith and not solely for delay or avoidance of penalties), may be assessed a civil penalty of not more than $7,000 for each day during which such failure or violation continues.

(e)
Pub. L. 98-473 Maximum criminal fines are increased by the Sentencing Reform Act of 1984, 18 U.S.C. Section 3551 et seq. See Historical notes.
Any employer who willfully violates any standard, rule, or order promulgated pursuant to section 6 of this Act, or of any regulations prescribed pursuant to this Act, and that violation caused death to any employee, shall, upon conviction, be punished by a fine of not more than $10,000 or by imprisonment for not more than six months, or by both; except that if the conviction is for a violation committed after a first conviction of such person, punishment shall be by a fine of not more than $20,000 or by imprisonment for not more than one year, or by both.

(f)

See historical notes.
Any person who gives advance notice of any inspection to be conducted under this Act, without authority from the Secretary or his designees, shall, upon conviction, be punished by a fine of not more than $1,000 or by imprisonment for not more than six months, or by both.

(g)
Whoever knowingly makes any false statement, representation, or certification in any application, record, report, plan, or other document filed or required to be maintained pursuant to this Act shall, upon conviction, be punished by a fine of not more than $10,000, or by imprisonment for not more than six months, or by both.

(h)

(1)
Section 1114 of title 18, United States Code, is hereby amended by striking out "designated by the Secretary of Health and Human Services to conduct investigations, or

inspections under the Federal Food, Drug, and Cosmetic Act" and inserting in lieu thereof "or of the Department of Labor assigned to perform investigative, inspection, or law enforcement functions."

(2)
Notwithstanding the provisions of sections 1111 and 1114 of title 18, United States Code, whoever, in violation of the provisions of section 1114 of such title, kills a person while engaged in or on account of the performance of investigative, inspection, or law enforcement functions added to such section 1114 by paragraph (1) of this subsection, and who would otherwise be subject to the penalty provisions of such section 1111, shall be punished by imprisonment for any term of years or for life.

(i)
Any employer who violates any of the posting requirements, as prescribed under the provisions of this Act, shall be assessed a civil penalty of up to $7,000 for each violation.

(j)
The Commission shall have authority to assess all civil penalties provided in this section, giving due consideration to the appropriateness of the penalty with respect to the size of the business of the employer being charged, the gravity of the violation, the good faith of the employer, and the history of previous violations.

(k)
For purposes of this section, a serious violation shall be deemed to exist in a place of employment if there is a substantial probability that death or serious physical harm could result from a condition which exists, or from one or more practices, means, methods, operations, or processes which have been adopted or are in use, in such place of employment unless the employer did not, and could not with the exercise of reasonable diligence, know of the presence of the violation.

(l)
Civil penalties owed under this Act shall be paid to the Secretary for deposit into the Treasury of the United States and shall accrue to the United States and may be recovered in a civil action in the name of the United States brought in the United States district court for the district where the violation is alleged to have occurred or where the employer has its principal office.

SEC.
18.
State Jurisdiction and State Plans

(a)
29 U.S.C. 667

Nothing in this Act shall prevent any State agency or court from asserting jurisdiction under State law over any occupational safety or health issue with respect to which no standard is in effect under section 6.

(b)
Any State which, at any time, desires to assume responsibility for development and enforcement therein of occupational safety and health standards relating to any occupational safety or health issue with respect to which a Federal standard has been promulgated under section 6 shall submit a State plan for the development of such standards and their enforcement.

(c)
The Secretary shall approve the plan submitted by a State under subsection (b), or any modification thereof, if such plan in his judgement—

(1)
designates a State agency or agencies as the agency or agencies responsible for administering the plan throughout the State,

(2)
provides for the development and enforcement of safety and health standards relating to one or more safety or health issues, which standards (and the enforcement of which standards) are or will be at least as effective in providing safe and healthful employment and places of employment as the standards promulgated under section 6 which relate to the same issues, and which standards, when applicable to products which are distributed or used in interstate commerce, are required by compelling local conditions and do not unduly burden interstate commerce,

(3)
provides for a right of entry and inspection of all workplaces subject to the Act which is at least as effective as that provided in section 8, and includes a prohibition on advance notice of inspections,

(4)
contains satisfactory assurances that such agency or agencies have or will have the legal authority and qualified personnel necessary for the enforcement of such standards,

(5)
gives satisfactory assurances that such State will devote adequate funds to the administration and enforcement of such standards,

(6)
contains satisfactory assurances that such State will, to the extent permitted by its law, establish and maintain an effective and comprehensive occupational safety and health program applicable to all employees of public agencies of the State and its political subdivisions, which program is as effective as the standards contained in an approved plan,

(7)
requires employers in the State to make reports to the Secretary in the same manner and to the same extent as if the plan were not in effect, and

(8)
provides that the State agency will make such reports to the Secretary in such form and containing such information, as the Secretary shall from time to time require.

(d)
If the Secretary rejects a plan submitted under subsection (b), he shall afford the State submitting the plan due notice and opportunity for a hearing before so doing.

(e)
After the Secretary approves a State plan submitted under subsection (b), he may, but shall not be required to, exercise his authority under sections 8, 9, 10, 13, and 17 with respect to comparable standards promulgated under section 6, for the period specified in the next sentence. The Secretary may exercise the authority referred to above until he determines, on the basis of actual operations under the State plan, that the criteria set forth in subsection (c) are being applied, but he shall not make such determination for at least three years after the plan's approval under subsection (c). Upon making the determination referred to in the preceding sentence, the provisions of sections 5(a)(2), 8 (except for the purpose of carrying out subsection (f) of this section), 9, 10, 13, and 17, and standards promulgated under section 6 of this Act, shall not apply with respect to any occupational safety or health issues covered under the plan, but the Secretary may retain jurisdiction under the above provisions in any proceeding commenced under section 9 or 10 before the date of determination.

(f)
The Secretary shall, on the basis of reports submitted by the State agency and his own inspections make a continuing evaluation of the manner in which each State having a plan approved under this section is carrying out such plan. Whenever the Secretary finds, after affording due notice and opportunity for a hearing, that in the administration of the State plan there is a failure to comply substantially with any provision of the State plan (or any assurance contained therein), he shall notify the State agency of his withdrawal of approval of such plan and upon receipt of such notice such plan shall cease to be in effect, but the State may retain jurisdiction in any case commenced before the withdrawal of the plan in order to enforce standards under the plan whenever the issues involved do not relate to the reasons for the withdrawal of the plan.

(g)
The State may obtain a review of a decision of the Secretary withdrawing approval of or rejecting its plan by the United States court of appeals for the circuit in which the State is located by filing in such court within thirty days following receipt of notice of such decision a petition to modify or set aside in whole or in part the action of the Secretary. A copy of such petition shall forthwith be served upon the Secretary, and thereupon the Secretary shall certify and file in the court the record upon which the decision complained of was issued as provided in section 2112 of title 28, United States Code. Unless the court finds that the Secretary's decision in rejecting a proposed State plan or withdrawing his approval of such a plan is not supported by substantial evidence the court shall affirm the Secretary's decision. The judgment of

the court shall be subject to review by the Supreme Court of the United States upon certiorari or certification as provided in section 1254 of title 28, United States Code.

(h)
The Secretary may enter into an agreement with a State under which the State will be permitted to continue to enforce one or more occupational health and safety standards in effect in such State until final action is taken by the Secretary with respect to a plan submitted by a State under subsection (b) of this section, or two years from the date of enactment of this Act, whichever is earlier.

SEC.
19.
Federal Agency Safety Programs and Responsibilities

(a)
29 U.S.C. 668

It shall be the responsibility of the head of each Federal agency (not including the United States Postal Service) to establish and maintain an effective and comprehensive occupational safety and health program which is consistent with the standards promulgated under section 6. The head of each agency shall (after consultation with representatives of the employees thereof)—

(1)
Pub. L. 50-241

provide safe and healthful places and conditions of employment, consistent with the standards set under section 6;

(2)
acquire, maintain, and require the use of safety equipment, personal protective equipment, and devices reasonably necessary to protect employees;

(3)
keep adequate records of all occupational accidents and illnesses for proper evaluation and necessary corrective action;

(4)
consult with the Secretary with regard to the adequacy as to form and content of records kept pursuant to subsection (a)(3) of this section; and

(5)
make an annual report to the Secretary with respect to occupational accidents and injuries and the agency's program under this section. Such report shall include any report submitted under section 7902(e)(2) of title 5, United States Code.

(b)
Pub. L. 97-375

The Secretary shall report to the President a summary or digest of reports submitted to him under subsection (a)(5) of this section, together with his evaluations of and recommendations derived from such reports.

(c)
Section 7902(c)(1) of title 5, United States Code, is amended by inserting after "agencies" the following: "and of labor organizations representing employees."

(d)
The Secretary shall have access to records and reports kept and filed by Federal agencies pursuant to subsections (a)(3) and (5) of this section unless those records and reports are specifically required by Executive order to be kept secret in the interest of the national defense or foreign policy, in which case the Secretary shall have access to such information as will not jeopardize national defense or foreign policy.

SEC.
20.
Research and Related Activities

(a)
(1)
29 U.S.C. 669

The Secretary of Health and Human Services, after consultation with the Secretary and with other appropriate Federal departments or agencies, shall conduct (directly or by grants or contracts) research, experiments, and demonstrations relating to occupational safety and health, including studies of psychological factors involved, and relating to innovative methods, techniques, and approaches for dealing with occupational safety and health problems.

(2)
The Secretary of Health and Human Services shall from time to time consult with the Secretary in order to develop specific plans for such research, demonstrations, and experiments as are necessary to produce criteria, including criteria identifying toxic substances, enabling the Secretary to meet his responsibility for the formulation of safety and health standards under this Act; and the Secretary of Health and Human Services, on the basis of such research, demonstrations, and experiments and any other information available to him, shall develop and publish at least annually such criteria as will effectuate the purposes of this Act.

(3)
The Secretary of Health and Human Services, on the basis of such research, demonstrations, and experiments, and any other information available to him, shall develop criteria dealing with toxic materials and harmful physical agents and substances which will describe exposure levels that are safe for various periods of employment, including but not limited to the exposure levels at which no employee will suffer impaired health or functional capacities or diminished life expectancy as a result of his work experience.

(4)
The Secretary of Health and Human Services shall also conduct special research, experiments, and demonstrations relating to occupational safety and health as are necessary to explore new problems, including those created by new technology in occupational safety and health, which may require ameliorative action beyond that which is otherwise provided for in the operating provisions of this Act. The Secretary of Health and Human Services shall also conduct research into the motivational and behavioral factors relating to the field of occupational safety and health.

(5)
The Secretary of Health and Human Services, in order to comply with his responsibilities under paragraph (2), and in order to develop needed information regarding potentially toxic substances or harmful physical agents, may prescribe regulations requiring employers to measure, record, and make reports on the exposure of employees to substances or physical agents which the Secretary of Health and Human Services reasonably believes may endanger the health or safety of employees. The Secretary of Health and Human Services also is authorized to establish such programs of medical examinations and tests as may be necessary for determining the incidence of occupational illnesses and the susceptibility of employees to such illnesses. Nothing in this or any other provision of this Act shall be deemed to authorize or require medical examination, immunization, or treatment for those who object thereto on religious grounds, except where such is necessary for the protection of the health or safety of others. Upon the request of any employer who is required to measure and record exposure of employees to substances or physical agents as provided under this subsection, the Secretary of Health and Human Services shall furnish full financial or other assistance to such employer for the purpose of defraying any additional expense incurred by him in carrying out the measuring and recording as provided in this subsection.

(6)
The Secretary of Health and Human Services shall publish within six months of enactment of this Act and thereafter as needed but at least annually a list of all known toxic substances by generic family or other useful grouping, and the concentrations at which such toxicity is known to occur. He shall determine following a written request by any employer or authorized representative of employees, specifying with reasonable particularity the grounds on which the request is made, whether any substance normally found in the place of employment has potentially toxic effects in such concentrations as used or found; and shall submit such determination both to employers and affected employees as soon as possible. If the Secretary of Health and Human Services determines that any substance is potentially toxic at the concentrations in which it is used or found in a place of employment, and such substance is not covered by an occupational safety or health standard promulgated under section 6, the Secretary of Health and Human Services shall immediately submit such determination to the Secretary, together with all pertinent criteria.

(7)
Within two years of enactment of the Act, and annually thereafter the Secretary of Health and Human Services shall conduct and publish industry wide studies of the effect of chronic or low-level exposure to industrial materials, processes, and stresses on the potential for illness, disease, or loss of functional capacity in aging adults.

(b)
The Secretary of Health and Human Services is authorized to make inspections and question employers and employees as provided in section 8 of this Act in order to carry out his functions and responsibilities under this section.

(c)
The Secretary is authorized to enter into contracts, agreements, or other arrangements with appropriate public agencies or private organizations for the purpose of conducting studies relating to his responsibilities under this Act. In carrying out his responsibilities under this subsection, the Secretary shall cooperate with the Secretary of Health and Human Services in order to avoid any duplication of efforts under this section.

(d)
Information obtained by the Secretary and the Secretary of Health and Human Services under this section shall be disseminated by the Secretary to employers and employees and organizations thereof.

(e)
The functions of the Secretary of Health and Human Services under this Act shall, to the extent feasible, be delegated to the Director of the National Institute for Occupational Safety and Health established by section 22 of this Act.

EXPANDED RESEARCH ON WORKER SAFETY AND HEALTH

29 U.S.C. 669a

Pub. L. 107-188, Title I, Section 153 added this text.

The Secretary of Health and Human Services (referred to in this section as the "Secretary"), acting through the Director of the National Institute of Occupational Safety and Health, shall enhance and expand research as deemed appropriate on the health and safety of workers who are at risk for bioterrorist threats or attacks in the workplace, including research on the health effects of measures taken to treat or protect such workers for diseases or disorders resulting from a bioterrorist threat or attack. Nothing in this section may be construed as establishing new regulatory authority for the Secretary or the Director to issue or modify any occupational safety and health rule or regulation.

SEC.
21.
Training and Employee Education

(a)
29 U.S.C. 670

The Secretary of Health and Human Services, after consultation with the Secretary and with other appropriate Federal departments and agencies, shall conduct, directly or by grants or contracts—

(1)
education programs to provide an adequate supply of qualified personnel to carry out the purposes of this Act, and

(2)
informational programs on the importance of and proper use of adequate safety and health equipment.

(b)
The Secretary is also authorized to conduct, directly or by grants or contracts, short-term training of personnel engaged in work related to his responsibilities under this Act.

(c)
The Secretary, in consultation with the Secretary of Health and Human Services, shall—

(1)
provide for the establishment and supervision of programs for the education and training of employers and employees in the recognition, avoidance, and prevention of unsafe or unhealthful working conditions in employments covered by this Act, and

(2)
Pub. L. 105-97, Section 2 added subsection (d). See Historical notes.
consult with and advise employers and employees, and organizations representing employers and employees as to effective means of preventing occupational injuries and illnesses.

(d)

(1)
The Secretary shall establish and support cooperative agreements with the States under which employers subject to this Act may consult with State personnel with respect to—

(A)
the application of occupational safety and health requirements under this Act or under State plans approved under section 18; and

(B)
voluntary efforts that employers may undertake to establish and maintain safe and healthful employment and places of employment. Such agreements may provide, as a condition of receiving funds under such agreements, for contributions by States towards meeting the costs of such agreements.

(2)
Pursuant to such agreements the State shall provide on-site consultation at the employer's worksite to employers who request such assistance. The State may also provide other education and training programs for employers and employees in the State. The State shall ensure that on-site consultations conducted pursuant to such agreements include provision for the participation by employees.

(3)
Activities under this subsection shall be conducted independently of any enforcement activity. If an employer fails to take immediate action to eliminate employee exposure to an imminent danger identified in a consultation or fails to correct a serious hazard so identified within a reasonable time, a report shall be made to the appropriate enforcement authority for such action as is appropriate.

(4)
The Secretary shall, by regulation after notice and opportunity for comment, establish rules under which an employer—

(A)
which requests and undergoes an on-site consultative visit provided under this subsection;

(B)
which corrects the hazards that have been identified during the visit within the time frames established by the State and agrees to request a subsequent consultative visit if major changes in working conditions or work processes occur which introduce new hazards in the workplace; and

(C)
which is implementing procedures for regularly identifying and preventing hazards regulated under this Act and maintains appropriate involvement of, and training for, management and non-management employees in achieving safe and healthful working conditions, may be exempt from an inspection (except an inspection requested under section 8(f) or an inspection to determine the cause of a workplace accident which resulted in the death of one or more employees or hospitalization for three or more employees) for a period of 1 year from the closing of the consultative visit.

(5)
A State shall provide worksite consultations under paragraph (2) at the request of an employer. Priority in scheduling such consultations shall be assigned to requests

from small businesses which are in higher hazard industries or have the most hazardous conditions at issue in the request.

SEC.
22.
National Institute for Occupational Safety and Health

(a)
29 U.S.C. 671

It is the purpose of this section to establish a National Institute for Occupational Safety and Health in the Department of Health and Human Services in order to carry out the policy set forth in section 2 of this Act and to perform the functions of the Secretary of Health and Human Services under sections 20 and 21 of this Act.

(b)
There is hereby established in the Department of Health and Human Services a National Institute for Occupational Safety and Health. The Institute shall be headed by a Director who shall be appointed by the Secretary of Health and Human Services, and who shall serve for a term of six years unless previously removed by the Secretary of Health and Human Services.

(c)
The Institute is authorized to—

(1)
develop and establish recommended occupational safety and health standards; and

(2)
perform all functions of the Secretary of Health and Human Services under sections 20 and 21 of this Act.

(d)
Upon his own initiative, or upon the request of the Secretary of Health and Human Services, the Director is authorized (1) to conduct such research and experimental programs as he determines are necessary for the development of criteria for new and improved occupational safety and health standards, and (2) after consideration of the results of such research and experimental programs make recommendations concerning new or improved occupational safety and health standards. Any occupational safety and health standard recommended pursuant to this section shall immediately be forwarded to the Secretary of Labor, and to the Secretary of Health and Human Services.

(e)
In addition to any authority vested in the Institute by other provisions of this section, the Director, in carrying out the functions of the Institute, is authorized to—

(1)
prescribe such regulations as he deems necessary governing the manner in which its functions shall be carried out;

(2)
receive money and other property donated, bequeathed, or devised, without condition or restriction other than that it be used for the purposes of the Institute and to use, sell, or otherwise dispose of such property for the purpose of carrying out its functions;

(3)
receive (and use, sell, or otherwise dispose of, in accordance with paragraph (2)), money and other property donated, bequeathed, or devised to the Institute with a condition or restriction, including a condition that the Institute use other funds of the Institute for the purposes of the gift;

(4)
in accordance with the civil service laws, appoint and fix the compensation of such personnel as may be necessary to carry out the provisions of this section;

(5)
obtain the services of experts and consultants in accordance with the provisions of section 3109 of title 5, United States Code;

(6)
accept and utilize the services of voluntary and noncompensated personnel and reimburse them for travel expenses, including per diem, as authorized by section 5703 of title 5, United States Code;

(7)
enter into contracts, grants or other arrangements, or modifications thereof to carry out the provisions of this section, and such contracts or modifications thereof may be entered into without performance or other bonds, and without regard to section 3709 of the Revised Statutes, as amended (41 U.S.C. 5), or any other provision of law relating to competitive bidding;

(8)
make advance, progress, and other payments which the Director deems necessary under this title without regard to the provisions of section 3324 (a) and (b) of Title 31; and

(9)

Pub. L. 97-258
make other necessary expenditures.

(f)
The Director shall submit to the Secretary of Health and Human Services, to the President, and to the Congress an annual report of the operations of the Institute

under this Act, which shall include a detailed statement of all private and public funds received and expended by it, and such recommendations as he deems appropriate.

(g)
Pub. L. 102-550 added subsection (g).
Lead-Based Paint Activities.

(1)

Training Grant Program.

(A)
The Institute, in conjunction with the Administrator of the Environmental Protection Agency, may make grants for the training and education of workers and supervisors who are or may be directly engaged in lead-based paint activities.

(B)
Grants referred to in subparagraph (A) shall be awarded to nonprofit organizations (including colleges and universities, joint labor-management trust funds, States, and nonprofit government employee organizations)—

(i)
which are engaged in the training and education of workers and supervisors who are or who may be directly engaged in lead-based paint activities (as defined in Title IV of the Toxic Substances Control Act),

(ii)
which have demonstrated experience in implementing and operating health and safety training and education programs, and

(iii)
with a demonstrated ability to reach, and involve in lead-based paint training programs, target populations of individuals who are or will be engaged in lead-based paint activities. Grants under this subsection shall be awarded only to those organizations that fund at least 30 percent of their lead-based paint activities training programs from non-Federal sources, excluding in-kind contributions. Grants may also be made to local governments to carry out such training and education for their employees.

(C)
There are authorized to be appropriated, a minimum, $10,000,000 to the Institute for each of the fiscal years 1994 through 1997 to make grants under this paragraph.

(2)
Evaluation of Programs. The Institute shall conduct periodic and comprehensive assessments of the efficacy of the worker and supervisor training programs developed and offered by those receiving grants under this section. The Director shall prepare reports on the results of these assessments addressed to the Administrator of the Environmental Protection Agency to include recommendations as may be

appropriate for the revision of these programs. The sum of $500,000 is authorized to be appropriated to the Institute for each of the fiscal years 1994 through 1997 to carry out this paragraph.

WORKERS' FAMILY PROTECTION

(a)
29 U.S.C. 671a

Short title
This section may be cited as the "Workers' Family Protection Act".

(b)
Findings and purpose

(1)
Pub. L. 102-522, Title II, Section 209 added this text.
Findings Congress finds that—

(A)
hazardous chemicals and substances that can threaten the health and safety of workers are being transported out of industries on workers' clothing and persons;

(B)
these chemicals and substances have the potential to pose an additional threat to the health and welfare of workers and their families;

(C)
additional information is needed concerning issues related to employee transported contaminant releases; and

(D)
additional regulations may be needed to prevent future releases of this type.

(2)

Purpose
It is the purpose of this section to—

(A)
increase understanding and awareness concerning the extent and possible health impacts of the problems and incidents described in paragraph (1);

(B)
prevent or mitigate future incidents of home contamination that could adversely affect the health and safety of workers and their families;

(C)
clarify regulatory authority for preventing and responding to such incidents; and

(D)
assist workers in redressing and responding to such incidents when they occur.

(c)
Evaluation of employee transported contaminant releases

(1)
Study

(A)
In general

Not later than 18 months after October 26, 1992, the Director of the National Institute for Occupational Safety and Health (hereafter in this section referred to as the "Director"), in cooperation with the Secretary of Labor, the Administrator of the Environmental Protection Agency, the Administrator of the Agency for Toxic Substances and Disease Registry, and the heads of other Federal Government agencies as determined to be appropriate by the Director, shall conduct a study to evaluate the potential for, the prevalence of, and the issues related to the contamination of workers' homes with hazardous chemicals and substances, including infectious agents, transported from the workplaces of such workers.

(B)
Matters to be evaluated
In conducting the study and evaluation under subparagraph (A), the Director shall—

(i)
conduct a review of past incidents of home contamination through the utilization of literature and of records concerning past investigations and enforcement actions undertaken by—

(I)
the National Institute for Occupational Safety and Health;

(II)
the Secretary of Labor to enforce the Occupational Safety and Health Act of 1970 (29 U.S.C. 651 et seq.);

(III)
States to enforce occupational safety and health standards in accordance with section 18 of such Act (29 U.S.C. 667); and

(IV)
other government agencies (including the Department of Energy and the Environmental Protection Agency), as the Director may determine to be appropriate;

(ii)
evaluate current statutory, regulatory, and voluntary industrial hygiene or other measures used by small, medium and large employers to prevent or remediate home contamination;

(iii)
compile a summary of the existing research and case histories conducted on incidents of employee transported contaminant releases, including—

(I)
the effectiveness of workplace housekeeping practices and personal protective equipment in preventing such incidents;

(II)
the health effects, if any, of the resulting exposure on workers and their families;

(III)
the effectiveness of normal house cleaning and laundry procedures for removing hazardous materials and agents from workers' homes and personal clothing;

(IV)
indoor air quality, as the research concerning such pertains to the fate of chemicals transported from a workplace into the home environment; and

(V)
methods for differentiating exposure health effects and relative risks associated with specific agents from other sources of exposure inside and outside the home;

(iv)
identify the role of Federal and State agencies in responding to incidents of home contamination;

(v)
prepare and submit to the Task Force established under paragraph (2) and to the appropriate committees of Congress, a report concerning the results of the matters studied or evaluated under clauses (i) through (iv); and

(vi)
study home contamination incidents and issues and worker and family protection policies and practices related to the special circumstances of firefighters and prepare and submit to the appropriate committees of Congress a report concerning the findings with respect to such study.

(2)
Development of investigative strategy

(A)
Task Force
Not later than 12 months after October 26, 1992, the Director shall establish a working group, to be known as the "Workers' Family Protection Task Force." The Task Force shall—

(i)
be composed of not more than 15 individuals to be appointed by the Director from among individuals who are representative of workers, industry, scientists, industrial hygienists, the National Research Council, and government agencies, except that not more than one such individual shall be from each appropriate government agency and the number of individuals appointed to represent industry and workers shall be equal in number;

(ii)
review the report submitted under paragraph (1)(B)(v);

(iii)
determine, with respect to such report, the additional data needs, if any, and the need for additional evaluation of the scientific issues related to and the feasibility of developing such additional data; and

(iv)
if additional data are determined by the Task Force to be needed, develop a recommended investigative strategy for use in obtaining such information.

(B)
Investigative strategy

(i)
Content
The investigative strategy developed under subparagraph (A)(iv) shall identify data gaps that can and cannot be filled, assumptions and uncertainties associated with various components of such strategy, a timetable for the implementation of such strategy, and methodologies used to gather any required data.

(ii)
Peer review

The Director shall publish the proposed investigative strategy under subparagraph (A)(iv) for public comment and utilize other methods, including technical conferences or seminars, for the purpose of obtaining comments concerning the proposed strategy.

(iii)
Final strategy
After the peer review and public comment is conducted under clause (ii), the Director, in consultation with the heads of other government agencies, shall propose

a final strategy for investigating issues related to home contamination that shall be implemented by the National Institute for Occupational Safety and Health and other Federal agencies for the period of time necessary to enable such agencies to obtain the information identified under subparagraph (A)(iii).

(C)
Construction
Nothing in this section shall be construed as precluding any government agency from investigating issues related to home contamination using existing procedures until such time as a final strategy is developed or from taking actions in addition to those proposed in the strategy after its completion.

(3)
Implementation of investigative strategy

Upon completion of the investigative strategy under subparagraph (B)(iii), each Federal agency or department shall fulfill the role assigned to it by the strategy.

(d)
Regulations

(1)
In general

Not later than 4 years after October 26, 1992, and periodically thereafter, the Secretary of Labor, based on the information developed under subsection (c) of this section and on other information available to the Secretary, shall—

(A)
determine if additional education about, emphasis on, or enforcement of existing regulations or standards is needed and will be sufficient, or if additional regulations or standards are needed with regard to employee transported releases of hazardous materials; and

(B)
prepare and submit to the appropriate committees of Congress a report concerning the result of such determination.

(2)
Additional regulations or standards If the Secretary of Labor determines that additional regulations or standards are needed under paragraph (1), the Secretary shall promulgate, pursuant to the Secretary's authority under the Occupational Safety and Health Act of 1970 (29 U.S.C. 651 et seq.), such regulations or standards as determined to be appropriate not later than 3 years after such determination.

(e)
Authorization of appropriations There are authorized to be appropriated from sums otherwise authorized to be appropriated, for each fiscal year such sums as may be necessary to carry out this section.

SEC.
23.
Grants to the States

(a)
29 U.S.C. 672

The Secretary is authorized, during the fiscal year ending June 30, 1971, and the two succeeding fiscal years, to make grants to the States which have designated a State agency under section 18 to assist them—

(1)
in identifying their needs and responsibilities in the area of occupational safety and health,

(2)
in developing State plans under section 18, or

(3)
in developing plans for—

(A)
establishing systems for the collection of information concerning the nature and frequency of occupational injuries and diseases;

(B)
increasing the expertise and enforcement capabilities of their personnel engaged in occupational safety and health programs; or

(C)
otherwise improving the administration and enforcement of State occupational safety and health laws, including standards thereunder, consistent with the objectives of this Act.

(b)
The Secretary is authorized, during the fiscal year ending June 30, 1971, and the two succeeding fiscal years, to make grants to the States for experimental and demonstration projects consistent with the objectives set forth in subsection (a) of this section.

(c)
The Governor of the State shall designate the appropriate State agency for receipt of any grant made by the Secretary under this section.

(d)
Any State agency designated by the Governor of the State desiring a grant under this section shall submit an application therefor to the Secretary.

(e)
The Secretary shall review the application, and shall, after consultation with the Secretary of Health and Human Services, approve or reject such application.

(f)
The Federal share for each State grant under subsection (a) or (b) of this section may not exceed 90 per centum of the total cost of the application. In the event the Federal share for all States under either such subsection is not the same, the differences among the States shall be established on the basis of objective criteria.

(g)
The Secretary is authorized to make grants to the States to assist them in administering and enforcing programs for occupational safety and health contained in State plans approved by the Secretary pursuant to section 18 of this Act. The Federal share for each State grant under this subsection may not exceed 50 per centum of the total cost to the State of such a program. The last sentence of subsection (f) shall be applicable in determining the Federal share under this subsection.

(h)
Prior to June 30, 1973, the Secretary shall, after consultation with the Secretary of Health and Human Services, transmit a report to the President and to the Congress, describing the experience under the grant programs authorized by this section and making any recommendations he may deem appropriate.

SEC.
24.
Statistics

(a)
In order to further the purposes of this Act, the Secretary, in consultation with the Secretary of Health and Human Services, shall develop and maintain an effective program of collection, compilation, and analysis of occupational safety and health statistics. Such program may cover all employments whether or not subject to any other provisions of this Act but shall not cover employments excluded by section 4 of the Act. The Secretary shall compile accurate statistics on work injuries and illnesses which shall include all disabling, serious, or significant injuries and illnesses, whether or not involving loss of time from work, other than minor injuries requiring only first aid treatment and which do not involve medical treatment, loss of consciousness, restriction of work or motion, or transfer to another job.

(b)
To carry out his duties under subsection (a) of this section, the Secretary may—

(1)
promote, encourage, or directly engage in programs of studies, information and communication concerning occupational safety and health statistics;

(2)
make grants to States or political subdivisions thereof in order to assist them in developing and administering programs dealing with occupational safety and health statistics; and

(3)
arrange, through grants or contracts, for the conduct of such research and investigations as give promise of furthering the objectives of this section.

(c)
The Federal share for each grant under subsection (b) of this section may be up to 50 per centum of the State's total cost.

(d)
The Secretary may, with the consent of any State or political subdivision thereof, accept and use the services, facilities, and employees of the agencies of such State or political subdivision, with or without reimbursement, in order to assist him in carrying out his functions under this section.

(e)
On the basis of the records made and kept pursuant to section 8(c) of this Act, employers shall file such reports with the Secretary as he shall prescribe by regulation, as necessary to carry out his functions under this Act.

(f)
Agreements between the Department of Labor and States pertaining to the collection of occupational safety and health statistics already in effect on the effective date of this Act shall remain in effect until superseded by grants or contracts made under this Act.

SEC.

25.

Audits

(a)

29 U.S.C. 674
Each recipient of a grant under this Act shall keep such records as the Secretary or the Secretary of Health and Human Services shall prescribe, including records which fully disclose the amount and disposition by such recipient of the proceeds of such grant, the total cost of the project or undertaking in connection with which such grant is made or used, and the amount of that portion of the cost of the project or undertaking supplied by other sources, and such other records as will facilitate an effective audit.

(b)
The Secretary or the Secretary of Health and Human Services, and the Comptroller General of the United States, or any of their duly authorized representatives, shall have access for the purpose of audit and examination to any books, documents, papers, and records of the recipients of any grant under this Act that are pertinent to any such grant.

SEC.
26.
Annual Report
29 U.S.C. 675 Pub. L. 104-66 Section 3003 terminated provision relating to transmittal of report to Congress.

Within one hundred and twenty days following the convening of each regular session of each Congress, the Secretary and the Secretary of Health and Human Services shall each prepare and submit to the President for transmittal to the Congress a report upon the subject matter of this Act, the progress toward achievement of the purpose of this Act, the needs and requirements in the field of occupational safety and health, and any other relevant information. Such reports shall include information regarding occupational safety and health standards, and criteria for such standards, developed during the preceding year; evaluation of standards and criteria previously developed under this Act, defining areas of emphasis for new criteria and standards; an evaluation of the degree of observance of applicable occupational safety and health standards, and a summary of inspection and enforcement activity undertaken; analysis and evaluation of research activities for which results have been obtained under governmental and nongovernmental sponsorship; an analysis of major occupational diseases; evaluation of available control and measurement technology for hazards for which standards or criteria have been developed during the preceding year; description of cooperative efforts undertaken between Government agencies and other interested parties in the implementation of this Act during the preceding year; a progress report on the development of an adequate supply of trained manpower in the field of occupational safety and health, including estimates of future needs and the efforts being made by Government and others to meet those needs; listing of all toxic substances in industrial usage for which labeling requirements, criteria, or standards have not yet been established; and such recommendations for additional legislation as are deemed necessary to protect the safety and health of the worker and improve the administration of this Act.

SEC.
27.
National Commission on State Workmen's Compensation Laws

29 U.S.C. 676
(Text omitted.)

SEC.
28.
Economic Assistance to Small Businesses
See notes on omitted text.

(Text omitted.)

SEC.
29.
Additional Assistant Secretary of Labor
See notes on omitted text.

(Text omitted.)

SEC.
30.
Additional Positions
See notes on omitted text.

(Text omitted.)

SEC.
31.
Emergency Locator Beacons
See notes on omitted text.

(Text omitted.)

SEC.
32.
Separability

29 U.S.C. 677
If any provision of this Act, or the application of such provision to any person or circumstance, shall be held invalid, the remainder of this Act, or the application of such provision to persons or circumstances other than those as to which it is held invalid, shall not be affected thereby.

SEC.
33.
Appropriations

29 U.S.C. 678
There are authorized to be appropriated to carry out this Act for each fiscal year such sums as the Congress shall deem necessary.
SEC.
34.
Effective Date
This Act shall take effect one hundred and twenty days after the date of its enactment.
Approved December 29, 1970.

As amended through January 1, 2004.

HISTORICAL NOTES

This reprint generally retains the section numbers originally created by Congress in the Occupational Safety and Health (OSH) Act of 1970, Pub. L. 91-596, 84 Stat 1590. This document includes some editorial changes, such as changing the format to make it easier to read, correcting typographical errors, and updating some of the margin notes. Because Congress enacted amendments to the Act since 1970, this version differs from the original version of the OSH Act. It also differs slightly from the version published in the United States Code at 29 U.S.C. 661 *et seq* . For example, this reprint refers to the statute as the "Act" rather than the "chapter." This reprint reflects the provisions of the OSH Act that are in effect as of January 1, 2004. Citations to Public Laws which made important amendments to the OSH Act since 1970 are set forth in the margins and explanatory notes are included below. NOTE: Some provisions of the OSH Act may be affected by the enactment of, or amendments to, other statutes. Section 17(h)(1), 29 U.S.C. 666, is an example. The original provision amended section 1114 of title 18 of the United States Code to include employees of "the Department of Labor assigned to perform investigative, inspection, or law enforcement functions" within the list of persons protected by the provisions to allow prosecution of persons who have killed or attempted to kill an officer or employee of the U.S. government while performing official duties. This reprint sets forth the text of section 17(h) as enacted in 1970. However, since 1970, Congress has enacted multiple amendments to 18 U.S.C. 1114. The current version does not specifically include the Department of Labor in a list; rather it states that "Whoever kills or attempts to kill any officer or employee of the United States or of any agency in any branch of the United States Government (including any member of the uniformed services) while such officer or employee is engaged in or on account of the performance of official duties, or any person assisting such an officer or employee in the performance of such duties or on account of that assistance shall be punished ..." as provided by the statute. Readers are reminded that the official version of statutes can be found in the current volumes of the United States Code, and more extensive historical notes can be found in the current volumes of the United States Code Annotated.

Amendments

On January 2, 1974, section 2(c) of Pub. L. 93-237 replaced the phrase "7(b)(6)" in section 28(d) of the OSH Act with "7(b)(5)." 87 Stat. 1023. Note: The text of Section 28 (Economic Assistance to Small Business) amended Sections 7(b) and Section 4(c)(1) of the Small Business Act. Because these amendments are no longer current, the text of section 28 is omitted in this reprint. For the current version, see 15 U.S.C. 636.

In 1977, the U.S. entered into the Panama Canal Treaty of 1977, Sept. 7, 1977, U.S.-Panama, T.I.A.S. 10030, 33 U.S.T. 39. In 1979, Congress enacted implementing legislation. Panama Canal Act of 1979, Pub. L. 96-70, 93 Stat. 452 (1979). Although no corresponding amendment to the OSH Act was enacted, the Canal Zone ceased to exist in 1979. The U.S. continued to manage, operate and facilitate the transit of ships through the Canal under the authority of the Panama Canal Treaty until December 31, 1999, at which time authority over the Canal was transferred to the Republic of Panama.

On March 27, 1978, Pub. L. 95-251, 92 Stat. 183, replaced the term "hearing examiner(s)" with "administrative law judge(s)" in all federal laws, including sections 12(e), 12(j), and 12(k) of the OSH Act, 29 U.S.C. 661.

On October 13, 1978, Pub. L. 95-454, 92 Stat. 1111, 1221, which redesignated section numbers concerning personnel matters and compensation, resulted in the substitution of section 5372 of Title 5 for section 5362 in section 12(e) of the OSH Act, 29 U.S.C. 661.

On October 17, 1979, Pub. L. 96-88, Title V, section 509(b), 93 Stat. 668, 695, redesignated references to the Department of Health, Education, and Welfare to the Department of Health and Human Services and redesignated references to the Secretary of Health, Education, and Welfare to the Secretary of Health and Human Services.

On September 13, 1982, Pub. L. 97-258, Section 4(b), 96 Stat. 877, 1067, effectively substituted "Section 3324(a) and (b) of Title 31" for "Section 3648 of the Revised Statutes, as amended (31 U.S.C. 529)" in section 22 (e)(8), 29 U.S.C. 671, relating to NIOSH procurement authority.

On December 21, 1982, Pub. L. 97-375, 96 Stat. 1819, deleted the sentence in section 19(b) of the Act, 29 U.S.C. 668, that directed the President of the United States to transmit annual reports of the activities of federal agencies to the House of Representatives and the Senate.

On October 12, 1984, Pub. L. 98-473, Chapter II, 98 Stat. 1837, 1987, (commonly referred to as the "Sentencing Reform Act of 1984") instituted a classification system for criminal offenses punishable under the United States Code. Under this system, an offense with imprisonment terms of "six months or less but more than thirty days," such as that found in 29 U.S.C. 666(e) for a willful violation of the OSH Act, is classified as a criminal "Class B misdemeanor." 18 U.S.C. 3559(a)(7).

The criminal code increases the monetary penalties for criminal misdemeanors beyond what is provided for in the OSH Act: a fine for a Class B misdemeanor resulting in death, for example, is not more than $250,000 for an individual, and is not more than $500,000 for an organization. 18 U.S.C. 3571(b)(4), (c)(4). The criminal code also provides for authorized terms of probation for both individuals and organizations. 18 U.S.C. 3551, 3561. The term of imprisonment for individuals is the same as that authorized by the OSH Act. 18 U.S.C. 3581(b)(7).

On November 8, 1984, Pub. L. 98-620, 98 Stat. 3335, deleted the last sentence in section 11(a) of the Act, 29 U.S.C. 660, that required petitions filed under the subsection to be heard expeditiously.

On November 5, 1990, Pub. L. 101-508, 104 Stat. 1388, amended section 17 of the Act, 29 U.S.C. 666, by increasing the penalties in section 17(a) from $10,000 for each violation to "$70,000 for each violation, but not less than $5,000 for each willful violation," and increased the limitation on penalties in sections (b), (c), (d), and (i) from $1,000 to $7,000.

On October 26, 1992, Pub. L. 102-522, 106 Stat. 3410, 3420, added to Title 29, section 671a "Workers' Family Protection" to grant authority to the Director of NIOSH to evaluate, investigate and if necessary, for the Secretary of Labor to regulate employee transported releases of hazardous material that result from contamination on the employee's clothing or person and may adversely affect the health and safety of workers and their families. Note: section 671a was enacted as section 209 of the Fire Administration Authorization Act of 1992, but it is reprinted here because it is codified within the chapter that comprises the OSH Act.

On October 28, 1992, the Housing and Community Development Act of 1992, Pub. L. 102-550, 106 Stat. 3672, 3924, amended section 22 of the Act, 29 U.S.C. 671, by adding subsection (g), which requires NIOSH to institute a training grant program for lead-based paint activities.

On July 5, 1994, section 7(b) of Pub. L. 103-272, 108 Stat. 745, repealed section 31 of the OSH Act, "Emergency Locator Beacons." Section 1(e) of the same Public Law, however, enacted a modified version of section 31 of the OSH Act. This provision, titled "Emergency Locator Transmitters," is codified at 49 U.S.C. 44712.

On December 21, 1995, Section 3003 of Pub. L. 104-66, 109 Stat. 707, as amended, effective May 15, 2000, terminated the provisions relating to the transmittal to Congress of reports under section 26 of the OSH Act. 29 U.S.C. 675.

On July 16, 1998, Pub. L. 105-197, 112 Stat. 638, amended section 21 of the Act, 29 U.S.C. 670, by adding subsection (d), which required the Secretary to establish a compliance assistance program by which employers can consult with state personnel regarding the application of and compliance with OSHA standards.

On July 16, 1998, Pub. L. 105-198, 112 Stat. 640, amended section 8 of the Act, 29 U.S.C. 657, by adding subsection (h), which forbids the Secretary to use the results of enforcement activities to evaluate the employees involved in such enforcement or to impose quotas or goals.

On September 29, 1998, Pub. L. 105-241, 112 Stat. 1572, amended sections 3(5) and 19(a) of the Act, 29 U.S.C. 652 and 668, to include the United States Postal Service as an "employer" subject to OSHA enforcement.

On June 12, 2002, Pub. L. 107-188, Title I, Section 153, 116 Stat. 631, Congress enacted 29 U.S.C. 669a, to expand research on the "health and safety of workers who are at risk for bioterrorist threats or attacks in the workplace."

Jurisdictional Note

Although no corresponding amendments to the OSH Act have been made, OSHA no longer exercises jurisdiction over the entity formerly known as the Trust Territory of the Pacific Islands. The Trust Territory, which consisted of the Former

Japanese Mandated Islands, was established in 1947 by the Security Council of the United Nations, and administered by the United States. *Trusteeship Agreement for the Former Japanese Mandated Islands*, Apr. 2-July 18, 1947, 61 Stat. 3301, T.I.A.S. 1665, 8 U.N.T.S. 189.

From 1947 to 1994, the people of these islands exercised the right of self-determination conveyed by the Trusteeship four times, resulting in the division of the Trust Territory into four separate entities. Three entities: the Republic of Palau, the Federated States of Micronesia, and the Republic of the Marshall Islands, became "Freely Associated States," to which U.S. Federal Law does not apply. Since the OSH Act is a generally applicable law that applies to Guam, it applies to the Commonwealth of Northern Mariana Islands, which elected to become a "Flag Territory" of the United States. *See Covenant to Establish a Commonwealth of the Northern Mariana Islands in Political Union with the United States of America*, Article V, section 502(a) as contained in Pub. L. 94-24, 90 Stat. 263 (Mar. 24, 1976)[citations to amendments omitted]; 48 U.S.C. 1801 and note (1976); *see also Saipan Stevedore Co., Inc. v. Director, Office of Workers' Compensation Programs*, 133 F.3d 717, 722 (9th Cir. 1998)(Longshore and Harbor Workers' Compensation Act applies to the Commonwealth of Northern Mariana Islands pursuant to section 502(a) of the Covenant because the Act has general application to the states and to Guam). For up-to-date information on the legal status of these freely associated states and territories, contact the Office of Insular Affairs of the Department of the Interior. (Web address: http://www.doi.gov/oia/)

Omitted Text. Reasons for textual deletions vary. Some deletions may result from amendments to the OSH Act; others to subsequent amendments to other statutes which the original provisions of the OSH Act may have amended in 1970. In some instances, the original provision of the OSH Act was date-limited and is no longer operative.

The text of section 12(c), 29 U.S.C. 661, is omitted. Subsection (c) amended sections 5314 and 5315 of Title 5, United States Code, to add the positions of Chairman and members of the Occupational Safety and Health Review Commission.

The text of section 27, 29 U.S.C. 676, is omitted. Section 27 listed Congressional findings on workers' compensation and established the National Commission on State Workmen's Compensation Laws, which ceased to exist ninety days after the submission of its final report, which was due no later than July 31, 1972.

The text of section 28 (Economic Assistance to Small Business) amended sections 7(b) and section 4(c)(1) of the Small Business Act to allow for small business loans in order to comply with applicable standards. Because these amendments are no longer current, the text is omitted here. For the current version see 15 U.S.C. 636.

The text of section 29, (Additional Assistant Secretary of Labor), created an Assistant Secretary for Occupational Safety and Health, and section 30 (Additional Positions) created additional positions within the Department of Labor and the Occupational

Safety and Health Review Commission in order to carry out the provisions of the OSH Act. The text of these sections is omitted here because it no longer reflects the current statutory provisions for staffing and pay. For current provisions, see 29 U.S.C. 553 and 5 U.S.C. 5108 (c).

Section 31 of the original OSH Act amended 49 U.S.C. 1421 by inserting a section entitled "Emergency Locator Beacons." The text of that section is omitted in this reprint because Pub. L. 103-272, 108 Stat.745, (July 5, 1994), repealed the text of section 31 and enacted a modified version of the provision, entitled "Emergency Locator Transmitters," which is codified at 49 U.S.C. 44712.

Notes on other legislation affecting the administration of the Occupational Safety and Health Act. Sometimes legislation does not directly amend the OSH Act, but does place requirements on the Secretary of Labor either to act or to refrain from acting under the authority of the OSH Act. Included below are some examples of such legislation. Please note that this is not intended to be a comprehensive list.

STANDARDS PROMULGATION

For example, legislation may require the Secretary to promulgate specific standards pursuant to authority under Section 6 of the OSH Act, 29 U.S.C. 655. Some examples include the following:

Hazardous Waste Operations. Pub. L. 99-499, Title I, section 126(a)-(f), 100 Stat. 1613 (1986), as amended by Pub. L. 100-202, section 101(f), Title II, section 201, 101 Stat. 1329 (1987), required the Secretary of Labor to promulgate standards concerning hazardous waste operations.

Chemical Process Safety Management. Pub. L. 101-549, Title III, section 304, 104 Stat. 2399 (1990), required the Secretary of Labor, in coordination with the Administrator of the Environmental Protection Agency, to promulgate a chemical process safety standard.

Hazardous Materials. Pub. L. 101-615, section 29, 104 Stat. 3244 (1990), required the Secretary of Labor, in consultation with the Secretaries of Transportation and Treasury, to issue specific standards concerning the handling of hazardous materials.

Bloodborne Pathogens Standard. Pub. L. 102-170, Title I, section 100, 105 Stat. 1107 (1991), required the Secretary of Labor to promulgate a final Bloodborne Pathogens standard.

Lead Standard. The Housing and Community Development Act of 1992, Pub. L. 102-550, Title X, sections 1031 and 1032, 106 Stat. 3672 (1992), required the Secretary of Labor to issue an interim final lead standard.

EXTENSION OF COVERAGE

Sometimes a statute may make some OSH Act provisions applicable to certain entities that are not subject to those provisions by the terms of the OSH Act. For example, the Congressional Accountability Act of 1995, Pub. L. 104-1, 109 Stat. 3, (1995), extended certain OSH Act coverage, such as the duty to comply with Section 5 of the OSH Act, to the Legislative Branch. Among other provisions, this legislation authorizes the General Counsel of the Office of Compliance within the Legislative Branch to exercise the authority granted to the Secretary of Labor in the OSH Act to inspect places of employment and issue a citation or notice to correct the violation found. This statute does not make all the provisions of the OSH Act applicable to the Legislative Branch. Another example is the Medicare Prescription Drug, Improvement, and Modernization Act of 2003, Title IX, Section 947, Pub. L. 108-173, 117 Stat. 2066 (2003), which requires public hospitals not otherwise subject to the OSH Act to comply with OSHA's Bloodborne Pathogens standard, 29 CFR 1910.1030. This statute provides for the imposition and collection of civil money penalties by the Department of Health and Human Services in the event that a hospital fails to comply with OSHA's Bloodborne Pathogens standard.

PROGRAM CHANGES ENACTED THROUGH APPROPRIATIONS LEGISLATION

Sometimes an appropriations statute may allow or restrict certain substantive actions by OSHA or the Secretary of Labor. For example, sometimes an appropriations statute may restrict the use of money appropriated to run the Occupational Safety and Health Administration or the Department of Labor. One example of such a restriction, that has been included in OSHA's appropriation for many years, limits the applicability of OSHA requirements with respect to farming operations that employ ten or fewer workers and do not maintain a temporary labor camp. Another example is a restriction that limits OSHA's authority to conduct certain enforcement activity with respect to employers of ten or fewer employees in low hazard industries. See Consolidated Appropriations Act, 2004, Pub. L. 108-199, Div. E - Labor, Health and Human Services, and Education, and Related Agencies Appropriations, 2004, Title I - Department of Labor, 118 Stat. 3 (2004). Sometimes an appropriations statute may allow OSHA to retain some money collected to use for occupational safety and health training or grants. For example, the Consolidated Appropriations Act, 2004, Div. E, Title I, cited above, allows OSHA to retain up to $750,000 of training institute course tuition fees per fiscal year for such uses. For the statutory text of currently applicable appropriations provisions, consult the OSHA appropriations statute for the fiscal year in question.

Index

D

E

G

M

N

O

P

Q

R

S

T

U

V

W

Z